Ecosystem Services, Green Infrastructure and Spatial Planning

Ecosystem Services, Green Infrastructure and Spatial Planning

Editor

Corrado Zoppi

MDPI • Basel • Beijing • Wuhan • Barcelona • Belgrade • Manchester • Tokyo • Cluj • Tianjin

Editor
Corrado Zoppi
Dipartimento di Ingegneria
Civile, Ambientale e Architettura
University of Cagliari
Cagliari
Italy

Editorial Office
MDPI
St. Alban-Anlage 66
4052 Basel, Switzerland

This is a reprint of articles from the Special Issue published online in the open access journal *Sustainability* (ISSN 2071-1050) (available at: www.mdpi.com/journal/sustainability/special_issues/green_planning).

For citation purposes, cite each article independently as indicated on the article page online and as indicated below:

LastName, A.A.; LastName, B.B.; LastName, C.C. Article Title. *Journal Name* **Year**, *Volume Number*, Page Range.

ISBN 978-3-0365-1716-2 (Hbk)
ISBN 978-3-0365-1715-5 (PDF)

Contents

About the Editor

Corrado Zoppi

Corrado Zoppi is a Doctor of Philosophy in Economics (USA, 1997), a Doctor of Research in Territorial Planning (Italy, 1992), and an M.Sc. in Economic Policy and Planning (USA, 1990). He is a Professor at the University of Cagliari (Sector ICAR/20 –Spatial planning). He is presently teaching at the Department of Civil and Environmental Engineering and Architecture of the University of Cagliari in the Undergraduate and Graduate Programs in Environmental and Territorial Engineering and in Sustainable Tourism Management and Monitoring (Regional and Urban Planning, Strategic Planning, and Environmental planning).

Preface to "Ecosystem Services, Green Infrastructure and Spatial Planning"

With regards the United Nations Convention on Biological Diversity, ratified by Italy by Law no. 1994/124, an ecosystem is "a dynamic complex of plant, animal and micro-organism communities and their nonliving environment interacting as a functional unit." Ecosystem goods and services, univocally defined as "ecosystem services" represent the benefits human populations derive, directly or indirectly, from ecosystem functions.

When trying to assess the ecosystem services of natural resources, the usual vision is always based on qualitative approaches. The importance of environmental services is generally recognized, as well as how much they are worth protecting and restoring. However, it is very difficult to compare the costs, which can be easily revealed in monetary terms, to the benefits, which are always in the abstract world of ideas. Actually, it is impossible to compare apples to oranges. It would be of huge utility for planning and managing to have tools that bridge this gap.

It has to be emphasized that the strategic environmental assessment (SEA) of management plans (MPs) for Natura 2000 sites has to be regarded as an assessment exercise concerning not merely a single node of the ecological network (that is, a single Natura 2000 site), but rather the network as a whole. SEA is intrinsically connected to sustainability because it establishes environmental protection-related objectives, and, therefore, it acts as a sustainability-oriented plan which becomes part of the planning process itself. This is of particular importance with reference to the definition of conservation measures, including the preparation of MPs, as SEA can help integrate sustainability within MP objectives and can be regarded as a real and effective learning path for the administrations in charge of Natura 2000 sites. Within the SEA, a fundamental issue is the assessment of the restoration of ecosystem services.

In the Communication from the Commission to the European Parliament, the Council, the European Economic and Social Committee and the Committee of the Regions (COM (2013) 249 final), a working definition of green infrastructure (GI) is proposed as follows: "A strategically planned network of natural and semi-natural areas with other environmental features designed and managed to deliver a wide range of ecosystem services. It incorporates green spaces (or blue if aquatic ecosystems are concerned) and other physical features in terrestrial (including coastal) and marine areas. On land, GI is present in rural and urban settings." Moreover, the Commission puts in evidence how and to what extent the issue of GI relates to the network of sites of community importance (SCI), special areas of conservation (SAC), and special protection areas (SPA): "The work done over the last 25 years to establish and consolidate the network means that the backbone of the EU's GI is already in place. It is a reservoir of biodiversity that can be drawn upon to repopulate and revitalize degraded environments and catalyze the development of GI. This will also help reduce the fragmentation of the ecosystem, improving the connectivity between sites in the Natura 2000 Network and thus achieving the objectives of Article 10 of the Habitats Directive."

Hence, it is evident that the definition of GI is strictly connected to the category of ecosystem services. Moreover, it has to be a planned network. Spatial planning, at the regional and urban levels, is an important and effective perspective to address the complex issue of defining, implementing, and managing networks of ecosystem services and GI.

As a consequence, GI has a decisive role in promoting the restoration of biodiversity and in reducing the fragmentation of ecosystems, and, eventually, in their capability of delivering ecosystem services. Therefore, a general goal of the SEA of MPs of sites of the Natura 2000 Network can be defined in order to address the issue of the role of GI in promoting and enhancing habitat restoration and the delivery of ecosystem services.

A wide range of issues related to ecosystem services and GI, as important points of reference for spatial planning, related to urban and rural contexts, with particular reference to the definition and implementation of planning policies aimed at protecting nature and natural resources, are discussed in the articles of the Special Issue.

Corrado Zoppi
Editor

sustainability

MDPI

Editorial

Ecosystem Services, Green Infrastructure and Spatial Planning

Corrado Zoppi

Department of Civil and Environmental Engineering and Architecture, University of Cagliari, Via Marengo 2, 09123 Cagliari, Italy; zoppi@unica.it

Received: 9 May 2020; Accepted: 26 May 2020; Published: 27 May 2020

Abstract: Ecosystem services and green infrastructure do not appear to inform spatial policies and plans. National governments hardly identify their ecological networks or make an effort to integrate them into their spatial policies and plans. Under this perspective, an important scientific and technical issue is to focus on preserving corridors for enabling species mobility and on achieving connectivity between natural protected areas. In this respect, this Special Issue takes a step forward insofar as it aims at proposing a theoretical and methodological discussion on the definition and implementation of ecological networks that, besides guaranteeing wildlife movements, also provide a wide range of ecosystem services. The social and economic profile of this question is also relevant since in the long run, savings in public spending (e.g., due to the reduced need for grey infrastructures aiming at contrasting soil erosion or at managing flood risk), savings in private spending (e.g., on water treatment costs) and the potential creation of green jobs are foreseeable. Moreover, indirect and less easily quantifiable social and health benefits (e.g., due to improved natural pollution abatement) are likely to occur as well.

Keywords: ecological corridors; ecosystem services; green infrastructure; landscape connectivity; landscape fragmentation; Natura 2000 Network

---

With regards the United Nations Convention on Biological Diversity, ratified by Italy by Law no. 1994/124, an ecosystem is "a dynamic complex of plant, animal and micro-organism communities and their nonliving environment interacting as a functional unit". Ecosystem goods and services, univocally defined as "ecosystem services", represent the benefits human populations derive, directly or indirectly, from ecosystem functions [1].

When trying to assess the ecosystem services of natural resources, the usual vision is always based on qualitative approaches. The importance of environmental services is generally recognized and as well as how much they are worth protecting and restoring. However, it is very difficult to compare the costs, which can be easily revealed in monetary terms, to the benefits, which are always in the abstract world of ideas. Actually, it is impossible to compare apples to oranges. It would be of huge utility for planning and managing to have tools that bridge this gap [2].

It has to be emphasized that the strategic environmental assessment (SEA) of management plans (MPs) for Natura 2000 sites has to be regarded as an assessment exercise concerning not merely a single node of the ecological network (that is, a single Natura 2000 site), but rather the network as a whole. SEA is intrinsically connected to sustainability because it establishes environmental protection-related objectives, and therefore it acts as a sustainability-oriented plan which becomes part of the planning process itself [3]. This is of particular importance with reference to the definition of conservation measures, including the preparation of MPs, as SEA can help integrate sustainability within MP objectives and can be regarded as a real and effective learning path for the administrations in charge of Natura 2000 sites [4]. Within the SEA, a fundamental issue is the assessment of the restoration of ecosystem services [5].

In the Communication from the Commission to the European Parliament, the Council, the European Economic and Social Committee and the Committee of the Regions (COM (2013) 249 final), a working definition of green infrastructure (GI) is proposed as follows: "A strategically planned network of natural and semi-natural areas with other environmental features designed and managed to deliver a wide range of ecosystem services. It incorporates green spaces (or blue if aquatic ecosystems are concerned) and other physical features in terrestrial (including coastal) and marine areas. On land, GI is present in rural and urban settings" [6]. Moreover, the Commission puts in evidence how and how much the issue of GI relates to the Network of Sites of community importance (SCIs), Special areas of conservation (SACs) and Special protection areas (SPAs): "The work done over the last 25 years to establish and consolidate the network means that the backbone of the EU's GI is already in place. It is a reservoir of biodiversity that can be drawn upon to repopulate and revitalize degraded environments and catalyze the development of GI [7]. This will also help reduce the fragmentation of the ecosystem, improving the connectivity between sites in the Natura 2000 Network and thus achieving the objectives of Article 10 of the Habitats Directive" [8].

Hence, it is evident that the definition of GI is strictly connected to the category of ecosystem services. Moreover, it has to be a planned network [9]. Spatial planning, at the regional and urban levels, is an important and effective perspective to address the complex issue of defining, implementing and managing networks of ecosystem services and GI [10].

As a consequence, GI has a decisive role in promoting the restoration of biodiversity and in reducing the fragmentation of ecosystems, and eventually, in their capability of delivering ecosystem services [11]. Therefore, a general goal of the SEA of MPs of Sites of the Natura 2000 Network can be defined in order to address the issue of the role of GI in promoting and enhancing habitat restoration and the delivery of ecosystem services [12].

A wide range of issues related to ecosystem services and GI, as important points of reference for spatial planning, related to urban and rural contexts [13], with particular reference to the definition and implementation of planning policies aimed at protecting nature and natural resources [14], are discussed in the articles of the Special Issue.

The published studies are related to four main questions, which can be highlighted as follows. A first point focuses on the different approaches to build a consistent Natura 2000 Network. As per Lai's article, outstanding differences are still in place, which generate relevant problems with regards to the completion of the establishment process of the Natura 2000 Sites. These inconsistencies can be generalized as potential hindrances concerning the implementation of the Network across the whole European Union.

A second relevant issue concerns the use of planning methodologies based on the definition of spatial policies which build on ecosystems and GI as supply sources of publicly available services in urban contexts. This theme is treated by: i. Colavitti, Floris and Serra with reference to the endowment standards of public services in urban plans; ii. Moura and Fonseca, with regards to the vulnerability of urban environments related to vegetation depletion, which can be effectively addressed through appropriate GI planning; iii. Santoro, Balena and Camarda, in relation to the comprehensive planning approach of the Master Plan of the Italian city of Bari; iv. Garau and Annunziata, with reference to the definition and implementation of a GI aimed at improving the attractiveness and livability of urban environments for children; v. Lai, Leone and Zoppi, with regards to the identification of spatial policies to support the implementation of a regional GI by enhancing its components at the local scale, represented by three urban areas of the Metropolitan City of Cagliari.

Thirdly, a clear cut research line is identified by the use of ecosystem services and GI to implement spatial policies aimed at protecting and enhancing nature, natural resources and environmental quality. The studies related to this issue are presented in the articles by: i. Scorza, Pilogallo, Saganeiti and Murgante on the features of the relationship between the protection of nature and environmental recovery concerning a Natura 2000 Site and a nationally recognized polluted site located in the Basilicata region; ii. Cialdea on the use of environmental matrices to define planning policies in a coastal area which surrounds the town of Termoli, in the Molise region, in the context of regional landscape planning-oriented spatial processes; iii. Floris, Gazale, Isola, Leccis, Pinna and Pira on the implementation of ecosystem services-based planning policies to define and develop the integration of the MPs of Natura 2000 Sites into the spatial regulations of marine protected areas; the proposed methodological approach is applied to the marine protected area of the Asinara Island; iv. Fallanca, Taccone and Corazziere on the identification of natural and historical heritage as fundamental ecosystem services-related features to promote local development processes based on the enhancement of environmental quality, natural capital, and historical and cultural endowment; the methodological approach is implemented into the Metropolitan City of Reggio Calabria; v. Leone and Zoppi on the definition of a methodology to support decision-making processes in the implementation of the Integrated Coastal Zone Management Protocol [15] at the local level; objectives deriving from different plans, such as coastal land use plans (CLUPs) and MPs of Natura 2000 Sites (MPs), are analyzed and compared in terms of reciprocal consistency in order to integrate their planning strategies and to identify the potentially negative impacts of CLUPs on MPs; the methodological approach is implemented into three spatial contexts of the Sulcis area, located in southwest Sardinia.

Finally, a couple of articles concern the use of ecosystem services and GI to set up spatial policies aimed at defining and implementing regional plans, which integrate objectives related to the protection of nature and natural resources as well as economic and social development goals. These studies are proposed by: i. Balletto, Milesi, Ladu and Borruso, who present a methodological approach to tourism planning based on a GI finalized to increase the tourist attractiveness of the Sardinian Sulcis area; ii. Magaudda, D'Ascanio, Muccitelli and Palazzo, who discuss the relations between agricultural production and the implementation of ecological networks and connectivity in the conceptual framework of the Common Agricultural Policy; iii. Campagna, Di Cesare and Cocco, who present an application of the geodesign methodology to a strategic planning framework related to the Metropolitan City of Cagliari, which identifies and integrates into the planning process a metropolitan GI; iv. De Pascali, Santangelo, Perrone and Bagaini, who identify some general guidelines to implement adequate relations between the supply of provisioning and regulating ecosystem services, and the decentralization process of the energy systems; v. Balletto, Milesi, Fenu, Borruso and Mundula, who target the valorization issue of military properties located in areas attractive for leisure and tourism which are made partially available to the public due to agreements between the National Defense Ministry and the local municipalities; the potential of these areas is assessed in terms of the supply of cultural and tourism-related (recreational) ecosystem services, and an analytical methodological approach is implemented with regards to a military coastal area located in southeast Sardinia.

Funding: This research was partially funded by the Research Project based on the Agreement between Dipartimento di Ingegneria Civile, Ambientale e Architettura (Department of Civil, Environmental Engineering and Architecture, DICAAR) of the University of Cagliari, Italy, and the Autonomous Region of Sardinia, Departmental Office of Environment Protection, finalized to the objectives of the Project "GIREPAM–Integrated Management of Ecological Networks through Parks and Marine Areas" (Programme INTERREG Marittimo Italy-France Maritime 2014–2020, Axis 2).

Conflicts of Interest: The author declares no conflict of interest.

References

1. Costanza, R.; Folke, C. Valuing ecosystem services with efficiency, fairness, and sustainability as goals. In *Nature Services. Societal Dependence on Natural Ecosystems*; Daily, G.C., Ed.; Island Press: Washington, DC, USA; Covelo, CA, USA, 1997; ISBN 1559634758.
2. Chaudhary, S.; McGregor, A.; Houston, A.; Chettri, N. The evolution of ecosystem services: A time series and discourse-centered analysis. *Environ. Sci. Policy* **2015**, *54*, 25–34. [CrossRef]
3. Geneletti, D. Reasons and options for integrating ecosystem services in Strategic environmental assessment of spatial planning. *Int. J. Biodivers. Sci. Ecosyst. Serv. Manag.* **2011**, *7*, 143–149. [CrossRef]
4. Mascarenhas, A.; Ramos, T.B.; Haase, D.; Santos, R. Ecosystem services in spatial planning and strategic environmental assessment—A European and Portuguese profile. *Land Use Policy* **2015**, *48*, 158–169. [CrossRef]

5. Zoppi, C. Integration of conservation measures concerning Natura 2000 Sites into Marine Protected Areas regulations: A study related to Sardinia. *Sustainability* **2018**, *10*, 3460. [CrossRef]
6. Liquete, C.; Kleeschulte, S.; Dige, G.; Maes, J.; Grizzetti, B.; Olah, B.; Zulian, G. Mapping green infrastructure based on ecosystem services, and ecological networks: A pan-European case study. *Environ. Sci. Policy* **2015**, *54*, 268–280. [CrossRef]
7. Anton, C.; Young, J.; Harrison, P.A.; Musche, M.; Bela, G.; Feld, C.K.; Harrington, R.; Haslett, J.R.; Pataki, G.; Rounsevell, M.D.A.; et al. Research needs for incorporating the ecosystem service approach into EU biodiversity conservation policy. *Biodivers. Conserv.* **2010**, *19*, 2979–2994. [CrossRef]
8. Hansen, R.; Pauleit, S. From multifunctionality to multiple ecosystem services? A conceptual framework for multifunctionality in Green Infrastructure planning for urban areas. *AMBIO* **2014**, *43*, 516–529. [CrossRef] [PubMed]
9. Cannas, I.; Lai, S.; Leone, F.; Zoppi, C. Green infrastructure and ecological corridors: A regional study concerning Sardinia. *Sustainability* **2018**, *10*, 1265. [CrossRef]
10. Arcidiacono, A.; Ronchi, S.; Salata, S. Managing multiple ecosystem services for landscape conservation: A green infrastructure in Lombardy Region. *Procedia Eng.* **2016**, *161*, 2297–2303. [CrossRef]
11. Taylor-Lovell, S.; Taylor, J.R. Supplying urban ecosystem services through multifunctional green infrastructure in the United States. *Landsc. Ecol.* **2013**, *28*, 1447–1463. [CrossRef]
12. Partidário, M.R.; Gomes, R.C. Ecosystem services inclusive strategic environmental assessment. *Environ. Impact Assess. Rev.* **2013**, *40*, 36–46. [CrossRef]
13. Vallecillo, S.; Polce, C.; Barbosa, A.; Castillo, C.P.; Vandecasteele, I.; Rusch, G.M.; Maes, J. Spatial alternatives for Green Infrastructure planning across the EU: An ecosystem service perspective. *Landsc. Urban. Plan.* **2018**, *174*, 41–54. [CrossRef]
14. De Groot, R.S.; Alkemade, R.; Braat, L.; Hein, L.; Willemen, L. Challenges in integrating the concept of ecosystem services and values in landscape planning, management and decision making. *Ecol. Complex.* **2010**, *3*, 260–272. [CrossRef]
15. The Protocol on Integrated Coastal Zone Management (ICZM Protocol) to the Barcelona Convention (Council Decision 2010/631/EU) Was Ratified by the EU on September 13, 2010. The Protocol Was Adopted by the Council on December 4, 2008 (2009/89/EC). 2010. Available online: http://ec.europa.eu/environment/iczm/barcelona.htm (accessed on 11 May 2020).

sustainability

MDPI

Article

Hindrances to Effective Implementation of the Habitats Directive in Italy: Regional Differences in Designating Special Areas of Conservation

Sabrina Lai

Dipartimento di Ingegneria Civile, Ambientale e Architettura, University of Cagliari, 09123 Cagliari CA, Italy; sabrinalai@unica.it

Received: 18 February 2020; Accepted: 16 March 2020; Published: 17 March 2020

Abstract: "Natura 2000" is a coordinated network of protected areas that stretches across the European Union in compliance with two directives (the so-called "Habitats Directive" and the "Birds Directive") that underpin the Union's policies on biodiversity conservation. This study is aimed at assessing the implementation of the network by qualitatively analyzing how Special Areas of Conservation are being designated. Such designation process, which is being implemented, although with great delay, in a number of member states, entails the establishment of site-specific conservation measures that may be included within appropriate management plans or other development plans. A systematic documental analysis of official acts establishing Special Areas of Conservation and approving conservation measures and management plans was performed by taking Italy as a case study. The analysis focuses on four key topics, as follows: use of conservation measures and appropriate management plans; multi-level governance of the Natura 2000 sites, in terms of involved institutions and tiers of government; stakeholders' inclusion in the designation process; and the relationship between conservation measures and the wider spatial planning system. The results show significant differences regarding the implementation of the Natura 2000 network and highlight potential general hindrances to completing the designation process in the European Union.

Keywords: Natura 2000 network; natural protected areas; environmental planning; multi-level governance

1. Introduction

Biodiversity protection in the European Union (EU) is grounded on two cornerstones: First, the EU Biodiversity Strategy, initiated in 1998 [1] and revised in 2011 [2]. Second, a legal framework whose main pillars are Council Directive 92/43/EEC of 21 May 1992 on the conservation of natural habitats and of wild fauna (the so-called "Habitats Directive"), Directive 2009/147/EC of the European Parliament and of the Council of 30 November 2009 on the conservation of wild birds, and the codified version of Council Directive 79/409/EEC of 2 April 1979 (the so-called "Birds Directive"). The legal framework establishes a strict protection regime for wildlife and their habitats; in addition, it creates an international, yet coordinated, network of protected areas aimed at maintaining or restoring biodiversity at a favorable conservation status both inland and at sea. Named "Natura 2000", this network comprises Sites of Community Importance (SCIs), Special Areas of Conservation (SACs), and Special Protection Areas (SPAs), and it spreads across more than 18% of the EU territory and over 530,000 square kilometers of sea waters [3].

If we were to follow Ostrom's view [4] (p. 1), Natura 2000 could be regarded as the EU tool designed to "govern natural resources used by many individuals in common" where an institution sets limits on the use of natural resources so as to ensure their long term maintenance and, by doing so, even their benefits (be they economic or non-economic) for people. It is considered in the literature as a champion example of an international network of protected areas [5,6] and as an outstanding implementation of biodiversity-related spatial policies [7], to the extent that Campagnaro et al. [8] state that it has made "unprecedented advances in implementing effective, evidence-based, internationally collaborative conservation policies and practices at continental scales".

But how can the effectiveness of biodiversity conservation policies be evaluated? Rauschmayer et al. [9] suggest that this can be done by assessing either their outcomes or their implementation processes.

As for outcome assessments, several studies have evaluated improvements in the conservation statuses of species and habitats since the establishment of the Natura 2000 network so as to assess its contribution to biodiversity protection. Some studies consider a specific country, or a set of countries: Among these, for instance, Maiorano et al. [10], McKenna et al. [11], and Trochet and Schmeller [12]. Other studies focus on a certain species, or a certain taxonomic rank (for instance, [13–15]) to analyze trends in their conservation statuses and understand whether such trends are linked to the establishment of the protection regime in force in Natura 2000 sites.

As for process evaluation, a number of interdisciplinary studies analyzed the establishment of the Natura 2000 network [16,17], its management [18], participation processes [19–22], and conflicts [23,24].

Under article 4.4 of the Habitats Directive, within six years from their establishment SCIs (identified by member states on scientific grounds only) are bound to be designated as SACs. The SAC designation process must be preceded by the establishment of site-specific "necessary conservation measures involving, if need be, appropriate management plans specifically designed for the site or integrated into other development plans" (article 6.1); such measures need to take "account of economic, social, and cultural requirements and regional and local characteristics" (article 2.3). Thus, as noted by Bouwma et al. [25], member states are provided with some degree of freedom as regards the management of the network, as they can choose either to "simply" identify conservation measures that should be implemented or to integrate them within planning tools (either existent ones or ones prepared ad-hoc—in the latter case, they are termed "management plans"). The voluntary character envisioned in the directive notwithstanding, ad-hoc plan-making processes (i.e., the definition of management plans containing the required conservation measures) are regarded in several studies as the way forward to mitigate conflicts [26,27] and as the optimal tool to include stakeholders' participation [16,28–30]; management plans have therefore been found to be preferred over stand-alone conservation measures in a number of member states and have been promoted by the EU [25]. Moreover, inclusive plan-making processes are regarded in the literature as a proper counterbalance to the top-down process whereby Natura 2000 sites were identified and designated [31].

However, the form that management plans take is heavily shaped by the way spatial planning is understood and implemented in the various member states as regards goal setting and regulation [25]; in addition, the practical implementation of management plans for biodiversity conservation can conflict with planning tools already in place at the regional and local levels, prominently land-use plans and territorial plans for economic development [32]. Hence, their effectiveness depends on the extent to which spatial planning integrates Natura 2000 sites [33], as land use regimes and policies should often be adjusted [34] to meet conservation goals, and therefore also on the planning system in force and the extent to which it allows for such integration.

Moreover, huge delays have generally been observed in setting conservation objectives and putting management plans in place [35,36]. This calls for more research to understand the reasons behind such low compliance with the directive, possibly signaling that competent authorities did not agree on what needs to be achieved on the sites [35], which, in turn, is highly likely to hinder progresses towards the broad objectives of biodiversity conservation.

So far, only a few studies have investigated the establishment of conservation measures (also including plan-making processes); one reason behind the low interest of the academic literature might possibly be attributed to the low involvement of the academic world in the establishment of conservation measures (eventually included within management plans). Among these studies,

Gil et al. [37], who propose a step-by-step methodology to prepare a Natura 2000 site management plan, tested in the Azores region in Portugal; Kovacs et al. [38], who analyze the way in which management plans were drafted for 25 Hungarian sites; and Cortina and Boggia [39], who define a methodology aimed at ascertaining whether, within a certain Natura 2000 site, the establishment of appropriate conservation measures is sufficient to pursue the directive's goals, or if a management plan is necessary.

Building upon current literature, this study is aimed at addressing the research gap identified above by analyzing the ongoing process through which SACs are being designated by means of an analysis of official documents that establish Natura 2000 conservation measures (CMs) or approve management plans (MPs).

Italy is selected as a case study because the SAC designation process has almost been completed, although with great delay with respect to the six-year deadline mandated by the Habitats Directive; such delay in October 2015 led to a formal notice (infringement procedure no. 2015-2163) from the European Commission, and, in January 2019, to an additional formal notice because of the still unachieved compliance. Similar formal notices, concerning issues in SAC designation, were also issued to other member states, including Portugal (no. 2015-2002), Spain (no. 2015-2003), Ireland (no. 2015-2006), Greece (no. 2015-2260), Germany (no. 2015-2262), Bulgaria (no. 2018-2352), and Slovakia (no. 2019-2141); none of these infringement procedures has been closed so far [40], which signals a general difficulty in fully implementing the Natura 2000 network, and the Italian case can shed some light on such issues.

The following section introduces the case study and provides the reader with both quantitative data on the Italian Natura 2000 network and a brief historic and normative description concerning the roles entrusted to the various institutions involved in the sites' identification as well as in the establishment and management of the network. In the third section, the qualitative methodology is presented, while results of the analysis are provided with reference to each Italian region or autonomous province in the fourth section. Next, the results are discussed in the fifth section, focusing on four topics as follows: the integration of CMs and MPs in ordinary spatial and territorial plans; differences, across regions, as to institutions and tiers of government involved in the management and planning of Natura 2000 sites; stakeholders' inclusion in the identification of CMs and preparation of MPs; and the nature and role of CMs and MPs in the Italian multilevel and hierarchical planning framework. Final conclusions are provided in the sixth section.

2. The Natura 2000 Network in Italy

The Natura 2000 network in Italy stretches over 64,124 km^2, of which land area amounts to 57,265 km^2, corresponding to approximately 19% of the national territory, and marine area amounts to 6859 km^2 [3]. To date, 2613 sites have been designated in Italy, of which 2335 have been identified as SCIs under the Habitats Directive. For 2307 SCIs, the six-year deadline for their designation as SACs has expired; as of December 31st 2019, 2261 sites have been designated as SACs (data retrieved from the institutional website of the Italian Ministry of the Environment and Land and Sea Protection http://www.minambiente.it/pagina/zsc-designate). Out of these 2261 SACs, only 403 had been designated before the infringement procedure was issued, thus well after the deadline had expired (it is therefore highly likely that such notice greatly accelerated the designation process), while the process is yet to be completed for the remaining SCIs.

Natura 2000 sites in Italy were initially identified within the "Bioitaly" project (1994-1998) by the Ministry for the Environment, supported by national scientific societies [41], and with the involvement of lower tiers of government [42], i.e., regions and the autonomous provinces of Trento and Bolzano, in turn supported by local universities and research centers [43]. On purely scientific bases, in compliance with the directive, around 2800 sites [44] were thus identified, which led to conflicts and resistance, both from stakeholders and municipalities [45], who complained about being left out of the process. A few years later, in April 2000, the Minister of the Environment issued a decree that also approved, besides the first list of SPAs, a preliminary list of proposed SCIs to be forwarded to the European Commission for designation (due to the encountered opposition, the two lists did not include all of the 2800 sites that had preliminarily been identified), which had to be finalized through a Commission Decision.

Following the sites' designation, each member state had to define the necessary site-specific CMs within six years of the lists being adopted (Commission Decisions approving lists of SCIs for the various biogeographic regions can be accessed from the EU Commission's website: https://ec.europa.eu/environment/nature/natura2000/biogeog_regions/index_en.htm); hence, as regards Italy, in 2010 for sites belonging to the Alpine and Continental biogeographic regions and in 2012 for the Mediterranean biogeographic region.

In Italy, the State (which is responsible for implementing the Habitats Directive) has delegated the identification of CMs and the preparation of MPs, as well as their approval, to its 19 regions and two autonomous provinces (as per the decree of the President of the Republic, no. 357/1997), while retaining, at the state level, the designation of SACs through a decree of the Ministry of the Environment and Land and Sea Protection.

Roles and responsibilities concerning the implementation of the Habitats Directive in Italy are therefore multilayered: First, the State, which proposes the lists of SCIs to the European Commission and designates the SACs; second, the European Commission, which adopts the lists of SCIs; third, the regions and autonomous provinces, which define and approve CMs and MPs as a prerequisite to SAC designation and are responsible for the management of the sites.

Figure 1 shows how the site designation process envisioned in the Habitats Directive works, from the preliminary identification of SCIs to the designation of SACs; moreover, it sums up the main aspects concerning the processes, approaches, and stakeholders' engagement. As portrayed in the image, the designation process is characterized by a rigid, hierarchical separation of the involved institutional tiers of government, and by technocratic approaches grounded only upon scientific data, in the absence of any participation of local communities and stakeholders who might possibly be involved in the sites' management. In other words, the first part of the process (SCI identification and designation), driven by scientific data and led by institutions, takes a top-down approach that leaves out local communities' and economic stakeholders' interests; such exclusion may—or may not—be compensated in the final part (SAC establishment and management), depending on how each member state designs the process.

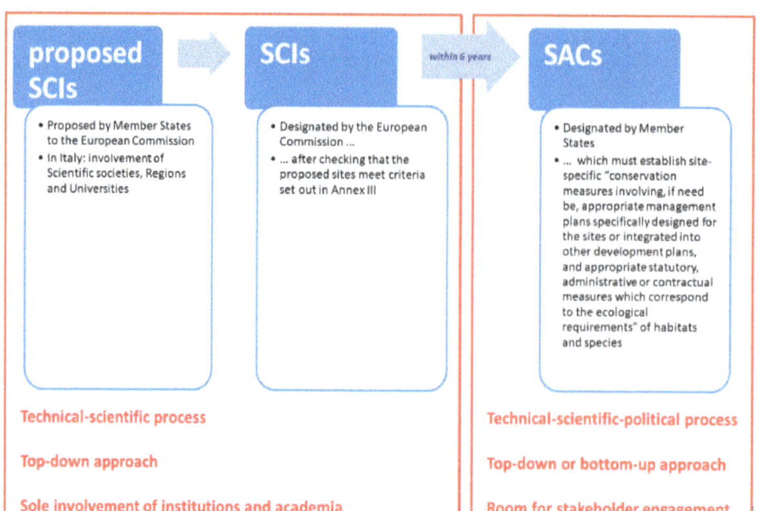

Figure 1. From preliminary identification of proposed Sites of Community Importance (SCIs) to designation of Special Areas of Conservation (SACs): roles, approaches, and stakeholders' involvement.

Figure 2 highlights the final part of the designation process shown in Figure 1 (hence, the transition from SCIs to SACs) and identifies the institutional tiers of government involved in the process in Italy, therefore highlighting the roles and responsibilities of the European Commissions, of the State, and of the regions and autonomous provinces. As previously mentioned, the Italian State has delegated the obligation to identify and approve CMs (sometimes included within MPs) to regions and autonomous provinces through the decree of the President of the Republic, no. 357/1997. Once CMs (or, optionally, an MP which includes them) are approved for a Natura 2000 site, the Ministry of the Environment and Land and Sea Protection checks them against the Habitats Directive's requirements, and, if the measures are deemed to be adequate, the site is established as a SAC through a ministerial decree. Finally, the State must provide information and official acts concerning SAC establishment to the European Commission, and it is also responsible for the six-year reporting on habitats' and species' conservation status, and therefore on the effectiveness of its Natura 2000 network in pursuing the ultimate goal of contributing to global biodiversity protection.

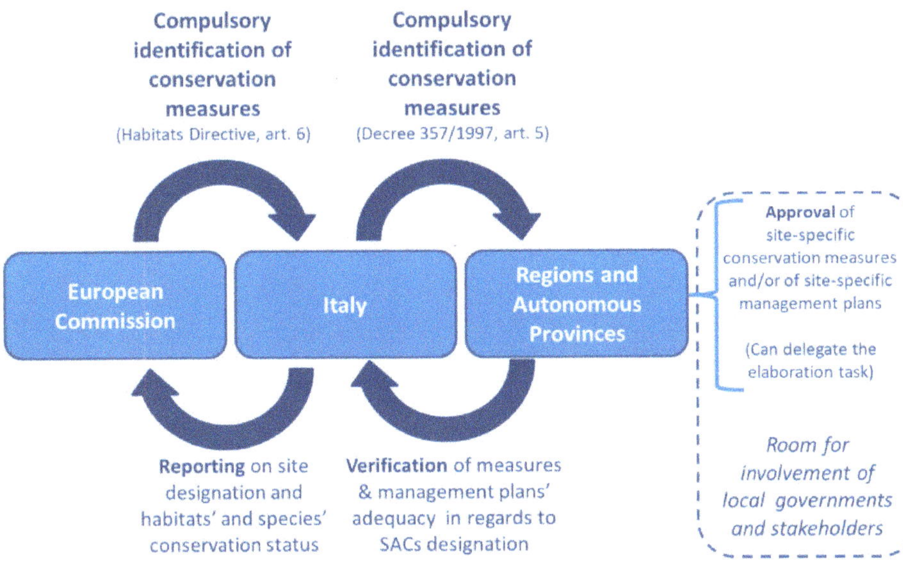

Figure 2. SAC designation in Italy: The State's obligations stemming from the Habitats Directive are transferred to the regions and autonomous provinces (upper arrows), which identify conservation measures and prepare management plans, while the State checks the measures, designate the SACs, and is accountable to the European commission (lower arrows).

3. Materials and Methods

For this research, a documentary case-study analysis was carried out. As stated in the Introduction, the SAC designation process in Italy has almost been completed, although with great delay with respect to the timeframe envisioned within the Habitats Directive. Thus, Italy is an "intrinsically interesting" [46] case study to analyze processes leading to SAC designation characterized by a high level of delay, because this case study can shed some light on institutional hindrances in designating SACs.

The documentary analysis was performed on official documents retrieved from the website of the Italian Ministry for the Environment and Land and Sea Protection [47] (monitored from July 2018 until December 2019), with a view to evaluating governance and inclusion mechanisms implemented in the management of the Natura 2000 Network in Italy. Since most of the official acts (i.e., the decrees of the State Minister for the Environment or deliberations of the regional executives) were available only as scanned documents, text mining techniques could not be applied. All of the official state and regional acts concerning SAC establishment were thoroughly read and analyzed, and their numbers are as follows:

- At the state level: approximately 70 ministerial decrees concerning SACs' designations;
- At the regional/local level: deliberations of local regional governments, or deliberations of natural parks' councils, approving site-specific CMs or MPs; a total of approximately 440 acts.

A preliminary "searching-out of underlying themes" of interest [48] (p. 557) was performed to double-check which information on issues emerging from the literature review, as presented in Section 1, could be retrieved. This search led to selecting the following four topics:

- Type of tool that was chosen, i.e., CMs or MPs, or both;
- Governance of the processes, i.e., institutions in charge of the definition of CMs and MPs;
- Level of inclusion within the processes, i.e., institutional and stakeholders' involvement in shaping CMs and MPs;
- Implications on spatial planning entailed by CMs and MPs.

Subsequently, once the topics of interest had been identified, all of the documents were scrutinized and fragments concerning the four topics were looked for, code, and noted at the site level.

According to Bryman [48] (p. 550), state-produced documents readily fulfill two of the four criteria that, according to Scott [49], should be looked for in social science research (i.e., authenticity and meaning), while issues of representativeness and credibility can sometimes be raised. Issue of representativeness (i.e., evidence that the documents analyzed are typical of their kind) can be excluded since, in this study, all of the available documents whereby SACs have been established in Italy were read and analyzed, without limiting the analysis to a sample. As for credibility, the official documents here analyzed do not provide any straight interpretations, opinions, or judgments, because they merely establish that a specific type of already-designated Natura 2000 site (namely, an SCI) is converted into another type (namely, an SAC) and enlist CMs to be enforced; however, the very fact that they approve a vision for the concerned sites, objectives to be pursued, and actions to be prioritized does provide an underlying official narrative about the significance or value of the sites.

4. Results

Results of the analysis are summarized in Table 1, which, for each region or autonomous province, provides: Progress towards designation completion (as: SCIs not yet designated as SACS, Column 2, and total number of SACs designated as of December 2019, Column 3); number of SACs designated based upon MPs or CMs (Columns 4 and 5); type of institution responsible for MP or CM preparation (Columns 6; note that their approval always lies with the region, or with the autonomous provinces of Trento and Bolzano in the case of Trentino-South Tirol); type of participation processes implemented, if any (Column 7); and whether any planning implication is entailed by MPs or CMs (Column 8).

Columns 2 and 3 in Table 1 show that the SAC designation process in Italy, although still incomplete, is nearing its end: If one takes into account that 20 out of the 74 SCIs still to be designated as SACs were established in 2017-2018 (which means that the six-year deadline has not passed yet for these sites), then conservation measures are in force in 97.67% of the sites, and the process has been completed in seven regions (Apulia, Basilicata, Campania, Liguria, Molise, Umbria, and Veneto). Figure 3 shows the geographical distribution of SCIs and SACs in Italy as of December 31, 2019, while Figure 4 shows how the process has progressed in time, by differentiating SACs on the basis of their designation year.

Table 1. The SAC designation process in Italy: Progress towards completion (number of SCIs still to be designated as SACs and number of SACs designated, Columns 2 and 3, respectively), number of SACs designated based on MPs and CMs (Columns 4 and 5), responsible institution for CM identification or MP preparation (Column 6), participatory processes carried out (Column 7), whether any planning relevance can be elicited from the official acts approving CMs and MPs (Column 8). All information is provided at the regional level. Source: Author's elaboration on documents retrieved from ftp://ftp.minambiente.it/PNM/Natura2000/Materiale%20Designazione%20ZSC - last access on December 31, 2019).

1) Regions and Autonomous Provinces	2) SCIs Not Yet SACs	3) No. SACs	4) No. SACs with MPs	5) No. SACs with CMs	6) Responsible Institution for CM/MP Preparation	7) Participatory Processes (*)	8) Planning Relevance
Abruzzo	3	193	---	193	CMs: Region MPs: Local authorities (MPs)	2	
Aosta Valley	1	76	32	44	Municipalities, Provinces, and Protected areas	2 (CMs and MPs), 4 (MPs)	
Apulia	---	85	60	25	Region, a Mountain community	2 (MPs by mountain communities), 2+3 (MPs by the region)	X (MPs only)
Basilicata	10	122	4	118	Region (CMs), Protected areas (MPs)	2, 4 (only in a few instances)	
Calabria	14	79	65	14	CMs: RegionMPs: Protected areas, Provinces, Municipalities	2, 4 (MPs); 1 (CMs)	X
Campania	16	207	203	4	Protected areas, Provinces, Regional forestry agency, University, non-governmental organizations	---	
Emilia-Romagna	4	40	---	40	Province	2	X
Friuli-Venezia Giulia	1	135	---	135	Province, Protected areas	2, 4	
Lazio	1	134	---	134	RegionFor one marine site overlapping one Marine Protected Area, the Ministry for the Environment	1	
Liguria	---	97	---	97	Region	2, 3	X (only as regards consultation)

Table 1. *Cont.*

1) Regions and Autonomous Provinces	2) SCIs Not Yet SACs	3) No. SACs	4) No. SACs with MPs	5) No. SACs with CMs	6) Responsible Institution for CM/MP Preparation	7) Participatory Processes (*)	8) Planning Relevance
Lombardy	---	104	---	104	Region	2	
Marche	3	193	---	193	CMs: RegionMPs: Local authorities (MPs)	2	
Molise	1	76	32	44	Municipalities, Provinces, and Protected areas	2 (CMs and MPs), 4 (MPs)	
Piedmont	---	85	60	25	Region, a Mountain community	2 (MPs by mountain communities), 2+3 (MPs by the region)	X (MPs only)
Sardinia	10	122	4	118	Region (CMs), Protected areas (MPs)	2, 4 (only in a few instances)	
Sicily	14	79	65	14	CMs: RegionMPs: Protected areas, Provinces, Municipalities	2, 4 (MPs); 1 (CMs)	X
Trentino-S.T. Bolzano province	16	207	203	4	Protected areas, Provinces, Regional forestry agency, University, non-governmental organizations	---	
Trentino-S.T. Trento province	4	40	---	40	Province	2	X
Tuscany	1	135	---	135	Province, Protected areas	2, 4	
Umbria	1	134	---	134	RegionFor one marine site overlapping one Marine Protected Area, the Ministry for the Environment	1	
Veneto	---	97	---	97	Region	2, 3	X (only as regards consultation)
TOTAL	74	2,261	460	1,801			

(*) 4: Meetings carried out during the plan-making or CM identification processes (both informative and participatory meetings; both with the general public and with selected stakeholders). 3: Public meetings following the plan-making or CM identification processes. 2: Written consultation on a pre-adopted and published version of CMs and MPs (the general public can submit written remarks and suggestions). 1: Consultation restricted to institutions only—no recorded participation.

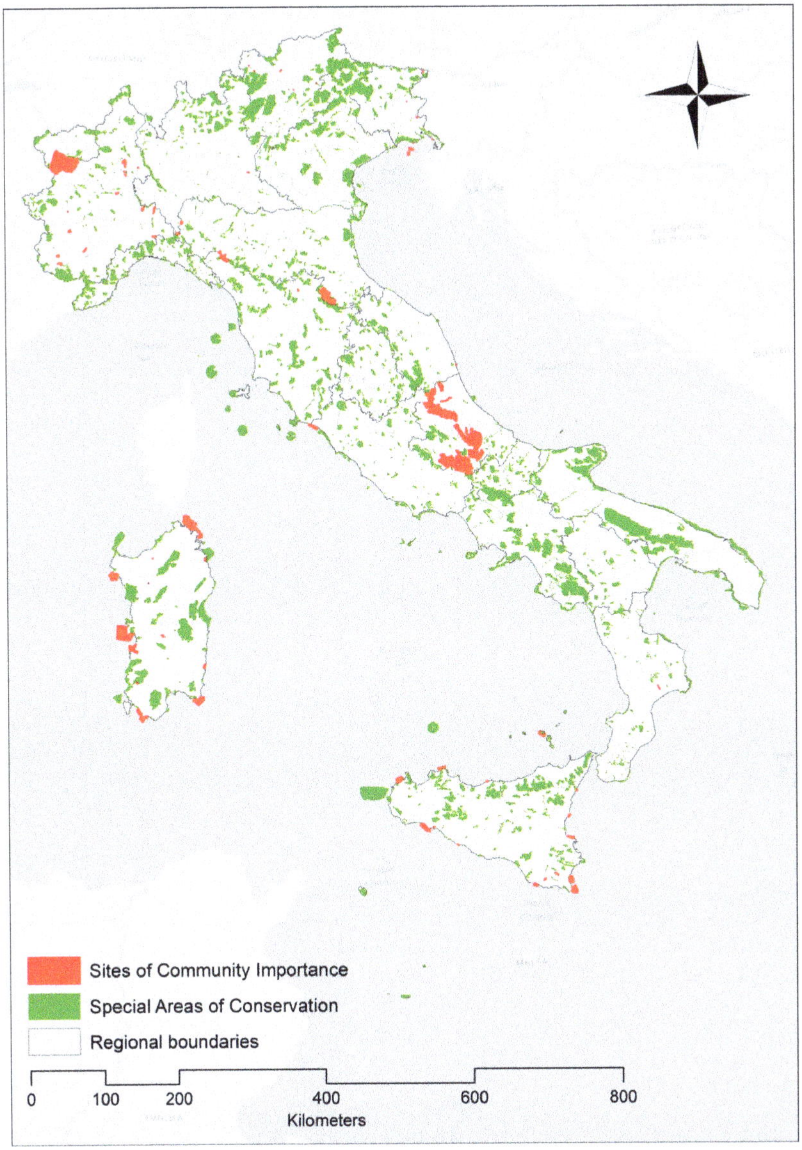

Figure 3. Sites of Community Importance and Special Areas of Conservation in Italy: designation process as of December 2019 (source: Author's own elaboration; Natura 2000 sites' boundaries retrieved from ftp://ftp.minambiente.it/PNM/Natura2000/TrasmissioneCE_dicembre2017/, last access on December 31, 2019).

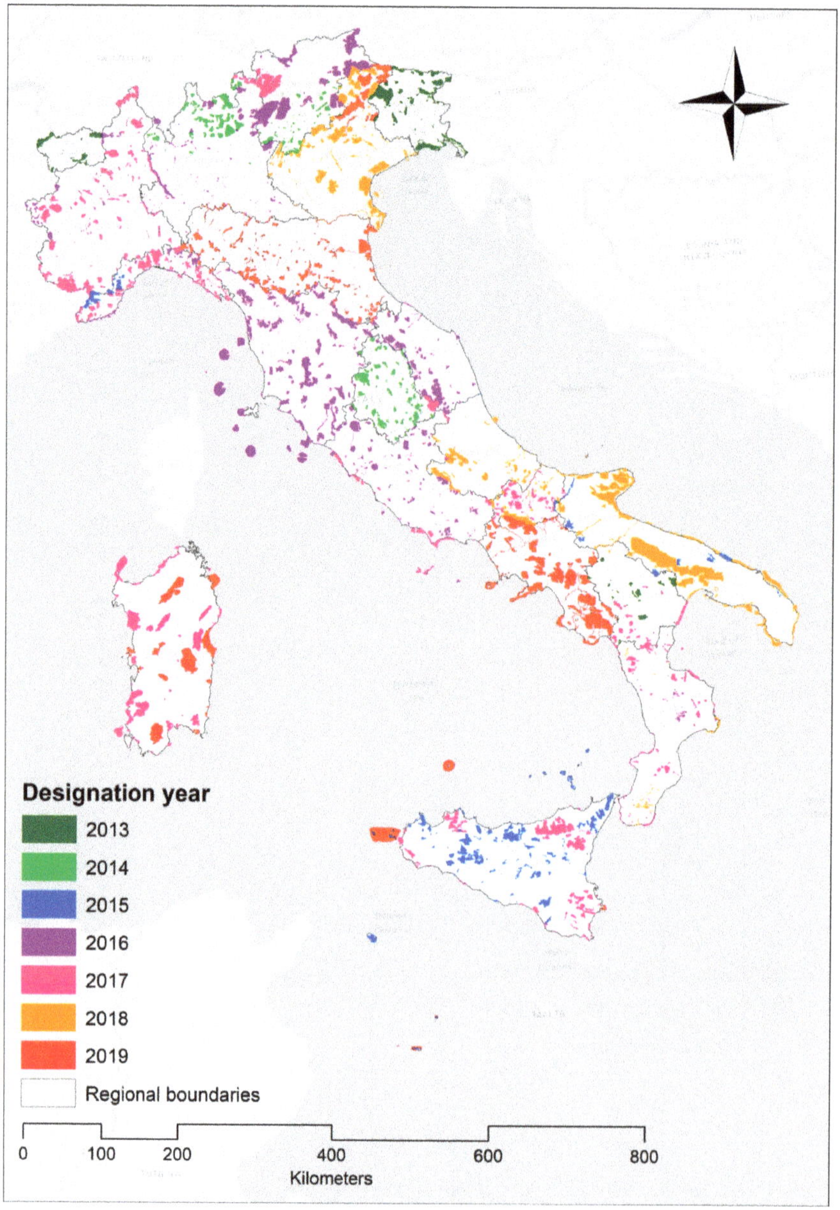

Figure 4. Special Areas of Conservation in Italy by designation year (source: Author's own elaboration; Natura 2000 sites' boundaries retrieved from ftp://ftp.minambiente.it/PNM/Natura2000/TrasmissioneCE_dicembre2017/, last access on December 31, 2019).

Column 4 and 5 in Table 1 provide quantitative information on the number of SACs that were designated based on MPs and CMs, respectively. Eleven regions have made use of both CMs and MPs; eight regions and the two autonomous provinces of Bolzano and Trento have approved only CMs, while in no region were all of the SACs designated based on MPs. A graphical representation is provided in Figure 5, where SACs are differentiated based on whether CMs or MPs were used for their designation; this figure also highlights that, for four marine sites overlapping as many Marine Protected Areas (MPAs) (one in Tuscany and three in Lazio), SACs where designated on the basis of their MPA regulatory tools only, defined at the state level (by the Ministry of the Environment and Land and Sea Protection), which are regarded as the sites' CMs.

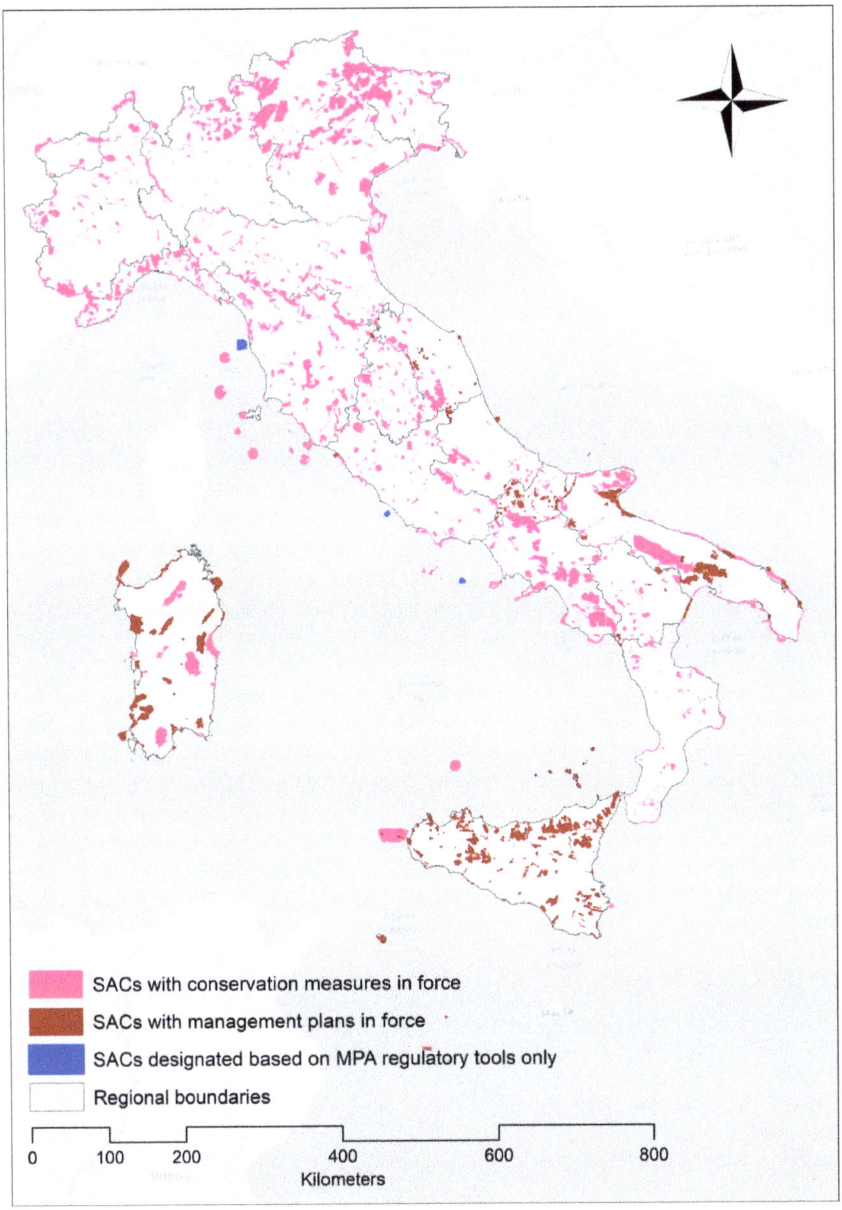

Figure 5. Special Areas of Conservation in Italy for which the official designation acts refer to conservation measures, management plans or (exclusively) MPA regulatory tools (source: Author's own elaboration; Natura 2000 sites' boundaries retrieved from ftp://ftp.minambiente.it/PNM/Natura2000/TrasmissioneCE_dicembre2017/, last access on December 31, 2019).

Column 6 provides information on which institution prepared CMs and MPs: While regions, or in the cases of Trentino Alto Adige and the autonomous provinces of Trento and Bolzano, retain the power to approve CMs and MP, they can delegate their preparation to other institutions. Each of the regions has therefore decided whether to carry out this task itself (some, such as Basilicata, have established internal offices or working groups while others, such as Veneto, have involved a local university, and others, like Molise or Apulia, have contracted external consultants) or whether to delegate lower tiers of government, such as provinces, municipalities, or mountain communities (which, in Italy, were statutory associations of mountain and hill municipalities). In the case of a Natura 2000 site overlapping a natural protected area such as a natural park or a nature reserve, some regions (for instance, Calabria, Lazio, Marche, Piedmont, and Sardinia), as well as the autonomous province of Trento, have delegated the task to the institution in charge of the area. Finally, a small number of other organizations appear to have been delegated, but this is a very uncommon choice, only made by two regions (Liguria: the Regional Environmental Agency; Sicily: the Regional forestry agency, a University, and some non-governmental organizations).

Column 7 in Table 1 summarizes information on whether, and how, participation was implemented in the making of CMs and MPs. Participation forms significantly vary among regions, with reference not only to the categories of institutional actors and stakeholders involved, but also to types and timing of participatory processes. In two cases only (Calabria and Sicily) neither participatory nor consultative process are recorded in the official acts here examined; in some other cases (Tuscany, Aosta Valley, and, to some extent, Emilia Romagna and Sardinia) consultation appears to have been restricted to institutions only. In most regions and autonomous provinces, official acts record higher-level participatory processes that took different forms, of which the most common is the written consultation on a pre-adopted and published version of CMs and MPs, which mirrors the traditional participatory process implemented in statutory planning in Italy. Truly participative processes that involve a Natura 2000 sites' key stakeholders (such as farmers, hunters, forest managers, and tourism businesses, as well as local communities) and that are carried out during the plan-making process or CM identification, which could in theory lead to better agreed-upon CMs, are rarer. Surprisingly, even rarer are informative meetings held after MP and CM approval.

Finally, Column 8 in Table 1 signals whether any planning-related content was detected in the official acts approving CMs and MPs. Such content can take various forms: references to the regional or provincial planning laws in force (e.g., Apulia, Friuli-Venezia Giulia, Molise, and the autonomous province of Bolzano) either as regards the approval procedure or the relationship between conservation measures and land-use plans and zoning schemes; call for integration of conservation measures within local planning tools (e.g., Basilicata); references to the Strategic Environmental Assessment (e.g., Apulia, Basilicata, Marche, Sardinia), a procedure envisioned in EU Directive 2001/42/EC "on the assessment of the effects of certain plans and programs on the environment", which some regions regard as mandatory for MPs and some explicitly rule out.

5. Discussion

The analysis has put in evidence remarkable differences concerning the SAC designation process among Italian regions; these differences concern a number of aspects, such as: the role played by the various tiers of government involved; the way socio-cultural-economic considerations, which must be taken into account according to the Habitats Directive, are incorporated in the process; and the nature of CMs and MPs.

First, as for the type of tool, SACs have mostly been established based on CMs in all of the regions and autonomous provinces; in some regions, MPs were used as well, and they only prevailed (as far as the number of sites is concerned) in four regions. This is not surprising and is consistent with the provision of the Habitats Directive because only site-specific CMs are required, while MPs are optional. In other EU member states, an MP is mandatory for each Natura 2000 site [45,50–53], while in Italy the literal provision of the directive has been transposed into the national law regulating

Natura 2000 network and retained within regional practices. In this regard, Ferranti et al. [52] have suggested that CMs were integrated within other development plans in most Italian regions. However, the analysis of official documents designating SACs, which enlist the measures in force in each site, has highlighted that none of the acts mention other development plans within which CMs could possibly have been integrated (such as, for instance, regional forestry plans, river basin MPs, or regional landscape plans). On the contrary, official regional acts only record a very limited number of cases where CMs were integrated within sectoral or territorial plans other than MPs, but even in such cases, integration only concerns planning tools specifically aimed at preserving biodiversity, such as plans for natural protected areas (i.e., national parks or nature reserves) or MPAs' regulatory tools. Indeed, this is consistent with Bouwma et al. [54,55], who have found that existing instruments are seldom used to implement EU policies, while new ad-hoc tools (such as, in this case, CMs or MPs) are usually developed, notwithstanding the absence of any explicit requirement to do so from the directive. This extremely low level of integration calls into question the extent to which nature and biodiversity are integrated within territorial plans, which in turns points to the extent to which human beings perceive the importance of biodiversity, which is irreplaceable in sustaining the natural processes needed to support human life and development. Such general low awareness is also at the root of perceived conflicts between biodiversity conservation and economic development. Hence, research that looks into nature's contributions to people, with particular reference to protected areas (e.g., [56–58]), on the inclusion of ecosystem services' tradeoffs and their assessment within planning processes [59–62], and on the ecosystem approach to spatial planning [63] could, in principle, contribute to raising awareness on the multiple benefits that nature and biodiversity provide to human beings, and at the same time provide new tools to manage the perceived conflicts.

Second, as regards the governance of the SAC designation processes, the European Commission's and member states' competences are clearly defined in the Habitats Directive. However, this study has highlighted that in Italy some biodiversity conservation imperatives that are key to achieving and efficiently managing the Natura 2000 network, such as species monitoring programs, conservation measures definition, and management plan preparation and implementation, have been transferred from the central state to the regions and autonomous provinces. Such transfer of competencies, which has now been in place for over two decades, has left plenty of room for different interpretations among regions. This is clearly shown, for instance, by the fact that some regions have identified CMs and prepared MPs themselves, while others have delegated this task to lower tiers of government, which are probably regarded as closer to local communities and therefore as a better choice to understand social and economic needs and expectations and account for them within CMs and MPs. In this regard, as direction for future research, this study points to the need to better understand whether there is any correlation between the tier of government at which CMs are identified (or MPs are prepared) and their effectiveness in contributing to maintaining habitats and species at a favorable conservation status and ultimately at halting biodiversity loss. To answer this question, further research that looks at whether the choice of institutional level impacts, on the one hand, on biodiversity health and, on the other hand, on the socio-economic conditions of the territories in which Natura 2000 sites have been established, would be needed.

Third, this study has shown that participation in the SAC designation process significantly varies among regions, as regards the form (e.g., written comments versus participation in meetings), the timing (e.g., within the CM or MP preparation, or at the very end), and the consulted actors (e.g., institutions, key stakeholders, local communities). It is worth remarking that, in this article, only the documented implementation of the participatory processes was analyzed. Two recent works [21,22] have proposed an assessment of the participatory processes carried out in some Italian regions; by surveying a small group of stakeholders, the authors have examined issues of inclusiveness, democracy, and conflicts concerning Natura 2000 management and planning, and have argued that only selected stakeholder groups were involved and that information, rather than participation, often took place, seldom allowing for effective integration of local knowledge within MPs. Hence, the authors have examined (not randomly or systematically, but through a snowball sampling technique) stakeholders' perceptions concerning their participation in the making of MPs, while this work has proposed a systematic assessment of official documents that have recorded such participation. Therefore, further research could bridge the two assessments and combine a systematic survey of stakeholders who took part (in whichever form) in SAC designation processes with the documental analysis here presented, so as to compare stakeholders' perceptions (on which focus, for instance, [31] and [64]), with the official narrative, hence ultimately assessing participation effectiveness.

Finally, implications on spatial planning only emerge in a limited number of cases and with contrasting perspectives. In the autonomous province of Bolzano, CMs are subject to the local planning law, similarly to what happens in Friuli Venezia Giulia, where, in compliance with the regional law no. 7/2008, the official acts approving CMs state that, in cases of differences or contradictions, CMs prevail over land use plans and regulations; moreover, the law also mandates that MPs are to be regarded as environmental territorial plans and that municipal land-use plans have to comply with MPs; this latter provision is also stated with reference to MPs approved by the Molise region. In the Apulian case, all of the acts whereby MPs are approved explicitly state that MPs constitute a sectoral planning tool, whose provisions integrate or replace those of the statutory and legally-binding land-use plans in force in the municipality whose territory overlaps the Natura 2000 site of concern; some acts also add that municipal land-use plans need to be revised in compliance with MPs. Moreover, for one Apulian site, the relevant act provides evidence that a Strategic Environmental Assessment procedure was carried out. In such cases, CMs (and MPs as well) are indeed regarded as territorial planning tools. Conversely, Basilicata Region's official acts, on the one hand, declare that CMs must be integrated within land use plans and sectoral plans; on the other hand, they state that MPs are not territorial plans and therefore they are not subject to the Strategic Environmental Assessment procedure. Similarly, MPs are not subject to the Strategic Environmental Assessment procedure also in the Marche region, but on different grounds: In this case, this procedure was regarded as not needed in the absence of actions for which an Environmental Impact Assessment procedure (first introduced in the EU by Council Directive 85/337/EEC) would have been mandatory, and in the absence of any significant adverse impact of the plan's action on the environment. On the contrary, a Strategic Environmental Assessment (also providing a legal framework for stakeholders' inclusion) was considered necessary in Sardinia, because most MPs also comprise interventions aimed at enhancing local assets and resources, or at supporting sustainable local economies, rather than at preserving biodiversity. Such differences concerning the very essence of CMs and MPs (as well as their binding or non-binding character) signal that an in-depth analysis of administrative and urban planning laws is required, possibly leading to a unified (national) legal framework across Italian regions. This would also reinforce the straightforward, yet very important, role played by Natura 2000 sites in spatial planning since the early 1990's, as such areas have contributed, through the Appropriate Assessment Procedure [65], to halting, preventing, or hindering urban and infrastructure development, which has ultimately resulted in supporting biodiversity conservation.

6. Conclusions

This study has examined the processes leading to SAC designation in Italy through a systematic qualitative documental analysis of official acts, issues of multi-level governance, inclusion, and the nature and role of CMs and MPs, including their relationship with the wider planning system.

Significant regional differences in the processes through which CMs and MPs have been defined and approved have emerged, which might be an Italian peculiarity because key functions and competences concerning the implementation of the Habitats Directive in Italy have been delegated from the state to regions and autonomous provinces. However, since general delays in establishing and managing SACs have been reported in the literature, the fragmented variety of Italian experiences (for instance, concerning participatory processes or the relationship between CMs and statutory planning and, ultimately, the perception and understanding concerning the very essence of CMs) in this regard could parallel that of a number of other EU countries.

Compared to other studies that have examined the implementation of the Habitats Directive, conflicts and participation in site designation, planning, and management, this study has taken a novel approach in that it has investigated the official institutional narrative through a thorough document analysis. Such narrative is usually neglected, and yet it should necessarily complement that of the involved communities and stakeholders, because Natura 2000 is an institutional, top-driven spatial policy that is grounded on a solid, multi-level legal framework comprising (at the very minimum, depending on each member state) directives and national laws. By doing so, this study also contributes to shedding some light on the establishment of conservation measures (eventually included within management plans), which is, so far, still an under-researched topic in the scientific literature, possibly due to the low involvement of the academic world in a process led by public institutions (at the national, regional, or local level, depending on the member state) and typically involving practitioners in natural sciences, forestry, and the marine environment, as far as the identification of measures is concerned.

According to Ferranti et al. [66], biodiversity conservation in the EU has historically been approached in various ways, consisting of a circular path; from an initial technocratic, science-driven, top-down approach (quite apparent in scientific grounded site designations) to a limited inclusion of selected local stakeholders such as farmers or hunters in site management, to the wider inclusion of other economic stakeholders less connected to the sites and yet still affected by site management and restrictions (for instance, tourism businesses), back to technocracy in the current phase, quite informed by the latest environmental economics research on natural capital and ecosystem services, which, according to the authors, results in a renewed marginalization of local communities. The fragmented and varied Italian experience concerning SAC designation suggests that, rather than a sequence of historical phases, the coexistence of various approaches, some more technocratic and some more democratic and inclusive, can be observed depending on the region.

Funding: This research received no external funding.

Conflicts of Interest: The author declares no conflict of interest.

References

1. European Commission. Communication from the Commission to the Council and the European Parliament on a European Community Biodiversity Strategy. COM (1998)42 final, 1998. Available online: https://eur-lex.europa.eu/legal-content/en/TXT/PDF/?uri=%20CELEX:51998DC0042 (accessed on 31 December 2019).
2. European Commission. Communication from the Commission to the European Parliament, the Council, the Economic and Social Committee and the Committee of the Regions Our Life Insurance, Our Natural Capital: An EU Biodiversity Strategy. COM (2011)0244, 2011. Available online: https://eur-lex.europa.eu/legal-content/en/TXT/PDF/?uri=CELEX:52011DC0244 (accessed on 31 December 2019).
3. European Environmental Agency. Natura 2000 Barometer, 2018. Available online: https://www.eea.europa.eu/data-and-maps/dashboards/natura-2000-barometer (accessed on 31 December 2019).
4. Ostrom, E. *Governing the Commons: In The Evolution of Institutions for Collective Action*; Cambridge University Press: Cambridge, UK, 1990.
5. Lockwood, M. Global protected area framework. In *Managing Protected Areas: A Global Guide*; Lockwood, M., Graeme, V., Kothari, A., Eds.; Cromwell Press: Trowbridge, UK, 2006.
6. Kukkala, A.S.; Arponen, A.; Maiorano, L.; Moilanen, A.; Thuiller, W.; Toivonen, T.; Zupan, L.; Brotons, L.; Cabeza, M. Matches and mismatches between national and EU-wide priorities: Examining the Natura 2000 network in vertebrate species conservation. *Biol. Conserv.* **2016**, *198*, 193–201. [CrossRef]
7. Popescu, V.D.; Rozylowicz, L.; Niculae, I.M.; Cucu, A.L.; Hartel, T. Species, habitats, society: An evaluation of research supporting EU's Natura 2000 network. *PLoS ONE* **2014**, *9*, e113648. [CrossRef] [PubMed]
8. Campagnaro, T.; Sitzia, T.; Bridgewater, P.; Evans, D.; Ellis, E.C. Half Earth or whole Earth: What can Natura 2000 teach us. *BioScience* **2019**, *69*, 117–124. [CrossRef]

9. Rauschmayer, F.; Berghöfer, A.; Omann, I.; Zikos, D. Examining processes or/and outcomes? Evaluation concepts in European governance of natural resources. *Environ. Policy Gov.* **2009**, *19*, 159–173. [CrossRef]
10. Maiorano, L.; Falcucci, A.; Garton, E.O.; Boitani, L. Contribution of the Natura 2000 Network to biodiversity conservation in Italy. *Conserv. Biol.* **2007**, *21*, 1433–1444. [CrossRef]
11. Mckenna, D.; Naumann, S.; Mcfarland, K.; Graf, A.; Evans, D. Literature Review. The Ecological Effectiveness of the Natura 2000 Network. Technical Paper n. 5/2014, 2014. Available online: https://bd.eionet.europa.eu/Reports/ETCBDTechnicalWorkingpapers/PDF/Ecologic_Effectiveness_of_Natura2000_LitReview_Report.pdf (accessed on 31 December 2019).
12. Trochet, A.; Schmeller, D. Effectiveness of the Natura 2000 network to cover threatened species. *Nat. Conserv.* **2013**, *4*, 35–53. [CrossRef]
13. D'Amen, M.; Bombi, P.; Campanaro, A.; Zapponi, L.; Bologna, M.A.; Mason, F. Protected areas and insect conservation: Questioning the effectiveness of Natura 2000 network for saproxylic beetles in Italy. *Anim. Conserv.* **2013**, *16*, 370–378. [CrossRef]
14. Zehetmair, T.; Müller, J.; Runkel, V.; Stahlschmidt, P.; Winter, S.; Zharov, A.; Gruppe, A. Poor effectiveness of Natura 2000 beech forests in protecting forest-dwelling bats. *J. Nat. Conserv.* **2015**, *23*, 53–60. [CrossRef]
15. Votsi, N.E.P.; Zomeni, M.S.; Pantis, J.D. Evaluating the effectiveness of Natura 2000 network for wolf conservation: A case-study in Greece. *Environ. Manag.* **2016**, *57*, 257–270. [CrossRef]
16. Alphandéry, P.; Fortier, A. Can a territorial policy be based on science alone? The system for creating the Natura 2000 network in France. *Sociol. Rural.* **2001**, *41*, 311–328. [CrossRef]
17. Haumont, F. Il recepimento della Direttiva negli Stati membri dell'Unione europea The Directive's transposition across the European Union's Member state. In *La Conservazione della Natura in Europa. La Direttiva Habitat ed il Processo di Costruzione della Rete Natura 2000 Nature Conservation in Europe. The Habitats Directive and the Making of the Natura 2000 Network*; Amirante, D., Ed.; FrancoAngeli: Milan, Italy, 2003.
18. Enengel, B.; Penker, M.; Muhar, A. Landscape co-management in Austria: The stakeholder's perspective on efforts, benefits and risks. *J. Rural. Stud.* **2014**, *34*, 223–234. [CrossRef]
19. Beunen, R.; van Assche, K.; Duineveld, M. Performing failure in conservation policy: The implementation of European Union directives in the Netherlands. *Land Use Policy* **2013**, *31*, 280–288. [CrossRef]
20. Cent, J.; Grodzińska-Jurczak, M.; Pietrzyk-Kaszyńska, A. Emerging multilevel environmental governance. A case of public participation in Poland. *J. Nat. Conserv.* **2014**, *22*, 93–102. [CrossRef]
21. De Meo, I.; Brescancin, F.; Graziani, A.; Paletto, A. Management of Natura 2000 sites in Italy: An exploratory study on stakeholders' opinions. *J. For. Sci.* **2016**, *62*, 511–520. [CrossRef]
22. Paletto, A.; Graziani, A.; Brescancin, F.; De Meo, I. Pubblica partecipazione nell'implementazione della rete Natura 2000 in Italia: Le esperienze dei portatori d'interessi Public participation in implementing Natura 2000 network in Italy: Stakeholders' experiences. *Forest* **2017**, *14*, 13–27. [CrossRef]
23. Bryan, S. Contested boundaries, contested places: The Natura 2000 network in Ireland. *J. Rural. Stud.* **2012**, *28*, 80–94. [CrossRef]
24. Gallo, M.; Malovrh, S.P.; Laktić, T.; De Meo, I.; Paletto, A. Collaboration and conflicts between stakeholders in drafting the Natura 2000 Management Programme (2015–2020) in Slovenia. *J. Nat. Conserv.* **2018**, *42*, 36–44. [CrossRef]
25. Bouwma, I.; Beunen, R.; Liefferink, D. Natura 2000 management plans in France and the Netherlands: Carrots, sticks, sermons and different problems. *J. Nat. Conserv.* **2018**, *46*, 56–65. [CrossRef]
26. Krott, M.; Julien, B.; Lammertz, M.; Barbier, J.M.; Jen, S.; Ballesteroz, M.; De Bovis, C. Voicing interests and concerns: Natura 2000: An ecological network in conflict with people. *For. Policy Econ.* **2000**, *1*, 357–366.
27. Kamphorst, D.A.; Bouwma, I.M.; Selnes, T.A. Societal engagement in Natura 2000 sites. A comparative analysis of the policies in three areas in England, Denmark and Germany. *Land Use Policy* **2017**, *61*, 379–388. [CrossRef]
28. Paavola, J. Protected areas governance and justice: Theory and the European Union's Habitats Directive. *Environ. Sci.* **2004**, *1*, 59–77. [CrossRef]
29. Rauschmayer, F.; Van Den Hove, S.; Koetz, T. Participation in EU biodiversity governance: How far beyond rhetoric? *Environ. Plan. C* **2009**, *27*, 42–58. [CrossRef]
30. Young, J.C.; Jordan, A.; Searle, K.R.; Butler, A.; Chapman, D.S.; Simmons, P.; Watt, A.D. Does stakeholder involvement really benefit biodiversity conservation? *Biol. Conserv.* **2013**, *158*, 359–370. [CrossRef]

31. Laktić, T.; Malovrh, S.P. Stakeholder participation in Natura 2000 management program: Case study of Slovenia. *Forests* **2018**, *9*, 599. [CrossRef]
32. Gibbs, D.; While, A.; Jonas, A.E.G. Governing nature conservation: The European Union Habitats Directive and conflict around estuary management. *Environ. Plan. A* **2007**, *39*, 339–358. [CrossRef]
33. Kati, V.; Hovardas, T.; Dieterich, M.; Ibisch, P.L.; Mihok, B.; Selva, N. The challenge of implementing the European network of protected areas Natura 2000. *Conserv. Biol.* **2015**, *29*, 260–270. [CrossRef]
34. Beunen, R.; de Vries, J.R. The governance of Natura 2000 sites: The importance of initial choices in the organisation of planning processes. *J. Environ. Plan. Man.* **2011**, *54*, 1041–1059. [CrossRef]
35. Langhout, W.; Brunner, A.L. The best idea Europe has ever had? Natura 2000: The largest network of protected areas in the world. *Georg. Wright Forum* **2017**, *34*, 86–95.
36. Schirpke, U.; Marino, D.; Marucci, A.; Palmieri, M. Positive effects of payments for ecosystem services on biodiversity and socio-economic development: Examples from Natura 2000 sites in Italy. *Ecosys. Serv.* **2018**, *34*, 96–105. [CrossRef]
37. Gil, A.; Calado, H.; Costa, L.T.; Bentz, J.; Fonseca, C.; Lobo, A.; Vergilio, M.; Benedicto, J. A methodological proposal for the development of Natura 2000 sites Management Plans. *J. Coastal. Res.* **2011**, *64*, 1326–1330.
38. Kovács, E.; Kelemen, E.; Kiss, G.; Kalóczkai, Á.; Fabók, V.; Mihók, B.; Megyesi, B.; Pataki, G.; Bodorkós, B.; Balázs, B.; et al. Evaluation of participatory planning: Lessons from Hungarian Natura 2000 management planning processes. *J. Environ. Manag.* **2017**, *204*, 540–550. [CrossRef] [PubMed]
39. Cortina, C.; Boggia, A. Development of policies for Natura 2000 sites: A multi-criteria approach to support decision makers. *J. Environ. Manag.* **2014**, *141*, 138–145. [CrossRef] [PubMed]
40. European Commission at Work: Infringement Decisions. Available online: https://ec.europa.eu/atwork/applying-eu-law/infringements-proceedings/infringement_decisions/index.cfm?lang_code=EN (accessed on 31 December 2019).
41. Postiglione, A. Impact of Natura 2000 Sites on Environmental Licensing. Italian Report, 2006. Available online: https://www.eufje.org/images/docConf/hel2006/IT%20hel2006.pdf (accessed on 31 December 2019).
42. Amirante, D. La Direttiva Habitat e la rete Natura 2000: Verso un modello europeo di conservazione integrata The Habitats directive and the Natura 2000 network: Towards a European model of integrated conservation. In *La Conservazione della Natura in Europa. La Direttiva Habitat ed il Processo di Costruzione della Rete Natura 2000 Nature Conservation in Europe. The Habitats Directive and the Making of the Natura 2000 Network*; Amirante, D., Ed.; FrancoAngeli: Milan, Italy, 2003.
43. Castorina, M.; Martimucci, D. Progetto BIOITALY (Biotopes inventory of Italy): Verso la Carta della Natura per la gestione del patrimonio naturale in Italia BIOITALY (Biotopes inventory of Italy) Project: Towards a Nature map to support natural heritage management in Italy. *Mondogis* **2000**, *2*, 45–48.
44. Blasi, C. BIOITALY: Natura 2000 in Italy. *Ann. Bot.* **1996**, *54*, 31–38.
45. Neven, M.G.G.; Kistenkas, F.H.; Van Apeldoorn, R.C.; Schröder, R.R.G.; Bruszick, A. Eurosites Insights. Image, Implementation, Interpretation and Integration of Natura 2000 in European Perspective (Alterra-rapport 1221-2), 2005. Available online: https://library.wur.nl/WebQuery/wurpubs/fulltext/39692 (accessed on 31 December 2019).
46. Denscombe, M. *The Good Research Guide for Small-Scale Social Research Projects*, 2nd ed.; Open University Press: Philadelphia, PA, USA, 2003.
47. Materiale Designazione ZSC. Available online: Ftp://ftp.minambiente.it/PNM/Natura2000/Materiale%20Designazione%20ZSC/ (accessed on 31 December 2019).
48. Bryman, A. *Social Research Methods*, 4th ed.; Oxford University Press: New York, NY, USA, 2012.
49. Scott, J. *A Matter of Record. Documentary Sources in Social Research*; Polity Press and Basil Blackwell: Cambridge, UK, 1990.
50. Beunen, R.; Van Assche, K. Contested delineations: Planning, law, and the governance of protected areas. *Environ. Plan. A* **2013**, *45*, 1285–1301. [CrossRef]
51. Bouwma, I.M.; Kamphorst, D.A.; Beunen, R.; van Apeldoorn, R.C. Natura 2000 Benchmark. A comparative analysis of the discussion on Natura 2000 management issues, 2008. Available online: http://edepot.wur.nl/3419 (accessed on 31 December 2019).
52. Ferranti, F.; Beunen, R.; Speranza, M. Natura 2000 Network. A comparison of the Italian and Dutch implementation experiences. *J. Environ. Pol. Plan.* **2010**, *12*, 293–314. [CrossRef]
53. Evans, D. Building the European Union's Natura 2000 network. *Nat. Conserv.* **2012**, *1*, 11–26. [CrossRef]

54. Bouwma, I.; Liefferink, D.; Van Apeldoorn, R.; Arts, B. Following old paths or shaping new ones in Natura 2000 implementation? Mapping path dependency in instrument choice. *J. Environ. Pol. Plan.* **2016**, *18*, 214–233. [CrossRef]
55. Bouwma, I.; Arts, B.; Liefferink, D. Cause, catalyst or conjunction? The influence of the Habitats Directive on policy instrument choice in Member States. *J. Environ. Plan. Man.* **2017**, *60*, 977–996. [CrossRef]
56. Castro, A.J.; Martín-López, B.; López, E.; Plieninger, T.; Alcaraz-Segura, D.; Vaughn, C.C.; Cabello, J. Do protected areas networks ensure the supply of ecosystem services? Spatial patterns of two nature reserve systems in semi-arid Spain. *Appl. Geogr.* **2015**, *60*, 1–9. [CrossRef]
57. Maczka, K.; Chmielewski, P.; Jeran, A.; Matczak, P.; van Riper, C.J. The ecosystem services concept as a tool for public participation in management of Poland's Natura 2000 network. *Ecosyst. Serv.* **2019**, *31*, 173–183. [CrossRef]
58. Lecina-Diaz, J.; Alvarez, A.; De Cáceres, M.; Herrando, S.; Vayreda, J.; Retana, J. Are protected areas preserving ecosystem services and biodiversity? Insights from Mediterranean forests and shrublands. *Landscape Ecol.* **2019**, *34*, 2307–2321. [CrossRef]
59. De Groot, R.S.; Alkemade, R.; Braat, L.; Hein, L.; Willemen, L. Challenges in integrating the concept of ecosystem services and values in landscape planning, management and decision making. *Ecol. Complex.* **2010**, *7*, 260–272. [CrossRef]
60. Geneletti, D. Assessing the impact of alternative land-use zoning policies on future ecosystem services. *Environ. Impact. Asses.* **2013**, *40*, 25–35. [CrossRef]
61. Gómez-Baggethun, E.; Barton, D.N. Classifying and valuing ecosystem services for urban planning. *Ecol. Econ.* **2013**, *86*, 235–245. [CrossRef]
62. García-Llorente, M.; Harrison, P.A.; Berry, P.; Palomo, I.; Gómez-Baggethun, E.; Iniesta-Arandia, I.; Montes, C.; García del Amo, D.; Martín-López, B. What can conservation strategies learn from the ecosystem services approach? Insights from ecosystem assessments in two Spanish protected areas. *Biodivers. Conserv.* **2018**, *27*, 1575–1597. [CrossRef]
63. Yigitcanlar, T.; Teriman, S. Rethinking sustainable urban development: Towards an integrated planning and development process. *Int. J. Environ. Sci. Technol.* **2014**, *12*, 341–352. [CrossRef]
64. Blondet, M.; de Koning, J.; Borrass, L.; Ferranti, F.; Geitzenauer, M.; Weiss, G.; Turnhout, E.; Winkel, G. Participation in the implementation of Natura 2000: A comparative study of six EU member states. *Land Use Policy* **2017**, *66*, 346–355. [CrossRef]
65. Zoppi, C.; Lai, S. An ontology of the appropriate assessment of Municipal master plans related to Sardinia (Italy). *Future Internet* **2014**, *6*, 223. [CrossRef]
66. Ferranti, F.; Turnhout, E.; Beunen, R.; Behagel, J.H. Shifting nature conservation approaches in Natura 2000 and the implications for the roles of stakeholders. *J. Environ. Pol. Plan.* **2014**, *57*, 1642–1657. [CrossRef]

sustainability

MDPI

Article

ESDA (Exploratory Spatial Data Analysis) of Vegetation Cover in Urban Areas—Recognition of Vulnerabilities for the Management of Resources in Urban Green Infrastructure

Ana Clara M. Moura [1,*] and Bráulio M. Fonseca [2]

1 Architecture School, Geoprocessing Laboratory, Federal University of Minas Gerais, Belo Horizonte, MG 30130-141, Brazil

2 IGC, Post-Graduate Program in Analysis and Modeling of Environmental Systems, Federal University of Minas Gerais, Belo Horizonte, MG 31270-901, Brazil; brauliomagalhaes@gmail.com

* Correspondence: anaclaramoura@yahoo.com; Tel.: +55-31-999595637

Received: 31 December 2019; Accepted: 29 February 2020; Published: 3 March 2020

Abstract: From the mapping of urban vegetation cover by high-resolution orthoimages, using IR band and NDVI classification (Normalized Difference Vegetation Index), added to three-dimensional representation obtained by LiDAR capture (Light Detection and Ranging), the volumetric values of vegetal cover are obtained as a base to construct spatial analysis in the district of Pampulha, in Belo Horizonte, investigating the role it plays in the neighborhood. The article aims to analyze the relationship between vegetation cover, income distribution and population density, as a support to urban environmental quality management. It applies Exploratory Spatial Data Analysis (ESDA) to identify the presence of clusters and patterns of spatial distribution and to examine spatial autocorrelation. The results confirm the concentration of vegetation cover in areas of high income and lower population density but the main contribution of the study is the use of a method to analyze the spatial behavior of this distribution. Calculating Moran global index and local index (LISA), these spatial combinations are mainly used to identify transformation pressures, which may result in the definition of priorities for public actions and the construction of proposals for parameterization of vegetation cover to support plans related to green infrastructure in urban areas.

Keywords: green infrastructure; ESDA and LISA; neighborhood spatial analysis

1. Introduction

High resolution data is each day more accessible to the construction of spatial analysis models that are the base for design and management of urban ecosystems. The municipality of Belo Horizonte is investing in the acquisition of data that allow high performance in the characterization of land use and land cover; with emphasis on high resolution spectral and spatial orthoimages (20 cm and the infrared band capture) and on the use of LIDAR (Light Detection and Ranging) that is used in the 3D representation of the city.

An initial production of data about vegetation cover, using the IR band and NDVI classification (Normalized Difference Vegetation Index) associated with 3D representation from vegetation, allowed us to calculate the volume of each vegetation plot and to present values per block and per census sector. Once the volumetric index of vegetation cover is calculated, the research that is described in this paper has the goal of analyzing the relationship between the presence of green areas, income distribution and population density.

The case study, the district of Pampulha, is known as a very green part of the city but it is also true that these areas are not well distributed, presenting inequalities. Moreover, it is under pressure of transformation, because the area projected by Oscar Niemeyer, the most famous Brazilian architect, was recently declared a World Heritage Site by UNESCO due to its notorious ensemble in which the vegetation cover is part of the genius loci. The position of the district is in the axis of city growth and densification. This explains the importance of the studies and the intention to support the proposal of green infrastructure parameters to the area. (Figure 1).

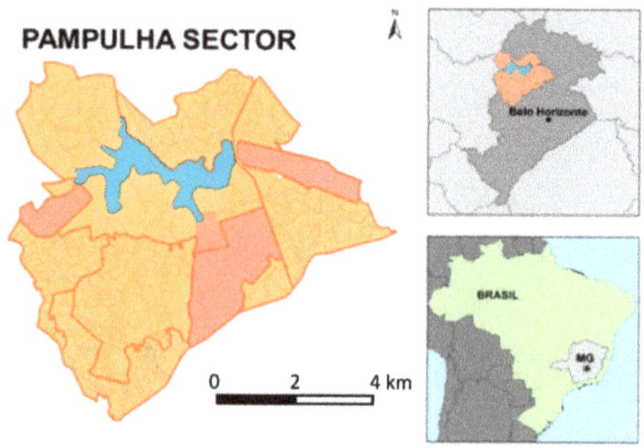

Figure 1. Pampulha Sector in the city of Belo Horizonte, state of Minas Gerais, in Brazil. Source: The authors.

The study explores the possibilities of spatial analysis with the support of GIS technologies, going beyond the simple analysis to prove the expected concentration of vegetation cover in areas of high income and low population density, which means that just a part of the population of the city is served by this good condition of green infrastructure. The use of GIS technologies and ESDA methods is a second goal, because the study can be used as a reference for further studies, as exploratory analysis is quite useful to help the researcher to understand spatial distributions and possible correlations.

Green areas in an urban environment have different functions, related to aesthetic quality, areas for socialization and leisure, protection against geotechnical problems, maintenance of aquifers and springs and also to ensure the maintenance of the environmental equilibrium related to the climate, the humidity of the environment, air quality and acoustic control [1–3]. Recognizing the importance of green areas and the need to plan green infrastructure, the study selected the robust vegetation that interferes more in urban quality when the goal is to compare some anthropic conditions of land use conditions (density and income).

The city of Belo Horizonte, where Pampulha is located, has around 2.5 million inhabitants, from which 145 thousand live in the 47 square kilometers of Pampulha region, what is considered low density in comparison to the rest of the city. Regarding income distribution, together with the South-Center region, it has the higher income of the city. However, as it can often be observed in Brazil, there is complexity in land use, as there are slums with very low income and portions with high density concentration of inhabitants in the area.

The selection of Pampulha was due to its expressive vegetation cover in comparison to the conditions of Belo Horizonte municipality area. The green areas are robust because of its environmental characteristics of gentle relief and declivity but also due to its formation history—it was projected by Niemeyer to be a reference of modernistic urban ensemble, with the contribution of Burle Marx, an important landscape planner.

Pampulha is just an example of interest to be presented but the same methods can be applied to any urban areas. Green areas are inseparable from the landscape that makes the cityscape and are related to cultural values and to the identity of a place in a broad degree of significance and coherence [4]. It is important to quality of life in physical and cultural senses.

The conformation of the original vegetal cover in Pampulha is characterized by a transition from the Atlantic Forest (that is more distributed from the center to the south of the city) to the Cerrado (Brazilian Savanna). The Atlantic Forest is constituted by high trees and dense composition, while the Cerrado is formed by medium size trees more adapted to harsh climates where water is scarce, with a less dense composition than the forest. With the construction of Pampulha in the 40s, with a modernistic style of urban planning, most of the original vegetation was replaced by exotic species, selected by designs of landscape compositions. In this sense, when the green areas of the city are studied, it is not possible to talk about original species but about volume of vegetation in a general sense.

The watershed of The Pampulha Dam, in Pampulha Region, is characterized by a depressed topography in relation to the geomorphological characteristics of the entire municipality of Belo Horizonte. It is a relatively flat area, with low relief, high density drainage and extensive river plains. The geological formation consists of gneiss granite from the Belo Horizonte Complex. Due to the intense weathering, deep soils were formed, however, with separated beds of sandy and clayey with poor cohesion, indicating susceptibility to erosion [5].

These characteristics also explain the dense vegetation cover, due to water, higher temperatures, flat topography and deep soils. Also because of these good conditions, there was a sprawl of urbanization and densification. The construction of the dam, in 1958, resulted in a "lake," which fostered the process of urbanization of the area and increased environmental conflicts.

The Pampulha Dam is formed by a body of water inserted in an urbanized hydrographic basin where the irregular occupation and the lack of sanitation and erosion control (mainly in the western portion, in the city of Contagem) caused the loss of approximately 50% of its volume, as well as an intense degradation of the water quality [5–7]. Environmental studies reported risks in biochemical parameters, such as cyanobacterial proliferation, including potentially toxic species, presence of heavy metals, eutrophication and low amount of dissolved oxygen in the water [5–8]. The problems of the water in The Pampulha Dam can be faced with the control of land use and density, the implementation of sanitation in the areas of low income occupation and the protection of green areas.

Green spaces in urban areas are responsible for helping regulate air quality and climate, reducing energy consumption by countering the warming effects of paved surfaces, recharging groundwater supplies, protecting lakes and streams from polluted runoff and maintaining the environmental balance that affects physical and psychological well-being through the influence of urban stressors and factors, such as overcrowded and polluted environment and reduced social support [2,9–11]. Green areas are reported as the most desired environment for relaxing and recovering from stress and sustaining mental efforts [12], therefore it is important to have these areas incorporated into the designs of the cities, in an interconnected network of natural areas and other spaces [13].

The main goal of this paper is, as a matter of fact, the use of a method based on the spatial analysis distribution of this observed condition. The spatial analysis technique used was ESDA (Exploratory Spatial Data Analysis), which identifies the existence of clusters of each land use typology, the existence of distribution patterns, the variety of spatial patterns and, more importantly, the spatial autocorrelation.

This case study, facing the relation between vegetation cover, income and population density with the use of ESDA, reveals spatial connections and tendencies that cannot be ignored, because spatial phenomena spread according to contiguity and neighborhood, changing behaviors, composing tendencies, defining new values and establishing a culture.

In this sense, studies that consider the risks and the tendencies in change to support the quality management in landscape are of great interest. This analysis can favor the understanding of the need for green infrastructure parameters in urban laws to be applied in an index of volumetric vegetation cover and in indexes of permeability per lot, per block and per any territorial unit. Using Moran global indexes, it is possible to understand each territorial unit in the sector, in this case study represented by units of census sectors, as reference to plan policies and projects for local scale in punctual interventions or to plan general projects for a broader area.

Geographic and territorial sciences that deal with the characterization, analysis, simulation of change and the proposals of alternative futures to a spatial composition are increasingly relying on the application of digital spatial models to support the studies of phenomena distribution. Among these geographic sciences are those dealing with physical phenomena but there are also those that investigate anthropic actions on space and the dynamics caused by man and environmental interaction. The concepts applied in this case study can be of wide application, being of interest to all those investigating variables related to anthropic issues (socioeconomic, population, values and cultures); to physical issues (vegetation, landscape typologies and many others); and those that study the relationships between anthropic and physical variables (for example, the relation between population density and vegetation cover and many other possibilities).

Geoinformation technologies have expanded significantly since the middle of the last century, associating knowledge about conceptual logics with the possibilities of representing and testing these logics with the technological support. Parallel to digital tools development, we also observe the developments in regulations and laws that favor the wide dissemination of information and facilitate access to data and applications, as support for planning and design.

As a legal point of view, it is important to mention INSPIRE European normative, that established the SDI (Spatial Data Infrastructure) in Europe and was used as a reference to Brazilian IDE in the scale of national data [14–16]. As a result, the access to georeferenced data is becoming possible in Brazil. In addition to that, the existence of free GIS software is a condition to make the use of the technology and methods in all scales possible, even in those municipalities with lack of financial conditions. The expectations for the next steps are the free web platforms, providing not only data but also services, increasing the possibilities for the users [17].

Spatial studies are always related to the process of analysis followed by synthesis. This means decomposing—composing—recomposing, according to the principles of systemic approach. The logic of producing analysis and synthesis to characterize the territory, to identify potentials and restrictions, has been in use before the advent of IT, based on analogic media. The overlay of maps, with the goal to find suitable areas for the equilibrium between environmental protection and anthropic use for the place, was proposed in the 60's and called "design with nature" [18].

The logic of decompose, compose and recompose is in the origin of science, when the investigator identifies the main component variables of a phenomenon or of a spatial occurrence and isolates these variables to study them. To understand the behavior of these variables, the researcher composes with them, to verify the significance of some aspects he observed. Once the behaviors are understood, he recomposes to define the insertion of each variable or of a group of variables in the system, to analyze the impacts of possible changes or transformations. Among the expressive literature about this theme, which is in the base of science, we chose to mention the studies in geography were also developed in the 60's, in discussions about models in geography [19].

An important field in science that studies the interrelationships among variables, phenomena, occurrences and territory is the System Approach. It defends that it is possible to isolate parts from the whole but without forgetting the existing relation among the parts. The approach was proposed by a biologist also in the 60's and it is applied to any spatial phenomena that consider mutual influence in anthropic and physical variables [20].

Once establishing the conceptual base for investigating and modeling the territory, the development of GIS (Geographic Information Systems) allowed the use of applications with the main known models but also allowed to plan and test new spatial models. To develop a model requires knowledge about logics and the capacity to translate these logics into algorithms. However, we are observing the indiscriminate use of spatial models as Apps (Applications) and GIS tools, just because they are available in software, without controlling the possible applications, limitations and suitability for that use. It is important to remember that spatial studies must begin with a clear definition of objectives, deep investigation about possible ways and the best way to achieve the objectives and tests on spatial models to define terms of use. It is also important to understand the specificities related to spatial analysis [21].

There are still few studies that approach three-dimensional studies applied to urban vegetation cover, especially using LiDAR (Light Detection and Ranging) associated with the NDVI (Normalized Difference Vegetation Index), in order to support the comparison between built and vegetation volumes to evaluate urban environmental quality [1,22,23].

Particularly in Brazil, this is very new, due to the cost of the data capture and to the lack of technicians that are able to work with geoprocessing tools but mainly because the consciousness about the importance of quality of people's environment and life is still very incipient in order to consider the importance of this variable in landscapes that are increasingly impacted by the processes of intense anthropization, especially with regard to the vegetation cover in the urban landscape. When the "green" variable is presented in any urban plan, it is calculated in bidimensional values, indirectly considered, as the index of permeability [24]. The parameters used, when applied by the municipality, are—rate of permeability per lot (a percentage of area that cannot be covered by impermeable materials but it does not really mean it will be vegetated); arborization plans to roads (that defines the need to plant a tree in the sidewalk when the proprietor implements the building); the general analysis that calculates the rate of green areas per inhabitant (that calculates the bidimensional surface of any green, without separating robust vegetation from shrubby vegetation and requires an average of 15 square meter per inhabitant) and the "environmental quota" (that accepts substitutes or compensation to the absence of the green, like rainwater collection box or green walls) [25,26]. In this sense, the present study chooses the volumetric values in vegetation as parameter of analysis, in comparison to the others that tell about anthropic use of urban land, understanding that this new parameter makes the difference in urban life quality [27].

Exploratory data analysis is not a new methodological resource, it can be observed in several works in the field of conventional econometrics [28,29]. In the same way, Exploratory Spatial Data Analysis is not new but its use is not well known in urban studies in Brazil, where, except for some big cities, there is no culture of strategic plans and, moreover, there are limitations in using spatial data analysis. In this sense, the Exploratory Spatial Data Analysis can play a role in studies about tendencies and behaviors in urban environments and can be considered an innovative contribution. This article presents conceptual bases about spatial analysis (state-of-the-art), as well as a case study to illustrate the applied methods (state-of-the-design), in order to contribute to new future researches.

2. Materials and Methods

One of the most important steps in geoprocessing is the definition of the territorial unit to develop the analysis, as space must be discretized in regular or non-regular units. Once representing the space in discrete units, they will be combined in spatial analysis models and in spatial representation. In the example of the present study, it is possible to understand the process of decompose—compose—recompose (Figure 2).

Initially, reality was represented in regular units of pixels (according to spatial resolution) and in spectral bands (spectral resolution). After this first decomposition, the NDVI (Normalized Difference Vegetation Index) model was chosen to compose the variables. This technique combines red and infrared bands, resulting in an index of vegetation, normalized from −1 to +1 [27,30]. Once the model is applied, the representation is recomposed according to the objective of investigation, which means—the identification of areas with robust vegetation cover, woody vegetation. It is also possible to recompose the results according to vegetation levels, from woody to scrubby and grass. In this case study, only the robust vegetation was selected and using the plots of this vegetation, a 3D model was composed, associating the footprints of the trees to the elevation values captured by LIDAR, from the DSM (Digital Surface Modelling). In this sense, we observe that many studies in spatial analysis can be done by processes of decompose—compose—recompose.

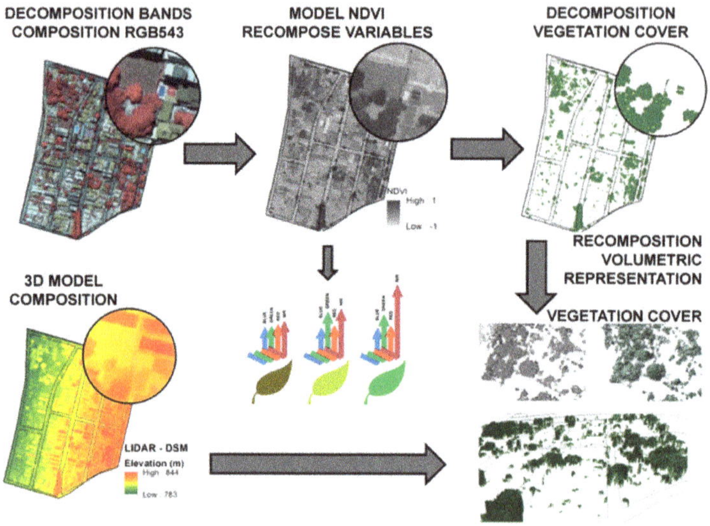

Figure 2. Example of decomposed reality in pixels and spectral bands, composed into an index of vegetation cover, recomposed in the identification of robust vegetation (woody and volumetric expression). Source: The authors.

The LIDAR data used was provided by the Municipality of Belo Horizonte, captured in 2015. The survey was performed through airborne LASER profiling. The parameters of capture were—FOV angle (field of view) of 20 degrees, flight height of 2388.1 m, stripe width of 1688.1 m, side overlap (between strips) of 36.4%, 76 strips in varied flight directions, with vertical accuracy of 25 cm and average density of 1.5 points per square meter. As the modelling considered the selection of robust vegetation, the number of points was enough to calculate the medium height of the trees, because none of the plots used had less than 36 square meters of projection.

The company responsible for the work performed the pre-processing of raw data and we performed the LIDAR data processing. The pre-processing was basically the use of GPS observations of the base station and of the aircraft, that were initially processed individually and subsequently concatenated in order to obtain a unique kinematic solution and adjusted to a known coordinate system. The first step on LIDAR data processing was also performed by the company, with the goal to separate the points of terrain from the points of surface land use. They used a tool that searches for the points with the lowest dimensions and builds a Triangular Irregular Network (TIN) grid. Most of the times the triangles of this initial model have the sides lower than the terrestrial surface, with few vertices touching the terrain and these irregularities are removed by the program. After that, the program starts modeling the terrain surface, adding more points to the model making it closer to the real shape of the terrain. The points that are added to the model are defined by iteration parameters of angles and distances. These parameters determine how close the points must be to the plane of a triangle to be accepted in the built model.

As the points were provided separated in terrain and surface data, the surface data was selected. They were in Multipoints format and it was necessary to convert to Singleparts to extract all the possible points and keeping the Z value (elevation). Using the shapes of robust vegetation plots, it was performed the Join by Location, which means to extract all the points values only to the projections of the existing trees. Using just the values correspondent to robust vegetation, it was constructed the raster surface model. Applying Zonal Statistics analysis, it was possible to get the Majority and the Range of the elevation values per shape of robust vegetation. The Range was used to control the difference from the lowest to the highest point of the plot element but the Majority value was used to calculate the volume of the vegetation (doing the multiplication of the area of the plot projection by the majority value in the height) [23,30].

An important step in the use of models is the definition of territorial unit analysis or spatial integration, that can be vector (decomposed according to graphic primitives of points, lines and polygons) or matrix (decomposed in pixels). In the specific case of neighborhood studies, in which ESDA (Exploratory Spatial Data Analysis) is included, the territorial unit or spatial discretization has a fundamental importance, because it interferes in the definition of what a "neighbor" is [31,32].

Searching for references in the past, when people were decoding the space into representation types, we can remember the pointillism impressionism (Figure 3). The question is—how many pixels must be selected from the point of a neighborhood so that the user is able to recognize an element or part of an element? Monet used to say that each color we see comes from the influence of its neighbor. Art had already anticipated the raster logic, that requires two conditions—the discretization of space into small parcels, that represents in spatial analysis the territorial units of integration (that can be pixels or polygons) and the fact that spatial relations are closely related to neighborhood conditions. Considering that elements and their influence in the territory can only be understood and characterized according to their position in the space, it is important to consider this condition while developing spatial analysis and applying spatial models, because it is very important to understand the context in which the elements are inserted.

Figure 3. Monet, pointillism impressionism. According to Monet, each color we see stems from the influence of its neighbor. Based on that, how many points should be grouped to form information? Source: Monet–Image flickriver–http://www.flickriver.com/photos/.

The definition of territorial units requires also the understanding that spatial parts are linked, to choose the best representation of these connections (Figure 4). Most geoprocessing models combine different layers that search to answer two main questions—"in this position, which are the characteristics?" or "these characteristics, where are they located?" [33,34]. It means that they think vertically, according to the combination of layers in a position but not considering the neighborhood (Figure 5).

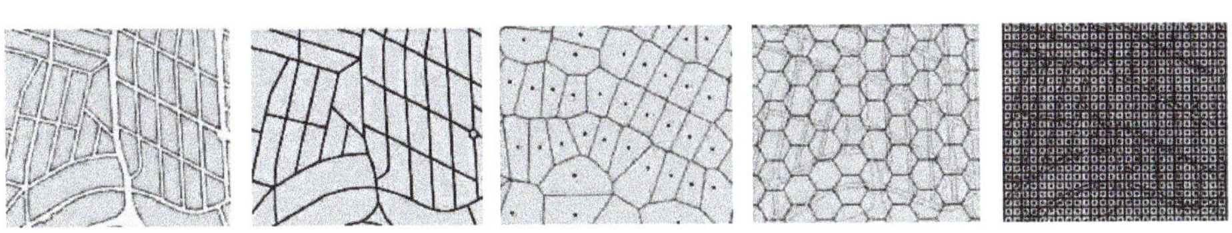

Figure 4. Limitations in the use of spatial representations—the need to construct connected patterns to work with the logic of a neighborhood. Blocks, tracks of roads, Voronoi or Thiessen from centroids, regular hexagonal tessellation and a regular grid of pixels. The different use supported by objectives. Source: The authors.

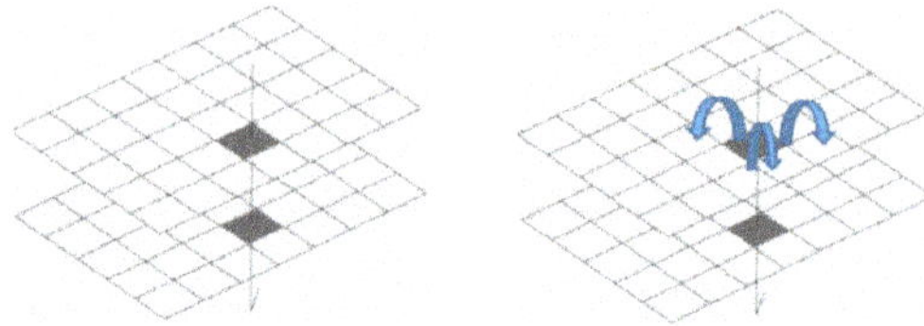

Figure 5. Models that integrate layers in vertical combinations, according to geographical position, compared to models that consider also horizontal combinations, recognizing the influence of the neighbors. From map algebra to the influence of the neighborhood. Source: The authors.

A classic application in spatial analysis is the geographic matrix [35–37] and the Multicriteria Analysis [38–41] that combine territorial information units (pixels or vector elements) that spatially coincide in a vertical way. The models that consider the neighborhoods must work with horizontal compositions, between elements in the same layer, according to any spatial definition or cutout.

Considering the models that work with neighborhood aspects, they can be separated into those that are used to classify typologies, as the image classification that applies filters or masks (Low-pass filters, High-pass filters, generalizations and groupings); those that work with topologic references (defining conditions of contiguity, connectivity and so on) and those that transform initial data (points or lines) into potential surfaces of spatial distribution of an occurrence or phenomenon (spatial interpolators) [41].

The potential surfaces can be composed by interpolators, each type of interpolators adapted to the behavior or characteristics of the data (if it has regular or irregular distribution), according to the function (if the model is applied to suggest values in parts of the territory in which there were no measures or data or to present the concentration of occurrences). There are many models that must be chosen according to the condition of the data and to the objective of the analysis, as the interpolator of Delauney example (to regular distribution phenomena), the mapping of density using Kernel or of concentration using Clusters (to distribution and concentration studies), the gravitational studies as IDW or Voronoi with friction and mass (to map the area of influence of occurrences), the probability of distribution in a territory simulated by Kriging (considers spatial tendencies and veins), among others. All of them aim to transform data into geographic information, considering the distribution of elements and composing numeric visualization of spatial distribution (Figure 6).

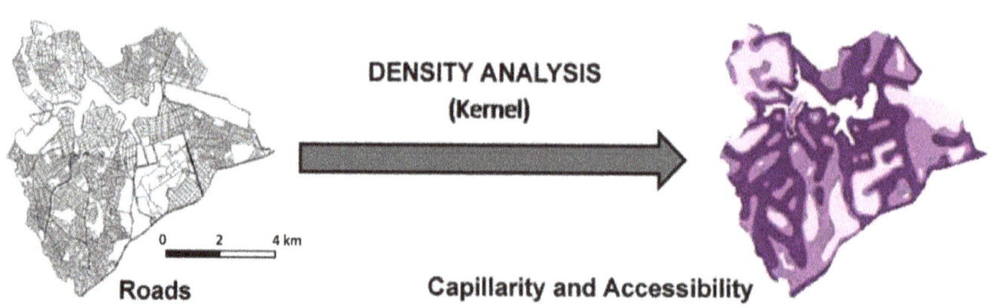

Figure 6. Composition examples of potential spatial surfaces. Source: The authors.

With the need to consider spatial groupings and the influence from the neighborhood in a process of change and transformation in a city, we must remember the first law of geography, from Tobler—"Everything is related to everything else but near things are more related than distant things" [42].

The present paper starts from the interest to study urban dynamics and to understand the distribution of life quality in the city, considering the surrounding effects. The variables that were chosen for the analysis were vegetation cover, income and population density distribution. The studies about the general distribution of each variable were done with the global Moran index, which tells if the area is of high value of the variable and the neighbors are also like this, if the area is of low value and the neighbors are also like this, if the area is of high value but the neighbors are of low value or if the area is of low value but the neighbors are of high value.

The studies can be done in a simple variable but also in a combination of variables. In this case, the global Moran index indicates the relation between high/low values of one variable surrounded by values of high/low of the other variable, in the 4 possibilities of combinations. In the case study, the goal was to investigate the relation between the vegetation and income variables, vegetation and population density and of income and population density.

It is also possible to analyze the behavior of one parcel of the territory, using local Moran index, the LISA (Local Index of Spatial Association). The local index allows to investigate, for example, if some territorial units with robust vegetation (woody) are surrounded by similar conditions or by neighbors that are very different in this aspect. This kind of investigation is important because of the risks of following the examples from the practices in the surroundings. The identification of territorial units that have some good conditions but are in the middle of others that do not have the same condition can be an alert to create policies or restrictions to discourage the change.

Also, in local studies, using LISA, it is possible to combine variables and analyze if in one territorial unit there is a good vegetation condition but surrounded by high density population or by low income population or the possible combinations. This kind of study can be strategic to plan the future of the area.

Most of the researches that apply ESDA (Exploratory Spatial Data Analysis) have the studies about equity as a goal. Most of them are interested to know, for example, if the distribution of an infrastructure is spatially coincident with income distribution, because the interest is to prove social inequalities. There are papers about indexes of human development and indexes of economic development or even about the distribution of services and facilities and the relation with incomes and poverty. Most of them are interested in analyzing the asymmetry in the distribution of opportunities [43–47].

In the present case study, it is possible to evaluate the equity concerned to green areas and socioeconomic or demographic aspects but the main interest was to use the analysis as a base for strategic studies that consider spatial phenomena that can be influenced by the behaviors of the surroundings. It is the logic that behaviors can conform tendencies, that can construct values that can establish a culture. In this sense, the way of living, the way of using the territory. The research has an innovative use of ESDA, because it is planned to be applied in strategic policies and designs.

The Moran Global (Global Spatial Autocorrelation Index–I-Moran Global) and Local (Local Indicators of the Space Association–LISA) indices were applied in this study using GeoDa software version 1.4.6 [31,48]. The I-Moran Global indicates the value of the spatial autocorrelation of a defined variable, regarding the entire data set, explaining if it is spatially concentrated or scattered. The positive values of I-Moran Global (between 0 and 1) mean the existence of direct correlation and the negative values (0 and −1) mean the presence of inverse correlation, functioning as a statistical test in which the null hypothesis is a spatial randomness.

An I-Moran Global statistic indicates a "strength" of spatial similarity or dissimilarity in neighboring regions. If $x_1, x_2, \ldots, x_n$ are places under observation, I-Moran Global for these data is:

$$I = \frac{N}{w_0} \frac{\sum_{i=1}^{N} \sum_{j=1, i \neq j}^{N} w_{ij}(x_i - \overline{x})(x_j - \overline{x})}{\sum_{i=1}^{n} (x_i - \overline{x})^2} \tag{1}$$

where

$$w_0 = \sum_{i=1}^{N} \sum\nolimits_{j=1} w\,(i.j) \tag{2}$$

LISA allows analyzing the spatial association for different locations of a variable distributed in space [35,48]. The I-Moran Local statistic measures the spatial autocorrelation of a specific location with its neighbors. Like the I-Moran Global, the significantly positive I-Moran Local indicates that the values of the place in question and its neighbors are similar, that is, there is positive autocorrelation (there are patterns of spatial similarity). The significantly negative I-Moran Local indicates that the value of the location under analysis is unequal in relation to its neighbors, that is, there are patterns of spatial dissimilarity. The Local I-Moran can be calculated for a location i and when the values of I_i are different from zero this indicates that the unit i is spatially associated with its neighbors, according to the equation below.

$$I_i = \frac{x_i - \overline{x}}{S_i^2} \sum\nolimits_{j=1}^{N} w_{ij}(x_i - \overline{x}) \quad (3)$$

where

$$S_i^2 = \frac{\sum\nolimits_{j=1, j \neq i}^{N} x_j^2}{N-1} - \overline{x}^2 \quad (4)$$

To summarize the steps that constitute the use of techniques related to the methods adopted, it is presented the main workflow of the study (Figure 7).

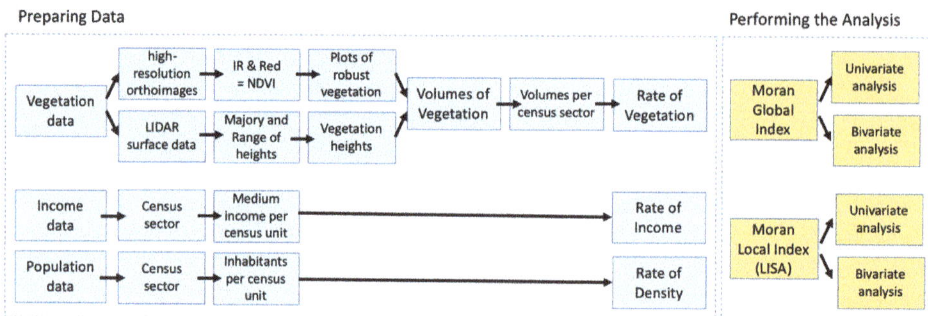

Figure 7. Workflow of the case study. Source: The authors.

3. Results

In order to use the ESDA Model, data from variables must be prepared to represent normalized and numeric distribution. It is not adequate, for example, to use data already segmented in value bands (grouped on discontinuous values) or even in categorical representation (typologies, also known as selective or nominal data). Data must be represented in real numbers, in continuous distribution and must be normalized to allow a comparison in variables.

For instance—if the objective is to compare income with vegetation cover, income cannot be grouped in bands. If we get the absolute numbers, income can go from R$0 to R$9691 (in Brazilian money) and green areas can go from 30 to 9655982 cubic meters (vegetation was associated with the volumetric value of each plot or fragment). To make it possible to compare these different variables, of different natures and with different scales of measurement, normalization is required. When data is normalized, it means that the initial and the last values are the same, while the internal distribution depends on the original distribution of the variable. It is possible to normalize, for example, from 0 to 1, from 0 to 100, according to the users' decision.

In the territorial units' decision, green vegetation was characterized by the footprint of the arborous body, the fragment projected in the surface but associated with the values about the volume (data captured from LIDAR and elaborated into volume values). With specific studies that we had done before, in which we were analyzing life quality, the values were grouped by blocks [23,30]. The ideal condition was to transform all the variables in distribution per block but in Brazil, socioeconomic data are available only per census sectors, which forced us to work with this territorial unit and to summarize the data from vegetation cover into the census sector unit (that covers many blocks but does not cut a block in the middle, because it respects the limits of the roads).

The neighborhood concept must be very clear when we work with ESDA. It interferes in the preparing of data. If we choose territorial units for the blocks, data must be treated so that the polygons' faces touch each other, covering the streets and conforming contiguous areas (Figure 8). In the case of grid representation (the logic of raster, matrix but organized in vectors), this contiguity is clear, because topologic conditions of a neighborhood are explicit but it is not in all cases that this representation can be interesting, as it is a regular grid that presents space as a homogeneous area and does not consider the variances in land cover (as roads, parks). It is also a problem when a variable measured by a zone is converted to grid (pixels) because the user is considering that the value is repeated in all the cells correspondent to that zone, ignoring that values were grouped in the initial representation.

Figure 8. Neighborhood studies. Blocks treated to ensure contiguity or adjacency. Source: The authors, adapted from Lloyd, Christopher. Spatial Data Analysis: an introduction for GIS users. Oxford University Press, 2010.

It is necessary to choose the scope of the neighborhood. In the use of ESDA it is necessary to define the relation of rook, bishop or queen. In the matrix of "rook" the neighbors are the cells or elements that have at least one face in contiguity or in contact with the cell or element of reference. In the matrix of "bishop" the neighbors are the elements that have at least one vertex in contact with the element of reference. In the matrix of "queen" the neighbors are those that have at least one vertex or one face in contact with the element of reference. Except for some specific motivation, to meet some objective, the ideal is to choose queen matrix, because it is a broader condition (Figure 9).

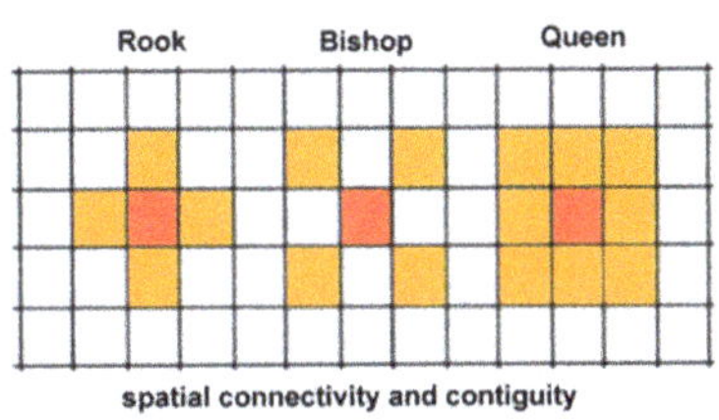

Figure 9. Definition of neighborhood: "rook," "bishop" and "queen." Source: The authors, adapted from Lloyd, Christopher. Spatial Data Analysis: an introduction for GIS users. Oxford University Press, 2010.

It is important to define the order of interest, that means the maximum number of closest neighbors (k-nearest neighbors) or even the maximum distance from the element of reference. If the k-nearest neighbors are chosen, it is important to define if they are going to be of first order, second order or third order (Figure 10). The choices are always related to the objectives of the analysis, and, mainly, related to the spatial dynamics of the place.

With the data already prepared and with the relation of the neighborhood already decided, the model is applied. The general Moran index goes from −1 to +1, which means perfect negative spatial autocorrelation and perfect positive spatial autocorrelation. It indicates if there is spatial randomness and presents an index and the visualization of the results in a scatter diagram. The interpretation is based on the direction and slope of the distribution line and by the four quadrants (Figure 11).

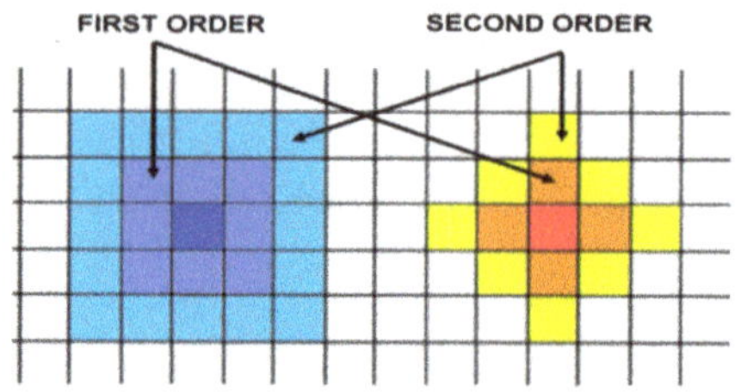

Figure 10. From a territorial unit, the neighbors of first order and the neighbors of second order, in matrix of "queen" and "rook." Source: The authors, adapted from ANSELIN, Luc. Local Indicators of Spatial Association–LISA. Geographical Analysis, Columbus, v.27, n.2, p. 93–115, 1995.

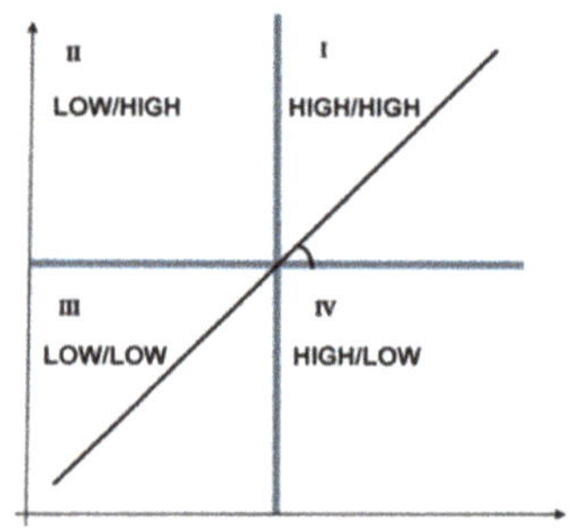

Figure 11. Scatter Plot–Moran Index representation. Source: The authors, adapted from ANSELIN, Luc. Local Indicators of Spatial Association–LISA. Geographical Analysis, Columbus, v.27, n.2, p. 93–115, 1995.

Randomness or autocorrelation interpretation is illustrated (Figure 12) [49]. The author explains that in image A, there is a spatial arrangement of perfect negative autocorrelation in Moran index (−1), because each black cell has only white cell neighbors, which means that all the neighbors are different from the cell of reference. In image B Moran index is +0.857, which is a high positive autocorrelation and could be +1 if all the cells were black or all the cells were white. The index was high and positive because in most of the cells the neighbors are the same, except for those in the frontiers of the two groups. In images D and E, we observe equal Moran indexes but one is positive and the other is negative. In the negative index (D), a great part of the image presents spatial regularity, forming a diagonal in which white cells have black cells as neighbors. In the positive index (E), most white cells have white cells as neighbors, which is also true for the black cells and with a spatial regularity in three horizontal lines of concentration. It is also interesting to observe that image C has a Moran index equal to 0 (zero) because there's no regularity in the distribution, it is completely random, it is not possible to find a spatial logic in the distribution or in the neighborhood.

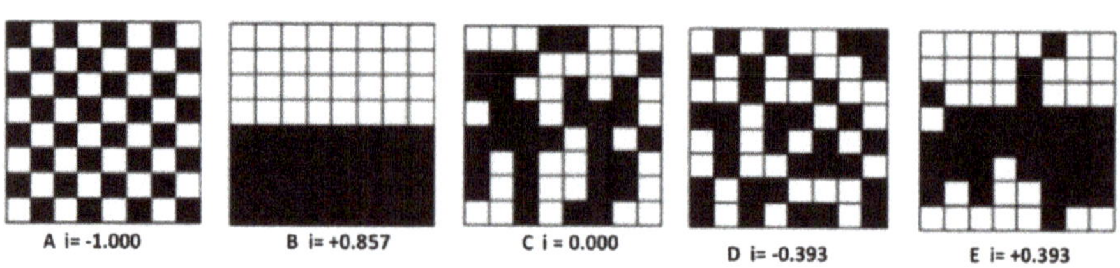

Figure 12. Moran index (that can go from −1 to +1) representing spatial autocorrelation. Source: Adapted from Goodchild, Michael F. Spatial Autocorrelation. Geo Books. 1986.

3.1. Preparing of Data

The first step was to calculate the volume of vegetation cover. Orthoimages with the resolution of 20 cm, divided into the visible spectral bands and the infrared band were used as a source of data. The NDVI (Normalized Different Vegetation Index) was composed to separate the robust vegetation (woody). From the classification, the footprints of the mass of trees were vectorized and fragment polygons of vegetation were composed. With the use of LIDAR data (Light Detection and Ranging), it was possible to get the elevation points and to calculate the vegetation volume from each footprint. The result was information about the volume of robust vegetation.

For the first studies, the values were grouped by block but for the studies in ESDA they were grouped by census sector, because the goal was to compare vegetation with socioeconomic and demographic data, that are made available by IBGE (Brazilian Institute of Geography and Statistics), from the census of 2010. Unfortunately, there is no more updated data on this scale. With all the data in the same territorial unit, the values were normalized from 0 to 1. This step was done in ArcGis©. From that point, a shapefile with polygons of census sectors with the attributes of vegetation, income and population density, all of them normalized from 0 to 1, started to be worked in Geoda [48].

3.2. Representation of Variables

The first step is the definition of the neighborhood matrix. The "queen" neighborhood was chosen and the level of first order. This means that we assumed as neighbors any census sector that had a vertex or a face in contact with the element in reference and that only the group from immediate proximity (first level) would be considered. This is because the case study, Pampulha, is divided into census sectors of large dimensions, which is different in other parts of the city. In other case studies, it must be considered the need to use a second or third order or even more.

The first maps produced were the distribution of each variable in the area, according to quintile parts (Figure 13). There are some similarities among the distribution of green areas and income (Figure 13a,b) and a relation that seems to be the opposite to population density (Figure 13c). This interpretation is in a general point of view but further investigations can be developed.

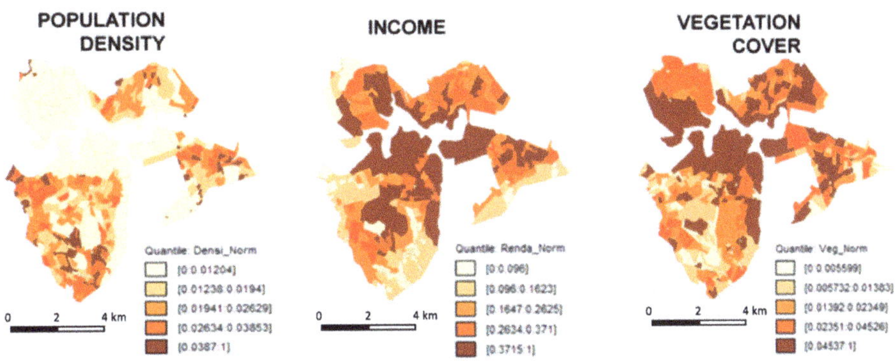

(a) (b) (c)

Figure 13. Population density (**a**), income distribution (**b**) and vegetation cover (**c**) in the Pampulha sector. Source: Socioeconomic data, IBGE, 2010 and vegetation data produced by the authors.

3.3. Moran Global Index

The study of Moran global index is applied to the whole territory and has the goal to evaluate the extension of groupings in the entire area, to measure if the grouping is significant, which allows us to think about the possibility of random or not random distribution of the occurrences in the region.

The evaluation can be done in a univariate way (only one variable at a time, analyzing its distribution in the territory) or in a bivariate way (analyzing the report between two variables in the area).

3.3.1. Moran Global Index per Univariate Analysis

The result of Moran global index for univariate analysis (just the isolated variable) shows that income is the variable that presents the highest positive autocorrelation (Figure 14a—Moran global index of 0.578), followed by population density (Figure 14c—Moran global index of 0.099) and vegetation (Moran global index of 0.057) (Figure 14b). In all variables, there is a positive autocorrelation index, which means that when the variable presents a high level in a part of the territory, its surroundings are also of high values, when there are low values, the neighborhood is of low value. But as all the indexes were not very high, there are also situations of high/low and low/high values. (Figure 14).

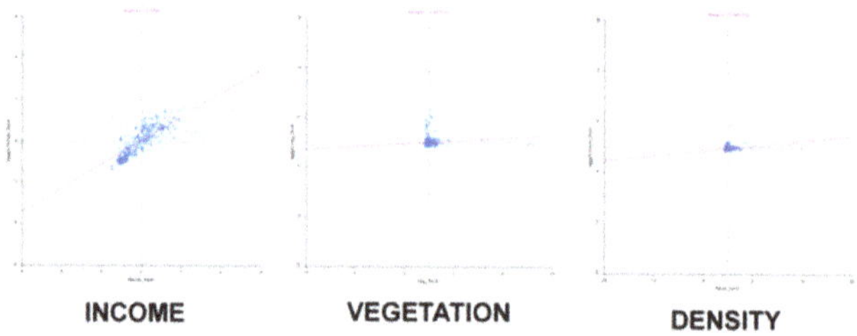

(a) (b) (c)

Figure 14. Moran Scatter Plot–Univariate analysis of income (**a**), vegetation (**b**) and population density (**c**). Source: The authors, using Geoda©.

In general, the values of the Moran Global index are not high in the positive sense, which reveals that there is no strong or total similarity between the analyzed phenomenon and its spatial occurrence. However, it is important to note that spatial randomness was not observed for any variable. Values close to zero must be observed, because they mean almost a condition of spatial randomness or spatial independence for the univariate analysis of the population density and vegetation variables. The income variable presents higher spatial correspondence between high values and its surroundings and low values and its surroundings, meaning spatial concentration of social conditions.

On the other hand, the same intensity of the Moran Global index is not observed for vegetation and population density. This means that the variable occurs in space but its spatial pattern of occurrence is tenuous. The result shows that there is not a large spatial concentration of vegetation and population density in the studied region.

3.3.2. Moran Global Index per Bivariate Analysis

The bivariate analysis evaluates whether there are spatial correlations between variable pairs. Population density was combined with income (Figure 15A), vegetation with population density (Figure 15B) and vegetation with income (Figure 15C), promoting all possible combinations (Figure 15). The result is that the relation between population density and income presents a negative autocorrelation (Moran global index of −0.092), which means that where income is high, population density is low. The relation between vegetation and population density also resulted as a negative spatial autocorrelation (Moran global index of −0.062), which means that the area is characterized by high concentration of vegetation in areas of low population density. In both cases, the indexes are not very expressive. The index is a little bit more expressive in the relation between income and vegetation, which presented a positive autocorrelation (Moran global index of +0.112).

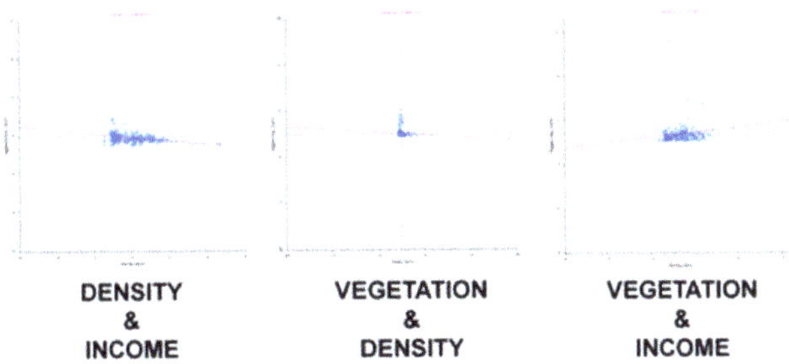

(**A**) (**B**) (**C**)

Figure 15. Scatter plot Moran global indexes–Bivariate analysis population density and income (**A**), vegetation and population density (**B**), vegetation and income (**C**). Source: The authors, using Geoda©.

In none of the autocorrelations the indexes were high, probably because of the high heterogeneity of the area. If some sectors were studied separately, according to homogenous areas, there would be a possibility of achieving higher values. Nonetheless, the indexes confirmed our initial perception and the next step is to study the area in detail, using local indexes (LISA).

Regarding the analysis of the bivariate Moran Global index, the values were close to zero, once again showing a low spatial trend of autocorrelation between the variables. Despite the low values, the expected relationship of negative spatial autocorrelation between population density and income was observed.

In the analysis of population density and the amount of vegetation, there was also negative spatial autocorrelation, showing that the increase in population density puts pressure on the vegetation fragments and their consequent suppression. However, the low value of the Moran index shows the low spatial expression of the relationship, that is, it occurs but it is not spatially evident.

The relationship between vegetation and income resulted in positive spatial autocorrelation, which demonstrates the tendency to preserve fragments of vegetation in areas of higher income and the risk of their suppression in areas of low income. Urban parameters can be used to explain this, since in high-income areas the lots are bigger and proportionally there will be more areas destined for vegetation cover. It was also observed a relationship between low income and the increase in population density and, consequently, greater pressure on vegetation cover.

3.4. Moran Local Analysis

Moran local analysis, LISA (Local Index of Spatial Association), has the goal to evaluate if there are clusters of high values with high values, of low values with low values or even the presence of outliers (high with low values or low with high values). The index also evaluates if these local clusters or the outliers are statistically significant. The results are maps that distribute the values in quadrants (high-high, low-high, low-low and high-low) and identify the territorial units in which the outcomes were not significant (in which there is no defined behavior) and maps that show, in the territorial units that presented expressive results in the previous step, how significant the result was. The studies can be done in univariate or in bivariate mode.

3.4.1. LISA—Moran Local Index, Univariate Analysis

In the case study, it was observed that the income variable presented spatial autocorrelation in two portions of the area, presenting territorial units of high income with neighbors of high income (red, high/high) and units of low income with neighbors of low income (blue, low/low). There are no situations of high income associated with neighbors of low income and there are very few units with low income in areas of high income and the existing ones are slums (light blue, low/high). Analyzing the significance of the observed combinations, the majority is high or medium to high, which supports the results of the analysis. However, most territorial units do not present positive spatial autocorrelation. Two agglomerations of high income and of low income in the area are observed but in the rest of the area there is a lot of heterogeneity. (Figure 16A–C).

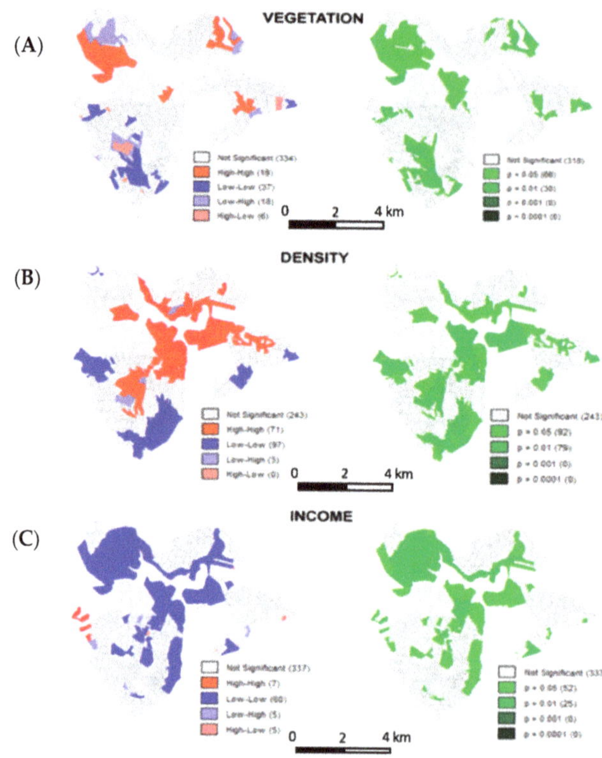

Figure 16. Local Index of Spatial Association (LISA), univariate analysis of vegetation (**A**), population density (**B**) and income (**C**) and a significance level of the identified relations (in green). Source: The authors, using Geoda©.

The vegetation distribution analysis (Figure 16A) presents an expressive number of territorial units with non-significant autocorrelation but it also presents two areas quite defined—a high vegetation level with neighbors with the same condition and another area of a low vegetation level with neighbors of a low level. The areas with low vegetation values call our attention, surrounded by areas of high values. In this case, investments in policies and projects can be done to these areas, so that they can follow local trends and make efforts to improve vegetation cover. Most of the classifications present a high significance.

In the population density analysis, the distribution that stands out is the presence of territorial units of low density inserted in neighborhoods of low density (Figure 16B) and an expressive number of units where there are no significant spatial autocorrelation and the spatial distribution is random. The significance of the results where spatial autocorrelation was observed is medium high to high.

3.4.2. LISA—Moran Local Index, Bivariate Analysis

The comparison between income and vegetation resulted that, even though we had many territorial units with non-significant results, there are parts of the area with high income surrounded by parts of high vegetation cover and parts with low income surrounded by parts with low vegetation cover (Figure 17A) and these two combinations presented significant results. This means that yes, the high income is associated with high presence of vegetation and low income is associated with low presence of vegetation cover.

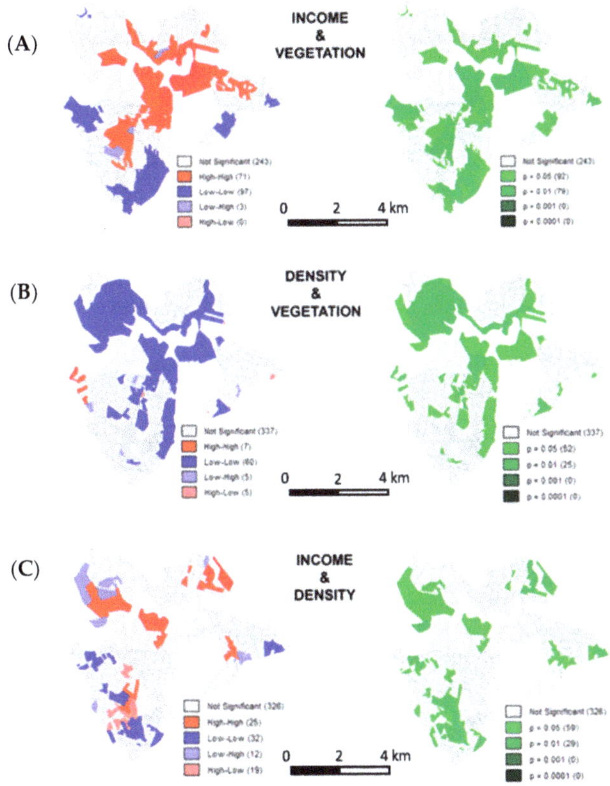

Figure 17. LISA, bivariate analysis of income and vegetation (**A**), population density and vegetation (**B**) and income and population density (**C**). Studies about the level of significance (in green). Source: The authors, using Geoda©.

4. Discussion

The first analysis about LISA bivariate, comparing income and vegetation, made us have the interest to investigate if population density was also correlated, in a negative sense, with vegetation cover (Figure 17B). The result is an expressive number of territorial units without autocorrelation, with non-significant results but there are parts of the territory with low population density surrounded by neighbors of high vegetation cover and parts of low population density with neighbors of low vegetation, both with significant results.

Pampulha is predominantly a low-density and very green area of the city. This is part of its genius loci and must be considered in public policies and in the management of land use of the area. The relation between income and population density shows that what predominates is the autocorrelation of high income surrounded by low population density followed by low income surrounded by low population density, with a level of significance from medium to high (Figure 17C).

From the analysis developed, it is possible to affirm that Pampulha has an expressive number of territorial units without spatial autocorrelation, with random characteristic compositions. However, from maps and indexes interpretation it was also possible to see that big zones in the area, especially those composed by big parcels, present significant spatial autocorrelation. These autocorrelations result in the main image of the area, its genius loci, which is expressive vegetation cover, low population density and high income. In face of these characteristics, it is important to maintain this quality of urban environment but also to create policies and projects that can extend this quality to the sectors of low income. It is also very important to observe the risks of transformation into high population density, because, if it happens, it will certainly change the genius loci of the area.

The main limitation about applying ESDA is related to existing limitations in spatial analysis in a general sense—the definition of territorial units of analysis and the aggregation of data, considered scale (resolution) and form (regularity) [50,51].

The territorial unit is a reference to collect, organize and analyze data. The ideal reference is a regular grid with high resolution, which means the discretization in very small portions of land cover. There are references to better define this resolution, that has to do with cartographic precision of the base map or of data capture. The smallest territorial unit is the minimum dimension of data capture or the resolution of 0.5mm in the scale of the base map, according to cartographic standard accuracy of class "A" maps [52,53]. As an example, using a base map produced in the scale of 1:5000, the minimum element can be 2,5m. Nevertheless, sometimes the users define the minimum grid value according to expectancy of territorial analysis, which means minimum spatial information. As an example, if the user needs to interpret the results by lot or block, he can use a regular grid measuring 10 m or 100 m.

In ESDA it is important to define not only the spatial resolution but also if it is possible to use a regular grid, because it depends on how the data was collected. The better the data collection was distributed, the more precise the analysis will be. Regular grids, considering also the importance of neighborhood definition, can be proposed in tessellations that cover the area of study and consider possible connections from spatial locations. In some cases, it is not possible to use a regular grid, because the data was captured and summarized according to territorial sample units that follow geographic references. Examples of these irregular grids are the use of blocks, districts, census tracts and municipalities. It is possible to convert these very irregular polygons into ones that reduce complexity but keep the proportional distribution of areas, considering a central point of the original polygon. Using the centroid of the original polygon, Thiessen polygons can be designed, with the goal to generalize spatial representation but keeping somehow the report of spatial dimension and neighborhood contacts.

In the case study discussed in this paper, data about vegetation cover was produced from images of high resolution, regular grid of pixels. It was possible to develop a very detailed analysis but the goal was to explore the relation through income and population density to vegetation cover, to predict vulnerabilities and potentialities to the maintenance of this value and to give support to policies in territorial management. Unfortunately, in Brazil, socioeconomic data about income and population are available summarized in census tracts and that defined the spatial resolution of the analysis.

In some cases, as it generally happens in Brazil, the definition of these irregular tracts follow administrative goals (the number of questionnaires by census taker) and not geographic goals (limits and regular dimensions of polygons) and result in lack of precision and possibilities of misunderstandings results. However, there is a possibility, in the future, to work with a better spatial resolution for data collection, by street segment, which will allow different aggregations and a more detailed interpretation.

5. Conclusions

The case study resulted as a useful instrument to work with quantitative and qualitative approach, because it presents the quantitative support to demonstrate a perception that inhabitants from Belo Horizonte have about the district of Pampulha. This makes the analyses supported by defensible and reproducible criteria.

The exploratory studies of spatial data have been used for over one decade in research about socioeconomic aspects, mainly to prove lack of distribution of facilities in the territory, which is being called equity. The contribution of this paper is to focus on the role that a neighborhood can have in behavior factors related to collective conscious or non-conscious decisions.

We defend that the observation of what is happening in the surroundings has an impact on the adoption of values and this can result in increment or decrement of common conditions of life quality to the group. We believe that behaviors produce tendencies, tendencies create values and values structure culture.

Observing those areas in the neighborhood that keep the vegetation cover is an incentive to adopt this practice as a value or as culture but it is important to say that policies and public projects must help to face the challenges of low income and the risks of population density growth. To plan policies and designs, it is very useful to observe each territorial unit, to identify those that have potential to receive interventions, according to their characteristics and neighborhoods. The expectation is that those units have the potential to spread the effects of an intervention, as catalysts, so that from punctual actions we can have effects of irradiation of results. With that knowledge, with less we can do more.

It is also important to consider the limitations and potentialities about ESDA studies. As ESDA is a model, represented from a mathematical and geometrical logic, it is a point of view, a simplification of reality. A model reduces the complexity of reality according to conceptual values, to methodological delimitation, to temporal restriction and to spatial delimitation. As a model, ESDA will never cover all of the reality complexity but it is a point of view. It works as a support to understand occurrences and phenomena, to make the users convert data into information and information into knowledge.

As it is "exploratory," the goal is to make users change references and parameters to test the behavior of main variables and to test the connections between variables. The main purpose is to provide a tool to explore decision making, considering generalizations. It must be understood as a relative reality portrait and not as an absolute index or report. If the model is understood just as this support to explore the behavior of variables, it is very useful in spatial analysis, mainly to decide about green infrastructure in urban planning.

The importance of identifying zones and spatial typologies is very clear to studies in physical geography, because it is the base of geographic classifications, applied in landscape descriptions and in spatial taxonomies. However, the awareness about this aspect of neighborhood is of interest, particularly, to research about behaviors and social values, because of spatial diffusion and spread of conditions, in which some parts of the territory influence its neighborhoods, which results in dynamic transformation. This means to identify behaviors that can result into tendencies that can construct new values that, already installed, will form a new culture (Figure 18). As a further development of this study, it is planned to understand what people think about green areas and how green infrastructure is seen in their daily life, with the goal to increase the behavior, tendencies, values and culture about environmental resources, as "nimby" behavior (*Not In My Back Yard*, expression to describe opposition to certain controversial projects or those that may be harmful to the environment) is still something to be overcome.

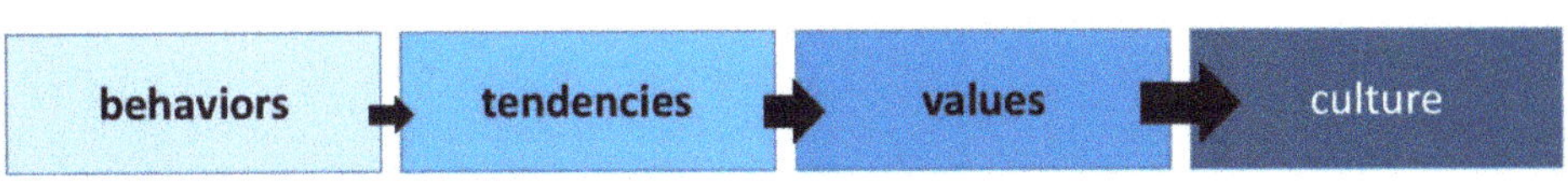

Figure 18. Diffusion effects of behaviors into culture in spatial phenomena. Source: The authors.

Author Contributions: Conceptualization, A.C.M.M. and B.M.F.; methodology, A.C.M.M. and B.M.F.; software, B.M.F.; validation, B.M.F.; formal analysis, A.C.M.M. and B.M.F.; investigation, A.C.M.M.; resources, A.C.M.M.; data curation, A.C.M.M.; writing—original draft preparation, A.C.M.M.; writing—review and editing, A.C.M.M. and B.M.F.; visualization, A.C.M.M.; supervision, A.C.M.M.; project administration, A.C.M.M.; funding acquisition, A.C.M.M. All authors have read and agreed to the published version of the manuscript.

Funding: The authors thank the Municipality of Belo Horizonte, PBH, the access to data and information, granted due to the agreement GGIU—Grupo de Gestão da Informação Urbana. The paper is a contribution to CNPq Project "Geodesign e Modelagem Paramétrica da Ocupação Territorial: Geoprocessamento para a proposição de um Plano Diretor da Paisagem para a região do Quadrilátero Ferrífero-MG," Process 401066/2016-9, Universal Call 2016, and Fapemig PPM-00368-18.

Conflicts of Interest: The authors declare no conflict of interest.

References

1. Rafiee, A.; Dias, E.; Koomen, E. Local impact of tree volume on nocturnal urban heat island: A case study in Amsterdam. *Urban For. Urban Green.* **2016**, *16*, 50–61. [CrossRef]
2. Loboda, C.R.; Angelis, B.L.D. Áreas Públicas Urbanas: Conceito, uso e funções. *Ambiência Guarapuava* **2005**, *1*, 125–139.
3. Antrop, M. Landscape change and the urbanization process in Europe. *Landsc. Urban Plan.* **2004**, *67*, 9–26. [CrossRef]
4. Benevolo, L. *A cidade e o Arquiteto: Debates*; Perspectiva: São Paulo, Brazil, 1984; p. 144.
5. Friese, K.; Schmidt, G.; De Lena, J.C.; Nalini, H.A., Jr.; Zachmann, D.W. Anthropogenic influence on the degradation of an urban lake—The Pampulha reservoir in Belo Horizonte, Minas Gerais, Brazil. *Limnologica* **2010**, *40*, 114–125. [CrossRef]
6. Lopes, A.M.M.B. Composição da Comunidade de Cianbactérias e Outros Grupos Microbianos em Dois Reservatórios Tropicais. Ph.D. Thesis, Conservação e Manejo da Vida Silvestre—Universidade Federal de Minas Gerais, Belo Horizonte, Brazil, 2013; p. 157.
7. Resck, R.P.; Bezerra, N.J.F.; Pinto-Coelho, R.M. Nova batimetria e avaliação de parâmetros morfométricos da Lagoa da Pampulha. *Geogr. Artig. Científicos* **2007**, *3*, 24–37.
8. Barbosa, F.; Garcia, F.; Marques, M.; Nascimiento, F. Balanço denitrogênio e fósforo em um reservatório eutrófico em Minas Gerais: Uma primeira abordagem. *Rev. Bras. Biol.* **1998**, *58*, 233–239.
9. Hergoz, C.P. *Cidades: (Re) Aprendendo a Conviver Com a Natureza*; Maud X Inverde: Rio de Janeiro, Brazil, 2013; p. 311.
10. Sirkis, A. O Desafio Ecológico das Cidades. In *Meio Ambiente No Século XXI*; Mendes, A.T., Ed.; Armazém do Ipê.: Campinas, Brazil, 2008; pp. 214–229.
11. Nucci, J.C. Metodologia para determinação da qualidade ambiental urbana. *Rev. Dep. Geogr.* **1998**, *12*, 209–224. [CrossRef]
12. Bajirao, B.N. Importance of Vegetation in Urban Environment. *Int. J. Sci. Res. Publ.* **2015**, *5*, 1–3.
13. Benedict, M.A.; McMahon, E.T. *Green Infrastructure: Linking Landscapes and Communities*; Island Press: Washington, DC, USA, 2006; p. 320.
14. Rhind, D. Funding an NGDI. In *Geospatial Data Infrastructures: Concepts, Cases, and Good Practice*; Groot, R., Mclaughlin, J., Eds.; Oxford University Press: New York, NY, USA, 2000; pp. 39–55.
15. Moura, A.C.M.; Magalhães, D.M.; Parizzi, M.G. Vocações e Conflitos de Interesse na Ocupação do Território da Região Metropolitana de Belo Horizonte—Apoio ao Plano Diretor de Desenvolvimento Integrado, PDDI. In Proceedings of the Anais XV Simpósio Brasileiro de Sensoriamento Remoto-SBSR, Curitiba, Brazil, 30 April 2011; pp. 690–697.
16. Andrade, F.G.; Baptista, C.S.; Davis Jr, C.A. Improving geographic information retrieval in spatial data infrastructures. *GeoInformatica* **2014**, *18*, 793–818. [CrossRef]
17. Freitas, C.R.; Freitas, C.R.; Ohata, A.; Rocha, G. SDI as an enterprise and sharing instrument: Datageo role in São Paulo's environmental system. *Rev. Bras. Cartogr.* **2017**, *69*, 1442–1455.
18. McHarg, I. *Design with Nature*; American Museum of Natural History: New York, NY, USA, 1969; p. 197.
19. Chorley, R.; Haggett, P. *Models in Geography*; Methuen: London, UK, 1967; p. 816.
20. Bertalanffy, L. *General System Theory.*; George Braziller: New York, NY, USA, 1968; p. 289.
21. Bonham-Carter, G.F. *Geographic Information Systems for Geoscientists: Modelling with GIS*; Pergamon: Ottawa, ON, Canada, 1994; p. 398.
22. Giongo, M.; Koehler, H.S.; Machado, S.A.; Kirchner, F.F.; Marchetti, M. LiDAR: Princípios e aplicações florestais. *Braz. J. For. Res.* **2010**, *30*, 231–244. [CrossRef]
23. Sena, I.S.; Casagrande, P.; Rocha, N.A.; Fonseca, B.M.; Moura, A.C.M. *Methodology for Green and Built Volume Analysis*; Mercator: Fortaleza, Brazil, 2018; pp. 1–13. [CrossRef]
24. Moura, A.C.M.; Rocha, N.A.; Sena, I.S.; Kechagioglou, X.; Campagna, M. Vegetation cover indexes in urban areas: Landscape Ecology studies with Sentinel2 images. *Rev. Bras. Cartogr.* **2018**, *70*, 1686–1714. [CrossRef]

25. Prefeitura Municipal de São Paulo. *Revisão da Lei de Parcelamento, Uso e Ocupação do Solo*; Prefeitura Municipal de São Paulo: Sao Paulo, Brazil, 2014.
26. Sociedade Brasileira de Arborização Urbana—SBAU. "Carta a Londrina e Ibiporã". *Bol. Inf.* **1996**, *3*, 3.
27. Jensen, J.R. *Remote Sensing of the Environment: An Earth Resource Perspective*; Pearson Prentice Hall: Minneapolis, MN, USA, 2007; p. 592.
28. Turkey, J.W. *Exploratory Data Analysis*; Addison Wesley Publishing Company: Don Mills, ON, Canada, 1977; p. 688.
29. Fotheringham, A.S.; Brunsdon, C.; Charlton, M.E. *Quantitative Geography*; Saga: London, UK, 2000; p. 288.
30. Rocha, N.A.; Sena, I.S.; Fonseca, B.M.; Moura, A.C.M. Association between a spectral index and a landscape index for mapping and analysis of urban vegetation cover. In *INPUT 2016*; SITI: Torino, Italy, 2016; pp. 526–531.
31. Anselin, L. The Moran scatter Plot as an ESDA tool to assess local instability in spatial association. In *Spatial Analytical Perspectives in GIS*; Fischer, M., Scholten, H.J., Unwin, D., Eds.; Taylor and Francis: London, UK, 1996; pp. 111–125.
32. Anselin, L. Interactive techniques and Exploratory Spatial Data Analysis. In *Geographical Information Systems: Principles, Techniques, Management and Applications*; Longley, P.A., Goodchild, M.F., Maguire, D.J., Rhind, D.W., Eds.; John Wiley and Sons: New York, NY, USA, 1998; pp. 251–264.
33. Bertin, J. *Semiologie Graphique: Les Diagrammes, Les Réseaux, Les Cartes*; Mouton et Gauthier-Villars: Paris, France, 1967; p. 431.
34. Bonin, S. *Initiation à la Graphique*; Epi: Paris, France, 1975; p. 172.
35. Berry, B.J.L. Approaches to regional analysis: A synthesis. *Ann. Assoc. Am. Geogr.* **1964**, *54*, 2–11. [CrossRef]
36. Menezes, A.C.F.; Faissol, S.; Ferreira, M.L. Análise da matriz geográfica: Estruturas e inter-relações. In *Tendências Atuais na Geografia Urbano-Regional: Teorização e Quantificação*; IBGE: Rio de Janeiro, Brazil, 1978; pp. 67–109.
37. Buzai, G.D. Análisis Espacial con Sistemas de Información Geográfica: Sus cinco conceptos fundamentales. In *Geografía y Sistemas de Información Geográfica, Aspectos Conceptuales y Aplicaciones*; Buzai, G.D., Ed.; Universidad Nacional de Luján–GESIG: Luján, Argentina, 2010; pp. 163–195.
38. Jankowski, P. Integrating geographical information systems and multiple criteria decision-making methods. *Int. J. Geogr. Inf. Syst.* **1995**, *9*, 251–273. [CrossRef]
39. Jankowski, P.; Nyerges, T. *GIS for Group Decision Making*; CRC Press: New York, NY, USA, 2001; p. 273.
40. Jankowski, P. Integrating geographical information systems and multiple criteria decision making methods: Ten years after. In *Classics from IJGIS: Twenty Years of the International Journal of Geographical Information Science and Systems*; Fisher, P., Ed.; CRC Press: New York, NY, USA, 2007; pp. 291–296.
41. Moura, A.C.M. *Geoprocessamento na Gestão e Planejamento Urbano*; Interciência: Rio de Janeiro, Brazil, 2003; p. 286.
42. Tobler, W. A computer movie simulating urban growth in the Detroit region. *Econ. Geogr.* **1970**, *46*, 234–240. [CrossRef]
43. Neves, M.C.; Ramos, F.R.; Camargo, E.C.G.; Câmara, G.; Monteiro, M.A. *Análise Exploratória Espacial de Dados Sócio-econôMicos de São Paulo*; GISBrasil: Salvador, Brazil, 2000; p. 11.
44. Ertur, C.; Le Gallo, J. An exploratory spatial data analysis of European disparities, 1980–1995. In *European Regional Growth*; Fingleton, B., Ed.; Springer: Berlin, Germany, 2003; pp. 55–97.
45. Haddad, M.A. Bolsa Família and the needy: Is allocation contributing to equity in Brazil? *J. Int. Dev.* **2008**, *20*, 654–669. [CrossRef]
46. Haddad, M.A. Examining the Spatial Distribution of Urban Indicators in São Paulo, Brazil: Do Spatial Effects Matter. In *Community Quality-of-Life Indicators: Best Cases III Community Quality-of Life Indicators*; Sirgy, J., Phillips, R., Rahtz, D., Eds.; Springer: Dordrecht, The Netherlands, 2009.
47. Celebioglu, F.; Dall'Erba, S. Spatial disparities across the regions of Turkey: An Exploratory Spatial Data Analysis. *Ann. Reg. Sci.* **2010**, *45*, 379–400. [CrossRef]
48. Anselin, L.; Syabri, I.; Kho, Y. GeoDa: An Introduction to Spatial Data Analysis. *Geogr. Anal.* **2006**, *38*, 5–22. [CrossRef]
49. Goodchild, M.F. *Spatial Autocorrelation*; University of Western Ontario: Canada, ON, Cadana, 1986; p. 56.
50. Anselin, L. Exploring Spatial Data with GeoDaTM: A Workbook. In *Spatial Analysis Laboratory Department of Geography*; University of Illinois: Champaign, IL, USA, 2005; p. 226.

51. Rodrigues, A.M.; Tenedório, J.A. Sensitivity Analysis of Spatial Autocorrelation Using Distinct Geometrical Settings: Guidelines for the Quantitative Geographer. *Int. J. Agric. Environ. Inf. Syst.* **2016**, *7*, 65–77. [CrossRef]
52. Decree N° 89.817. *Establishes the Regulatory Instructions for the Technical Standards of National Cartography*; Presidency of the Republic Civil House Legal Affairs Subsection: Brasilia, Brasil, 20 June 1984.
53. Monico, J.F.G.; Póz, A.P.D.; Galo, M.; Santos, M.C.; Oliveira, L.C. *Acurácia e Precisão: Revendo os Conceitos de Forma Acurada*; Boletim de Ciências Geodésicas: Curitiba, Brazil, 2009; pp. 469–480.

sustainability MDPI

Article

Knowledge Models for Spatial Planning: Ecosystem Services Awareness in the New Plan of Bari (Italy)

Stefania Santoro, Pasquale Balena and Domenico Camarda *

Department of Civil, Environmental, Land, Construction and Chemistry (DICATECh), Polytechnic University of Bari, 70125 Bari, Italy; stefania.santoro@poliba.it (S.S.); pasquale.balena@poliba.it (P.B.)
* Correspondence: domenico.camarda@poliba.it

Received: 26 December 2019; Accepted: 9 February 2020; Published: 18 February 2020

Abstract: The concept of ecosystem services (ES) arises as a formal outcome of historical processes of understanding and interpreting settlements as complex ecological systems. Because of a straightforward, bottom-up demand for environment enhancement, this concept increasingly occurs in discourses, in narratives, in the demands of common people, triggering a new urban environmental awareness. This is now often arising spontaneously in the protocols of participatory plan processes, especially when planning for the future of complex environments such as city areas. The present study tries to elicit reflections around the significance of ES issues awareness in the case study of Bari (Italy), which is experiencing an inclusive and participatory process of construction of shared knowledge for the new master plan. Starting from an initial campaign of civic walks (CWs) along the urban neighborhoods and a subsequent semi-structured interview to the community, the paper carries out comparative analyses using problem-structuring methods (PMs), in order to evaluate and reflect on community behaviors and expectations about ES. Then the paper ends by emphasizing the role of structured knowledge-raising approaches, as critical activities to enhance ecosystem awareness in planning settlements as complex ecological systems.

Keywords: knowledge modelling; spatial planning; problem structuring methods

1. Introduction

In the Italian season of so-called *third-generation plans* following earlier post-war planning experiences [1], issues of qualitative (as well as interstitially speculative) transformation of cities appear, apparently in terms of urban facilities and services.

This type of approach is generally considered as extended until the 1980s, with some medium-sized cities often cited as examples, such as Pavia, Pistoia, Arezzo.

This is also the period, however, of an eruption of the environmental question in scientific debates. New reflections focus on the limits of dissipative growth especially within the residential settlements. What emerges is the need for progressively increased attention to natural resources and their regeneration cycles, especially in urban areas. Some observers even deduce from this circumstance an emerging *fourth generation* of spatial plans, contaminated by new increasing socio-environmental operational programs, such as Agenda 21 [2].

Certainly, a new awareness is growing around the need for closing natural cycles, to avoid problems of liveability, health, consumption of ecological resources. Settlement areas are increasingly considered, planned and managed as complex ecological systems and not as simple territories to be transformed.

The hand of public administration and policymaking can do much in this framework, in its role as a service provider to support the life and welfare of communities. By the new millennium, the new and simple reading of this commitment is thus immediately turned into operationally considering the role played by service places as also resource regenerators.

Also, owing to this simple, natural evolution, the new concept of ecosystem services develops in planning practices, as a lexical outcome of a historical process of understanding and interpreting settlements as complex ecological systems [3] (p. 43). It is therefore a formal name that corresponds to a straightforward, bottom-up demand for environmental enhancement. It increasingly occurs in discourses, in narratives, in the demands of common people, triggering a new urban environmental awareness [4] (p. 161). This is now arising spontaneously in the protocols of participatory plan constructions, especially when planning the future of areas at environmental risk [5].

The present study starts from these considerations, trying to elicit reflections around the weight of ecosystem instances through inclusive processes of cognitive planning, with the aim of verifying their final policy enhancement. This research objective in the present work is not oriented to discuss features, roles, history of ES in planning as such. Rather, the purpose is to reason about the effectiveness of some knowledge-based planning models in embedding and valorizing ES significance in spatial plans. The work refers to the case study of Bari (Italy), in which a multi-faceted process of construction of shared knowledge is in progress for the preparation of the new urban plan of the city. After the present introduction, Section 2 shows a basic literature review about ES research in a knowledge-raising perspective, while Section 3 presents the case study layout. Section 4 shows and describes the applied methodology and Section 5 outlines and discusses some results of the case study. Final remarks and future developments close the paper (Section 6).

2. Ecosystem Services in a Knowledge-Oriented Perspective: A Relevant Background

According to a literature review, a large variety of ecosystem service (ES) definitions and classification approaches exist [6–12]. However, it is recognized by the scientific community that ESs offer benefits to the human ecosystem [13]. According to the Common International Classification of Ecosystem Services (CICES) [14], ESs is defined as the effects of ecosystems on people's well-being. The CICES framework uses and classifies ESs into three categories affecting directly the human ecosystem: (i) *provisioning* (e.g., food and fresh water), (ii) *regulating and maintenance* (e.g., water purification), and (iii) *cultural* (e.g., recreation and aesthetics).

In the present study, that started from a thesis work at Polytechnic University of Bari, the CICES framework has been developed according to the Environmental Protection Agency (EPA) [15] as (i) Natural resources: water, land, soil, and air; (ii) drivers of change: policy, land use, climate, pollution; (iii) benefits: economy, well-being, food-water and materials and integrated into the methodology (Table 1).

Table 1. Ecosystem services. Adapted from Environmental Protection Agency (EPA).

ECOSYSTEM SERVICES		
NATURAL RESOURCES	**DRIVERS OF CHANGE**	**BENEFITS**
Water	Policy	Economy
Land	Land Use	Well-being
Soil	Climate	Food- Water and Materials
Air	Pollution	Public Health

Natural resources are the resources offered by the planet without any human intervention (UNESCO); according to Millennium Ecosystem Assessment (MA), drivers of change are defined as natural or artificial factors that impact directly or indirectly on the ecosystems, causing a change. While the first kind of drivers influence directly ecosystem processes, an indirect driver alters one or more direct drivers. For this work, direct drivers have been considered; the benefits refer to the advantage deriving from ESs in different fields [16].

A literature review shows a growing interest of ES in the past decades [17–19]. However, in order to deal with the complexity of ES, few approaches exist [20]. A brief review on this topic still shows the lack of practical suggestions for the implementation of ES in spatial planning [10,21,22]. The result is a deep gap between science and political decisions [23]. In the attempt of reducing this gap, literature shows the role of the participatory processes (PP) as critical integration of expert knowledge—insufficient to support decisions in relation to specific problem situations [24,25]. PP could facilitate information acquisition, contributing to the knowledge sharing and decision implementation [26]. There is no straightforward strategy to elicit, collect, and structure citizen knowledge, experience, and perception, nor to improve their awareness of a problematic situation. Several approaches and methodologies is still being explored, although growing debates exist toward promoting stakeholders' engagement, with mixed results [27]. These suggestions are increasing effectively today addressed within the field of participatory modelling techniques (PMT) (see a brief review in the Section 4).

In order to structure the citizen knowledge about ES, the present study tries to apply a methodology based on a combination of PMT and PSM. This methodology is experimentally applied to the case study of Bari in Apulia region (Italy).

3. The Case Study of Bari

The capital of Apulia in Southern Italy, Bari city is extended over 117,38 km^2 with 324,198 citizens (ISTAT 2011). It is subdivided into five districts characterized by the presence of natural, cultural, and social resources to be reactivated, able to offer numerous ES (Figure 1).

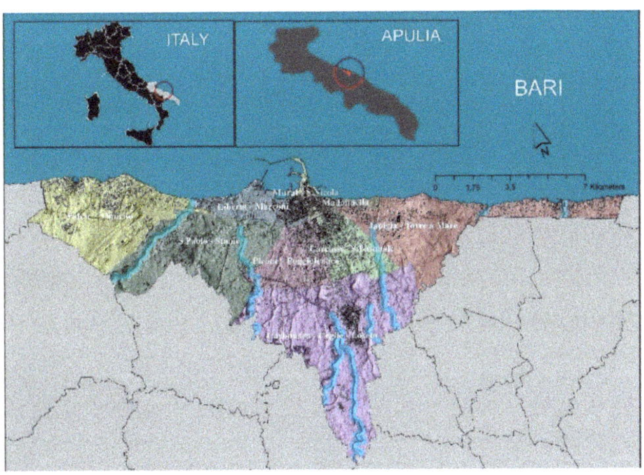

Figure 1. Location of the city, municipality, and nine blades.

The natural capital of Bari can be referred to a number of consolidated aspects, such as (i) The sea, which is embedded in the culture and economy of the city; (ii) the nine blades, organized in a complex ecological system (Figure 1); (iii) areas of archaeological interest and public parks located in different districts. The cultural and social capital can be represented by (iv) the historical pre-existences offered in the city center by houses dating back to the Byzantine era (Figure 2); (v) the lictor architecture developed during fascism in the peripheral center and on the waterfront (Figure 3), and the architecture of twentieth-century rationalism in the peripheral districts (PPTR Puglia 2009) [28].

While it is easier to trace the benefits for the well-being of humans and of the ecosystems offered by the natural factors (such as, for example, the role of urban green that can regulate the climate by absorbing carbon dioxide and supporting the recycling of nutrients and the formation of soil or attract people for leisure and tourism), it is harder to associate the benefits offered by the built landscape, despite building issues are central to the public discontent with natural resource management decisions [29]. Therefore, setting up a participatory process in the identification and evaluation of ES is an increasing issue in planning agendas. Starting from resolution nr. 565/2015 in 2015 Bari Department of urban planning started a participatory process at several levels, asking citizens to participate in the drafting of the master plan of the city (PUG) [30]. The path was designed in different time steps, performed between May and November 2016, through (i) nr.30 *urban front offices* (UFOs) activated in the municipal area (Figure 4), (ii) nr.9 *civic walks* (CWs) to single out peculiar aspects and features of relevant areas (Table 2; Figures 5 and 6), and (iii) nr.5 *public workshops* (PW). Located in different parts of the city (Figure 4), UFOs have supported the participatory process through the distribution of surveys regarding the themes of (i) public space, (ii) landscape, and (iii) mobility; CWs aim to share new visions of the places [31]. Analysts have chosen nine CWs in order to highlight the peculiarities of each area (Table 2); *experts knowledge* (institutional referents and technicians) and *non-expert knowledge* (citizens) have been included in order to present observations, questions, and desires. *PWs* aimed at expanding the dissemination of participation culture in urban policies, through the proposal of some open meetings to citizenship on relevant urban topics.

Figure 2. Archaeological area of San Pietro, (Baritoday, 2013).

Figure 3. The lictor architecture developed during fascism on the waterfront.

Figure 4. Allocation of urban front office (Report PUG 2016).

Table 2. Civic walk description.

Subject of Civic Walks	Areas
1. Minor historical centers and hypogea	Carbonara e Ceglie
2. Relations between the city and the sea	Libertà e Marconi
3. The places of the "gigantism" of the Quaroni Plan and the city of children	Carrassi e Poggiofranco
4. The reuse and redevelopment of historical fabrics in the multiethnic city	Madonnella e Libertà
5. Archaeological protection as a natural barrier to the consumption of soil and a resource for development	Torre a Mare
6. Recovery of brownfield areas	Santo Spirito
7. The historical nucleuses of public buildings and the landscape resources in the suburbs	San Paolo
8. The quality of living in the large public housing districts: public space and private space in the different neighborhood settlements	Japigia
9 The public housing districts to be reconnected	Stanic

Figure 5. Civic walks (CW) n°1.

Figure 6. CW n°5.

The present study starts from a participatory process carried out by the Municipality of Bari, trying to highlight the limitation of the methodology adopted and to elicit some reflections on the weight of ES in citizens awareness. Section 4 describes the proposed methodology.

4. Methodology

Historically, building knowledge models for spatial planning purposes has a very recent genesis. This happened essentially because of a research area, dominant from the second half of the 1900s until around the 1990s, characterized by the qualitative and discretionary data management, often referred to as "participatory planning." The rationale behind this position can be sought mainly as a response to a strong demand for bottom-up democracy in decisions, after long traditions of top-down decision-making centralism [32] (p.27). Toward the end of this exciting season, dangerous demagogic, rhetorical, and speculative pitfalls ended up characterizing and often degrading the ethical and social effectiveness of participatory planning. The increasing awareness that has fortunately followed has been paralleled by an emerging research on new methodologies offered by information and communication technology for the management of complex and widespread data. ICT-based statistical-mathematical platforms have increasingly allowed the management of multisource and multiagent data in a quantitative or quali-quantitative way [33]. A growing scientific research has emerged regarding the construction of quantitative bottom-up knowledge management models, to support more informed, real-time decisions toward more effective spatial policy and planning. For the social management of environmental resources and components of the territory, the problem is particularly complex as it is linked to an inherently embedded social and environmental complexity. In addition to mainly qualitative methods, e.g., structured in terms of building future scenarios, ontological and relational knowledge management models have evolved in an attempt to preserve such inherent complexity while maintaining a synthesis necessary to allow its management [34–38]. In our paper, the context is characterized by a hybrid methodological approach for which a quantitative knowledge management model accompanies and criticizes a mainly qualitative pre-existing approach.

In particular, the present study proposes an approach based on knowledge structuring to (i) overcome the limits emerging during the participatory process adopted by the municipality, and (ii) to investigate the weight of citizens' awareness about ES through the construction of knowledge models.

Specifically, the reflection on CWs raised three critical issues: (i) Numerical predominance of considerations by expert knowledge on non-expert knowledge; (ii) the lack of structured knowledge broadly following narrative patterns; (iii) a small number of participants, never exceeding 30 units.

In the present study, in order to overcome these limits, information emerged in narrative patterns deriving from CWs has been recorded and formalized using ad-hoc structuring platforms, particularly relevant to PSMs modelling area. Specifically, a qualitative analysis of the information deriving from CWs through knowledge discovery in text (KDT) was needed to build causal loop diagrams (CLDs) and semi-structured interviews (SSIs). KDT was oriented to structure problem framework [39] and build CLDs, whereas SSIs held a dual function of validating CLDs and involving a more significant sample of citizens. Figure 7 shows the developed methodology.

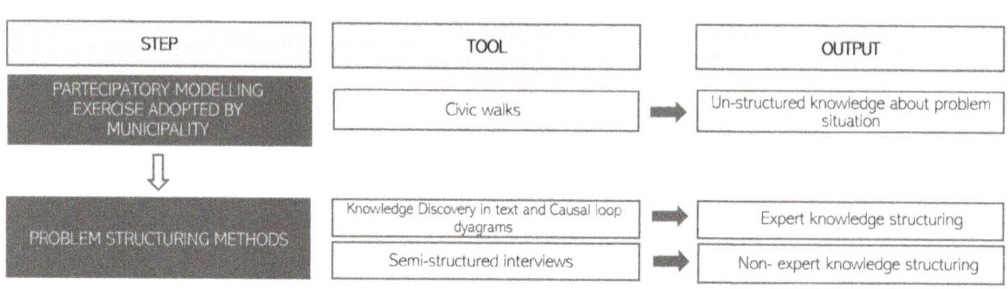

Figure 7. The developed methodology.

The choice to use CLDs and SSIs arise from an analysis of existing PMT. Levels of involvement could be enhanced to strengthen the strategic knowledge on the environment, scientific literacy, and the empowerment of citizens in helping to inform and monitor policies and management efforts related to the ecosystem services [40]. Some authors [41–44] have recently proposed literature reviews on PMT (Table 3).

Table 3. Type of participatory modelling technique.

Typology	Description	Tool
Group Model Building (GMB)	It is based on the involvement of a group of stakeholders in order to create a conceptual model. The model building process can start from a story telling, a set of interviews or narratives and it is supported by a facilitator.	Causal cycle diagrams and similar visual tools
Mediated Modelling (MM)	It is based on the collect of information by a group of stakeholders in order to create a system dynamic model. The model building process use the computer models in order to create scenarios.	System dynamic model (Stella)
Companion Modelling (CM)	It is based on involving stakeholders through role-plays in order to make them aware of the variety of points of view and their consequences in terms of actions.	Role-playing games
Participatory Simulation (PS)	It is based on involving stakeholder through mobile devices and physical activities. The rules of the games cannot be modified by the stakeholders. Every decision and every interaction are registered for further analyses.	Agent-based model
Shared Vision Planning (SVP)	It is based on computer simulation in order to identify alternatives and trade-offs in a manner where stakeholders without modelling experience can actively participate in the modelling process.	Scenario simulation
Collaborative Learning (CL)	It is based on the idea that learning is a naturally social act that takes place through communication. The aim is to teach and learn in groups to work together on problems, complete a task, or create a product.	Workshop, seminar, and similar tool

For this work, a group model building (GMB) has been used and the model has been built by the analyst following the narratives deriving from the CWs and validated by the SSIs. The analysis led on one of the nine CWs is described below, as an explicatory example.

4.1. *Participatory Modelling Exercise: Civic Walks (CWs)*

Civil walks are an early and widespread instrument of citizens' participation in the field of urban policies in order to activate new forms of knowledge about the city [31]. They are typically collective walks along the streets of urban neighborhoods, developed by citizens who accompany an expert scholar describing and discussing the features and issues related to the area. Through questions and answers, a mutual interaction of knowledge is established between the participants and the expert, which is oriented to enrich the knowledge base useful for planning purposes. CWs tell the dynamics, spaces, and urban regeneration through the eyes of citizens. CWs aim to enhance the environmental aspects, the old and new forms of the urban space and economy, the quality of public space in terms of beauty and sociality to calibrate the urban redevelopment actions.

According to the literature analysis, CWs seem to ensure some important features: (i) The implementation of *de-professionalization* visions, i.e., not only professionals shape the future of the districts; (ii) a *demystification* of problems, turning territorial planning into real and concrete perspectives, away from a virtual or mediatized knowledge; (iii) the *democratization* of knowledge and decisions, as many citizens are directly involved in the process of reflection and decision, especially those that represent an interest in the future of the districts [32].

In several Italian cities CWs have been used (see the case of Bologna, Ravenna, Modena, Livorno e Valdagno).

Despite these assumptions, the analyzed CWs present a preponderance of interventions by the expert knowledge. The CW analyzed here has crossed areas coming from two planning seasons of the city of Bari. The first one comes from the urban plan drafted by Calza-Bini and Piacentini in 1954 characterized by a traditional urban design (concentric and equidistant road links, called *mediane,* connected by radial roads). Figure 8 taken from a Google satellite shows the average morning traffic flow of a road that does not relate to the expanding Bari of the post-war years (Figure 8(1a,1b)).

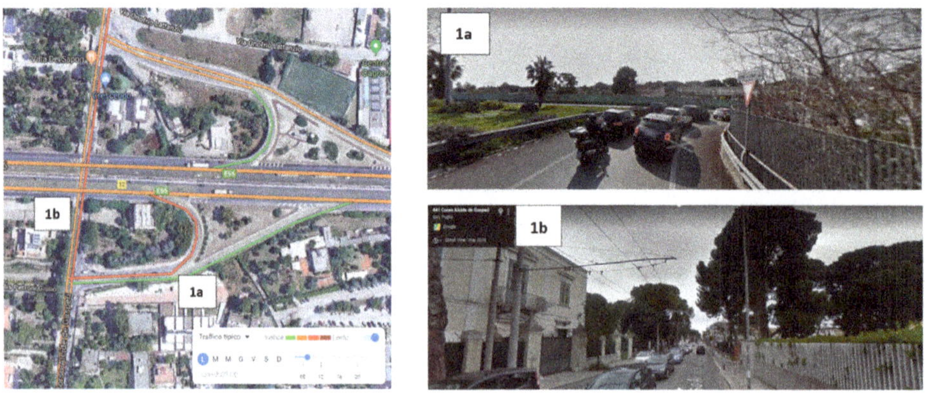

Figure 8. Ring road connection to Alcide De Gasperi street (**1a**); Alcide De Gasperi street (**1b**).

The second one is Quaroni's urban plan (1976), characterized by the gigantism of roads and buildings in view of demographic growth and city flows (Figures 9 and 10).

Figure 9. Mother Teresa of Calcutta street, Quaroni's urban plan.

During the CW analyzed, 22 citizens and 3 technicians were involved. The CW was focused on three places: Alcide De Gasperi street (Place 1); Gandhi Mohandas street (Place 2); Mother Teresa of Calcutta street (Place 3) (Figure 11).

Figure 10. Gandhi Mohandas street, Quaroni's urban plan.

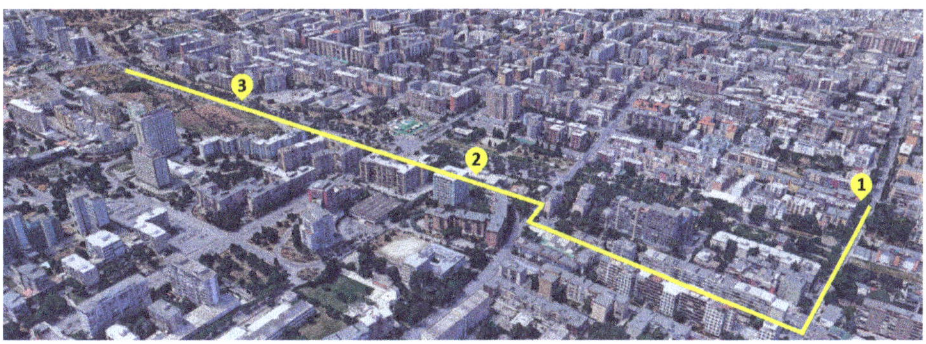

Figure 11. CW n°3 The places of the "gigantism" of the Quaroni Plan and the city of children.

4.2. *Knowledge Discovery in Text (KDT) and Causal Loop Diagram (CLD)*

The analysis of the information that emerged from the conversations during CWs was carried out through the knowledge discovery in text (KDT) approach [45]. The selection of this approach depends on its offering the possibility, using automated intelligent systems, to extract knowledge from unstructured texts. Specifically, through KDT it is possible to extrapolate contents from an unstructured text, despite the lexical difficulties inherent in a conversation [46].

The application to the present case study took place with the use of Rapid Miner studio software [47] and is divided into two phases: the first one, in which the text is "cleaned up," and the second one in which the analysis of contents and the construction of cause-effect links are carried out.

The first "cleaning" phase follows the following steps: (i) Tokenization, able to select the main words (tokens) included in a document according to a frequency logic; (ii) the identification of "Stopwords" in order to delete all the irrelevant words listed in the dictionaries provided by the system (deriving from the Italian dictionary); (iii) stemming, able to reduce the number of words that share the same root as a given token. The pieces of information emerging from the cleaning process are grouped into a correlation matrix (Figure 12) representing the second phase.

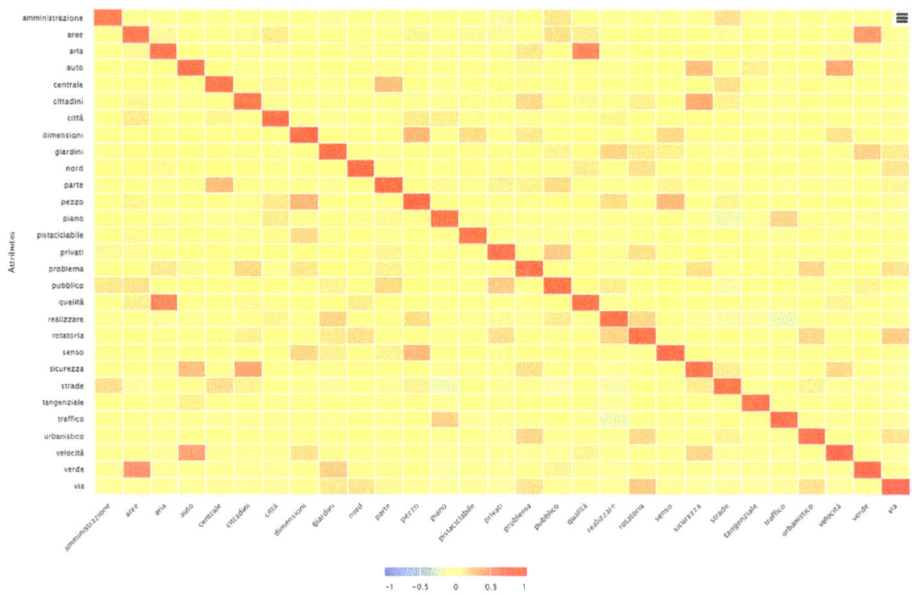

Figure 12. Correlation matrix (excerpt).

Correlation (1) is a statistical technique that shows the extent to which the pairs of variables X and Y are related:

$$r = \frac{1}{n-1}\sum((X-\overline{X})/Sx)((Y-\overline{Y})/Sy) \quad (1)$$

where, X and Y are the frequency of the single words with the relative means and letter S indicate the standard deviation.

Correlation is expressed by a number between -1 and 1 that measures the degree of association between the two words. A positive value indicates the presence of a positive association. It is obtained when high values of Y tend to be associated with high values of X and low values of X tend to be associated with low values of Y. An inverse association, on the other hand, is expressed by a negative value. That is, high values of X tend to be associated with the minimum of the values of Y and vice versa. These relations refer to the instances of the text, thus allowing the creation of a structured cause–effect relationship [48].

The analysis just described was fundamental in order to reduce the ambiguity of interpretation deriving from the reading of a textual document. Also, it allowed us to reconstruct the causal links between the variables in a more structured way and overcoming the limit of the necessary presence of the actors involved, as required by group model building technique (GMB) theory.

CLDs are a formal modelling tool of GMB. It is a method used mostly in business applications but also for natural resource management [49]. It consists of the involvement of a group of agents, stakeholders, in one or more sessions to build the conceptual model or problem situation represented by CLDs and similar visual tools, starting from a history or even from interviews, facts, and narratives [50]. The facilitator helps the group in the model construction, remaining generally neutral with respect to the content. CLDs is symbolized by variables and links with polarity representing the effect of one variable on another [51].

The elaboration process described above took place by dividing the text into three sections, relating to the three Places of reference (Figure 11).

As a matter of simplicity, the present section shows the methodology applied to only Place 1.

The KDT process filtered 826 words and selected the most common word pairs (29 relationships between variables with weight included, for negative values between −1 and −0.5 and for positive values between 0.5 and 1). The relationships have been represented using a *Force-Directed Graph Drawing Algorithm* (FDGA) (Figure 13) [52].

The FDGA is represented by nodes and links between nodes. The size of the node is associated with the characteristic of each of them; the distance depends on the repulsive or attractive force between nodes. The FDGA allows to build the framework of CLDs representing the knowledge model of citizen involved.

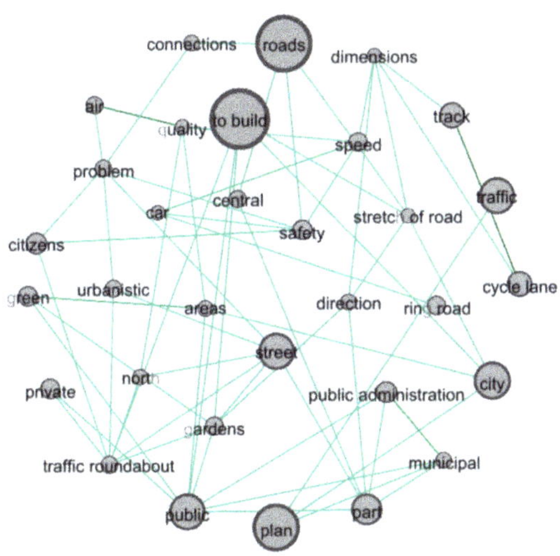

Figure 13. Relevant concepts emerged at Place 1 using a force-directed graph drawing algorithm (FDGA). (The Fruchterman-Reingold algorithm implemented in the "Gephi v. 0.9").

From the CLD referring to Place 1, two main themes emerged: (i) excessive vehicular traffic and (ii) the lack of public green spaces (Figure 14).

Figure 14. Causal loop diagrams (CLD) referring to Place 1.

Specifically, on one hand, the road section is claimed to be unable to meet contemporary mobility demands. On the other hand, the problem of lacking green public spaces is due to intensive buildings and possibly worsened by the misappropriation of the few remaining areas by some private owners. A re-building of the traffic plan for the management of vehicular flows on the one hand, and the supervision by the public administration on the other, are the solutions proposed by the *expert* knowledge in response to the issues raised.

The second CLDs referring to Place 2 and 3 was carried out using the same procedure (Figure 15).

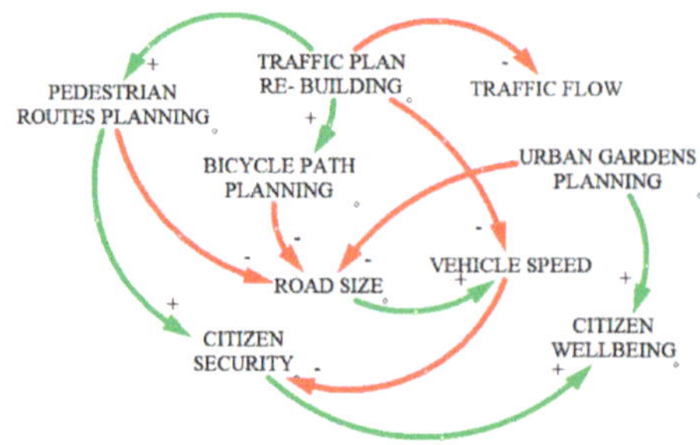

Figure 15. CLD referring to Place 3.

The CLDs show the causes and the effects that these variables entail. Following the same procedure, the other map has been built, in which the CLDs of Place 2 and 3 have been aggregated referring to the same issues. The use of land, deriving from the reduction of some road sections first conceived as urban highway and never completed, was a central theme referring to the Place 2 and 3. Specifically, two suggestions have been proposed: (i) Urban gardens for community along the roads and (ii) the reorganization with partial pedestrianization and bicycle path of the street to reduce the speed of traffic. The need to expand public spaces, by redeveloping the underutilized areas, was claimed in a different part of the district.

In the next section the validation of CLDs is explained through SSIs.

4.3. Problem Structuring Method: Semi-Structured Interviews

SSIs are commonly used in policy research and are applicable to many research questions [53]. They combine some structured, formalized questions with some unstructured exploration. They are useful when dealing with complex systems, owing to the use of spontaneous approaches able to better explore, understand, clarify answers to questions [54]. In this context, SSIs have been carried out and submitted to citizens, retracing the same places of the CWs, also aiming at checking the relevance and consistency of issue previously raised by *expert* knowledge.

SSIs have been structured in three sections: (i) citizen profiles, (ii) mobility issues, and (iii) public space issues. Citizens could express own preferences on a Likert (1932) 1–5 scale of agreement and to insert a free contribution on the actions to be addressed on the issues raised (Figure 16). A sample of 88 citizens, divided into 53 women and 35 men, aged between 35 and 50 years, were interviewed.

MOBILITY	1	2	3	4	5
Alcide De Gasperi street is undersized in relation to daily vehicular traffic					
On Calcutta street the cars speed reduces the pedestrians' safety					
Lack of signage and lighting on pedestrian pathways reduce pedestrian safety					
The crossroads should be managed with the integration of roundabouts					
Car parks are adapted to the traffic flow					
In order to improve mobility in your neighbourhood, please suggest some actions					

PUBLIC AREAS	1	2	3	4	5
Children and elders have adequate entertainment facilities					
Don Tonino Bello park is adequately maintained					
There are many empty spaces that could be redesigned for other public functions					
There are public facilities for citizens' support					
I feel safe in my neighbourhood					
In order to improve growth of public areas, please suggest some actions					

Figure 16. Semi-structured interview.

5. Results and Discussion

A qualitative analysis of the information deriving from CWs through KDT allowed us to build the CLDs in order to structure *expert* knowledge and SSI in order to validate *expert* knowledge and at the same time to structure *non-expert* knowledge.

The results emerging from the application of the methodology are described below. CLDs building tried to overcome the limit relating the unstructured approach emerged during CWs. The information thus emerging was subsequently connected to ES (EPA) classes, so making it possible to draw out considerations regarding the issues related to ES, by observing Table 4.

It can be noted that the most common drivers of change are the land use and policy. The benefits related to well-being and public health are connected through cause–effect relationships.

The "traffic plan re-building" variable emerged in all three places. Referring to Place 1, it was suggested to act on the traffic flow, through a study of vehicular flows, not being able to physically modify the undersized road section. Referring to Place 2 and 3, the construction of cycle paths, pedestrian route and urban gardens was suggested. The latter seems to meet a dual function of reducing the road section and (consequently) vehicle speed, while promoting sustainable mobility and equipping the district with urban gardens. In terms of benefits, these actions induce an improvement in the well-being of citizenship owing to the presence of areas for leisure, a decrease in vehicular traffic with more safety for pedestrians, an increase in health and clean air-related benefits.

The above statements have been submitted to citizens' opinion and degree of validation through SSIs. On the one hand, this allowed a general validation by the citizens on the issues emerged from expert knowledge, thus somehow balancing the preponderance of interventions by expert knowledge. On the other hand, it helped to bring out new issues such as waste management, the inclusion of public lighting, and the planting of new plant species. Variables have been relocated to relevant ES categories (Table 5).

Table 4. Ecosystem services (ES) emerged from CW *expert* knowledge.

NATURAL RESOURCES	PLACE	DRIVERS OF CHANGE		BENEFITS	
		Policy	**Land Use**	**Well-Being**	**Public Health**
LAND	1	Municipality supervision	traffic plan re-building	decrease traffic flow	increase air quality
			use of public green area	areas for children	
	2/3		traffic plan re-building	increase bicycle path planning	decrease vehicles speed
				increase pedestrian route planning	
				decrease traffic flow	
			urban gardens planning	increase citizen well-being	

Table 5. ES emerged from semi- structured interviews by *non-expert* knowledge.

NATURAL RESOURCES	PLACE	DRIVERS OF CHANGE		BENEFITS	
		Policy	**Land Use**	**Well-Being**	**Public Health**
Land	1		Planting of different tree species		Decrease of allergies
	2/3		public lighting	pedestrian security	
			waste management	neighborhood cleaning	increase air quality

The issues emerged, which are added to those already known to be derived from expert knowledge are: the planting of new tree species in order to reduce the problems linked to allergies that characterize children residing in Alcide De Gasperi street (Place 1); the strengthening of public lighting at Place 2 in order to increase pedestrian safety and finally, at Place 3, the need of improving the waste management system to guarantee adequate hygienic conditions of spaces and healthiness of air. The validation of CLDs took place through the analysis of semi-structured interviews. The preference of citizens in relation to questions have been analyzed. For the sake of simplicity one example is shown "Corso Alcide De Gasperi is undersized" (Figure 17a).

A total of 48% of citizens involved expressed an agreement at Likert scale 4, whereas 34% of citizens agreed at grade 5 (Figure 17a). Most citizens involved acknowledges that Alcide De Gasperi street is undersized. It is possible to summarize the results obtained from the questionnaire through a histogram in which the abscissas represent the questions, and the ordinates represent the average of citizens' preference for each question (Figure 17b).

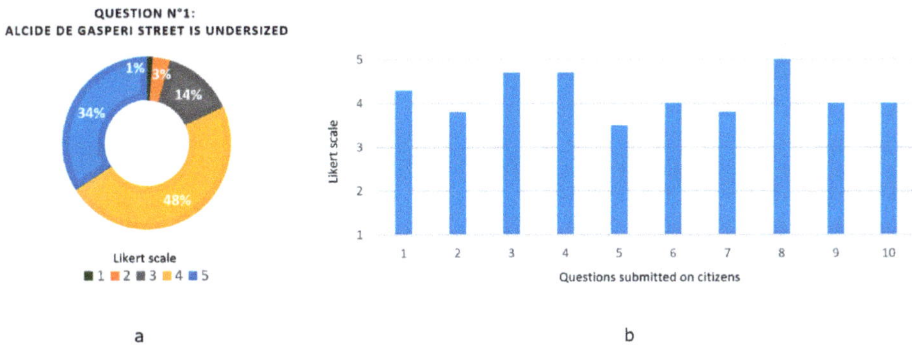

Figure 17. (**a**) Percentage of citizens' agreement with the question n°1; (**b**) average of 88 citizens' preferences on each question of the semi- structured interviews.

6. Conclusions

The application of knowledge structuring models through PSMs aims to challenge some limits of PMT and to investigate the level of citizens' knowledge and then, the awareness about ES.

The study has brought about some general considerations, that can be synthesized as follows.

First, CWs seem to be not completely able to lay out, analyze, and understand issues and problem situations emerging along walking discourses. An integration offered by other methods, such as extended SSIs, seems to be effectively integrative of the knowledge building process, being also possible to involve a greater number of people.

Second, the concept of ES seems to be now somehow inherent in the culture of citizens. In fact, virtually every action that emerged from CWs and SSIs can be assigned to a category of ES. However, some limitations still appear, such as: (i) Emerging ESs consider only the natural resource *Land*; (ii) the drivers used are only *Policy* and *Land* use and the benefits arising are only related to *Well-being* and *Public Health*.

Interestingly, CLDs seem to usefully integrate future-modelling activities, such as scenario-building models. For example, they seem to be useful to investigate the implications of citizen potential decisions on areas, as well as to facilitate citizens' knowledge about ES and, more broadly, to support the construction of collective futures. In this perspective, more work will be devoted to check such issues on different case studies.

Collective futures show up here as being certainly based on expert knowledge, but also largely non-expert, common, bottom-level knowledge. As a matter of facts, for the first time non-expert knowledge is considered in a computationally structured way and not only in qualitative sense, through the search and identification of knowledge database management models.

Actually, we must say that the results of the present work do parallel a constant evolution of attempts to structure informal data [55,56]. Stimulated by a well-known urgency of the environmental problem both at the global and at the local level, spatial decision-making and planning processes tend to confront complexity, rather than reducing or even denying it as occurred in the past [57].

Yet for this this knowledge-inclusive confrontation to be operational, intelligent and operationally refined instruments are needed. This represents a clear limitation of the hybrid approach used here, which represents just an explorative effort to evaluate an applied qualitative methodology, rather than propose an original one. This is in fact a path with a large experimental and area-based component, able to provide the details and the experimental assortment that guarantee the usefulness, quality, and effectiveness of the path itself. But the availability and/or construction of these instruments is still far from being in step with theoretical approaches and methodologies and requires continuous and reiterated efforts.

In many ways this recalls a famous scientific approach of *Simonian* memory, whose trial-and-error steps allow just the necessary clarification and operational fine-tuning to problem management [58]. It is therefore a still wide and interesting research path in perspective to which our next research efforts will be oriented as follow-up activities.

Author Contributions: The paper is the result of a joint research work carried out by authors together. Nonetheless, D.C. wrote Sections 1 and 6, S.S. wrote Sections 2, 3 and 5 whereas P.B. wrote Section 4. All authors have read and agree to the published version of the manuscript.

Funding: This research received no external funding.

Acknowledgments: The authors thank Raffaele Giordano, Alessandro Pagano and Irene Pluchinotta for their insights and support in developing part of the applied methodology. Also, P.B. thanks S.S. in particular, for her collaboration in laying out Section 4.

Conflicts of Interest: The authors declare no conflict of interest.

References

1. Venuti, G.C. *La Terza Generazione Dell'urbanistica*; Angeli: Milano, Italy, 1990.
2. Lucci, P.; Ruzzon, D. *Pianificazione Locale: L'Agenda 21 e la Nuova Generazione di Piani in Quaderni–Ambiente e Società*; ISPRA: Rome, Italy, 2015; Volume 1, ISBN 978-88-448-0728-3.
3. Sassen, S. *Human Settlement Development*; EOLSS Publications: Oxford, UK, 2009.
4. Newman, P.; Jennings, I. *Cities as Sustainable Ecosystems: Principles and Practices*; Island Press: Washington, DC, USA, 2012.
5. Chilvers, J. Environmental Risk, Uncertainty, and Participation: Mapping an Emergent Epistemic Community. *Environ. Plan.* **2008**, *40*, 2990–3008. [CrossRef]
6. Binning, C.E.; Cork, S.J.; Parry, R.; Shelton, D. *Natural Assets: An Inventory of Ecosystem Goods and Services in the Goulburn Broken Catchment*; CSIRO Sustainable Ecosystems: Canberra, Australia, 2001.
7. Millennium, M.A. *Ecosystem Assessment. Ecosystems and Human Well-being: Synthesis*; Island Press: Washington, DC, USA, 2005.
8. Wallace, K.J. Classification of ecosystem services: Problems and solutions. *Biol. Conserv.* **2007**, *139*, 235–246. [CrossRef]
9. Boyd, J.; Banzhaf, S. Ecosystem Services and Government Accountability: The need for a new way of judging Nature's Value. *Resources* **2005**, *158*, 16–19.
10. Boyd, J.; Banzhaf, S. What are ecosystem services? The need for standardized environmental accounting units. *Ecol. Econ.* **2007**, *63*, 616–626. [CrossRef]
11. Landers, D.; Nahlik, A. *Final Ecosystem Goods and Services Classification System (Fegs-cs). U.S.*; Environmental Protection Agency: Washington, DC, USA, 2013.
12. Lee, H.; Lautenbach, S. A quantitative review of relationships between ecosystem services. *Ecol. Indic.* **2016**, *66*, 340–351. [CrossRef]
13. Holt, A.R.; Mears, M.; Maltby, L.; Warren, P. Understanding spatial patterns in the production of multiple urban ecosystem services. *Ecosyst. Serv.* **2015**, *16*, 33–46. [CrossRef]
14. CICES. Available online: https://biodiversity.europa.eu/maes/ecosystem-services-categories-in-millennium-ecosystemassessment-ma-the-economics-of-ecosystem-and-biodiversity-teeb-and-common-international-classification-of-ecosystem-services-cices (accessed on 26 May 2019).
15. EPA. United States Environmental Protection Agency. Available online: https://www.epa.gov/enviroatlas/ecosystem-services-enviroatlas (accessed on 15 April 2019).
16. Vogdrup-Schmidt, M.; Strange, N.; Olsen, S.B.; Thorsen, B.J. Trade-off analysis of ecosystem service provision in nature networks. *Ecosyst. Serv.* **2017**, *23*, 165–173. [CrossRef]
17. Fisher, B.; Turner, R.; Morling, P. Defining and classifying ecosystem services for decision making. *Ecol. Econ.* **2009**, *68*, 643–653. [CrossRef]
18. Pedersen Zari, M. Ecosystem services analysis for the design of regenerative built environments. *Build. Res. Inf.* **2012**, *40*, 54–64. [CrossRef]
19. Raymond, C.M.; Kenter Jasper, O. Transcendental values and the valuation and management of ecosystem services. *Ecosyst. Serv.* **2016**, *21*, 241–257. [CrossRef]

20. Holling, C.S. Understanding the complexity of economic, ecological and social systems. *Ecosystems* **2001**, *4*, 390–405. [CrossRef]
21. Armsworth, P.R.; Chan, K.M.A.; Daily, G.C.; Ehrlich, P.R.; Kremen, C.; Ricketts, T.H.; Sanjayan, M.A. Ecosystem service science and the way forward for conservation. *Conserv. Biol.* **2007**, *21*, 1383–1384. [CrossRef] [PubMed]
22. Ascough, J.C.; Maier, H.R.; Ravalico, J.K.; Strudley, M.W. Future research challenges for incorporation of uncertainty in environmental and ecological decision-making. *Ecol. Model.* **2008**, *219*, 383–399. [CrossRef]
23. Hansen, R.; Frantzeskaki, N.; McPhearson, T.; Rall, E.; Kabisch, N.; Kaczorowska, A.; Kain, J.H.; Artmann, M.; Pauleit, S. The uptake of the ecosystem services concept in planning discourses of European and American cities. *Ecosyst. Serv.* **2015**, *12*, 228–246. [CrossRef]
24. Fischer, F. *Citizens, Experts, and the Environment: The Politics of Local Knowledge*; Duke University Press: Durham, NC, USA, 2000.
25. Barreteau, O.; Bots, P.W.G.; Daniell, K.A. A framework for clarifying "participation" in participatory research to prevent its rejection for the wrong reasons. *Ecol. Soc.* **2010**, *15*, 1–22. [CrossRef]
26. Maskrey, S.A.; Mount, N.J.; Thorne, C.R.; Dryden, I. Participatory modelling for stakeholder involvement in the development of flood risk management intervention options. *Environ. Model. Softw.* **2016**, *82*, 275–294. [CrossRef]
27. Pagano, A.; Pluchinotta, I.; Pengal, P.; Cokan, B.; Giordano, R. Engaging stakeholders in the assessment of NBS effectiveness in flood risk reduction: A participatory System Dynamics Model for benefits and co-benefits evaluation. *Sci. Total Environ.* **2019**, *690*, 543–555. [CrossRef]
28. PPTR. *Piano Paesaggistico Territoriale Regionale*; General Report; Regione Puglia: Puglia, Italy; Available online: http://paesaggio.regione.puglia.it/PPTR_2015/1_Relazione%20Generale/01_Relazione%20Generale.pdf (accessed on 12 February 2020).
29. Satterfield, T.; Roberts, M. Incommensurate risks and the regulator's dilemma: Considering culture in the governance of genetically modified organisms. *New Genet. Soc.* **2008**, *27*, 201–216. [CrossRef]
30. PUG Report. Available online: http://pug.comune.bari.it/report/ (accessed on 20 June 2019).
31. Jones, B. *Neighborhood Planning. A Guide for Citizens and Planner*; American Planning Association: Chicago, IL, USA, 1990.
32. Fagence, M. *Citizen Participation in Planning*; Elsevier: Amsterdam, The Netherlands, 2014.
33. Borri, D.; Camarda, D.; De Liddo, A. Envisioning environmental futures: Multi-agent knowledge generation, frame problem, cognitive mapping. *Lect. Notes Comput. Sci.* **2004**, *3190*, 230–237. [CrossRef]
34. Brody, S.D.; Godschalk, D.R.; Burby, R.J. Mandating citizen participation in plan making: Six strategic planning choices. *J. Am. Plan. Assoc.* **2003**, *69*, 245–264. [CrossRef]
35. Berthet, E.T.A.; Barnaud, C.; Girard, N. How to foster agroecological innovations? A comparison of participatory design methods. *J. Environ. Plan. Manag.* **2016**, *59*, 280–301. [CrossRef]
36. Hatchuel, A.; Weil, B. C-K design theory: An advanced formulation. *Res. Eng. Des.* **2009**, *19*, 181. [CrossRef]
37. Borri, D.; Camarda, D. Spatial ontologies in multi-agent environmental planning. In *Technologies for Supporting Reasoning Communities and Collaborative Decision Making: Cooperative Approaches*; Yearwood, J., Stranieri, A., Eds.; IGI Global Information Science: Hershey, PA, USA, 2010; pp. 272–295.
38. Tsoukiàs, A.; Pluchinotta, I.; Ferretti, V. Studying the generation of alternatives in public policy making processes. *Eur. J. Oper. Res.* **2019**, *273*, 353–363. [CrossRef]
39. Homer, J.; Oliva, R. Maps and models in system dynamics: A response to Coyle. *Syst. Dyn. Rev.* **2001**, *17*, 347–355. [CrossRef]
40. Schröter, M.; Kraemer, R.; Mantel, M.; Kabisch, N.; Hecker, S.; Richter, A.; Neumeier, V.; Bonn, A. Citizen science for assessing ecosystem services: Status, challenges and opportunities. *Ecosyst. Serv.* **2017**, *28*, 80–94. [CrossRef]
41. Lynam, T.; De Jong, W.; Sheil, D.; Kusumanto, T.; Evans, K. A review of tools for incorporating community knowledge, preferences, and values into decision making in natural resources management. *Ecol. Soc.* **2007**, *12*, 5. [CrossRef]
42. Daniell, K.A. Co-Engineering Participatory Modelling Processes for Water Planning and Management. Ph.D. Thesis, Agro Paris Tech. and the Australian National University, Canberra, Australia, 2008.
43. Voinov, A.; Bousquet, F. Modelling with stakeholders. *Environ. Model. Softw.* **2010**, *25*, 1268–1281. [CrossRef]

44. Hare, M. Forms of Participatory Modelling and its Potential for Widespread Adoption in the Water Sector. *Environ. Policy Gov.* **2011**, *21*, 386–402. [CrossRef]
45. Dulli, S.; Polpettini, P.; Trotta, M. *Text Mining: Teoria e Applicazioni*; Franco Angeli: Milano, Italy, 2004.
46. Bolasco, S.; Canzonetti, A.; Capo, F. *Text Mining—Uno Strumento Strategico per Imprese e Istituzioni*; Cisu Editore: Roma, Italy, 2005.
47. Rapid Miner Studio Software. Available online: https://rapidminer.com/get-started/ (accessed on 30 November 2019).
48. Verma, T. Processo di tokenizzazione e filtro in RapidMiner. *IJAIS* **2014**, *7*, 16–18. [CrossRef]
49. Exter, K.; den Specht, A. Assisting stakeholder decision making using system dynamics group model-building. In *Proceedings of APEN National Forum*; The Regional Institute: Hobart, Australia, 2003; p. 43.
50. Vennix, J.A.M.; Akkermans, H.A.; Rouwette, E.A.J.A. Group model-building to facilitate organizational change: An exploratory study. *Syst. Dyn. Rev.* **1996**, *12*, 39–58. [CrossRef]
51. Richardson, G.P. Loop polarity, loop dominance, and the concept of dominant polarity. *Syst. Dyn. Rev.* **1995**, *11*, 67–88. [CrossRef]
52. Fruchterman, T.M.J.; Reingold, E.M. Graph Drawing by Force-Directed Placement. *Softw. Pract. Exp.* **1991**, *21*, 1129–1164. [CrossRef]
53. Barriball, K.L.; While, A. Collecting data using a semi-structured interview: A discussion paper. *J. Adv. Nurs.* **1994**, *19*, 328–335. [CrossRef] [PubMed]
54. Wilson, C. *Interview Techniques for UX Practitioners: A User-Centered Design Method*; Newnes Publications: Newton, MA, USA, 2013.
55. Freitag, D. *Machine Learning for Information Extraction in Informal Domains*; Carnegie Mellon University Press: Pittsburgh, PA, USA, 1998.
56. Liu, B. *Sentiment Analysis and Opinion Mining*; Morgan & Claypool: London, UK, 2012.
57. De Roo, P.G.; Silva, E.A. *A Planner's Encounter with Complexity*; Ashgate Publishing Limited: Farnham, Surrey, UK, 2010.
58. Simon, H. Trial and error search in solving difficult problems. *Behav. Sci.* **1962**, *7*, 425–429. [CrossRef]

sustainability

MDPI

Article

Urban Standards and Ecosystem Services: The Evolution of the Services Planning in Italy from Theory to Practice

Anna Maria Colavitti *, Alessio Floris * and Sergio Serra *

Department of Civil and Environmental Engineering and Architecture, via Santa Croce 67, 09124 Cagliari, Italy
* Correspondence: amcolavt@unica.it (A.M.C.); alessio.floris@unica.it (A.F.); sergioserra@unica.it (S.S.)

Received: 4 February 2020; Accepted: 12 March 2020; Published: 20 March 2020

check for updates

Abstract: Human well-being is determined by multiple factors related to health, social relations, safety, environment, landscape, cultural heritage, and quality of services. The Italian planning system provided a set of "urban standards", in terms of threshold values of areas per inhabitant destined for public services and facilities. The application of urban standards, for a period of more than fifty years, did not result in a broad improvement of life quality in the urban areas. This paper discusses the issue of urban facilities in Italy in order to evaluate the opportunity to innovate traditional standards according to the environmental and ecological paradigm, focusing on the benefits provided to humans by natural ecosystems, the so-called ecosystem services (ESs). The paper investigates the evolution of the Italian planning practice through the introduction of quality standards and innovative tools able to meet the ever-changing social demand. The research aims to verify if the ES concept is really implemented in the Italian planning practice and if the ecosystem approach has a real impact on political decision-making. Using a comparative method, four case-studies of urban municipal plans are selected and analyzed in order to identify different approaches and possible fields of innovation. The research highlighted a lack of integration of ecosystem services approach in the land use decisions, although there is an in-depth survey on the state of conservation of ecological and environmental resources. The local experiments of qualitative standards represent an attempt to deal with specific ecological emergencies, namely flood risk, air, water, and soil pollution, and loss of biodiversity. Conclusions discuss, from an international perspective, the need to revise the traditional planning approach in the field of public services and facilities, taking into account the influence of ecosystem services on human well-being.

Keywords: urban standards; ecosystem services; urban facilities

1. Introduction

The system of common goods, services and infrastructures strongly affects the quality of life for communities living in urban areas. The social aspirations associated with urban life must be addressed inasmuch as they overcome the individual needs expressed by the consumer society and meet the higher "right to the city" [1]. The city should always enable the free expression of collective and individual needs for complementary or contrasting activities, including gathering and socializing places, creativity and cultural events locations, job opportunities, sports and entertainment facilities where the communities can enjoy tangible and intangible services. The aim to achieve a fair and sustainable human well-being must lead the regeneration processes of the existing urban fabrics, without new developments on natural land [2]. The current challenge for urban planning is to integrate the environmental paradigm into the issue of collective services, according to the principles of social equity and environmental justice [3]. The application of this assumption allows the supply of public services, transportation, recreational and cultural spaces to be expanded, while ensuring greater flexibility on the adaptation to the ever-changing social needs [4]. The city is indeed the place where the quality of life, the economic productivity, the safety and health of the population and the environment are deeply intertwined and sometimes underestimated [4,5].

In Italy, the traditional planning approach is based on a rational and regulative definition of strict land use zoning and parameters. Since 1968, the provision of public services and facilities is regulated by "urban standards", intended as the minimum provision of public space for each inhabitant (present or future) or new function. Several studies highlighted the need to move from the traditional approach, typically based on "a priori" quantitative definition of types and minimal areas for common services and infrastructures, to a more innovative solution ensuring an adequate qualitative level of public services and contextually guarantee the preservation and accessibility of the environmental system [4,6–8].

Accordingly, the definition of criteria that take into account the benefits provided to humans by natural ecosystems [9–11] can increase the effectiveness of the traditional urban-environmental parameters, based exclusively on quantitative principles for the spatial identification of areas intended for public services, neglecting to consider the effective contribution in terms of well-being of the inhabitants.

This statement is linked to the concept of ecosystem services (ESs) [12,13], together with the so-called "ecological print", connected to the human consumption of natural resources in relation to the Earth's ability to reproduce them [14]. The potentialities offered by ESs clarified exhaustively the relation between human welfare and ecosystem functionality [13,15] and the relevant contribution toward a sustainable and resilient urban development [15–18].

The paper analyzes a few recent experiences of Italian cities' land use plans to uncover several innovative elements to apply to the urban standards and, in particular, new models of ecological/environmental and quality standards. In Italy, the state of integration of ESs in spatial and land use planning and its effective impact on political decision-making have been investigated.

While the scientific research focuses on the methods for mapping of ESs and their specific performance, the planning practice shows the difficulty to integrate them into a comprehensive evaluation of the demand and supply of public services into the decisional processes for the governance of the territory [4].

According to the scientific literature, urban plans should take into account the benefits derived from appropriate management of the ecosystems, in order to reach satisfactory levels of well-being but also to provide the maintenance and protection of ESs [13,15,18–21].The study of a model for the analysis of urban ESs is a research field that grew considerably in the last two decades, although the techniques used to implement the services and the choices underlying the planning are still immature [22–25]. Some planning theories and ideas, such as the "garden city" and the "intelligence growth", have a similar conceptual foundation [4,26], which integrate the concept of green infrastructure as a systematic method to contain the expansion of the urban fabric, ensuring a coherent relationship between the natural and the built environment. Due to the lack of attention to the protection of ecosystems and biodiversity, the Italian urban and regional planning often targets just the protection of the environmental values, regarded as natural, identity, or cultural elements [3]. Given that changes in land use play a central role in the delivery of ESs and urban planning, in turn, has a significant influence on their conservation or enhancement [21,27,28], the paper has adopted a comparative case study approach to evaluate the current state of integration of the concept of ES in land use plans and the opportunity to consider the ESs in a broad sense, as common goods that influence the communities' quality of life.

2. From the Quantitative Model to the Qualitative and Ecological Standards

In the second half of the 1950s, in order to improve the quality of general urban plans, the Italian government realized the necessity to define standards and parameters on the supply of public services and facilities, applicable to often extremely different contexts [7,8]. The issue of service supply assessment, influencing the relationship between public and private spaces, has also been managed by public bodies involved in the field of social housing, such as the "INA Casa" (abbreviation for the Italian housing program financed by INA (National Insurance Agency). The debate on the urban standards led to the draft of a national law (L. n. 765/1967, known as "Legge Ponte") [29]. It took place in a historical context that dedicated great attention to the quick expansion of Italian cities, planned with simplistic and inadequate tools [30].The concept of "urban planning standard", was definitively introduced with Inter-ministerial Decree 1444/1968. Over the decades, each region contributed to a better definition of the national urban standards by promulgating reforms that adjusted the quantitative thresholds. This provision is mandatory, regardless of the local management choices and the spending capacity of the institution in charge [31]. As described in Table 1, the urban standards differ, in terms of quantity and type, according to the zoning [32].

Table 1. Urban standards defined by Inter-ministerial Decree 1444/1968.

Zoning	Criteria	Parameters
Residential areas	18 square meters of urban standards per capita (100 cubic meters per capita of building provisions)	education areas—4.5 sq. m/inh.
		common facilities and services—2 sq. m/inh.
		public spaces such as parks, playgrounds, sports fields—9 sq. m/inh.
		public parking lots—2.5 sq. m/inh.
Industrial areas	Percentage of the total area	10% of the area for the industrial settlements
Commercial or offices areas	Parameter referred to the gross floor area of expected buildings	80 sq. meters of public facilities per each 100 sq. meters of the gross floor area of the expected buildings. 50% of them destined to parking lots

The determination of a standard value can be considered as a compromise between economic forces—involved in the transformation of the territory—and political forces, safeguarding collective interests. This value guaranteed basic satisfactory requirements that were capable of meeting the social needs and were consistent with the interests negotiated by a specific class [6,32].Therefore, the national criteria were clearly applicable only in a certain historical period characterized by a remarkable expansion of urban areas, no longer acceptable considering the changed conditions of the economic and cultural context [4]. Many studies have focused on developing indicators to evaluate factors that directly affect human well-being, also assessing adequate levels of quality of life in terms of services, facilities, and infrastructures [33–35]. For example, the World Health Organization recommended to guarantee a minimum threshold of green spaces (nine square meters pro capita) but considered an ideal amount of 50 square meters of urban green spaces for each inhabitant [33,36].

Today, cities show a shortage of services and infrastructures due to the scant commitment of the administrators, the insufficient public funds, the high expropriation cost to acquire areas, the tendency to ensure the mere availability of public surfaces, with no concern for the performance of services provided to the community [4,37]. The costs relating to the expropriation allowance, the realization of the service and its management cannot be sustained by the local authorities. This results in the unfair treatment of private property, subjected to indefinitely reiterated constraints, without compensation. After the sentence n.348/2007 of the National Constitutional Court, that required to pay market price for the areas for public facilities instead of the agricultural value, the expropriation became increasingly unused [37,38].

The plans often remained unapplied in the provision of public services and facilities, leading to inadequate and variable implementation of the urban standards [39]. For example, the survey 2018 of the Italian National Institute of Statistics (ISTAT) showed encouraging data about the available urban green spaces in Italy, with an average of 32.8 square meters per inhabitant [40]. However, in the main provincial towns, the data reflect a remarkable heterogeneity, namely the provision of urban green spaces per capita is notably different from town to town, from the low value of Crotone (3.6 sq. m./inh.) to the wider green areas of Matera (997.2 sq. m./inh). Since the first applications, urban standards have shown excessive rigidity, which overlooks the social structural changes, the heterogeneity and complexity of the geographical context, the availability of the resources and the different strategies enacted by the political choices [8]. For example, the traditional criteria overlook that the volume per capita is generally higher in low-density urban areas than in central areas [41], the demand for educational services is connected to the expected increase in the school-age population [42,43], the need for religious buildings is not limited to Catholic churches [44,45] and the provision of parking lots is linked with urban mobility system and hubs. For these reasons, a review of the criteria, used for the analysis of needs and supply, is clearly required in order to consider the settlement conditions and the demographic and social trends, monitoring the actual necessities expressed by the stable population and by the daily or seasonal flows of non-resident users [31,39,46].

In a general view of urban regeneration, the concept of "urban standard" is replaced by "territorial provision", which includes the services requested by individuals, families, and companies, regardless of the providers, either public or private and mixed [47]. Moreover, the reuse of existing buildings and areas can contribute to achieving the dual purpose of renovating the urban fabric and improving the local system of services and facilities [31].

The quality of the service system is particularly important in the disadvantaged territories, namely in the so-called inner areas, that cannot enjoy essential services. In this context, due to the high costs/benefits ratio for the realization of new public facilities, the provision of services may take place at regional/provincial (supra-municipal) level, guided by quality and efficiency criteria to limit the public expenditure and maximize the benefits for the local communities [48]. Unfortunately, this goal clashes with poor integration between regional policies and local planning [39,49].

If urban standards can easily control the amount of service provisions, the quality criteria are difficult to translate into standard values referred to more debatable aspects, depending on the specific physical, temporal, and economic features of the context, such as functionality, accessibility, spatial distribution, sphere of influence, usability, and efficiency [39,50].For example, the accessibility to urban green spaces can positively affect public health and the opportunities for physical and social activities, acting as a qualifying factor for the city and a driver of the real estate demand [51,52]. An unfair and irregular distribution of green areas in the urban areas mirrors the characteristics and conditions of the inhabitants (i.e., income, social level, age, and gender) and can be an indicator of racial, ethnic and socio-economic disparities [53,54].

The reform of the regional planning laws reflected the debate on the qualitative/quantitative dimension of the standard. The Italian regions of Lombardy and Basilicata focused on the qualitative aspects of the standards which they put under the control of the municipal plans, while other regions, such as Emilia Romagna, Veneto, Calabria, Valle d'Aosta, and Lazio, focused on both the quantitative and the qualitative requirements. Few regions, for example Umbria and Lombardy, chose a specific programmatic plan to address the management of the collective services, known as "Plan of Services" [55]. The remaining regions have confirmed the national model, sometimes expanding the thresholds and requirements, without questioning the underlying logic [31].

The Basilicata Region required an assessment of the system of services and infrastructures, including a quantitative and qualitative analysis of the territorial distribution, the state of use, the management costs and the performance (in terms of accessibility, suitability for different age groups, functionality, and technological adequacy).

A recent law proposal of the Piedmont Region takes into account, in addition to the national urban standards, the provision of additional surfaces for environmental, ecological, and ecosystem services (Law Proposal Piedmont Region n. 302, June 2018).

The adaptation of the service supply is achieved through the creation of new quality standards aiming at the safeguard of the spatial structures that guarantee functions and advantages provided by the ESs, improving the quality of life of the communities and the territory resilience, according to both natural science and socio-economic studies [56–58].Cities and urbanizing regions take advantage from the ESs produced in the urban areas, but also are at the center of material and energy flows and natural goods availability that expands on much larger areas [59,60].

Although urban citizens are dependent on global ecosystems, the locally generated ESs have a particularly high value due to the great number of beneficiaries. In this direction, the research had sometimes focused on the evaluation of the consequences of changes in ecosystems on human well-being, in terms of social-economic costs, in order to support decision-makers in the definition of ESs conservation and enhancement actions [61–65].

Over the years, many authors proposed a variety of ESs classifications that the environmental analysis can suggest [12,66–69]. However, every system of classification is based on an unrealistic simplification, because "ecosystems are complex, dynamic, and adaptive systems with non-linear feedbacks, thresholds, hysteresis effects" [69] (p. 351).

The most popular is the Millennium Ecosystem Assessment (MEA), which divides the ESs into four categories: Supporting services, provisioning services, regulating services, and cultural services [12]. Nowadays, it has been overcome by more effective theoretical systems that clarified the distinction between ends and means [70].

Services exist if a specific benefit or beneficiary can be identified. It is something that people can enjoy, and it must be considered only in relation to people's needs [67,68]. According to Costanza, ESs are means to human well-being and it is difficult to categorize and distinguish processes, benefits, and services (intermediate and/or final) [68,69,71]. This is the assumption of the Cascade framework, proposed by Haines-Young and Potschin, that analyzed the connection between ecological structures, processes generated by ecosystems, services and benefits gained by society, following a pattern similar to a production chain [66,67,72].

A more recent classification, the Common International Classification of Ecosystem Services (CICES), has been developed by the European Environment Agency (EEA) with the support of other institutions [66]. It represents a revision of the MEA classification, based on the recognition of the need to standardize the criteria of ESs analysis and description. Figure 1 represents the conceptual background for CICES, using the Cascade model with the distinction between benefits and values.

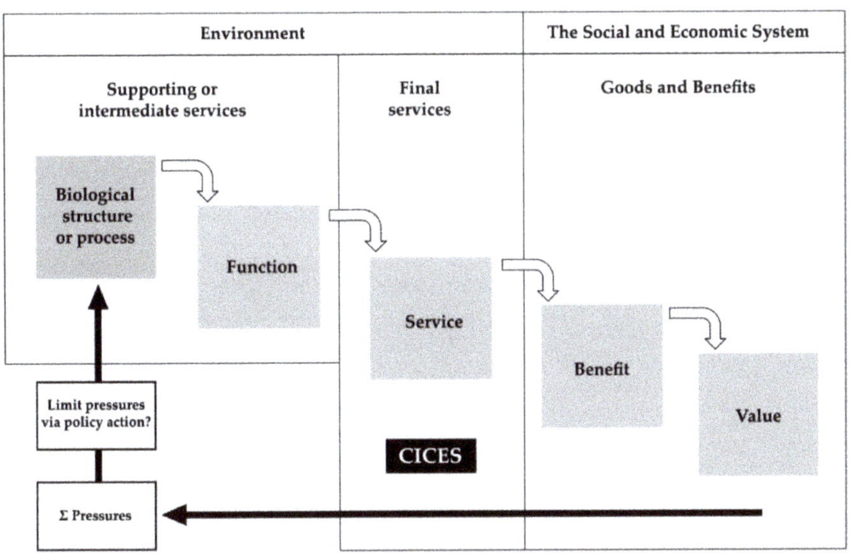

Figure 1. The ecosystem service cascade model. Processed by the authors based on Haines-Yong, R., Potschin, M. CICES Version 5.1, 2018, available at: www.cices.eu [66].

The CICES excluded the so-called supporting services and focused only on the provisioning, regulating, and cultural ones. The reason is related to the difficulty of identifying and describing the final outputs of supporting services, also known as intermediate services, from ecosystems that people use and value [66,67,73].

The classification proposed by the Intergovernmental Science-Policy Platform on Biodiversity and Ecosystem Service program (IPBES) recognized that many services fit in more than one of the MEA categories. In the last proposal, dated May 2019, ecosystem goods and services are defined "Nature's contributions to people (NCP)", including all the contributions of nature, both positive and negative, to the quality of life of humans as individuals, societies, or humanity as a whole [74]. In the IPBES conceptual framework, ESs are delivered by a category of nature, biodiversity, and ecosystems and contribute to a good quality of life, thus separating out means and ends.

The paper aims to verify if the ES concept is really implemented in the Italian planning practice and if the ecosystem approach has a real impact on political decision-making. The state of integration of ESs in spatial and land use planning in the Italian context has been investigated to understand whether or not there is a lack of consideration of ESs approach in the land use decisions.

3. Materials and Methods

The methodology is based on a comparative case study approach, considered an effective strategy in the research field of urban and regional studies [75,76]. It is defined to understand the path of different regional contexts and the points of convergence or divergence across these [77].

The framework of the qualitative and comparative analysis, summarized in Table 2, has been applied to the review of a few recent examples of Italian land use plans, focusing on the integration of ES concept into a more comprehensive system of services and infrastructures able to achieve a higher level of life quality.

The first section of the comparative framework focuses on the analytical part of the plans concerning the ES knowledge. In particular, the paper examines in depth the recognition and classification of the spatial components that are part or can become part of municipal ESs networks, understood as the system of components that make up the environmental structure of a limited urban fabric. According to the points identified in the cascade model, the paper investigates the description of the environmental, cultural, and ecological structures and their attributes (biophysical structure and processes). The other points concern the evaluation of ecosystem functions, services, and their influence on human well-being, in terms of benefits and values for the community. The second part of the analysis concerns the investigation of strategies and actions to improve human wellbeing through the conservation, rehabilitation, or enhancement of ecosystem service in urban areas [77].

Table 2. Analysis framework. Processed by the authors based on literature review [24,66–77].

ANALYSIS	Biophysical structure and processes	Identification of the spatial components of existing or potential ecological networks. Analysis of the structure, the processes, and the biophysical function of ecosystems.
	Functions	Description of ecosystem properties or urban areas' potentialities in terms of ecosystem functions.
	Services	Identification of the final services and products provided to the community by ecosystems.
	Benefits and values	Data on the influence of ecosystems on human well-being in terms of benefits and values provided.
STRATEGIES AND ACTIONS	Conservation	Strategies and actions aimed at preserving the health and the functionality of ecosystems.
	Regeneration	Strategies and actions aimed at restoring the health and the functionality of ecosystems.
	Enhancement	Strategies or actions aimed at enhancing the existing provision of ecosystem services or creating new ecosystems.

The focal points of the investigation are four case studies of urban municipal plans adopted by Bologna, Rome, Milan, and Florence (Figure 2), that are the capital cities of Italian regions characterized by different sets of rules in terms of regional planning tools and environmental policies. The selection is based on the criterion of achieving a wider range of approaches used to deal with the issue of environmental, ecological standards and ESs: Bologna and Rome, as they represent interesting examples of municipal ecological networks that are still not connected to ESs; Milan, because it is experimenting with an innovative model of sectorial public services planning that includes the ecological network; Florence, that has attempted the integration of the ESs in the management of the soil usage, with an emphasis on building an ecological network that becomes the driver of city transformation. It allowed us to uncover several innovative elements to apply to the urban planning standards and, in particular, new models of ecological/environmental and quality standards.

Figure 2. Map of case studies (Processed by the authors based on ISTAT open data 2019).

The main documents of general urban plans of each case study have been collected into the official websites of the municipalities (the latest updated version available at December 2019). Appendix A provides a full list of the documents used in the comparative case studies analysis, in particular:

- the plan reports, both general or specialized (e.g., the description of the methods and strategies adopted in the planning instrument);
- the regulative sections (e.g., the system of rules to manage urban development, the measures to preserve ecological functions);
- the figures and tables (e.g., the identification of services and green infrastructures, the evaluation of ecological potentialities, the description of the environmental system).

In the discussion section, an assessment of the degree of knowledge and consideration of each point of the analysis framework was provided in order to allow the comparison of the case studies. If the plan contains data, descriptions, or adopts criteria relating to a specific point of the analysis framework, the level of detail is assessed (inadequate, moderate, or good).

4. Results

4.1. Bologna

Bologna ranks at the top of the list of Italian cities for its vast heritage of gardens and urban parks, which covers an area of over 350 hectares. In the last decades, the policies of environmental protection have contributed to the conservation of the natural, historical, and rural landscape, which corresponds to almost half of the municipal administrative area. However, it does not automatically guarantee a sufficient supply of ESs to control the anthropic pressure on the environmental resources.

With the aim of improving the overall quality of the urban environment, the 2008 Structural Municipal Plan (PSC) focuses on the strengthening of the ecological network and preserving and recovering the natural habitat. The municipal ecological network is organized into three hierarchical levels (main, secondary, and urban level) constituted by a polyvalent system of "knots", which are eco-systemic elements acting as reservoirs of the local biodiversity of fauna and flora.

The main ecological network consists of a series of complex ecological hubs and passageways that connect the different parts of the city with the protected wide areas of the provincial territory. Sometimes it also includes agricultural areas, such as vegetable gardens, sports areas with or without the relative public services, and facilities that do not compromise its ecological value.

The secondary ecological network is defined by simple hubs, local ecological passageways, and spread-out areas that guarantee the connections among the parts of the city. The PSC enlists the linear elements, both natural and semi-natural (box hedges and tree lines, arboreal strips, waterways, reclamation drains, linear lawns, salvaged green scarps, etc.) that either show an ecological value or can acquire it through recovery actions. The local ecological passageways are nature vehicles leading, through connective areas, to the urban ecological hubs that are city sectors with a relevant ecological value, often only potential, containing at the same time rural and urban public green elements. Usually destined to playful recreational areas and pedestrians and bikers, they represent, with the urban ecological connective green, a set of open spaces that have actual or potential ecological value, often allocated for public use. The urban ecological connective green is represented by a set of private and public gardens and parks, boulevards and decorative greenery, vegetation, and natural soils. Spread all over the city, it constitutes an ecological reservoir of permeable soil that helps to reduce air pollution.

In addition to the guidelines for safeguarding and strengthening the ecological network, the plan includes further measures aimed at mitigating the impacts produced by mobility infrastructures on the natural environment. In the development of new settlements, environmental compensation measures are provided in order to increase the vegetation cover in green areas and the permeable soil(at least 75% of the total area).

The plan adopts market-based tools, such as non-financial compensation, to freely acquire more than 200 hectares of natural land by transferring the development rights to other suitable areas.

The main results of the analysis are summarized in the table below (Table 3).

Table 3. Results of the analysis. The case of the Urban General Plan of Bologna.

ANALYSIS	
Biophysical structure and processes	Identification of three ecological networks based on the specific environmental and landscape value or the ecological potentiality: Mainecological network(complex ecological knots and corridors); Secondary ecological network(simple ecological knots, local ecological corridors, ecological and landscape connective tissue); Ecological urban network (urban ecological knots, local ecological tissue)
Functions	Connection between urban and external areas, preservation of environmental system and protected areas, soil permeability.
Services	Food production, periurban agriculture, environmental and natural goods production, reduction of air and electromagnetic pollution, climate regulation, preservation of existing and potential biodiversity, mitigation of the infrastructure's impact on the landscape and environment.
Benefits and values	Not considered.
STRATEGIES AND ACTIONS	
Conservation	Identification of new ecosystems and river areas to protect.
Regeneration	Revitalization and naturalization of compromised public properties, in order to create new ecosystem services at a regional scale (i.e., planting intervention).

Table 3. *Cont.*

Enhancement	Valorization of the agricultural land near the urban area, through the creation of new public green areas and the promotion of collective use of agricultural lots. Increase in the permeable surfaces. Valorization and improvement of existing tree areas to realize a new important biological and ecological reserve, connected to rural and river areas, with a public recreational function. Realization of a green lung to mitigate the impact of road and train infrastructures to became tree-lined roads. Realization of an important river park of supra-municipal level through the creation of path along the river, the naturalization of large brownfields and the recovery of degraded areas.

4.2. Rome

The 2008 General City Plan of Rome (PRG) paid specific attention to the ecological and environmental issues, although the city registered record values for the increase of newly built areas in the last years [78]. The cartographic documents identified the green and ecological network that includes primary, secondary, and new additional components. The primary elements are the important natural ecosystem components, such as environmental protected areas, agricultural parks, and areas of naturalistic relevance. The secondary elements connect other areas of the hydrographic and agricultural network and some parts of the residential zones, also including services and infrastructures. The primary elements usually concern areas subjected to measures of environmental protection, whereas the secondary components are intended to recovery and restoration actions, in particular on rundown areas.

The plan introduces some restrictions on the use of green areas included in the ecological network and in the local public services system. In particular, it mandates a public destination for park or sport areas and a compulsory requirement of a permeable surface larger than 80% of the total area. To strengthen the ecological network, in the "Renovating City" zone, new building constructions in the free areas are forbidden, while the development rights can be transferred to other suitable locations [79]. The executive planning allocates the areas included in the ecological network either to the public green, in order to satisfy the urban standard requirements, or to the private green, supporting the ecological strategies. The plan increases the number of urban standards defined by national law, up to 22 square meters for each inhabitant.

Moreover, it established incentives for the demolition of the incompatible buildings included in the ecological network, encouraging the transfer of the development rights and the conversion into a green area [39]. The municipal authority has the possibility to adopt integrated plans to coordinate and promote safeguard measures and interventions, recovery and enhancement of the ecological network through the increase of spontaneous greenery, maintenance of the watercourses, reconversion of parts of the hydrographic network, promotion of the archeological, historic and monumental heritage, preservation of the panoramic vantage points and belvederes, protection of the landscape integrity and continuity, limitation of the soil sealing, and reduction of the environmental pollution. The private and equipped green areas are subjected to specific rules aimed to guarantee a minimum amount of vegetation (40 trees and 80 shrubs/hectare) or to provide sports facilities and relative services.

The main results of the analysis are summarized in Table 4.

Table 4. Results of the analysis. The case of the Urban General Plan of Rome.

ANALYSIS	
Biophysical structure and processes	Identification of ecological networks(structured on three levels of components), natural protected areas, hydrographic network, agricultural areas and parks, green belts.
Functions	Ecological and hydrological functions provided to the urban area thanks to the connection to the natural areas and the ecological network. Agricultural parks as a land supply for the economic activities.
Services	Biological agricultural production.
Benefits and values	Not considered.
STRATEGIES AND ACTIONS	
Conservation	Preservation of agricultural soils (agricultural parks). Strategies and actions of safeguard for the municipal ecological network. Monitoring and periodic evaluation of the municipal ecological network conservation in terms of naturality and functionality.
Regeneration	Rehabilitation and naturalization of the hydrographic network, through the creation of agricultural parks. Restoration and environmental requalification of compromised or degraded areas.
Enhancement	Realization of new green belts and natural components to integrate the ecological network.

4.3. Milan

Collective services and infrastructures represent an important field of innovation in the planning system of the Lombardy Region. The introduction of a new municipal sectorial instrument, the Plan of Services (PS), aimed at the identification and implementation of an adequate qualitative and quantitative provision of public services (Lombardy Regional Law n.1 of 2001). According to the other plans or programs in force, this instrument establishes the desired quality level and the methods to be used to implement and manage the service system by encouraging forms of public and private partnerships to share economic resources, functions, and goals [55].

The PS has to take into account the environmental function of the green areas, the organization of parking lots in a broad approach to the management of the mobility system, and the coordination with the other municipal plans.

A few years later, the Region of Lombardy strengthened the role of the PS, which is entrusted with the task of "[...] ensuring a global provision of areas for public facilities of general interest, of any area intended for public residential housing, vegetation, ecological passageways, and green areas connecting the rural with the urban territory"(art.9, Lombardy Regional Law n.12/2005). Moreover, the law promotes the realization of green connections between infrastructures and urbanized areas and their rational distribution within the municipal territory, to support the current use and the possible future applications. A detailed program, concerning the assessment of costs, timing, priorities, and methods to realize the services, has been established by the law but often overlooked in the plan drafting.

The Milan municipality has drawn up a Plan of Services (March 2019), as it is stated in the Regional Law, concerning the identification and the regulation of the existing and future services, such as urban green, mobility infrastructures, public transportation, and social housing.

It analyzes in detail the ecological network and makes reference to the ESs as part of the collective services and infrastructure system, for example, the regulation services and the mitigation services to tackle climate change, promote and strengthen biodiversity, decontaminate waterways, and enhance the circular economy. The system of the green spaces strongly favors the improvement of the urban quality as it encompasses ecological, environmental, and social needs.

The identification, in the cartographic documents, of the existing and new elements that characterize the green network, should take into account the numerous types of public green spaces, harmonically merge with the private areas destined to agricultural activities. For example, the category of the "environmental green" also includes areas for the mitigation and decontamination of roads and infrastructures, not directly enjoyed by the inhabitants but nonetheless essential for the maintenance of the land permeability, the control of thermoregulation and other ESs. Finally, an innovative point lies in the so-called Local Identity Nuclei (NIL), namely city sections or neighborhoods that share common identity features. Each NIL has to be equipped with a local park or a system of common gardens and directly connected with the municipal green network.

The achievement of predefined levels of ESs production is encouraged by a ten percent reduction of the urban standards requirements. Moreover, the interventions of afforestation on private areas and of waterways regeneration are considered as a part of the amount of the urban standards.

The main results of the analysis are summarized in the table below (Table 5).

Table 5. Results of the analysis. The case of the Urban General Plan of Milan.

	ANALYSIS
Biophysical structure and processes	Identification of the green areas, subdivided into: Existing urban green areas, environmental green areas, new urban green areas, and planting areas. These components constitute the municipal ecological network, where the equipment of ecosystem services is considered as an environmental infrastructure. Identification of historical urban fabrics and of the existing or new religious buildings, recognized as services for the local communities.
Functions	Mitigation and naturalization of the areas affected by the infrastructural network. Recognition of religious buildings to support the cohabitation between different communities and to improve integration processes. Identification of the historic urban fabrics in order to preserve the local identities.
Services	The environmental green areas guarantee the permeability of the soil, the reduction of climate pollution and greenhouse gas, and the improvement of ecosystem services, contributing to the connection of the ecological network.
Benefits and values	Not considered.
	STRATEGIES AND ACTIONS
Conservation	Protection of agricultural uses in order to safeguard the environment. Protection of traditional cultures and knowledge through the improvement of public areas of the historic urban fabrics.
Regeneration	Revitalization of watercourses and of the brown fields, in order to reduce the amount of low-quality soils.
Enhancement	Improvement of the amount of natural soils. Identification of areas in which to provide the afforestation. Promotion and strengthening biodiversity. Guarantee a high percentage of natural soils in case of new intervention.

4.4. Florence

The Structural Plan (PS) and the Urban Regulation (RU) of Florence are very interesting examples of planning instruments that focus on the ecological network in order to preserve the ecosystemic value of the territory. After the plan was adopted, in 2010, the many revisions that took place did not question the general strategy of land-take reduction based on initiatives of urban regeneration. The project proposals consisted of building replacement and restoring techniques that allow the improvement of the urban and environmental quality and the enhancement of collective services and infrastructures.

The plan provided a dynamic network of public spaces and services that can be adapted to the new social needs. The enhancement of the infrastructural and service supply is supported by private agreements to gives up freely 50% of the development areas to make parks and other public spaces.

The plan redefines the ecological network by including ecological corridors, ecological knots and other parts of the ecological network within the city. The large green infrastructure connects linear elements (corridors) and areas (knots) that act as "biodiversity reservoirs" for which actions for the conservation of the fauna and flora biodiversity are provided. The ecological knots are differentiated into new or pre-existing knots, that may or may not need requalification. The elements of the urban ecological network (parks and urban gardens, green areas, tree lines) are often quantitatively adequate but not accessible, making it necessary to create new connections. The enhancement of the environmental quality is achieved through the improvement of the ecological functions of the existing green areas, through interventions of urban renewal and development of the ecological network with the support of public or private investments.

Urban green spaces are classified according to the following categories: The green garden, including all urban or peri-urban green spaces, both public or private, the green parks, including large areas, often next to historical buildings, the tree-lined boulevards, namely the continuous linear rows of trees placed along the main city roads.

In the analysis of the municipal green areas, the whole surface has been divided into pre-sized cells that were subsequently classified depending on their type, size, quality, accessibility, and spatial relation with the anthropic background. The density, the characteristics, the composition, and the spatial distribution of the vegetation represent useful factors to identify the areas fit for purposes of biodiversity conservation, climate control, air purification, and micro-habitat support.

The ecological network has been designed to connect the urban area to the peripheral zones, allowing the transit and migration of the fauna. It is involved in a wide set of plan actions, such as the management of existing trees and vegetation (amelioration, completion, and/or substitution of specific plant essences) or the enhancement of areas already characterized by ecological potentialities. The project of the ecological network is based on the following criteria: Avoid the fragmentation and isolation, increase the functionality of the existing corridors used for the fauna transferring and migration (including the watercourses, the vegetation and the tree-lines), create new knots or linear elements of the ecological network and modify the existing tracts according to the distribution of trees and vegetation in order to guarantee the connection of flora and fauna to the city.

The integration of the ESsinto the urban planning is based on specialized studies that are part of research on the distribution of the zoological and botanical diversity in the green areas.

The chart of the ecological classes introduces a set of indicators related to the flora and fauna biodiversity. Another section focused on the 64 urban green spaces, spread out in the whole municipal territory and intended for a garden or public park, with special emphasis on their capacity to accept animal and vegetation species. The gathered data allowed to define a strategy of improvement of those areas, ecological potential, and a final synthetic value expressing their ecological suitability.

The indicators of the municipal green areas concerned five pivotal aspects: Flower quality, habitat quality, soil quality, extension of the green area, quality of the surroundings of the green area. The overall value of the green area is the result of the summation of these five parameters.

The plan examines the existing ecological areas in need to become more efficient, the so-called "areas for the environmental upgrading". It also defines rules and criteria to address the public interventions in those areas, with specific provisions regarding the botanical and zoological species to use and how to ensure their fruition. Moreover, the new public and private developments represent opportunities to improve the blue and green infrastructures and to complete or regenerate the ecological network.

The RU introduces the concept of "ecological class", refereed to the ecological potential determined by a synthesis between botanic and the zoological components. The ecological classes and the continuity of the urban network have allowed us to pinpoint possible defects or potential value points that can become a basis for future projects to guarantee and develop the functionality of the ecological network.

The main results of the analysis are summarized in the table below (Table 6).

Table 6. Results of the analysis. The case of the Urban General Plan of Florence.

ANALYSIS	
Biophysical structure and processes	Identification of the ecological network, subdivided into core areas, ecological corridors, stepping zones, and buffer zones. Mapping of green urban areas and green periurban areas. Definition of the concept of ecological class, which refers to the ecological potential determined by a synthesis between botanical and the zoological components, setting up qualitative indicators for biodiversity. Census of the UNESCO historic settlement and recognition of the significant post-war architectures.
Functions	Ecological network aimed to build connections of natural and semi-natural environments, in order to support species in transit and host microenvironments in critical situations. Identification of the historical heritage in order to preserve the local identities.
Services	Guarantees of the graduality of the habitats, the mobility of the species, and the genetic exchange.
Benefits and values	Benefits are implicitly defined but not assessed in terms of their results and quality.
STRATEGIES AND ACTIONS	
Conservation	Safeguard of existing ecological network elements. Environmental monitoring. Historical heritage preservation.
Regeneration	Redevelopment of some existing sections of the ecological network.
Enhancement	Completion and strengthening of the territorial ecological network and improvement of natural dynamics of renewal of resources. Strengthening of internal and external connections to the ecological network. Development of compatible economic activities, in order to contribute to the protection of ecological values. Realization of new ecological knots.

The plan provides the enhancement of the service system with the application of non-financial compensation and expropriation tools for the acquisition of private areas and the achievement of a standard of 30.9 square meters for each inhabitant. In the new developments project, the urban standards required by the national legislation must be guaranteed anyway.

5. Discussion

In Italy, the explicit and voluntary use of the ES concept in spatial planning is still limited [21,80–82]. As shown by the review of Italian land use plans analyzed, the model of the ecological network integrates settlements, infrastructures, and natural components, following a new approach to the environmental preservation to achieve the goals of high-quality landscape and sustainable development [82].

In this direction, the urban planning focuses on the improvement of the urban green system in the new development projects, recognizing the ecological values and benefits provided by private green areas, adjacent and connected to the new settlements, used as public parks and gardens, although directly managed by private owners.

The widespread diffusion of green areas in the city and the construction of ecological networks, which combine public green areas with new private areas of ecological interest, can be an effective tool to tackle the problems generated by ongoing development [39], going beyond the concept of urban standard of green areas, recognizing their contribution in terms of benefits provided regardless of ownership.

The protection of the natural heritage and the ecological functionality of the territory does not require the acquisition of the property by the public authorities.

In Table 7, the results of the comparative case study analysis are summarized according to the theoretical framework described in the methodological section.

Table 7. Summary of the comparative case study analysis.

		BOL	ROM	MIL	FLO
	ANALYSIS				
Biophysical structure and processes	Identification of the spatial components of existing or potential ecological networks	•••	•••	•••	•••
	Analysis of the structure, the processes and the biophysical function of ecosystems	••	•	••	•••
Functions	Description of ecosystem properties or urban areas potentialities in terms of ecosystem functions	••	•	••	•••
Services	Identification of the final services and products provided to the community by ecosystems	••	•	••	••
Benefits and values	Data on the influence of ecosystems on human well-being in terms of benefits and values provided	–	–	–	•
	STRATEGIES AND ACTIONS				
Conservation	Strategies and actions aimed at preserving the health and the functionality of ecosystems	•	••	••	•••
Regeneration	Strategies and actions aimed at restoring the health and the functionality of ecosystems	••	••	••	••
Enhancement	Strategies or actions aimed at enhancing the existing provision of ecosystem services or creating new ecosystems	••	••	••	•••

– not considered or not clearly developed; • inadequate level of detail •• moderate level of detail ••• good level of detail.

All the plans identify in detail the spatial components of the municipal ecological networks, both existing or potential. Starting from similar and good levels of detail in the description of biophysical structures and processes, there is a progressive reduction in the quality of analysis and information according to the sequence described in the Cascade model. This assumption is confirmed by an inadequate analysis of the ecosystem properties and functions, excepted Florence that shows in-depth knowledge of urban ecosystem potentialities. The plans appear even more deficient in the identification of final ecosystem services and products for the community, while the data on the assessment of benefits and values for the human well-being are almost absent. The plan of Florence represents an example of partial integration of the issue of ESs in spatial planning to extend the benefits derived from the environmental protection policies to the system of common goods and infrastructures. In the case of Milan, the Plan of Services is structured by a dynamic path of identification of the real demand for the satisfaction of the community needs, but it did not developed a quantitative and qualitative assessment of the benefits provided by ESs. It is particularly innovative in terms of evaluation of the accessibility, usability, and feasibility of the collective services, but it is only an analytical tool, mostly inefficient as regards the realization, adaptation, and improvement of the public services [79,83].

Research highlights a defective awareness in defining the meaning of ESs outside the scientific and academic field and a difficult transfer of specialized knowledge in practices [27,84–86]. The mapping of the ESs should contribute to the territorial governance and is pivotal for the full acknowledgment of the environmental sustainability of the actions of land use and transformation provided by plans and to evaluate their consequences [86,87].

The ESs offer the opportunity to improve the land use planning, considering their direct contribution to the amelioration of the life quality and the welfare of the community [88,89]. In this direction, new environmental ecological standards have generally been recognized in the analyzed plans, but they are not linked to the services and benefits provided by the ecosystems. These parameters usually refer to strategies and actions for the conservation, regeneration, and enhancement of the environmental and ecological system (soil permeability, carrying capacity, non-renewable resources consumption, size and composition of green biomass, urban microclimate, air quality, noise reduction) [31,39]. Strategies and actions usually refer to the conservation and regeneration of the ecological networks without awareness of the positive consequences in terms of ecosystem services provisions. Only Florence evaluates the state of conservation of ESs in the urban area and the expected effects of development projects.

6. Conclusions

Over the last several decades, in the international context, the urban planning research has innovated and perfected methods and tools used to analyze and evaluate the indispensable functions that the ecosystem is capable of supplying to a large number of stakeholders. ESs becomes fundamental to foster the preservation of biodiversity and landscape that are seen as crucial contributors to human welfare. The paper demonstrates that the integration of ESs into planning enables connecting environmental protection with other community goals, such as the realization of green and blue infrastructures and public spaces. For this reason, it is fundamental to intensify the studies about the ESs and set up protocols to incorporate them into the spatial planning, avoiding the application of undifferentiated parameters disconnected from the local identity.

The scientific literature has provided plenty of indicators to measure the benefit generated by the ESs for the life quality of settled communities. In the Italian example, the previous planning experiences used strategies of sustainable development that, although kept in consideration the ecological function of the land, often used ineffective preservation and monitoring measures. The traditional model of urban standards, introduced to guarantee a fair and minimum ratio between public and private spaces in the city, needs a substantial revision to consider ESs in the urban environment as a field of options available to society. ESs should represent services to inhabitants to be guaranteed, therefore, resources to safeguard or restore in the local planning strategies.

The definition of the new quality standards has to take into account the ever-changing and heterogeneous factors affecting the urban welfare directly. We recommend that it is fundamental to define a model of collective services and infrastructures in order to improve the life quality of the community and, at the same time, to ensure the preservation of the common goods. In the territorial and urban plans, the quantitative and qualitative criteria (namely "urban quality standards") should control the human pressure on the natural environment, as well as improve the health of the cities, through lower consumption of non-renewable resources, a rebalance and mitigation of the anthropic impact and the enhancement of the ecological and environmental system.

Moreover, the definition of innovative thresholds and qualitative standards referring to human well-being, shared at the European and international level, can help prevent the speculation and exploitation of the common goods, the trampling of the rights of the people and the consequent decrease of the community welfare. At the same time, the necessary flexibility and adaptability to the specific territorial situations can be guaranteed.

Author Contributions: Conceptualization, A.M.C., A.F. and S.S.; methodology, A.M.C., A.F. and S.S.; investigation, A.M.C., A.F. and S.S. All sections are jointly written by the authors, except to Section 1, written by A.M.C., Sections 4.1 and 4.4. written by S.S., Sections 4.2 and 4.3, written by A.F. All authors have read and agreed to the published version of the manuscript.

Conflicts of Interest: The authors declare no conflict of interest.

Appendix A List of the Analyzed Official Documents

Plans	Documents (*Official Name*)
Municipality of Bologna, Piano StrutturaleComunale, 2008, available at http://www.comune.bologna.it/psc/documenti/848 (accessed on14 January 2020).	Relazione illustrativa Quadro normativo Valutazione di sostenibilità ambientale e territoriale Quadro conoscitivo Tables: Figure della ristrutturazione Strategie per la qualità Regole
Municipality of Rome, Piano Regolatore Generale, 2008, available at http://www.urbanistica.comune.roma.it/images/uo_urban/prg_vigente/prg_nta.pdf (accessed on 14 January 2020).	Norme tecniche di attuazione Sistemi e regole Rete Ecologica Standard Urbanistici Tables and drawings
Municipality of Milan, Piano dei Servizi, 2019, available at: https://www.comune.milano.it/aree-tematiche/urbanistica-ed-edilizia/pgt-vigente/piano-dei-servizi (accessed on14 January 2020).	Relazione Generale Norme e ambiti a servizi Catalogo, NIL e servizi Accessibilità Rete ecologica Tables and drawings
Municipality of Florence, Piano Strutturale Comunale. Relazione Illustrativa, 2014, available at http://pianostrutturale.comune.fi.it/documenti_del_piano/piano_strutturale_2014_approvata.html (accessed on14 January 2020).	Allegato A—Relazione Allegato B—Tavole Allegato C—Quadro conoscitivo Tables and drawings
Municipality of Florence, Regolamento Urbanistico. Relazione, 2015, available at: http://regolamentourbanistico.comune.fi.it/documenti/RU_vigente.html (accessed on 14 January 2020).	Allegato A—Relazione Allegato B—Norme tecniche di attuazione Allegato F—Rete ecologica Allegato H—Quadro conoscitivo Tables and drawings

References

1. Lefebvre, H. *Le Droit à la Ville*; Edizione Economica: Paris, France, 1968.
2. Gardi, C. Toward zero net land take by 2050: An EU perspective. In *Growing Compact*; Fregolent, L., Tonin, S., Eds.; FrancoAngeli: Milano, Italy, 2015; pp. 177–190.
3. Wolch, J.R.; Byrne, J.; Newell, J.P. Urban green space, public health, and environmental justice: The challenge of making cities just green enough. *Landsc. Urban Plan.* **2014**, *125*, 234–244. [CrossRef]
4. Colavitti, A.M.; Serra, S. Fifty years of service planning in Italy (1968–2018). The evolution of standard toward the efficiency of governance. In Proceedings of the International Conference on Changing Cities IV: Spatial, Design, Landscape & Socio-Economic dimensions, Chania, Crete Island, Greece, 24–29 June 2019; Gospodini, A., Ed.; University of Thessaly: Volos, Greece, 2019.
5. Salzman, J.; Arnold, C.A.; Garcia, R.; Hirokawa, K.; Jowers, K.; Lejava, J.; Peloso, M.; Olander, L. The Most Important Current Research Questions in Urban Ecosystem Services. *Duke Environ. Law Policy Forum* **2014**, *25*, 1–47.
6. Renzoni, C. Cinquant'anni di standard urbanistici (1968–2018). *Radici. Territorio* **2018**, *84*, 21–23. [CrossRef]
7. Falco, L. *I Nuovi Standard Urbanistici*; Edizioni delle Autonomie: Rome, Italy, 1987.
8. Contardi, L. Cinque ragioni per tornare a ragionare sugli standard. *Urban. Doss.* **1999**, *21*, 2–5.
9. Daily, G.C. (Ed.) Introduction: What Are Ecosystem Services? In *Nature's Services: Societal Dependence on Natural Ecosystems*; Island Press: Washington, DC, USA, 1997; pp. 1–10.

10. Costanza, R.; d'Arge, R.; de Groot, R.; Farber, S.; Grasso, M.; Hannon, B.; Limburg, K.; Naeem, S.; O'Neill, R.V.; Paruelo, J.; et al. The value of the world's ecosystem services and natural capital. *Nature* **1997**, *387*, 253–260. [CrossRef]
11. Daily, G.C. *Nature's Services: Societal Dependence on Natural Ecosystems*; Island Press: Washington, DC, USA, 2008.
12. Millennium Ecosystem Assessment (MEA). *Ecosystems and Human Well-Being*; Island Press: Washington, DC, USA, 2005.
13. De Groot, R.S. *Functions of Nature, Evaluation of Nature in Environmental Planning, Management and Decision Making*; Wolters-Noordhoff: Groningen, The Netherlands, 1992.
14. Rees, W.; Wackernagel, M. Urban Ecological Footprints: Why Cities Cannot be Sustainable and Why They are a Key to Sustainability. In *Urban Ecology. An International Perspective on the Interaction between Humans and Nature*; Marzluff, J., Shulenberger, E., Endlicher, W., Alberti, M., Bradley, G., Ryan, C., ZumBrunnen, C., Simon, U., Eds.; Urban Ecology; Springer: Boston, MA, USA, 2008; pp. 537–555.
15. Cortinovis, C.; Zardo, L.; Geneletti, D. Servizi ecosistemici: Nuovi strumenti per la pianificazione urbana. *Sentieri Urbani* **2016**, *19*, 27–31.
16. Giaimo, C.; Santolini, R.; Salata, S. Performance urbane e servizi ecosistemici. Verso nuovi standard? In *Dopo 50 Anni di Standard Urbanistici in Italia—Verso Percorsi di Riforma*; Giaimo, C., Ed.; INU Edizioni: Rome, Italy, 2019; pp. 63–69.
17. Saxer, S.R.; Rosenbloom, J. *Social-Ecological Resilience and Sustainability*; Wolters Kluwer: New York, NY, USA, 2018.
18. Brauman, K.A.; Daily, G.C.; Duarte, T.K.; Mooney, H.A. The nature and value of ecosystem services: An overview highlighting hydrologic services. *Annu. Rev. Environ. Resour.* **2007**, *32*, 67–98. [CrossRef]
19. Grêt-Regamey, A.; Celio, E.; Klein, T.M.; Hayek, U.W. Understanding ecosystem services trade-offs with interactive procedural modeling for sustainable urban planning. *Landsc. Urban Plan.* **2013**, *109*, 107–116. [CrossRef]
20. Biggs, R.; Schlüter, M.; Schoon, M.L. *Principles for Building Resilience: Sustaining Ecosystem Services in Social-Ecological Systems*; University Press: Cambridge, UK, 2015.
21. Geneletti, D. Reasons and options for integrating ecosystem services in strategic environmental assessment of spatial planning. *Int. J. Biodivers. Sci. Ecosyst. Serv. Manag.* **2011**, *7*, 143–149. [CrossRef]
22. Lopes, R.; Videira, N. Valuing marine and coastal ecosystem services: An integrated participatory framework. *Ocean Coast. Manag.* **2013**, *84*, 153–162. [CrossRef]
23. Albert, C.; Aronson, J.; Fürst, C.; Opdam, P. Integrating ecosystem services in landscape planning: Requirements approaches, and impacts. *Landsc. Ecol.* **2014**, *29*, 1277–1285. [CrossRef]
24. Mascarenhas, A.; Ramos, T.B.; Haase, D.; Santos, R. Integration of ecosystem services in spatial planning: A survey on regional planners' views. *Landsc. Ecol.* **2014**, *29*, 1287–1300. [CrossRef]
25. De Groot, R.S.; Alkemade, R.; Braat, L.; Hein, L.; Willemen, L. Challenges in integrating the concept of ecosystem services and values in landscape planning: Management and decision making. *Ecol. Complex* **2010**, *7*, 260–272. [CrossRef]
26. Colding, J. The role of ecosystem services in contemporary urban planning. In *Urban Ecology: Patterns, Processes and Applications*; Niemelä, J., Ed.; Oxford University Press: Oxford, UK, 2011; pp. 228–237.
27. Salata, S.; Ronchi, S.; Ghirardelli, F. I servizi ecosistemici a supporto della pianificazione paesaggistica. *Territorio* **2016**, *77*, 45–52. [CrossRef]
28. Delphin, S.; Escobedo, F.J.; Abd-Elrahman, A.; Cropper, W.P. Urbanization as a land use change driver of forest ecosystem services. *Land Use Policy* **2016**, *54*, 188–199. [CrossRef]
29. Mocine, C.R. The Legge Ponte, a Step toward More Effective City Planning in Italy. *Urban Law Annu.* **1970**, *187*, 187–195. Available online: https://openscholarship.wustl.edu/law_urbanlaw/vol1970/iss1/11 (accessed on 14 January 2020).
30. Odorisio, C. Il dibattito del '68 ed il bilancio attuale. *Urban. Doss.* **1999**, *21*, 6–7.
31. Francini, M.; Colucci, M. La ridefinizione degli standard urbanistici quale strategia di rigenerazione urbana. *Urban. Doss.* **2013**, *4*, 439–441.
32. Guzzo, G. *La Pianificazione Urbanistica. Soggetti, Contenuti e Ambiti Applicativi*; Giuffrè Editore: Milano, Italy, 2012.

33. World Health Organization. *Health Indicators of Sustainable Cities in the Context of the Rio+20 UN Conference on Sustainable Development*; WHO: Geneva, Switzerland, 2012.
34. Science for Environment Policy. 2018 Indicators for Sustainable Cities. In-Depth Report 12. Produced for the European Commission DG Environment by the Science Communication Unit, UWE, Bristol. Available online: http://ec.europa.eu/science-environment-policy (accessed on 14 January 2020).
35. Istituto Nazionale di Statistica, BES 2019. Il Benessere Equo e Sostenibile in Italia. ISTAT, Roma, 2019. Available online: https://www.istat.it/it/files//2019/12/Bes_2019.pdf (accessed on 18 January 2020).
36. Russo, A.; Cirella, G.T. Modern Compact Cities: How Much Greenery Do We Need? *Int. J. Environ. Res. Public Health* **2018**, *15*, 2180. [CrossRef]
37. Urbani, P. *Urbanistica Solidale. Alla Ricerca Della Giustizia Perequativa tra Proprietà e Interessi Pubblici*; Bollati Boringhieri: Torino, Italy, 2011.
38. Centofanti, N. *Diritto di Costruire, Pianificazione Urbanistica, Espropriazione*; Giuffrè Editore: Milano, Italy, 2010.
39. Giallanella, F. Standard urbanistici e piano locale. Indicatori quantitativi e riferimenti prestazionali. In *Piano Locale e ... Nuove Regole, Nuovi Strumenti, Nuovi Meccanismi Attuativi*; Ricci, L., Ed.; Franco Angeli: Milan, Italy, 2009.
40. ISTAT, Report Ambiente Urbano. Anno 2018. Rilevazione Dati Ambientali Nelle Città. Tavola 11.1—Disponibilità di Verde Urbano nei Comuni Capoluogo di Provincia/Città Metropolitana—Anni 2014–2018 (m2 per Abitante), 2018. Available online: https://www.istat.it/it/archivio/236912 (accessed on 18 January 2020).
41. Contardi, L. Nuovi standard urbanistici e procedure negoziali. *Urban. Inf.* **1999**, *167*, 31.
42. Agger, S.G. *Autogestione Urbana: L'urbanistica per una Nuova Società*; Dedalo Libri: Bari, Italy, 1977.
43. Giaimo, C. La trama. Dopo 50 anni, ripartire dagli standard. In *Dopo 50 Anni di Standard Urbanistici in Italia—Verso Percorsi di Riforma*; Giaimo, C., Ed.; INU Edizioni: Rome, Italy, 2019; pp. 30–40.
44. Bolgiani, I. Attrezzature religiose e pianificazione urbanistica: Luci e ombre. *Stato Chiese Plur. Confess.* **2013**, *28*, 1–23.
45. Moroni, S.; Chiodelli, F. Urbanistica, moschee e altri luoghi di culto. Riflessioni a partire da una recente legge della Regione Lombardia. *Crios* **2016**, *11*, 21–34. [CrossRef]
46. Ghiloni, M.; Karrer, F. Verso standard urbanistici convenzionali? *Urban. Inf.* **1999**, *167*, 38–39.
47. Karrer, F.; Ricci, M. *Città e Contratto: Il Piano dei Servizi Tra Programmazione Urbana e Gestione*; Officina Edizioni: Rome, Italy, 2006.
48. Gerundo, R.; Fasolino, I.; Grimaldi, M.; Graziuso, G. The performance of urban standards as a way of evaluating the efficiency of facilities in the municipalities of inland areas. *Plurimondi VII* **2015**, *16*, 133–141.
49. Colavitti, A.M.; Serra, S.; Usai, A. Le città metropolitane come motori dello sviluppo: Le risposte della pianificazione territoriale di area vasta alle recenti riforme istituzionali ed amministrative. *Planum* **2014**, *2*, 438–446.
50. La Rosa, D. Accessibility to greenspaces: GIS based indicators for sustainable planning in dense urban context. *Ecol. Indic.* **2014**, *42*, 122–134. [CrossRef]
51. Pafi, M.; Siragusa, A.; Ferri, S.; Halkia, M. *Measuring the Accessibility of Urban Green Areas. A Comparison of the Green ESM with Other Datasets in Four European Cities*; JRC Technical Reports, EUR 28068 EN; Publications Office of the European Union: Luxembourg, 2016.
52. Kabisch, N.; Strohbach, M.; Haase, D.; Kronenberg, J. Urban green space availability in European cities. *Ecol. Indic.* **2016**, *70*, 586–596. [CrossRef]
53. Dai, D. Ratial/ethnic and socio economic disparities in urban green space accessibility: Where to intervene? *Landsc. Urban Plan.* **2011**, *102*, 234–244. [CrossRef]
54. European Environment Agency. *Spatial Analysis of Green Infrastructure in Europe*; European Environment Agency: Luxembourg, 2014.
55. Rossetti, M. La nuova politica dei servizi urbani della Regione Lombardia. Urbanistica Informazioni 2001, 176. Arcidiacono, A., Viviani, S. Nuovi standard per la pianificazione urbanistica. In *Rapporto Consumo di Suolo, Dinamiche Territoriali e Servizi Ecosistemici*; ISPRA: Rome, Italy, 2016; pp. 77–78.
56. Mussinelli, E.; Tartaglia, A.; Fanzini, D.; Riva, R.; Cerati, D.; Castaldo, G. New Paradigms for the Urban Regeneration Project between Green Economy and Resilience. In *Regeneration of the Built Environment from a Circular Economy Perspective*; Della Torre, S., Cattaneo, S., Lenzi, C., Zanelli, A., Eds.; Springer: Berlin/Heidelberg, Germany, 2020; pp. 59–68.

57. Bastian OHaase, D.; Grunewald, K. Ecosystem properties, potentials and services—The EPPS conceptual framework and an urban application example. *Ecol. Indic.* **2012**, *21*, 7–16. [CrossRef]
58. Liang, Y.; Liu, L.; Huang, J. Integrated Ecosystem Services Assessment in Urbanizing Regions. In *Integrated Modelling of Ecosystem Services and Land-Use Change*; Springer Geography; Springer: Singapore, 2020; pp. 153–167.
59. La Notte, A.; Rhodes, C. The theoretical frameworks behind integrated environmental, ecosystem, and economic accounting systems and their classifications. *Environ. Impact Assess. Rev.* **2020**, *80*, 106–117. [CrossRef]
60. Franchina, A. La Città e la Dimensione Ambientale: Il Paradigma dei Servizi Ecosistemici Nella Pianificazione Alla Scala Urbana. Ph.D. Thesis, University of Palermo, Palermo, Italy, 2018.
61. Gómez-Baggethun, E.; Barton, D.N. Classifying and valuing ecosystem services for urban planning. *Ecol. Econ.* **2013**, *86*, 235–245. [CrossRef]
62. Bottero, M.; Bravi, M.; Giaimo, C.; Barbieri, C.A. Ecosystem Services: From Bio-physical to Economic Values. In *Values and Functions for Future Cities. Green Energy and Technology*; Mondini, G., Oppio, A., Stanghellini, S., Bottero, M., Abastante, F., Eds.; Springer: Berlin/Heidelberg, Germany, 2020; pp. 37–50.
63. Bolund, P.; Hunhammar, S. Ecosystem services in urban areas. *Ecol. Econ.* **1999**, *29*, 293–301. [CrossRef]
64. Syrbe, R.U.; Grunewald, K. Ecosystem service supply and demand. The challenge to balance spatial mismatches. *Int. J. Biodivers. Sci. Ecosyst. Serv. Manag.* **2017**, *13*, 148–161. [CrossRef]
65. Lilburne, L.; Eger, A.; Mudge, P.; Ausseil, A.-G.; Stevenson, B.; Herzig, A.; Beare, M. The Land Resource Circle: Supporting land-use decision making with an ecosystem-service-based framework of soil functions. *Geoderma* **2020**, *363*, 114–134. [CrossRef]
66. Haines-Yong, R.; Potschin, M. Common Classification of Ecosystem Services CICES Version 5.1 and Guidance on the Application of the Revised Structure, 2018. Available online: www.cices.eu (accessed on 14 January 2020).
67. Potschin, M.; Haines-Young, R. Defining and measuring ecosystem services. In *Routledge Handbook of Ecosystem Services*; Potschin, M., Haines-Young, R., Fish, R., Turne, R.K., Eds.; Routledge: London, UK; New York, NY, USA, 2016; pp. 25–44.
68. Costanza, R. Ecosystem services: Multiple classification systems are needed. *Biol. Conserv.* **2008**, *141*, 350–352. [CrossRef]
69. Wallace, K.J. Classification of ecosystem services: Problems and solutions. *Biol. Conserv.* **2007**, *139*, 235–246. [CrossRef]
70. Costanza, R.; de Groot, R.; Braat, L.; Kubiszewski, I.; Fioramonti, L.; Sutton, P.; Farber, S.; Grasso, M. Twenty years of ecosystem services: How far we have we come and how far do we still need to go? *Ecosyst. Serv.* **2017**, *28*, 1–16. [CrossRef]
71. Fisher, B.; Turner, R.K. Ecosystem services: Classification for valuation. *Biol. Conserv.* **2008**, *141*, 1167–1169. [CrossRef]
72. La Notte, A.; D'Amato, D.; Mäkinen, H.; Paracchini, M.L.; Liquete, C.; Egoh, B.; Geneletti, D.; Crossman, N.D. Ecosystem services classification: A system ecology perspective of the cascade framework. *Ecol. Indic.* **2017**, *74*, 392–402. [CrossRef]
73. Díaz, S.; Pascual, U.; Stenseke, M.; Martín-López, B.; Watson, R.T.; Molnár, Z.; Hill, R.; Chan, K.M.A.; Baste, I.A.; Brauman, K.A.; et al. Assessing nature's contributions to people. *Science* **2018**, *359*, 270–272. [CrossRef]
74. Adhikari, K.; Hartemink, A.E. Linking soils to ecosystem services: A global review. *Geoderma* **2016**, *262*, 101–111. [CrossRef]
75. Krehl, A.; Weck, S. Doing comparative case study research in urban and regional studies: What can be learnt from practice? *Eur. Plan. Stud.* **2019**, 1–17. [CrossRef]
76. Nadin, V.; Stead, D. Opening up the compendium: An Evaluation of International Comparative Planning Research Methodologies. *Eur. Plan. Stud.* **2013**, *21*, 1542–1561. [CrossRef]
77. Geneletti, D.; Cortinovis, C.; Zardo, L.; Esmail, B.A. *Planning for Ecosystem Services in Cities*; Springer: Cham, Switzerland, 2020.
78. Corrado, R. *L'urbanistica Italiana Dopo le Sentenze del TAR sul PRG di Roma*; Gangemi Editore: Roma, Italy, 2010.
79. Pogliani, L. L'evoluzione del piano dei servizi. *Territorio* **2009**, *49*, 68–73.

80. La Rosa, D. Why is the inclusion of the ecosystem services concept in urban planning so limited? A knowledge implementation and impact analysis of the Italian urban plans. *Socio-Ecol. Pract. Res.* **2019**, *1*, 83–91. [CrossRef]
81. Gibelli, G.; Santolini, R. Reti ecologiche e governo del territorio. *Territorio* **2011**, *58*, 61–74.
82. Woodruff, S.C.; BenDor, T.K. Ecosystem services in urban planning: Comparative paradigms and guidelines for high quality plans. *Landsc. Urban Plan.* **2016**, *152*, 90–100. [CrossRef]
83. Caldarice, O. La pianificazione dei servizi in Lombardia. Tentativi di innovazione. *Urban. Doss.* **2013**, *4*, 379–381.
84. Scolozzi, R.; Morri, E.; Santolini, R. Pianificare territori sostenibili e resilienti: La prospettiva dei servizi ecosistemici. *Territorio* **2012**, *60*, 167–175.
85. Calzolari, C.; Ungaro, F.; Filippi, N.; Guermandi, M.; Malucelli, F.; Marchi, N.; Staffilani, F.; Tarocco, P. A methodological framework to assess the multiple contributions of soils to ecosystem services delivery at regional scale. *Geoderma* **2016**, *261*, 190–203. [CrossRef]
86. Burkhard, B.; Maes, J. *Mapping Ecosystem Services*; Pensoft Publishers: Sofia, Bulgaria, 2017; 374p.
87. Banerjee, O.; Crossman, N.D.; de Groot, R.S. Ecological Processes, Functions and Ecosystem Services: Inextricable Linkages between Wetlands and Agricultural Systems. In *Ecosystem Services in Agricultural and Urban Landscapes*; Wratten, S., Sandhu, H., Cullen, R., Costanza, R., Eds.; John Wiley & Sons: Hoboken, NJ, USA, 2013; pp. 16–27.
88. Teixera da Silva, R.; Fleskens, L.; van Delden, H.; van der Ploeg, M. Incorporating soil ecosystem services into urban planning: Status, challenges and opportunities. *Landsc. Ecol.* **2018**, *33*, 1087–1102. [CrossRef]
89. Fugazza, B.; Ronchi, S.; Salata, S. La ricomposizione degli assetti ecosistemici a partire dalla valutazione delle funzioni dei suoli: Una proposta di green infrastructure per il territorio lodigiano. *Reticula* **2014**, *7*, 103–109.

sustainability

MDPI

Article

Smart City Governance and Children's Agency: An Assessment of the Green Infrastructure Impact on Children's Activities in Cagliari (Italy) with the Tool "Opportunities for Children in Urban Spaces (OCUS)"

Chiara Garau * and Alfonso Annunziata

Department of Civil and Environmental Engineering and Architecture, University of Cagliari, Cagliari 09123, Italy

* Correspondence: cgarau@unica.it; Tel.: +39-070-675-5565

Received: 22 July 2019; Accepted: 30 August 2019; Published: 5 September 2019

Abstract: The increases in urbanization, pollution, resource depletion, and climate change underline the need for urban planning policies that incorporate blue–green infrastructure (BGI) and ecosystem services. This paper proposes a framework for assessing BGI's effect on children's outdoor activities. This effect, called meaningful usefulness, is a central issue due to the influence of experiences with nature on children's development and the global trend of concentration of children in urban areas. Based on the concept of affordance, the methodology formalizes meaningful usefulness in terms of an index of usefulness of individual settings (I_{UIS}) and a synthetic index of usefulness of BGI in a specific area (I_{SGI}). These are determined via an audit protocol, Opportunities for Children in Urban Spaces (OCUS), which incorporates a set of indicators measuring micro-scale properties of individual places and contextual macro-scale factors. The methodology is applied to BGI components in Cagliari, Sardinia, Italy, which was selected for its superior density of urban green spaces. The application of the OCUS tool confirms its usefulness for investigating functional affordances incorporated into the trans-scalar structures of BGIs. The analytic protocol further contributes to the implementation of urban planning strategies within the smart city paradigm.

Keywords: green infrastructure; smart city; affordance; children

1. Introduction

As Martintotti [1] observed, the city needs to be celebrated as the product and heart of the most advanced manifestation of civilization, the center for commerce, learning, and culture, and the production of the most significant scientific and technological advancements.

The emergence—and triumph—of cities manifests in the global process of mass urbanization, a trend determining that "today, half the world's population lives in urban areas and, by 2050, all regions will be predominantly urban. According to current projections, virtually the whole of the world's population growth over the next 30 years will be concentrated in urban areas" (p. IX) [2]. This process is associated with the phenomena of pollution, resource depletion, climate change, rising energy costs, decrease in biodiversity, growing inequality, and deterioration of microclimatic conditions [3]. These phenomena, alongside the "preoccupation with preventing and minimizing the effects of the next natural or manmade disaster" and the need to preserve the centrality of cities as places of excellence, will determine the formulation and implementation of urban paradigms that orient the transformation of the built environment, including the smart city paradigm. In fact, the latter calls for governance practices and planning strategies that activate synergies among traditional infrastructures, information and communication technology (ICT) infrastructures, and socio-economic structures aimed at supporting sustainable economic development and a high quality of life, with proper management of natural resources through participatory action and engagement [4]. In this respect, the green infrastructure-based approach emerges as a key tool for re-configuring the city as an inclusive, healthy, anti-fragile, connected urban ecosystem by achieving a balance among urban and ecological processes and systems, increasing biodiversity, and incorporating resilience measures for preventing and minimizing effects of storm, flood, heat, drought, and pollution [3].

The benefits provided by blue–green infrastructures extend beyond the ecological and environmental dimensions and likewise affect the social and economic structures of the city. These benefits are comprehensively understood within the conceptual framework based on the notion of ecosystem services (ES). A comprehensive definition of urban blue–green infrastructure (BGI) and ecosystem service is presented in Section 2.

Focusing on the social dimension of ecosystem services, the purpose of this paper is to structure the theoretical framework and provide an analytic protocol for assessing the ways in which green infrastructures affect children's opportunities to engage in independent meaningful outdoor activities. These are defined as activities and practices motivated by intentions, goals, and purposes, thereby significantly contributing to the emergence of psychological place experience [5]. The potential of urban BGIs to enable children's mobility, agency, and meaningful engagement with places and objects by supporting their independent functional, optional, recreational, and social outdoor activities is referred to as meaningful usefulness.

The theoretical framework is based on the concepts of capability, affordance, behavior setting, and functional description of places, introduced by Sen [6], Gibson [7], Barker and Wright [8], and Heft [9], respectively. In particular, the affordance concept refers to the functional, social, and emotional opportunities and constraints for a specific individual, incorporated into environmental features. For this reason, it is considered as a central category for describing the public space potential to promote people's activities. The methodology builds on these concepts by operationalizing the concept of meaningful usefulness in terms of an index of usefulness of individual natural settings (I_{UIS}) and of a synthetic index of usefulness of the urban blue–green infrastructure in a specific area (I_{SGI}). These indexes are determined via the application of an audit protocol, the Opportunities for Children in Urban Spaces and Natural Settings (OCUS_NS). This analytic tool, created by the authors, incorporates a set of qualitative and quantitative indicators that measure micro-scale properties of individual places and contextual macro-scale factors. This research focuses on the urban context for two reasons: the potential of building on the findings from the literature on children's independent mobility, physical activity, and walkability, as well as the emergence of the contemporary city as the most common milieu of children's development determined by the global trend toward urbanization. As United Nations Children's Fund (UNICEF) [10] observed, more than one billion children live in urban settings around the world; in the future, the majority of children will grow up in towns and cities. This research fills a gap in the literature on blue–green infrastructure and on children's experience of the public space by underlining the increase in opportunities of children's engagement and transactions with natural settings as a central aspect of the social dimension of ecosystem services.

In fact, children's contact with nature emerges as a central issue for two intertwined reasons: the influence of experiences with nature on children's cognitive, bodily, social, and emotional development [11–14], and the correlation between early nature experiences and the development of human nature connections, which in turn affect the possibility of the "trans-generational establishment of sustainable futures" (p. 2) [15]. The human–nature connection emerges as a central leverage point for enabling the transition of a socio-ecological system to a sustainable and resilient future [16].

The methodology is applied to the assessment of a trans-scalar mosaic of blue–green infrastructure components identified across the city of Cagliari in Sardinia, Italy. The city of Cagliari was selected as a subject for the case study due to its significant availability and density of urban green spaces, which are superior to the average values measured for major urban areas on a national scale [17]. The paper is divided into in five sections. In Section 2, a literature review on urban blue–green infrastructure and on children's experience of outdoor spaces outlines the theoretical framework by defining the concepts of affordance, behavior setting, green infrastructure, and ecosystem service; afterward, the methodological framework and the case study are described. The results of the application of the audit protocol to the case study are outlined in Section 3, while the most relevant findings are discussed in Section 4. Finally, Section 5 discusses the relevance and the limitations of this research and outlines its development in the future.

2. Materials and Methods

2.1. Urban Blue–Green Infrastructures

Benedict and McMahon [18] define green infrastructure as "an interconnected network of waterways, wetlands, woodlands, wildlife habitats, and other natural areas; greenways, parks, and other conservation lands; working farms, ranches, and forests; and wilderness and other open spaces that support native species, maintain natural ecological processes, sustain air and water resources, and contribute to the health and quality of life of communities and people". Consequently, an urban green infrastructure (Urban BGI) can be conceptualized as an interconnected network of open spaces, natural areas, urban woodland, and parks; green streets, squares, and public realm; sustainable drainage systems, rivers, and waterways; cycleways and pedestrian routes; and smaller-scale interventions such as green roofs, walls, and facades that contribute to people's wellbeing and to the balance between city and nature by providing ecological, economic, and social benefits, including water purification, retention, and drainage, biodiversity, local food production, recreation carbon storage, social cohesion, and identity building [3,19–22]. Hence, the green infrastructure-based approach implies planning, designing, realizing, regenerating, and connecting the components of the green and blue networks into a trans-scalar, capillary, contiguously connected infrastructure.

The benefits that humans obtain from green infrastructures are referred to as ecosystem services. The concept of an ecosystem service is widely utilized since it incorporates and underlines the notion that natural systems are socially valuable and are present in forms that are not immediately intuitable [23]. Nevertheless, several definitions of ecosystem services and different classification protocols can be traced in the current literature [21,22,24–26]. In general terms, ES can be defined as "the benefits of nature to households, communities, and economies" [22]. Several studies underline that ES determines trans-scalar neutral, synergic, and trade-off relationships among households and communities, thus emphasizing the significance of understanding these interactions as a pre-condition for informed decisions on politics and interventions related to environment, economics, and land use [27–29].

2.2. Concepts for Assessing Urban BGI Components

As stated in the introduction, this paper focuses on the social dimension of ecosystem services and highlights the relevance of urban BGI components in terms of children's nature experiences and independent outdoor activities. Children's independent activities are herein defined as the complex of children's practices carried out across public space without adult supervision. These include independent mobility—the freedom and/or ability of children to travel across the urban space and play outdoors. This is conceptualized both as a vector for physical activity [30] and as a creative act of spatial appropriation and of meaningful engagement with spaces and objects, conducive to dwelling and enchantment [31]. Within this framework of reference, nature experiences can be conceptualized as the complex of children's transactions with natural settings. The benefits to children's wellbeing brought about by nature experiences and outdoor activities are better understood through the capability approach. Capability is described as a valuable state of being or a condition that a person can access [6,14,32]. According to Chawla [14], children's transactions with natural settings are associated with the realization of 10 general capabilities: life; bodily health; bodily integrity; affiliation; practical reason; play; senses, imagination, and thought; emotions; connection to nature and other species; control over one's environment. These capabilities are pre-conditions to the development of foundational capability. In particular, direct nature experiences during childhood emerge as crucial moments of "sustainable enculturation, with long-lasting consequences for sustainable social–ecological systems" (p. 2) [15]. In fact, early nature experience is central to structuring the psychological traits of human–nature connection, which in turn emerges as the most relevant leverage point for modifying society's mindset and for structuring resilient and sustainable social–ecological structures [33,34].

The concept of affordance is, thus, introduced here as a central category for understanding and describing the transactions among individuals and spaces and for investigating environmental factors that influence practices and behaviors. Building on definitions proposed by Gibson [7] and Heft [9], an affordance can be conceptualized as a functionally significant character of the environment, considered in relation to an individual (p. 20). Kyttä [35] observed that the concept of affordance also includes the emotional and social opportunities and restrictions incorporated in an environment in relation to a specific individual. Kyttä [36] likewise introduced a distinction between potential and actualized affordances. The former refer to the infinite number of generic possibilities incorporated into the structures of a setting, while the latter refer to opportunities determined by the attributes of the spatial, material, and sociocultural features of a setting that are congruent with the corporality of individuals, as defined by the complex of their necessities, abilities, and physical, social, and psychological characteristics. Thus, the concept of affordance is relational in nature and overcomes the subject–object dichotomy.

A concept related to that of affordance is the notion of behavior setting, introduced by Barker and Wright [8]. According to Kyttä [35], a behavior setting can be defined as a social and cultural context where a dynamic, yet stable pattern of actions is generated by joint participation of two or more individuals with the support of affordances. The relationship between activity and the milieu is synomorphic, that is, it implies consistency among actions and environmental features.

These concepts are the structural categories for investigating the activities across a given space. Building on Barker and Wright's "one boy's day" [37], Heft outlines a taxonomy of environmental features alternative to that "of the standard classification by form". In Heft's words, environmental elements within this taxonomy are classified in terms of distinctive functional properties—being ride-on-able, climb-on-able, sit-on-able, run-on-able, etc.—that constitute the environmental counterparts of observed activities [9].

The observation of different hierarchical relationships in affordance categories results in the establishment of a synthetic functional taxonomy of environmental features, which are structured into 10 categories: (i) flat, relatively smooth surface; (ii) relatively smooth slope; (iii) graspable/detached object; (iv) attached object; (v) non-rigid attached object; (vi) climbable feature; (vii) aperture; (viii) shelter; (ix) moldable material; and (x) water [9].

The functional perspective incorporated into Heft's taxonomy suggests a conceptualization of environmental features that is commensurate to the intentional and active characteristic of psychological functioning. Moreover, it emphasizes the developmental dimension of places and of interactions within places; the functional possibilities incorporated into environmental features change according to the developmental status of the individual or group.

Building on these considerations, Lerstrup and Konijnendijk van den Bosch [38] clarified Heft's definitions and revised the hierarchical relationships among the functional categories, defining an alternative functional taxonomy structured in 10 classes: open ground, sloping terrain, shielded places; rigid fixtures; moving fixtures; loose objects; loose material, water; creatures; fire. Two fundamental considerations emerge from this study. The first concerns the individuation of transversal characteristics of environmental features that significantly influence children's patterns of activities; these qualitative properties are variation and uniqueness, abundance, novelty and change, and size and gradation. These properties affect the significance and attractiveness of environmental features, the availability of the affordance for a particular activity to all users, and the creation of a testing environment, which in turn enables the development of children's abilities and competencies.

The second observation concerns the constitutive ambiguity of the notion of affordance, whose meaning depends on the particular perspective adopted. Hence, the term affordance can be interpreted as "affordances of a setting", referring to the action possibilities incorporated into a setting, as "affordances for an activity", thus identifying features that support specific actions, and sometimes as "affordances for someone", referring to either of the two. Nevertheless, according to the definition of affordance as a multi-dimensional concept, which includes social, emotional, cognitive, and cultural properties, an analysis on the affordances incorporated into a setting should extend beyond the dimension of instrumental functionality. According to Broeberg et al. [39], social affordances, thus, refer to the opportunities "to be with adults", "meet friends", and "make new friends", as well as including the unpleasant possibilities of contact with social milieu fragilities, i.e., antisocial practices, such as "unpleasant gangs" and "scary adults", or of negative social interactions including "feeling like an outsider" or "being lonely". Moreover, social affordances also refer to conditions conducive to sense of privacy and territoriality, affected by adults' practices and appropriation of spaces, including "kids not allowed" or "strict control". On the other hand, emotional/contextual affordances refer to possibilities, incorporated into the spatial and social structures of a place, of experiencing positive or negative emotions and stimuli, including the possibilities of perceiving a place as beautiful, calm, quiet, safe, exciting, and clean or, contrastingly, noisy, dirty, dangerous, and stressful.

Finally, accessibility is defined either as potential for interaction or as the actual freedom to participate in different activities, thus partially overlapping with the definition of capability. Geurs and Van Wee [40] observed that the concept of accessibility incorporates four dimensions: a transport component, a land-use component, a temporal component, and an individual component. The concepts of capability, affordance, accessibility, and behavior setting are central to the construction of the theoretical model of the proposed audit tool. The complex of functional, emotional, social affordances, and accessibility conditions incorporated in environmental features determines the potential of urban BGI components to support children's nature experiences and meaningful outdoor activities, which in turn enable children's realization of foundational capabilities. This potential is referred to as meaningful usefulness.

The concept of meaningful usefulness, hence, incorporates the conceptualization of child-friendliness as the potential of the public space to support intense engagement with the environment, leading to experiences of enchantment, learning, and gaining competence through experience, as well as the claiming, interpreting, and appropriation of spaces and environmental features. Nevertheless, while the concept of child-friendliness, as observed by Whitzman et al. [41], is pre-eminently associated with a social and health planning perspective, the introduction of the concept of meaningful usefulness reflects the intention to focus on material and spatial conditions of the environment that incorporate the functional, social, and emotional opportunities that influence children's experience of the urban space.

2.3. *Review of Methodological Approaches*

Finally, a central contribution for structuring a methodological framework is derived from different methodological approaches incorporated in models which operationalized the concepts of child-friendliness and walkability.

In particular, the Bullerby model, structured by Kyttä [36,42] and revisited by Broberg et al. [39] and Kyttä et al. [35,43], combines children's experiences, operationalized in terms of actualized affordances and of their diversity, collected through public participation geographic information systems (PPGIS), and objectively measures structural properties of the built environment. PPGISs are also central in the web-map-based survey structured by Lopes et al. [44]. A different observational protocol is incorporated in audit tools. These tools—including the Survey on Conditions of Practicable Environments (SCOPE), the Public Open Space Desktop Audit Tool (POSDAT), the Pedestrian Environment Data Scan (PEDS), the Environmental Assessment of Public Recreation Spaces (EAPRS), and the Procedure and the Quality Index of Parks for Youth (QUINPY)) [45–48]—incorporate qualitative and quantitative indicators and sub-indicators and aggregate data collected from direct on-site observation and/or secondary data retrieved from informative territorial services, internet-based street-level imagery services (Google Street View), and territorial imagery services (Google Maps, Google Earth, Bing Maps). The combination of primary and secondary data, and of qualitative and quantitative indicators, is also central to multicriteria evaluation models [49–52]. These tools integrate spatial information related to macro-scale built environment factors and evaluation of micro-scale site-specific urban design features, reflecting a conceptualization of a place's inclusivity or usefulness as the product of both inherent and endowed contextual properties. Finally, questionnaires focus on reporting pedestrians' perceptions and preferences related to significant spatial, environmental, and social properties of the urban space (Neighborhood Environment Walkability Survey (NEWS)) [53]. In Section 3, the methodological framework for assessing the meaningful usefulness of natural settings does not incorporate the urban blue/green infrastructures and describes the "Opportunities for Children in Urban Spaces" tool.

3. Methodology

The OCUS tool is considered an audit tool that integrates quantitative and qualitative indicators for evaluating and measuring the availability of functional, emotional, and social affordances and the conditions of access to open public spaces, determined by the configurational, compositional, and material conditions of the urban blue/green infrastructures at different scales (Table 1). Indicators and sub-indicators are, hence, organized into four categories: (i) functional opportunities; (ii) social opportunities; (iii) emotional/contextual opportunities; and (iv) independent accessibility opportunities. With respect to the functional opportunities dimension, each indicator refers to a specific category of functional properties that evaluates and measures the availability, variety, size, gradation, and uniqueness of environmental features incorporating a specific class of affordances. For instance, the indicator potential appropriation of open grounds measures the presence of flat void surfaces, the number of available regions of open ground, the size and dimensions of the most favorable surface, and variety in form, size, and surface among the regions of open ground. Moreover, the developmental characteristic of transactional interactions among users and settings, incorporated in the concept of affordance, compels the definition of the users considered and the determination of the pertinent indicators of meaningful usefulness. Consequently, the research focuses on users ranging from 9–13 years of age. According to Shaw et al. [54], this range is consistent with existing studies on children's experience of places and reflects sensible variations in children's level of independent mobility. The lower limit of this age range is associated with a relevant increase in the number of children in Italy who are allowed to cross major roads and go to relevant places within walking distance. On the other hand, age 13 is associated with the threshold beyond which the majority of children exercise the ability to cross major roads, travel alone on local buses, and independently go to school and to relevant places within walking distance.

Indicators included in the emotional and social opportunities dimensions assess and measure conditions and materials of the public space that incorporate specific social and emotional opportunities and restrictions for children. The social and emotional affordances, which were defined through building on findings from studies conducted by Kyttä [42], Broberg et al. [39], and Kyttä et al. [35], are associated with specific configurational, compositional, and material characteristics of the public space, operationalized in terms of representative indicators and sub-indicators. The correlation among features of natural settings, emotional and social affordances, and the individuation of pertinent indicators is based on the literature on built environment. This study correlates walkability, children's independent mobility, and children's physical activity.

Table 1. Indicators incorporated into the Opportunities for Children in Urban Spaces (OCUS) tool. POS—public open space.

Affordance	Environmental Features	Indicator (Ind.)
Functional affordances		
Walking, running, cycling, playing ball	Open ground	Ind. potential appropriation of open grounds (presence, size, quantity, variety)
Rolling/sliding/running down, rolling objects down, jumping down, jumping over, sitting in	Sloping terrain	Ind. potential appropriation of sloping terrains (presence, gradation, quantity, variety)
Hiding, as frame, microclimate	Repaired space	Ind. potential appropriation of repaired spaces (presence, quantity, variety)
Sitting-on, jumping, running around, hiding behind	Rigid features	Ind. potential transaction with rigid features (presence, quantity, variety)
Climbing, balancing on, hanging by arms, hanging in legs	Rigid climbable features	Ind. potential transaction with rigid climbable features (presence, quantity, variety)
Swinging on, dangling swaying in, seesawing on, spinning	Non-rigid, moving features	Ind. potential transaction with non-rigid moving features (presence, quantity, variety)
Drawing, scratching, throwing, hammering, batting, building	Loose objects	Ind. potential manipulation of loose objects (presence, quantity, variety)
Construction of objects, pouring, modification of surface features	Loose materials	Ind. potential manipulation of moldable materials (presence, quantity, variety)
Splashing, pouring, floating objects, drinking	Water	Availability of water (presence, quantity, variety, size)
Following, catching, caring for	Creatures	Presence of animals, insects, birds
Ensuring hygienic conditions	Lavatories	Availability of lavatories
Accessing web-based apps, communicating	Internet access	Internet coverage
Feeling safe	Natural control of the POS	Eyes on the POS
Pleasant place	Conspicuousness	Imageability of the POS
Quiet	Acoustic Environment	Quality of the acoustic environment
Breathing clean air	Pollution	Concentration of particulate matter
Clean public space	Maintenance of POS	Cleanliness of surfaces, equipment
Luminous	Illumination	Ind. of potential usability of the POS during night hours
Meeting friends	Meeting places	Presence or visibility of meeting places
Privacy/control	Sense of privacy and territoriality	Degree of children's appropriation of spaces
Being with adults	Intergenerational activities	Presence of intergenerational activities
Make new friends	Anchor places	Presence of anchor places within a 400-m buffer (sports facilities, educational institutions, shopping malls, formal sites for play)
Lively	Presence of people	Degree of liveliness of the public space (density of retail activities and services; presence of outdoor activities)
Unpleasant/scary people/antisocial practices	Signs of neglect	Broken window
Access by collective transport	Access to mass transit	Proximity of collective transport nodes
Access by walking alone or with friends	Accessible pedestrian network	Category of contiguous pedestrian facilities
Dealing with vehicular traffic	Priority of vulnerable users and soft mobility modes	Barrier effect (main entrance or worst condition)
Access by cycling alone or with friends	Accessible bicycle facilities	Category of contiguous bicycle facilities
Access to other relevant places	Land-use diversity	Walk score
Frequent access	Residential density	(Prevailing typology/segments surrounding the POS)

Moreover, according to Garau and Pavan [51], Garau et al. [45], and Abis et al. [55], indicators and sub-indicators are selected as a function of criteria of objectivity (indicators must be clear, unambiguous, and precise); relevance (significance and pertinence to the focus of the analysis, i.e., the concept of practicability); measurability and reproducibility (indicators must be quantitative and systematically observable); validity (implying possibility of verification and data quality control); representativeness, comparability over time, and applicability (possibility of use of findings from previous research and existence of accessible databases); and understanding (indicators must be easily understandable by the target audience, i.e., by decision-makers and the general public) [45,51,55].

For instance, a repaired space is an enclosed region of space incorporating a refuge, affording prospects for hiding, lying in, or sitting in. An enclosed region is defined as a space bounded on three sides, or as a space covered by a ceiling of height not superior than 2.15 m. The indicator potential appropriation of repaired spaces is calculated as the mean of the scores assigned to sub-indicators assessing the presence, number, and variety of enclosed regions of space. Similarly, the indicator potential transaction with rigid features assesses the presence, quantity, and variety of rigid, non-movable features (including trees, retaining walls, benches, stairs) that afford sitting on, jumping on/over/down from, running around, hiding behind, and building on [9,38]. The presence of creatures, determining the affordances of observing, following, and caring for creatures, is considered as a condition that affects the central children's capability to live with concern for and in relation to animals, plants, and the world of nature [32].

Moreover, according to Jacobs's [56] seminal work, the emotional affordance of feeling safe is associated with the opportunities of natural surveillance of the public space. This environmental parameter is operationalized in terms of the sub-indicators "visibility of the nearest buildings", "interactivity of façades", and "presence of activity with extended open hours", introduced by Saelens et al. [47], Gehl [57], and Moura et al. [49], respectively. These sub-indicators are then aggregated into the indicator "eyes on the street". Furthermore, the social affordance for making new friends is associated with the proximity of anchor places and, thus, to the presence of primary functions conducive to the concentration of people. This environmental factor is operationalized in terms of the indicator "existence and visibility of anchor places", introduced by Moura et al. [49]. The conspicuousness of a setting, related to the emotional affordance of being in a pleasant space, is conceptualized as the resultant of the geometric layout of a surface, captured by the sub-indicator articulation of edges, and of the presence of unique elements, including natural structures, artifacts, major landscape features, and distinctive buildings. These distinct aspects of a setting are associated with the coherence and imageability of a space and to its effects on re-orientation and individuation of goal locations. In fact, a study by Lee, Shusterman, and Spelke [58] revealed that children's navigation behaviors depend on two distinct processes: a modular process of re-orientation based on the geometry of the surrounding surface layout and an associative process that relates unique elements and landmarks to locations.

In addition, developmental processes extend core abilities, permitting more flexible strategies, emerging from the combination of geometric and non-geometric information [58]. Opportunities for independent accessibility refer to a multi-dimensional [40,59] concept of accessibility, encompassing transport, land use, temporal, and individual components within the conceptual framework of the OCUS tool. This concept is related to macro-scale contextual factors, including access to mass transit [45], access to pedestrian facilities [59], access to bicycle facilities, priority of vulnerable users [49], residential density, and land-use diversity [45].

The determination of indicators related to micro-scale features incorporating functional, emotional, and social affordances is based on primary data, collected during on-site observation and integrated with secondary data retrieved from informative territorial services and internet-based street-level imagery services. Moreover, on-site observations are conducted according to a form, containing a list of items related to specific environmental features and properties to assess. Information retrieved from the forms is integrated with images taken during the observation. Finally, accessing internet is considered a relevant opportunity for children in order to interact, via their mobile phones, with peers and parents, for instance, exchanging images, audio files, and text messages.

The combination of qualitative and quantitative indicators, measured via specific scales, poses the problem of the normalization of results in homogeneous quantitative terms in order to aggregate the single indicators, related to specific affordances, into the global index of meaningful usefulness. Consequently, with respect to the qualitative indicators, a score ranging from 0–4 is determined for each scale level. For quantitative indicators, levels of performance corresponding to a specific range of values are defined.

Afterward, a score ranging from 0–4 is assigned to each value band. Each indicator incorporating a set of sub-indicators is expressed via a score calculated as the mean of the score determined for the single sub-indicators (Table 2).

Table 2. Examples of indicators and scoring procedures.

Indicator	Type	Measurement	Scale	Score
Ind. potential appropriation of open grounds	Quantitative	Presence (binary evaluation)	Yes	4
			No	0
		Size (L_1 and/or L_2)	70 m $\geq L_1$ and $L_2 >$ 25 m	4
			100 m $\geq L_1$ and/or $L_2 >$ 70 m	3
			25 m $\geq L_1$ and $L_2 >$ 15 m	3
			15 m $\geq L_1$ and/or $L_2 >$ 6 m	2
			(L_1 and/or L_2) > 100 m	1
			6 m $\geq$ (L_1 and/or L_2)	0
		Quantity	=1 region of open ground	0
			=2 regions of open ground	2
			>2 regions of open ground	4
		Variety in terms of surface, size, geometry	Yes	0
			No	4
		Average	$(S_1 + S_2 + S_3 + S_4)/4$	
Density of retail activities and services	Quantitative	r/100 m; r = number of retail activities	20 ≥ r/100 m > 16	4
			16 ≥ r/100 m > 12	3
			12 ≥ r/100 m > 8	2
			8 ≥ r/100 m > 4	1
			4 ≥ r/100 m ≥ 0	0
Proximity of collective transport nodes	Quantitative	Distance from the nearest transport node	Dct ≤ 100 m	4
			200 m ≥ Dct > 100 m	3
			300 m ≥ Dct > 200 m	2
			400 m ≥ Dct > 300 m	1
			Dct > 400 m	0
Degree of children's appropriation of spaces	Qualitative	Degree of adults control of POSs	Restrictions on access	0
			Constraints on uses	1
			Competition for space	2
			Time/coupling constraints	3
			Manicured spaces	3
			No constraints on activities	4

The indicators representative of specific micro- and macro-scale environmental factors are, thus, aggregated into a global index of usefulness of individual natural settings, formalized as a score ranging from 0–120. This score is then converted to a value ranging from 0–1 by dividing the actual level of meaningful usefulness n by the potential level (120) (Table 3).

Moreover, four additional synthetic indexes are determined: (i) a functional affordances index, I_f; (ii) an emotional affordances index, I_e; (iii) a social affordances index, I_s; and (iv) an access conditions index, I_a. Hence, each index aggregates the single indicators related to a specific category of affordances. The synthetic category indexes are expressed by a value ranging from 0–1, defined as the ratio of the sum of the scores attributed to single indicators (S_{II}) compared to a potential score (P_j) (Table 2).

Table 3. Examples of determination of the index of usefulness of individual natural settings (I_{UIS}), of the continuity factor f, and of the synthetic index of usefulness of urban blue–green infrastructure (BGI) in a specific area (I_{SGI}).

Index	Measurement	Scale	Rating
I_{UIS}	$(\Sigma I_{Ci})/P$ (P = 120)	$1.00 \geq I_{UIS} \geq 0.85$	Optimal
		$0.84 \geq I_{UIS} \geq 0.75$	Good
		$0.74 \geq I_{UIS} \geq 0.65$	Fair
		$0.64 \geq I_{UIS} \geq 0.55$	Adequate
		$0.54 \geq I_{UIS} \geq 0.35$	Inadequate
		$0.34 \geq I_{UIS} \geq 0$	Poor
I_f; I_e; I_s; I_a	$(\Sigma S_{II})/Pj$, Pf = 48, P_e = 24, P_s = 24, P_a = 24	$1.00 \geq I_{UIS} \geq 0.85$	Optimal
		$0.84 \geq I_{UIS} \geq 0.75$	Good
		$0.74 \geq I_{UIS} \geq 0.65$	Fair
		$0.64 \geq I_{UIS} \geq 0.55$	Adequate
		$0.54 \geq I_{UIS} \geq 0.35$	Inadequate
		$0.34 \geq I_{UIS} \geq 0$	Poor
f	$A_{pce}/(\Sigma A_i)$	$1.00 \geq I_{UIS} \geq 0.80$	Optimal
		$0.79 \geq I_{UIS} \geq 0.60$	Fair
		$0.59 \geq I_{UIS} \geq 0.45$	Inadequate
		$0.44 \geq I_{UIS} \geq 0$	Poor
I_{SGI}	$[(\Sigma I_{UISi} \times A_i)/(\Sigma A_i)] \times f$	$1.00 \geq I_{UIS} \geq 0.85$	Optimal
		$0.84 \geq I_{UIS} \geq 0.75$	Good
		$0.74 \geq I_{UIS} \geq 0.65$	Fair
		$0.64 \geq I_{UIS} \geq 0.55$	Adequate
		$0.54 \geq I_{UIS} \geq 0.35$	Inadequate
		$0.34 \geq I_{UIS} \geq 0$	Poor

Note: P is the Potential Performance of a Natural Settings, Represented by a Score of 120; Apce = Surface Area of the Primary Connected Element; Ai = Surface Area of the *i*-th BGI Component.

The layout of the audit tool is expressed via a matrix that organizes the single indicators and sub-indicators into the four dimensions (functional, social, emotional, and independent access opportunities) and formalizes the relationships among dimensions, functional categories or affordances, environmental features, and indicators. Moreover, for each indicator and sub-indicator, there are specified scales, levels of performance, bands of values, definitions of conditions related to each scale level for qualitative indicators, and, if required, queries for measuring the value of quantitative indicators via GIS platform data processing tools. The subsequent stage consists of the determination of a continuity factor f that measures whether the urban blue–green infrastructure components in a specific area are dispersed or connected in a continuous spatial structure. The continuity factor, thus, measures the ratio of the surface area of the urban BGI primary component compared to the total surface area of natural settings constituting the components of the urban BGI in a determined area. The primary component is defined as the segment of contiguously connected urban BGI components that comprises the largest surface area of natural settings. Multiplying the weighted average index of usefulness of urban BGI components by the continuity factor f, both considered for a specific area, determines the value of the synthetic index of usefulness of urban BGI in a specific area (I_{SGI}). This index, thus, formalizes the extent to which the urban blue–green infrastructure constitutes a continuous spatial structure that reinforces children's rights to the city by enabling their agency and independent mobility, and that promotes children's meaningful nature experiences and transactions with the environment. Moreover, the I_{SGI} index, as a synthetic indicator representative of the conditions of continuity, inclusivity, and friendliness of the network of natural settings in a specific area, enables the comparison between distinct areas. The audit tool incorporates a procedure structured on eight stages: (i) selection and characterization of the case study; (ii) selection of pertinent indicators; (iii) individuation of publicly available datasets; (iv) definition of measurement scales for single indicators as a function of the availability of secondary data and definition of observational protocols for collecting data from on-site direct observations; (v) data collection and evaluation of single indicators; (vi) normalization of measurements and aggregation of partial results into the synthetic index of meaningful usefulness for single urban BGI components; (vii) determination of the continuity factor f; and (viii) measurement of the global index of usefulness of urban BGI in a specific area.

4. Selection of the Case Study

The city of Cagliari was selected as a case study for three reasons: firstly, the tradition in terms of policies aimed at promoting children's mobility and agency (Figure 1); secondly, the recognition of the strategic relevance of the networks of natural and green areas within territorial and urban planning; and thirdly, the density of green and natural areas, superior to the national average. With respect to the second point, both the municipal urban plan [60] and the plan for the historic center [61] recognize the strategic relevance of the network of natural spaces, parks, safeguard areas, and edges as a carrying structure of the urban fabric and as an infrastructure that provides aesthetic, social, ecological, and climatic benefits, as well as fundamental services, in terms of environmental education and quality of life. In particular, the plan for the historic center [61] individuates the creation of an urban historic park as an intervention aimed at reinforcing the distinctive and environmental characteristics of the compact city and public city.

Furthermore, Cagliari is among the greenest provincial capitals and metropolitan city municipalities in Italy. Approximately 61.6% of its surface area constitutes green areas, including preserved natural areas and urban green spaces; this value is superior to the 19.2% national average ratio of green areas compared to municipal surface areas [17]. The OCUS tool was used to compare the meaningful usefulness of urban BGIs within two distinct districts of the urban area. The characterization of the case study was based on a consideration grounded in the literature on children's practices across public spaces; the possibility of children's outdoor independent activities is associated with the availability of areas close to home and not dominated by adults [14,62]. According to Tonucci [63], the proximity of places available for spontaneous practices emerges as a pre-condition for children's experience of autonomy and for the development of competencies and abilities. Consequently, the application of the OCUS tool aims to evaluate the opportunities for children's agency, autonomy, and nature experience by assessing the extent to which natural settings, including parks, gardens, buffer areas, and planted squares, to constitute a capillary and continuous structure of meaningful and stimulating places, which, at the local scale, support children's independent functional, recreational, and social practices.

Figure 1. Cagliari: the red line delimits the boundary of the municipality of Cagliari; the orange field delimits the area of study.

The analysis compared two districts that differ in terms of morphology, building typology, and population density (Figure 2): the historic center, characterized by a population density of 12,013.76 inhabitants per km^2, a dense and compact urban fabric delimited by and intertwined with different forms of public open spaces, including gardens, parks, specialized green areas, and fringes that are individuated as the components of a continuous structure of natural spaces, the historic urban park, and environments instrumental to the reconstruction of an inclusive public city; and the Bonaria district, a modern residential district with a population density equal to 5022.72 inhabitants per km^2, structured by a regular grid of secondary and local streets serving detached houses and apartment buildings. The result of the application of the OCUS tool and the comparison of the selected districts in terms of usefulness and continuity of the local urban blue–green infrastructure are discussed in the Section 5.

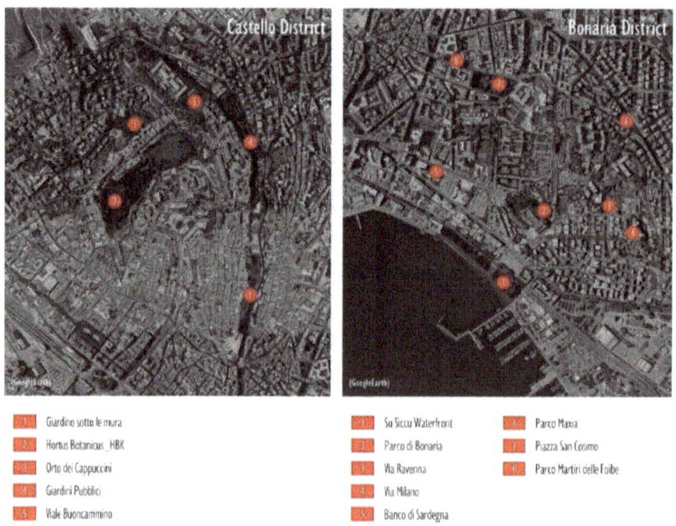

Figure 2. Representation of urban blue–green infrastructure (BGI) components explored during the study.

5. Results

Results obtained via the application of the audit tool revealed that the meaningful usefulness of the urban BGIs identified in the areas of study was marginal for the district of Castello (I_{SGI} equal to 0.34) and for the Bonaria district (I_{SGI} equal to 0.30).

The weighted average index of practicability of single spaces was equal to 0.70 in the Castello district and 0.62 in the Bonaria district (see Figure 3 and Table 4). These values were different from the results determined for the indexes of meaningful usefulness of single natural settings (I_{UIS}). The levels of usefulness for the BGI components within the Castello district ranged from inadequate (I_{UIS} equal to 0.52 for the fringe area in Viale Buoncammino) to fair (I_{UIS} equal to 0.72, 0.73, and 0.74 for Giardino Sotto le Mura, Giardini Pubblici, and Orto dei Cappuccini, respectively) and good (I_{UIS} equal to 0.76 for the vast area including the Roman Amphitheatre and the Hortus Botanicus Karlitanus); conversely, the values of the I_{UIS} for the BGI components within the Bonaria district ranged from to 0.50 for the Garden in via Milano (indicative of an inadequate level of usefulness) to 0.69 for Piazza San Cosimo, 0.65 for Parco Martiri delle Foibe and for the garden area along via Ravenna, and 0.64 for Parco Maxia (indicative of a fair level of usefulness). Nevertheless, the other individuated available BGI components revealed adequate levels of usefulness, corresponding to values of the I_{UIS} ranging from 0.60 for Parco di Bonaria to 0.62 for the Su Siccu waterfront.

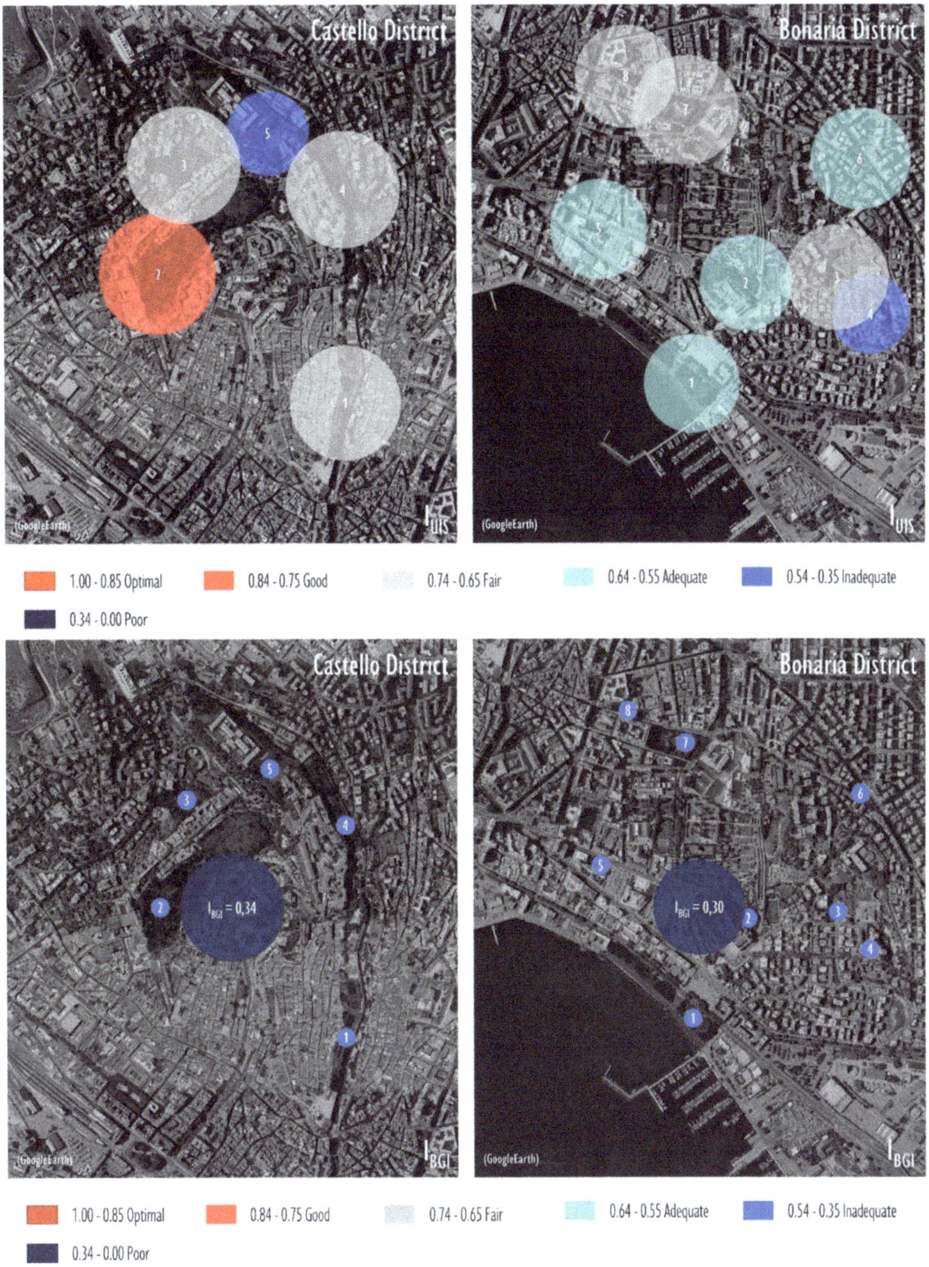

Figure 3. Distribution of values of the index of usefulness of individual settings (I_{UIS}) and the synthetic index of usefulness of BGI in a specific area (I_{SGI}): varying levels of meaningful usefulness of the selected BGI components.

Table 4. Values of the I_{SGI} and of I_{UIS} indexes for the selected area of study. HBK—Hortus Botanicus Karlitanus.

Urban BGI Components	Indexes of Meaningful Usefulness		
	I_{UIS}	Continuity Factor f	I_{SGI}
Giardino Sotto le Mura (1)	0.72		
HBK (2)	0.76		
Orto dei Cappuccini (3)	0.74		
Giardini Pubblici (4)	0.74		
Viale Buoncammino (5)	0.53		
Castello district	0.71	0.48	0.34
Su Siccu (1)	0.62		
Parco Bonaria (2)	0.60		
Via Ravenna (3)	0.66		
Via Milano (4)	0.50		
Banco di Sardegna (5)	0.63		
Parco Maxia (6)	0.64		
Piazza San Cosimo (7)	0.69		
Parco Martiri Foibe (8)	0.65		
Bonaria district	0.62	0.48	0.30

Further insights could be derived from the review of partial indexes (I_f, I_e, I_s, I_a), which aggregated the indicators related to the specific dimensions of functional affordance, emotional affordance, social affordance, and accessibility (Table 5).

Table 5. Values of the I_f, I_e, I_s, and I_a indicators.

Urban BGI Components	Indexes of Meaningful Usefulness			
	I_f (Functional)	I_e (Emotional)	I_s (Social)	I_a (Access)
Giardino Sotto le Mura (1)	0.67	0.78	0.81	0.64
HBK (2)	0.86	0.83	0.67	0.60
Orto dei Cappuccini (3)	0.77	0.83	0.81	0.52
Giardini Pubblici (4)	0.79	0.89	0.81	0.43
Viale Buoncammino (5)	0.71	0.49	0.25	0.50
Castello district	0.79	0.76	0.63	0.55
Su Siccu (1)	0.63	0.42	0.83	0.59
Parco Bonaria (2)	0.69	0.78	0.27	0.56
Via Ravenna (3)	0.61	0.81	0.67	0.59
Via Milano (4)	0.52	0.64	0.23	0.59
Banco di Sardegna (5)	0.40	0.88	0.85	0.60
Parco Maxia (6)	0.42	0.93	0.83	0.61
Piazza San Cosimo (7)	0.72	0.82	0.69	0.52
Parco Martiri Foibe (8)	0.62	0.77	0.69	0.54
Bonaria district	0.63	0.61	0.66	0.57

Natural settings within the Castello district revealed fair to optimal levels of availability of functional affordances (I_f ranging from to 0.67 for Giardino Sotto le Mura, 0.79 for Giardini Pubblici, and 0.86 for the Hortus Botanicus), and good to optimal levels of availability of emotional and social affordances (see Figures 4–6). In particular, a value of the index of availability of emotional affordances superior to 0.80 was observed for Giardini Pubblici and the Hortus Botanicus. Furthermore, the index of availability of social affordances was equal to 0.81 for Giardini Pubblici, Giardini Sotto le Mura, and Orto dei Cappuccini.

A significant exception was represented by the fringe area along Viale Buoncammino, connoted by an inadequate level of availability of positive emotional affordances (I_e = 0.49) and by a poor level of availability of social opportunities (I_s = 0.25). Conditions of accessibility were likewise observed to be inadequate for Giardini Pubblici, Orto dei Cappuccini, and the fringe area along Viale Buoncammino (I_a equal to 0.43, 0.52, and 0.5, respectively) and adequate (I_a equal to 0.60 and 0.64) for Hortus Botanicus and Giardino Sotto le Mura. Finally, conditions of accessibility were inadequate for Giardini Pubblici, the fringe area along Viale Buoncammino, and Orto dei Cappuccini, while they were adequate for the remaining settings (Figure 7).

The values of the I_f, I_e, I_s, and I_a indexes for the districts were determined as the weighted average of the indexes calculated for each BGI component.

Contrastingly, natural settings within the Bonaria district revealed the lesser availability of functional affordances (Figure 4). The most significant space in terms of functional affordances was Piazza San Cosimo (I_f = 0.72), while three settings (Via Milano garden, Parco Maxia, and Banco di Sardegna garden) revealed marginal levels of availability of functional affordance (I_f equal to 0.52, 0.42, and 0.40, respectively).

The potential to actualize emotional affordances (Figure 5) was optimal in both Parco Maxia (I_e = 0.93) and underneath the Banco di Sardegna Building (I_e = 0.88), and good in Piazza San Cosimo (I_e = 0.82), Parco di Bonaria (I_e = 0.78), in the public gardens along via Ravenna (I_e = 0.81), and in Parco Martiri delle Foibe (I_e = 0.77), while it was marginal in the Su Siccu waterfront (I_e equal to 0.42).

Moreover, the availability of social affordances (Figure 6) was particularly variable. A scarce potential to actualize social affordances was observed in Parco di Bonaria and Via Milano gardens, a good potential was estimated in the Su Siccu waterfront and Parco Maxia (I_s = 0.83), and an optimal one was found in the Banco di Sardegna gardens (I_s = 0.85).

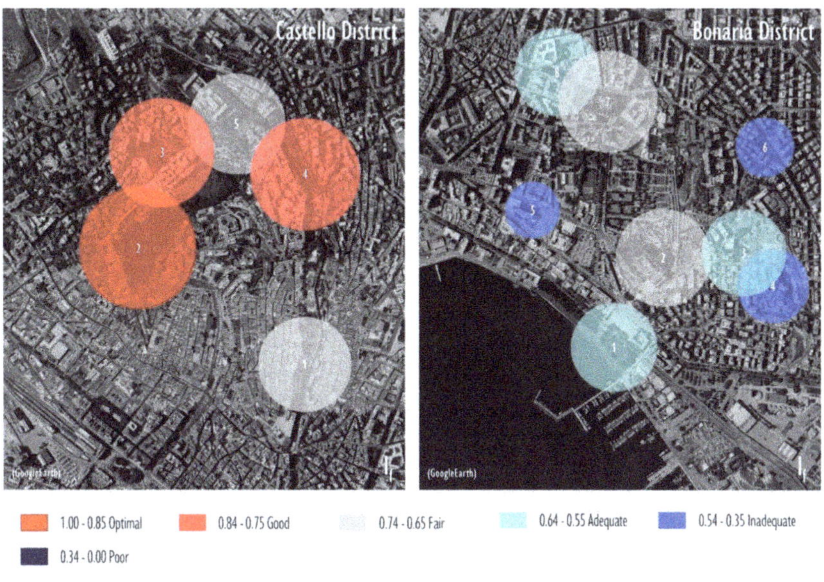

Figure 4. Distribution of values of the I_f functional indicator among urban BGI components across the Castello and the Bonaria districts.

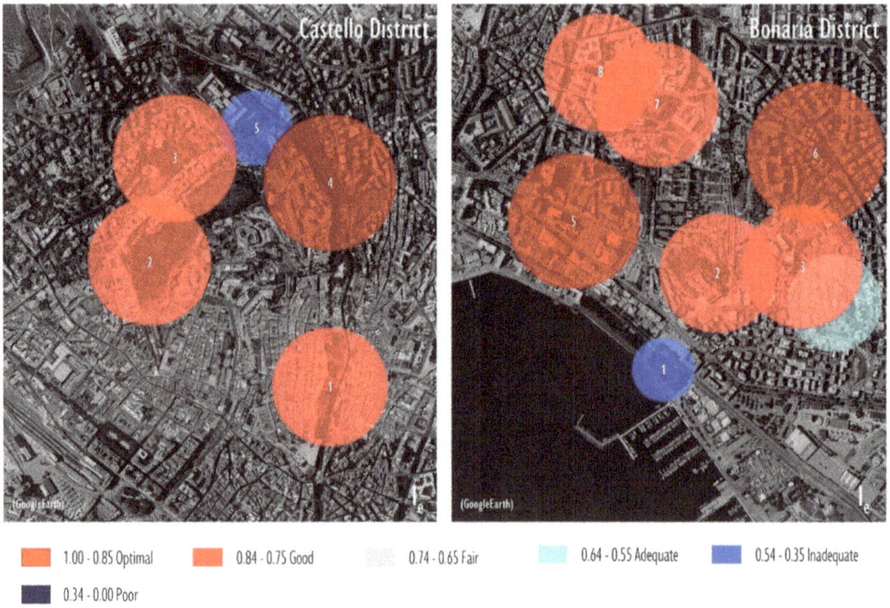

Figure 5. Distribution of values of the I_e emotional indicator among urban BGI components across the Castello and the Bonaria districts.

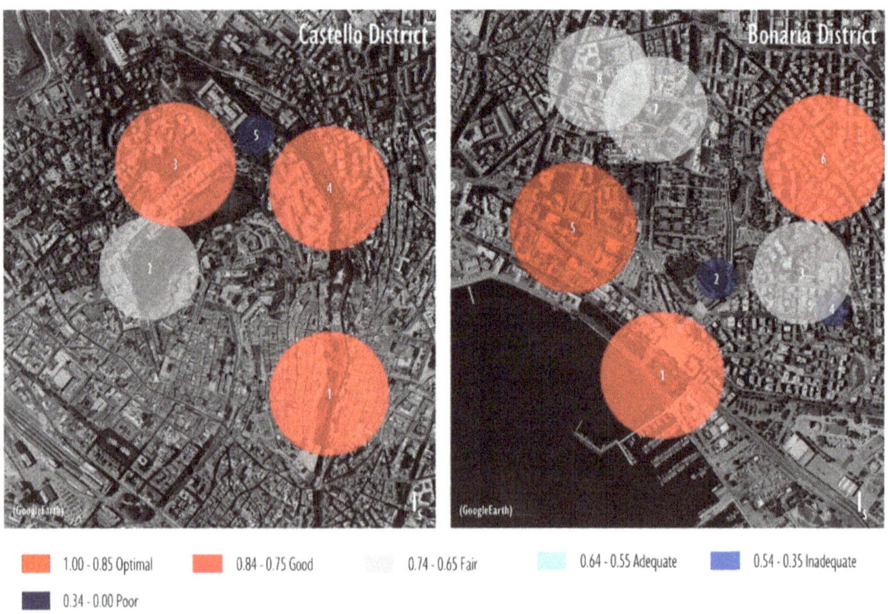

Figure 6. Distribution of values of the I_s social indicator among urban BGI components across the Castello and the Bonaria districts.

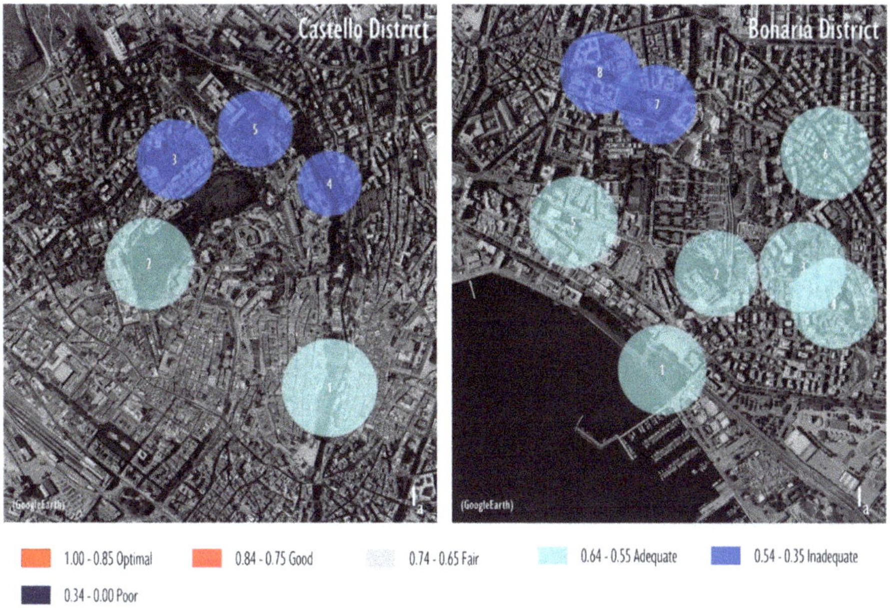

Figure 7. Distribution of values of the I_a access indicator among urban BGI components across the Castello and the Bonaria districts.

Finally, the condition of accessibility (Figure 7) was assessed as inadequate only for Parco Martiri delle Foibe (I_a equal to 0.50) and adequate for the remaining considered settings. In conclusion, the continuity factor f of the system of natural spaces was equal to 0.48 both in the Castello district and in the Bonaria district.

A comprehensive discussion of these results and a synthetic description of the relevant criticalities observed and related to the modest level of practicability measured are presented in Section 6.

6. Discussion

The results described in the previous section revealed a set of common criticalities, both at the scale of the specific setting and at the scale of the system of urban BGIs. In particular, the continuity factor revealed the scarce arteriality of the network of natural settings, which is fragmented and configured as a patchwork of dispersed natural and semi-natural settings.

This characteristic affects both the ecological and the social dimension of ecosystem services; it reduces the potential of the urban BGI to constitute a reserve for biodiversity and its potential to constitute a continuous structure of walkable and meaningful spaces, which are conducive to children's outdoor practices and independent mobility across the urban space.

Moreover, the analysis of the case study revealed that, in the case of the Bonaria district, the regeneration of residual spaces and specialized areas (the monumental cemetery, the buffer zone around it, the surface area of a disused industrial facility, the surface of an abandoned military facility) could lead to the configuration of an urban blue/green infrastructure with a continuity factor f equal to 0.79 (see Figure 8).

Figure 8. Possible reconfiguration of the BGIs across the Bonaria district.

This value indicates the potential to structure a contiguously connected subnetwork comprising 79% of the surface area of public spaces constituting urban BGI components in the area of study.

Focusing on site-specific factors, a relevant issue underlined by the evaluation of functional properties of urban green areas is the constraint on children's social and recreational practices determined by over-designed and by neatly maintained—or manicured—surfaces, reflecting adults' control and possession of public spaces. Vilanueva et al. [64] observed that natural play environments, incorporating natural elements and vegetation, appear more conducive to children's cognitive and physical development rather than physical manmade play areas.

In particular, the modest size and the uniformity in terms of surface textures and materials of regions of open ground limit opportunities for group practices and physical activities (riding a bike, running, playing football). Moreover, the morphological uniformity and regularity of surfaces reduces the variety of spatial conditions, consequently reducing the opportunity to explore, experiment, and appropriate spaces, and to combine intense physical activities with balancing acts.

A further related issue is the limited quantity and variety of loose objects and moldable materials which can constitute enabling materials for creative and imaginative games, including construction, manipulation, drawing, creation of patterns, and make-believe play. In her seminal work, Jacobs [56] observed that diversity at the ground level, determined by variations in the morphology of surfaces, concentration and dispersion of vegetation, and openings toward focal points, increases the availability of a space to different uses, rhythms, and purposes, and reinforces its complexity and attractiveness. Inversely, if a space, such as a billboard, can be embraced and understood at a glance, with its parts appearing to be homogeneous, then it barely supports diversity of uses or frequency of visits.

As for emotional affordances, a relevant issue is represented by the contextual conditions promoting the natural control of the space and determining the perceived safety of the public space. These conditions are positively related to children's outdoor activities, independent mobility, and to the perception of a space as an enabling place [13,64]. In particular, the partial or non-visibility of buildings and the absence of activities with extended service hours affect the natural surveillance of Giardini Sotto le Mura and Viale Buoncammino in the Castello district, as well as of the Su Siccu waterfront and

Parco di Bonaria in the Bonaria district. A related issue is the presence of signs of neglect and physical decay, including the quantity and content of graffiti, traces of drugs or alcohol consumption, and the condition of surfaces, seating facilities, illumination, and furniture.

These factors are individuated as a proxy of the occurring anti-social practices within the public space and are negatively related to levels of children's independent mobility and to perceptions about the child-friendliness of the urban space; furthermore, they are associated with parental restrictions on children's independent activities [13,65–67]. The analysis reveals the critical conditions of the fringe area along Viale Buoncammino, Via Milano garden, Piazza San Cosimo, and Parco Martiri delle Foibe. Thus, maintenance, reinforcement of perceptual relationships among natural settings, the built environment, and the collocation of specialized activities and functions could increase the natural surveillance, as well as the vitality, liveliness, and attractiveness of natural spaces. Another relevant aspect affecting emotional affordances is the degree of illumination of natural settings; this factor is related to perceptions about safety, comfort, and usability of spaces during night hours. In particular, the analysis revealed good to optimal condition of lampposts, underlined by the broken window indicator, and, more significantly, the non-uniform distribution of lampposts in larger natural settings. In fact, illumination is concentrated along paths, thus limiting the possibility for children's exploration, spatial appropriation, and structured group activities to daylight hours.

A further relevant issue related to social affordances is adults' control of spaces, resulting in lesser sense of privacy and territoriality, which affects children's spatial appropriation and the attribution of conceptional and use values to spaces [13,41,68]. The application of the audit tool revealed different forms of adult interference, including manicuring of spaces, time constraints, competition for spaces among adults and children, and practices and restrictions on children's outdoor activities. Moreover, the tokenistic and marginal involvement of children in the planning, design, and maintenance of green areas reinforces the adult-centered approach to the construction of the public space, hindering children's right to participation and reducing their influence on the environment. Finally, the most significant issues are related to the conditions for independent access to urban BGI components. In particular, the analysis revealed three criticalities: the barrier effect, which is the fragmentation of the public space determined by the pervasiveness of surfaces reserved for vehicular mobility and by the discrepancy between formal intersections and total road crossings, including informal crossings; the tokenistic integration between the system of urban BGIs and the network of bicycle facilities; and the contiguity of BGIs components to pedestrian facilities that are inadequate, in terms of separation from spaces for vehicular mobility, functionality of surfaces, and geometric characteristics, to support children's independent mobility, recreational activities, and assimilation. The OCUS tool, thus, underlined an organization of the public space oriented by political and economic interests that do not respond to, and instead marginalize, children's needs. Children's interests and their right to participate in the city's life are considered peripheral elements in the planning and design process. The public space formalizes the segregation of children into controlled environments and their conceptualization as "incomplete individuals" [65], thus reinforcing their psychological distance from the adults. A general question concerns the fragmented and tokenistic approach to the construction of a child-friendly city. This appears to be focused on specific issues (such as active travel to school) and implemented through interventions that affect the social and health planning dimension [41]. On the contrary, an effective approach should affect the structural characters of the built environment. In this regard, a significant opportunity is encompassed in the construction of a continuous trans-scalar network of urban BGIs, incorporating different functional, social, and emotional affordances for children and connecting to the networks of public spaces, pedestrian and bicycle mobility, collective transport, and communication infrastructures. This strategy actualizes the shift to a citizenship approach that acknowledges children's right to the city and broadens the concept of meaningful place to the entire urban realm.

This approach would integrate, as central issues of the planning process, the effects of independent mobility and spontaneous exploration of everyday spaces on children's physical wellbeing, acquisition of spatial and environmental skills, emotion regulation, intellectual and creative development, and, ultimately, the construction of their social and individual identity.

7. Conclusions

This research underlines a perspective in urban planning founded on the intersection among research on ecosystem services, blue–green infrastructures, and child-friendly cities. In particular, this study addresses a fundamental yet neglected issue: the significance of urban BGI components as enabling and meaningful places supporting children's spontaneous outdoor activities and affecting their wellbeing and integral development. This issue emerges as a central aspect of the social dimension of ecosystem service. At the same time, it emphasizes the need for a radical shift in policies and strategies aimed at constructing an inclusive child-friendly city. In addition to actions focused on the cultural, social, and health planning dimension, there emerges the necessity for strategies that account for children's needs and interests in the planning, design, and management of the different networks that comprise the contemporary city.

Furthermore, this research underlines the centrality of the concept of affordance for understanding and describing the intentional transactions among children and public open spaces. The relevance of this concept, derived from research in the field of environmental psychology, is still overlooked in the field of urban planning. This concept is operationalized in terms of indicators for the assessment of the meaningful usefulness of BGI components from children's perspectives. In this respect, the analysis of the case study revealed that the OCUS tool and the indexes of usefulness of individual natural settings and of BGIs can support different stages of the planning process: (i) the individuation of structural criticalities of the systems of BGIs, expressed by the continuity factor f; (ii) individuation of criticalities related to functional, social, emotional, and access conditions of individual spaces; (iii) evaluation of alternative scenarios of urban regeneration at different scales, in terms of their impact on children's opportunities to independently and meaningfully engage with natural settings; (iv) comparison among different parts of urban areas; and (v) monitoring of interventions of regeneration of open spaces integrated into urban BGIs, via comparison of levels of usefulness over time. Moreover, the proposed methodological framework and its theoretical premises, based on the concepts of affordance and of capability, can be adapted in order to assess the inclusivity and meaningful usefulness of natural settings and of other elements of the public space from the point of view of other categories of vulnerable users, particularly the elderly.

Yet, the research revealed two limitations. Primarily, it underlined the need to determine the relative importance of environmental features and of pertinent indicators. Environmental features and indicators need to be weighted according to contextual factors and children's individual characteristics. Different studies [64,69–71] revealed that children's patterns of activities across public space are influenced by cultural constructs, parents' socio-economic status, individual abilities and purposes, age, and gender. Secondly, the research revealed the need to validate results and to assess the congruency of results with children's actual patterns of activities and perception of the public space. Moreover, Spelke's research underlined the significance of geometrical properties of spaces and of distinctive elements in supporting re-orientation and individuation of locations. Finally, a further consideration concerns the possibility of comparing natural settings differing in terms of scale, morphology, and function. This results from the conceptualization of the index of meaningful usefulness of a single natural setting I_{UIS}, as an index of performance, determined in relation to a potential level of performance. Consequently, the future development of the research will focus on four aspects: (i) determining procedures, founded on the Delphi method, for supporting the participation of experts and stakeholders in the process of establishing the relative importance of environmental correlates of meaningful usefulness of natural settings; (ii) defining procedures based on surveys or direct observations, structured according to the saturation principle and the phenomenological approach, for the validation of results; (iii) reformulating the indicator imageability of the public open spaces to assess the extent to which the geometry of a setting and the presence and organization of its distinctive elements enable children's comprehension and understanding of the surrounding space; and (iv) increasing the trans-scalarity of the OCUS tool by conceptualizing quantitative indicators related to specific environmental features as indicators of performance or by structuring a taxonomy of BGI components based on morphology, function, and scale and establishing a potential level of performance specific for each category of natural setting. Consequently, the OCUS tool will be increasingly configured as a flexible, modifiable, protocol, whose layout, indicators, and coefficients of relative importance are adapted to the specificity of contextual factors and characteristics of a particular category of users. Thus, consistent with the principle that, with respect to methodological framework for evaluating the public spaces, one size does not fit all, the future development of this research will be aimed at reinforcing the adaptability of the OCUS tool to different contexts and its integrability into the planning process, particularly within the smart city paradigm.

Author Contributions: This paper is the result of the joint work of the authors. In particular, "Review of Methodological approaches", "Methodology", "Selection of the Case Study", "Results", and "Discussion" were written jointly by the authors. C.G. wrote "Introduction" and "Conclusions". A.A. wrote "Urban Blue–Green Infrastructures" and "Concepts for Assessing Urban BGI Components".

Funding: This study was supported by the MIUR (Ministry of Education, Universities, and Research (Italy)) through the project "Governing the smart city: A governance-centered approach to smart urbanism—GHOST" (Project code: RBSI14FDPF; CUP Code: F22I15000070008), financed with the SIR (Scientific Independence of Young Researchers) program. We authorize the MIUR to reproduce and distribute reprints for governmental purposes, notwithstanding any copyright notations thereon. Any opinions, findings, and conclusions or recommendations expressed in this material are those of the authors, and do not necessarily reflect the views of the MIUR. This study was also supported by the project "Healthy Cities and Smart Territories", founded by the Foundation of Sardinia and Autonomous Region of Sardinia (Fondazione di Sardegna—Convenzione triennale tra la Fondazione di Sardegna e gli Atenei Sardi Regione Sardegna 2016).

Conflicts of Interest: The authors declare no conflicts of interest.

References

1. Martinotti, G. *Sei Lezioni Sulla Città*; Feltrinelli Editore: Milan, Italy, 2017; ISBN 978-880710530-2.
2. Glaeser, E. *Triumph of the City*; Pan: Basingstoke, UK, 2011; ISBN 978-110147567-6.
3. Armour, T.; Armour, S.; Hargrave, J.; Revell, T. *Cities Alive: Rethinking Green Infrastructure*; Arup: London, UK, 2014.
4. Caragliu, A.; Del Bo, C.; Nijkamp, P. Smart cities in Europe. *J. Urban Technol.* **2011**, *18*, 65–82. [CrossRef]
5. Min, B.; Lee, J. Children's neighborhood place as a psychological and behavioral domain. *J. Environ. Psychol.* **2006**, *26*, 51–71. [CrossRef]
6. Sen, A. Capability and Well-being. In *The Quality of Life*; Nussbaum, M., Sen, A., Eds.; Clarendon Press: Oxford, UK, 1993; pp. 30–53.
7. Gibson, J.J. *The Ecological Approach to Visual Perception*; Lawrence Erlbaum: Hillsdale, NJ, USA, 1979.
8. Barker, R.G.; Wright, H.F. *Midwest and Its Children: The Psychological Ecology of an American Town*; Row, Peterson: Evanston, IL, USA, 1954.
9. Heft, H. Affordances of children's environments: A functional approach to environmental description. *Child. Environ. Q.* **1988**, *5*, 29–37.
10. UNICEF. Children in an Increasingly Urban World. 2012. Available online: https://www.unicef.org/sowc2012/pdfs/SOWC-2012-Chapter-1-Children-in-an-increasingly-urban-world.pdf (accessed on 31 March 2019).
11. Goltsman, S.; Kelly, L.; McKay, S.; Algara, P.; Wight, L. Raising "free range kids": Creating neighborhood parks that promote environmental stewardship. *J. Green Build.* **2009**, *4*, 90–106. [CrossRef]
12. Lin, E.Y.; Witten, K.; Oliver, M.; Carroll, P.; Asiasiga, L.; Badland, H.; Parker, K. Social and built-environment factors related to children's independent mobility: The importance of neighbourhood cohesion and connectedness. *Health Place* **2017**, *46*, 107–113. [CrossRef]
13. Witten, K.; Kearns, R.; Carroll, P.; Asiasiga, L. Children's Everyday Encounters and Affective Relations with Place: Experiences of Hyperdiversity in Auckland Neighbourhoods. 2017. Available online: https://www.tandfonline.com/doi/abs/10.1080/14649365.2017.1347700 (accessed on 10 August 2019).
14. Chawla, L. Benefits of nature contact for children. *J. Plan. Lit.* **2015**, *30*, 433–452. [CrossRef]

15. Giusti, M.; Svane, U.; Raymond, C.M.; Beery, T.H. A Framework to Assess Where and How Children Connect to Nature. *Front. Psychol.* **2018**, *8*, 2283. [CrossRef]
16. Abson, D.J.; Fischer, J.; Leventon, J.; Newig, J.; Schomerus, T.; Vilsmaier, U.; von Wehrden, H.; Abernethy, P.; Ives, C.D.; Jager, N.D.; et al. Leverage points for sustainability transformation. *Ambio* **2017**, *46*, 30–39. [CrossRef]
17. ISTAT. Ambiente Urbano. 2018. Available online: https://www.istat.it/it/archivio/225505 (accessed on 31 March 2019).
18. Benedict, M.; McMahon, E. *Green Infrastructure: Linking Landscapes and Communities*; Island Press: Washington, DC, USA, 2006.
19. Cannas, I.; Lai, S.; Leone, F.; Zoppi, C. Green Infrastructure and Ecological Corridors: A Regional Study Concerning Sardinia. *Sustainability* **2018**, *10*, 1265. [CrossRef]
20. Tjallingii, S. Planning with water and traffic networks—Carrying structures of the urban landscape. In *Flowscape—Designing Infrastructure as Landscape*; Nijhuis, S., Jauslin, D., van Der Hoeven, F., Eds.; Delft University of Technology: Delft, NL, USA, 2015; pp. 57–80, ISBN 978-946186472-7.
21. Boyd, J.; Banzhaf, S. What are ecosystem services? The need for standardized environmental accounting units. *Ecol. Econ.* **2007**, *63*, 616–626. [CrossRef]
22. Millennium Ecosystem Assessment. *Ecosystems and Human Well-Being*; Island Press: Washington, DC, USA, 2005.
23. Daily, G. *Nature's Services: Societal Dependence on Natural Ecosystems*; Island Press: Washington, DC, USA, 1997; ISBN 1-55963-476-6.
24. Landers, D.H.; Nahlik, A.M. Final Ecosystem Goods and Services Classification System (FEGS-CS). Available online: https://gispub4.epa.gov/FEGS/FEGS-CS%20FINAL%20V.2.8a.pdf (accessed on 27 May 2019).
25. Wallace, K.J. Classification of ecosystem services: Problems and solutions. *Biol Conserv.* **2007**, *139*, 235–246. [CrossRef]
26. Binning, C.; Cork, S.; Parry, R.; Shelton, D. *Natural Assets: An Inventory of Ecosystem Goods and Services in the Goulburn Broken Catchment*; Ecosystem Services Project: Canberra, Australia, 2001.
27. Holt, A.R.; Mears, M.; Maltby, L.; Warren, P. Understanding spatial patterns in the production of multiple urban ecosystem services. *Ecosyst. Serv.* **2015**, *16*, 33–46. [CrossRef]
28. Dennis, M.; James, P. Ecosystem services of collectively managed urban gardens: Exploring factors affecting synergies and trade-offs at the site level. *Ecosyst. Serv.* **2017**, *26*, 17–26. [CrossRef]
29. Vogdrup-Schmidt, M.; Strange, N.; Olsen, S.B.; Thorsen, B.J. Trade-off analysis of ecosystem service provision in nature networks. *Ecosyst. Serv.* **2017**, *23*, 165–173. [CrossRef]
30. Aarts, M.J.; Wendel-Vos, W.; van Oers, H.A.; van de Goor, I.A.; Schuit, A.J. Environmental determinants of outdoor play in children: A large-scale cross-sectional study. *Am. J. Prev. Med.* **2010**, *39*, 212–219. [CrossRef]
31. Pyyry, N. Thinking with broken glass: Making pedagogical spaces of enchantment in the city. *Environ. Educ. Res.* **2017**, *23*, 1391–1401. [CrossRef]
32. Nussbaum, M.C. *Creating Capabilities*; Harvard University Press: Cambridge, MA, USA, 2011; ISBN 978-067407235-0.
33. Conrad, E. Human and social dimension of landscape stewardship. In *The Science and Practice of Landscape Stewardship*; Bieling, C., Plieninger, T., Eds.; Cambridge University Press: Cambridge, UK, 2017; pp. 38–53, ISBN 978-110714226-8.
34. Matsuba, M.K.; Pratt, M.W.; Norris, J.E.; Mohle, E.; Alisat, S.; McAdams, D.P. Environmentalism as a context for expressing identity and generativity: Patterns among activists and uninvolved youth and midlife adults. *J. Personal.* **2012**, *80*, 1091–1115. [CrossRef]
35. Kyttä, M.; Oliver, M.; Ikeda, E.; Ahmadi, E.; Omiya, I.; Laatikainen, T. Children as urbanites: Mapping the affordances and behavior settings of urban environments for Finnish and Japanese children. *Child. Geogr.* **2018**, *16*, 319–332. [CrossRef]
36. Kyttä, M. The extent of children's independent mobility and the number of actualized affordances as criteria for child-friendly environments. *J. Environ. Psychol.* **2004**, *24*, 179–198. [CrossRef]
37. Barker, R.G.; Wright, H.F. *One Boy's Day: A Specimen Record of Behavior*; Archon Books: Hamden, CT, USA, 1951; ISBN 978-0208004857.
38. Lerstrup, I.; van den Bosch, C.K. Affordances of outdoor settings for children in preschool: Revisiting Heft's functional taxonomy. *Landsc. Res.* **2017**, *42*, 47–62. [CrossRef]

39. Broberg, A.; Kyttä, M.; Fagerholm, N. Child-friendly urban structures: Bullerby revisited. *J. Environ. Psychol.* **2013**, *35*, 110–120. [CrossRef]
40. Geurs, K.T.; van Wee, B. Accessibility evaluation of land-use and transport strategies: Review and research directions. *J. Transp. Geogr.* **2004**, *12*, 127–140. [CrossRef]
41. Whitzman, C.; Mizrachi, D. Creating child-friendly high-rise environments: Beyond wastelands and glasshouses. *Urban Policy Res.* **2012**, *30*, 233–249. [CrossRef]
42. Kyttä, M. Children in Outdoor Contexts: Affordances and Independent Mobility in the Assessment of Environmental Child Friendliness. Ph.D. Thesis, Helsinki University of Technology, Helsinki, Finland, 2003.
43. Kyttä, A.M.; Broberg, A.K.; Kahila, M.H. Urban environment and children's active lifestyle: SoftGIS revealing children's behavioral patterns and meaningful places. *Am. J. Health Promot.* **2012**, *26*, e137–e148. [CrossRef]
44. Lopes, F.; Cordovil, R.; Neto, C. Independent Mobility and Social Affordances of Places for Urban Neighborhoods: A Youth-Friendly Perspective. *Front. Psychol.* **2018**, *9*, 2198. [CrossRef]
45. Garau, C.; Annunziata, A.; Coni, M. A Methodological Framework for Assessing Practicability of the Urban Space: The Survey on Conditions of Practicable Environments (SCOPE) Procedure Applied in the Case Study of Cagliari (Italy). *Sustainability* **2018**, *10*, 4189. [CrossRef]
46. Mygind, L.; Bentsen, P.; Badland, H.; Edwards, N.; Hooper, P.; Villanueva, K. Public open space desktop auditing tool—Establishing appropriateness for use in Australian regional and urban settings. *Urban For. Urban Green.* **2016**, *20*, 65–70. [CrossRef]
47. Saelens, B.E.; Frank, L.D.; Auffrey, C.; Whitaker, R.C.; Burdette, H.L.; Colabianchi, N. Measuring physical environments of parks and playgrounds: EAPRS instrument development and inter-rater reliability. *J. Phys. Act. Health* **2006**, *3* (Suppl. 1), S190–S207. [CrossRef] [PubMed]
48. Rigolon, A.; Németh, J. A QUality INdex of Parks for Youth (QUINPY): Evaluating urban parks through geographic information systems. *Environ. Plan. B Urban Anal. City Sci.* **2016**, *45*, 275–294. [CrossRef]
49. Moura, F.; Cambra, P.; Gonçalves, A.B. Measuring walkability for distinct pedestrian groups with a participatory assessment method: A case study in Lisbon. *Landsc. Urban Plan.* **2017**, *157*, 282–296. [CrossRef]
50. Taleai, M.; Amiri, E.T. Spatial multi-criteria and multi-scale evaluation of walkability potential at street segment level: A case study of Tehran. *Sustain. Cities Soc.* **2017**, *31*, 37–50. [CrossRef]
51. Garau, C.; Pavan, V.M. Evaluating Urban Quality: Indicators and Assessment Tools for Smart Sustainable Cities. *Sustainability* **2018**, *10*, 575. [CrossRef]
52. Jabbari, M.; Fonseca, F.; Ramos, R. Combining multi-criteria and space syntax analysis to assess a pedestrian net-work: The case of Oporto. *J. Urban Des.* **2018**, *23*, 23–41. [CrossRef]
53. Rosenberg, D.; Ding, D.; Sallis, J.F.; Kerr, J.; Norman, G.J.; Durant, N.; Harris, S.K.; Saelens, B.E. Neighborhood environment walkability scale for youth (NEWS-Y): Reliability and relationship with physical activity. *Prev. Med.* **2009**, *49*, 213–218. [CrossRef]
54. Shaw, B.; Bicket, M.; Elliott, B.; Fagan-Watson, B.; Mocca, E.; Hillman, M. *Children's Independent Mobility: An International Comparison and Recommendations for Action*; Policy Studies Institute: London, UK, 2015.
55. Abis, E.; Garau, C. An assessment of the effectiveness of strategic spatial planning: A study of Sardinian municipalities. *Eur. Plan. Stud.* **2016**, *24*, 139–162. [CrossRef]
56. Jacobs, J. *The Death and Life of Great American Cities*; Random House: New York, NY, USA, 1961; ISBN 978-067964433-0.
57. Gehl, J. *Cities for People*; Island Press: Washington, DC, USA, 2013; ISBN 9781597265737.
58. Lee, S.A.; Shusterman, A.; Spelke, E.S. Reorientation and landmark-guided search by young children: Evidence for two systems. *Psychol. Sci.* **2006**, *17*, 577–582. [CrossRef]
59. Blečić, I.; Cecchini, A.; Congiu, T.; Fancello, G.; Trunfio, G.A. Evaluating walkability: A capability-wise planning and design support system. *Int. J. Geogr. Inf. Sci.* **2015**, *29*, 1350–1374. [CrossRef]
60. Comune di Cagliari. *Piano Urbanistico Comunale—Indirizzi Programmatici Operativi*; Comune di Cagliari: Cagliari, Italy, 2017. (In Italian)
61. Comune di Cagliari. *Piano Particolareggiato Per il Centro storico, Progetti Strategici Per lo Spazio Pubblico*; Comune di Cagliari: Cagliari, Italy, 2015. (In Italian)
62. Hart, R. *Children's Experience of Place*; Irvington: New York, NY, USA, 1979.
63. Tonucci, F. Citizen child: Play as welfare parameter for urban life. *Topoi* **2015**, *24*, 183–195. [CrossRef]

64. Villanueva, K.; Giles-Corti, B.; Bulsara, M.; Trapp, G.; Timperio, A.; McCormack, G.; Van Niel, K. Does the walkability of neighbourhoods affect children's independent mobility, independent of parental, socio-cultural and individual factors? *Child. Geogr.* **2014**, *12*, 393–411. [CrossRef]
65. Davis, A.; Jones, L.J. Children in the urban environment: An issue for the new public health agenda. *Health Place* **1996**, *2*, 107–113. [CrossRef]
66. Nordström, M. Children's views on child-friendly environments in different geographical, cultural and social neighbourhoods. *Urban Stud.* **2010**, *47*, 514–528. [CrossRef]
67. Chaudhury, M.; Hinckson, E.; Badland, H.; Oliver, M. Children's independence and affordances experienced in the context of public open spaces: A study of diverse inner-city and suburban neighbourhoods in Auckland, New Zealand. *Child. Geogr.* **2017**, *17*, 49–63. [CrossRef]
68. Furneaux, A.; Manaugh, K. Eyes on the alley: children's appropriation of alley space in Riverdale, Toronto. *Child. Geogr.* **2018**, *17*, 204–216. [CrossRef]
69. Galvez, M.P.; McGovern, K.; Knuff, C.; Resnick, S.; Brenner, B.; Teitelbaum, S.L.; Wolff, M.S. Associations between neighborhood resources and physical activity in inner-city minority children. *Acad. Pediatr.* **2013**, *13*, 20–26. [CrossRef]
70. Gillespie, J. Being and becoming: Writing children into planning theory. *Plan. Theory* **2013**, *12*, 64–80. [CrossRef]
71. Villanueva, K.; Badland, H.; Kvalsvig, A.; O'Connor, M.; Christian, H.; Woolcock, G.; Giles-Corti, B.; Goldfeld, S. Can the neighborhood built environment make a difference in children's development? Building the research agenda to create evidence for place-based children's policy. *Acad. Pediatr.* **2016**, *16*, 10–19. [CrossRef]

sustainability

MDPI

Article

Assessment of Municipal Masterplans Aimed at Identifying and Fostering Green Infrastructure: A Study Concerning Three Towns of the Metropolitan Area of Cagliari, Italy

Sabrina Lai [1], **Federica Leone** [2,*] **and Corrado Zoppi** [2]

[1] Regional Administration of Sardinia, Department for the Environment, via Roma 80, 09123 Cagliari, Italy; slai@regione.sardegna.it
[2] Department of Civil and Environmental Engineering and Architecture, University of Cagliari, Via Marengo 2, 09123 Cagliari, Italy; zoppi@unica.it
* Correspondence: federicaleone@unica.it; Tel.: +39-070-6755213

Received: 29 January 2019; Accepted: 6 March 2019; Published: 10 March 2019

Abstract: Building upon a recent piece of research that maps a regional green infrastructure (RGI) in relation to four components (natural value, conservation value, landscape value, and recreational value), this study aims at identifying planning policies that can foster the enhancement of the RGI by increasing one or more of its components at the sub-regional scale. To this end, the RGI suitability map is overlaid with the planning schemes of the municipal masterplans (MMPs) of three towns belonging to the Metropolitan City of Cagliari (Italy), and multiple linear regressions are performed. The outcomes of the study imply that the eligibility of a land parcel to be part of the RGI depends on several factors related to planning policies entailed by the zoning schemes of the MMPs, such as presence and spreading of conservation and safeguard areas within urban fabrics, improved accessibility of historic and natural landmarks, planned use of nature-based solutions within the regulating codes of MMPs, improvement of habitat quality in the spatial context of rural areas. Main limitations of the proposed methodology concern the fragile theoretical foundations concerning the assessment of the recreational value, and the need for structured integration of nature-based solutions into the proposed methodology.

Keywords: green infrastructure; ecosystem services; Natura 2000 Network; environmental planning

1. Introduction

The concept of green infrastructure (GI) arises within the international debate at the end of the 1990s as a distinctive approach to landscape planning [1]. GI is considered as a reference category in the contexts of several disciplines, e.g., landscape ecology [2], greenway planning [3], and management of water resources [4]. Moreover, different functions of GI are identified, e.g., biodiversity conservation [5], or benefits provided to local communities and to civil society as a whole [6]. Therefore, several definitions of GI are available in the literature. Among many, Benedict and McMahon's [5], Wright's [7], Weber et al.'s [8] and the European Commission's [9] are the most relevant. Benedict and McMahon [5] define GI as the ecological system that supports environmental, social and economic health, emphasizing the socio-economic approach to GI. According to Wright [7], although connectivity, multifunctionality and green areas represent the core ideas as regards the category of GI, a deterministic definition is somewhat questionable because, on the one hand, such definition would be inconsistent with a progressively evolving conceptual framework concerning GI, and, on the other hand, its intrinsic interoperability would imply the opportunity of using the GI conceptual framework in a number of research and technical fields related to environmental and spatial studies, which would entail a preference to a flexible, non-deterministic definition. Weber et al. [8] stress the environment-related character of the GI concept, conceived as a system of natural and semi-natural areas spread over the landscape. Broadly speaking, from the above-cited literature GI can be understood as a network of natural and semi-natural areas that play a key role in supporting ecological, social and economic activities.

This study takes as a reference the definition by the European Commission [9] (p. 3), which describes GI as "a strategically planned network of high quality natural and semi-natural areas with other environmental features, which is designed and managed to deliver a wide range of ecosystem services and protect biodiversity in both rural and urban settings." This definition stresses two fundamental characteristics of GI, that is, provision of ecosystem services and protection of biodiversity [10].

Habitats located in the countries of the European Union are characterized by growing fragmentation and deterioration due to human pressures on the environment and the ecosystems [11,12]. The spatial identification of GI is a key element of the European Union biodiversity strategy to 2020 since the second target states that "By 2020, ecosystems and their services are maintained and enhanced by establishing green infrastructure and restoring at least 15% of degraded ecosystems" [13] (p. 5). Consequently, GI represents a spatial planning tool that significantly enhances the quality of life through environmental, social and economic values generated by multifunctional uses of the ecosystems [12].

A number of authors highlight the importance of the inclusion of GI within spatial planning practice [14,15]. Lennon and Scott [15] target the identification and the expansion of GI as one of the most important strategies to implement the ecosystem-based approach into spatial planning. The use of GI in urban contexts entails the development of spatial processes whose planning and management involve particular attention to the impact of non-ecological benefits generated by urban contexts on natural (non-urbanized) environments [16]. According to Kambites and Owen [6], GI planning should be (1) holistic, related to both human and ecological functions; (2) strategic, reflecting a forward-looking view that goes beyond administrative boundaries; (3) inclusive, involving different stakeholders; and, (4) qualitative, assessing biodiversity value and natural resources quality.

However, the implementation of the GI category and related principles is still limited in planning practices, partially due to the limited understanding of the GI concept by planners [17], whose knowledge and information concerning GI is not supported by adequate scientific foundations [18]. Moreover, the implementation of the GI concept into policy-making is problematic because of the complexity of planning processes, which deal with environmental, social, and political contexts and the related regulatory framework [17].

Several studies discuss benefits generated by GI, but they seldom take account of the critical relationships between theoretical assumptions and practices derived thereof as regards the planning framework at stake [19]. Davies et al. [20] investigate the implementation of GI at the urban and regional levels, in relation to twenty European case studies, by using data derived from a number of sources such as official planning documents, questionnaires, Internet analysis, and statistical archives. Davies et al.'s study reports that only seven out of 32 analyzed plans mention the term GI and that only in four cases (Edinburgh, Liverpool, Bristol, and Barcelona) references to connections between GI and spatial planning can be identified. In the work of Artmann et al. [21], GI is the key concept to propose guidelines on how landscape plans can make urban networks of green spaces consistent with compact city spatial frameworks. From this point of view, particular attention should be paid to how local plans should address GI planning [22], and to the lack of models that assess synergies and tradeoffs between ecological and social benefits generated by the spatial identification and planning of GI [23,24]. Under this perspective, this study aims at proposing a methodological approach to include and implement GI within spatial planning at the city level, hence it addresses an outstanding gap concerning scientific and technical research on GI.

As regards these questions, this study builds upon a few recent articles, related to Sardinia, concerning the identification of a spatial taxonomy of areas eligible to be part of a regional green infrastructure (RGI) [25,26] on the basis of four factors, namely the natural, conservation, landscape and recreational values. This study aims at defining and analyzing the relationship between the RGI, identified through the implementation of the methodology proposed in the above-cited articles, and the rules of municipal masterplans (MMPs).

In order to achieve this goal, a methodology based on the overlay mapping of the spatial taxonomy of areas eligible to be part of the RGI and the zoning layouts of MMPs, and on the analysis of correlations between the spatial taxonomy and the zoning rules, is proposed. Correlations are identified through regression analysis. The methodology is applied to the MMPs of three municipalities belonging to the Metropolitan City of Cagliari (MCC, Sardinia, Italy). The outcomes of the study offer important suggestions as regards the definition and implementation of the planning policies of the MCC, based on the general goal of strengthening the GI-related characteristics of the towns located within the metropolitan boundaries, with a view to a future expansion of the RGI within the MCC. The proposed methodology can be easily applied to other national and international urban contexts, and the results of its implementation into the towns of the MCC are important comparative references as regards analogous studies.

This study is structured as follows. This introduction (Section 1) has identified the wider context and the debates that the study is contributing to. Next, Section 2 describes the proposed methodological approach and the spatial context for the implementation of the case study, that is, the towns of Cagliari, Assemini and Capoterra. The results coming from the regression analysis which explores and detects correlations between the RGI and the spatial zoning rules of the MMPs of the three towns are presented in Section 3. In Section 4, implications for spatial planning policies related to the urban contexts of the MCC are discussed. Finally, directions for future research and concluding remarks are proposed in the Section 5.

2. Materials and Methods

2.1. Case Study

Municipalities are, in Italy, in charge of programs and plans, ruling on land development and land-use changes, hence they draft, adopt and approve their own MMPs, which simultaneously lay down a strategic policy for the concerned territory and provide the setting for the management of small scale land-use transformation [27]. Due to the hierarchic nature of the Italian planning system [27] (p. 35), MMPs must conform to a number of higher level plans, the most prominent of which are regional plans, and especially the Regional Landscape Plan (RLP).

This study takes the towns of Assemini, Cagliari and Capoterra, in Sardinia (Italy) as case studies. Each town is a municipality, with its own elected local government and mayor, and it is also part of the MCC, recently established under national law no. 2014/56 and regional law no. 2016/2 (Figure 1). Cagliari, with its approximately 150,000 inhabitants and 85 km^2 in size, is the regional capital and the metropolitan center; Assemini (having around 27,000 inhabitants and 118 km^2 in size) and Capoterra (with about 23,000 inhabitants and 69 km^2 in size) are two medium-sized towns both geographically and economically close to the regional capital, since they belong to the same travel-to-work area, in that a good share (approximately 30 percent [28]) of their populations commutes to Cagliari on a daily basis.

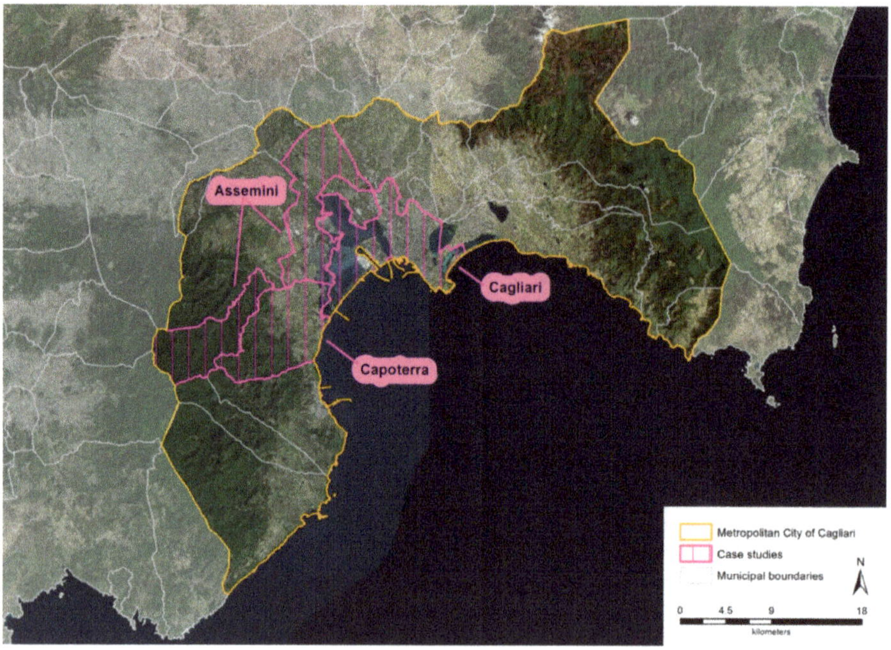

Figure 1. The study area.

The three above-mentioned towns were deemed useful as case studies because, notwithstanding their geographic proximity (which implies similarities in natural, social, and landscape features which lay the basis for the identification of the RGI) and their common planning framework (due to their belonging to the same region, their MMPs are subject to the same regional planning laws and must conform to the same RLP), some outstanding differences can be found that make the comparison interesting. The first difference has to do with the share vegetated (hence, natural and semi-natural areas, which are a necessary, but not sufficient, condition for the presence of a GI as defined by the European Commission [9]) versus urbanized area, since in Assemini and in Capoterra a large portion of the municipal land cover is either natural or agricultural, contrary to what happens in Cagliari, where artificial land covers dominate, if wetlands are not taken into account. A second difference concerns the concentration of the resident population, which in Cagliari amounts to about 1,760 residents per square kilometer, about seven times as much as that of Capoterra (339.5) and eight times as much as that of Assemini (225.3), which in terms of GI implies significant differences in municipal demand for ecosystem services, and, possibly, delocalization of the supplying areas. A third difference regards the land-use plans in force: only two of them (Assemini's and Capoterra's) conform to the RLP in force, while the third (Cagliari's) has very recently started the adjustment process [29] to the RLP, hence in principle for Cagliari there might be more room for improving nature and landscape protection policies and therefore for enhancing the supply of ecosystem services and ultimately the GI. Finally, a common feature of the three municipalities is the presence of several natural protected areas, including various Natura 2000 sites and two regional natural parks, which can constitute core areas for an RGI provided that their inter-connections are properly identified, safeguarded and managed through appropriate planning actions.

2.2. Zoning Schemes

Within both the municipalities of Assemini and Capoterra, an MMP recently approved and compliant with the Sardinian RLP is in force; their planning documents and zoning schemes, approved in August 2015 and May 2016, respectively, are available on the municipalities' official web pages [30,31]. As for the municipality of Cagliari, a much older MMP, dating back to 2004, is in force; such plan was approved under the former landscape planning system, hence the complex and conflictual process of adjustment to the RLP [29] has not taken place yet. The planning documents and zoning scheme for the municipality of Cagliari are available on its official webpage [32] and geoportal [33].

For each of the three municipalities, the zoning schemes were retrieved and analyzed in the light of their respective technical implementation norms. Next, the schemes were simplified on the basis of the provisions contained in the norms, so as to reduce as much as possible the number of zone types, for instance, by joining together sub-zones belonging to the same zone type, or by merging zones with similar planning or building rules. This simplification led to identifying ten types of planning zones; out of the ten types, listed in Table 1, type "E" is not included in Cagliari's zoning scheme, while types "GS" and "IC" are not included in Assemini's and Capoterra's ones.

Table 1. Homogeneous zones identified by the zoning rules of the municipal masterplans (MMPs) of Cagliari, Assemini and Capoterra: simplified and detailed zone types.

Simplified Zone Types		Detailed Zone Types		
Type	**Description**	**Cagliari**	**Assemini**	**Capoterra**
A	Historic districts.	PQCS	A1; A2	A1; A2.1
B	Residential completion zones.	B1; B1*; B2; B2*; B2R1; B2R2; B2R3; B3; B3*; B3R1; B3R2; B3R3; B4; B4*; B4R1; B4R2; B4R3; B5; B5*; B5R1; B5R2; B6; B6*; B6R1; B6R2; B7; B7R1; B7R2; B8; B8*; B8R2; B8R3; B9; B9R1	B1.a/B1.b; B2; B3; B3	B1; B2
C	Residential expansion zones.	C*; C1; C1*; C2*; C3*; C4*; C5*; ER	C1; C2.1; C2.2; C3	C1a; C1b; C1c; C1d; C2; C3
D	Industrial and commercial zones.	D; DG; DR	D1; D2.1; D2.2; D2.3; D2.4; D2.5; D2.6	D2a; D2b; D2c; D2PIP
E	Agricultural zones.	—	E1; E2; E5	E1; E2; E3; E4; E5
G	Collective service zones.	G1; G2; G3; G4; G5I; G6; G7; GA1; GA2; GH; GI; GI/S; GIT; GM; GM*; GM*/GI/S; GOM; GP1; GT; GT*/GA2; GT*/GI; GT*/GS; GTS; D*/G	G1.1; G1.2a; G1.2b; G1.2c; G1.2d; G1.2e; G1.3; G1.4; G1.5; G1.6; G1.7; G1.8; G1.9; G1.10; G1.11; G2.1; G2.2; G2.3; G2.4a; G2.4b; G2.4c; G2.5; G3.1; G4.1a; G4.1b; G4.2; G4.3; G4.4; G4.5; G4.6	G1a; G1b; G1c; G1d; G2a; G4
GS	Collective service zones: green parks significant at the city level.	GS	—	—
H	Conservation and safeguard zones.	AR; AS; H; HG; HS	H1; H2a; H2b; H3	H1; H2; H3
EZ	Enterprise zones.	BS3*; IC; PI1; PI2; D*/GI; D*/GI1; S3*	—	—
S	Public spaces reserved for collective activities, green areas, or parking lots at the district level.	S1; S1*; S1-S2; S2; S2*; S2/E; S3; S4	S1; S1/S2; S2; S2/S3; S3; S4; SBpr	S; S2; S3; S3*; S4

"A" zones coincide with historic districts (sometimes, as in Cagliari, featuring buildings dating back to the Middle Age, and mostly comprising built-up areas developed before World War II), although sometimes they also include other built-up areas with distinctive historic or artistic features.

"B" zones are built-up areas generally developed from the 1950s onwards. They are usually completely developed (meaning that the full housing capacity in terms of cubic meters per square meter of land has already been built), although some partially developed areas can also be included, provided that roads and other infrastructures have already been realized and that at least 10 percent of the allowed housing volume has been developed.

"C" zones are bound to be residential areas; they comprise both undeveloped or partially developed parts of a town that do not qualify as "B" zones (either because they lack infrastructures or because the built volume does not reach the minimum threshold).

"D" zones are reserved for new commercial and industrial buildings, also including processing, storage and selling of agricultural and fishing products.

"E" zones are allocated for agricultural uses, and comprise five sub-zones depending on the importance of agricultural and farming activities that can be carried out, hence ultimately depending on the land suitability for crop production, grazing, or farming. In these areas, residential buildings are allowed only if connected to, and necessary for, the maintenance of the agricultural plot of land in which they are built.

"G" zones are meant to host buildings and facilities, both public and private, of collective interest; they include, among others, high schools, universities, hospitals, sports facilities, water treatment plants, waste incinerators.

"GS" zones identify large green areas that are reserved for urban parks, in which only low-density public facilities (for sports or recreation) can be built.

"H" zones comprise areas where development is almost completely forbidden; they include fragile and sensitive areas, such as beaches or buffer zones around wetlands and rivers, as well as areas to be preserved for their archaeological or speleological or landscape relevance, and also buffer zones around major infrastructures (e.g., roads or railways) that must be left undeveloped.

"EZ", zones, the so-called "enterprise zones" [34,35], are undeveloped or partially-developed parts of the city of Cagliari where integration of different functions (residential buildings, public facilities and recreational areas) must be guaranteed.

"S" zones are areas that must be surrounded by the developer to the municipality when a private detailed development plan (be it for housing, or for industrial or commercial development) is implemented; such areas are therefore public and can be turned into primary schools, public facilities or green spaces for the district, or parking lots.

2.3. Methodology

This study builds upon a methodology applied in previous studies [25,26,36,37], where a potential RGI is mapped taking an Italian region as a case study: Lombardy in [36], and Sardinia in [25,26,37]. In the Sardinian case, the suitability of each patch of land to belong to an RGI is assessed based upon four factors expressing as many functions provided by a GI, as follows:

- natural value (NatVal), which represents habitats' quality notwithstanding pressures and threats exerted on biodiversity, and hence the GI's capacity to provide ecosystem services. NatVal is assessed and mapped through the tool "Habitat quality" of the software "InVEST" [38] that uses as input: i. data on land covers, retrieved from the regional geoportal [39]; ii. threats to biodiversity identified in the Natura 2000 Standard Data Forms of the European Environment Agency [40] and weighed by experts in the field; iii. a sensitivity matrix of each land cover type to each threat;
- conservation value (ConVal), which accounts for the fact that green infrastructures are, in the definition provided by the European Commission [9] and quoted in Section 1, "a .... network of high quality natural and semi-natural areas". Hence ConVal accounts for the presence of natural and semi-natural habitats protected under the European Union legislation because rare, or in danger of disappearance, or providing outstanding examples of typical characteristics of one of the European biogeographical regions;

- recreation value (RecVal), which provides an indication of the extent to which landscapes are attractive for recreational uses and hence provide recreational ecosystem services; RecVal is assessed and mapped through the tool "Visitation: Recreation and Tourism" of the software "InVEST" [38] which retrieves spatial and quantitative information from geotagged pictures uploaded by users on the social media Flickr;
- landscape value (LandVal), which accounts for the quality of landscapes as implied in the RLP's normative framework. In compliance with the European Landscape Convention of the Council of Europe [41], landscape in the plan is regarded as "an area, as perceived by people, whose character is the result of the action and interaction of natural and/or human factor" (article 1) and a "foundation of people's identity" (article 5); therefore, depending on their quality, landscapes are providers of cultural ecosystem services. From the spatial dataset of the plan, available on the regional geoportal [42], the location of each landscape good was retrieved and, through an expert-based approach, a score was assigned to each good; the higher the score, the stricter the rules contained in the plan so as to protect and preserve the good, and ultimately, the higher the quality of that good.

The suitability of each patch of land to belong to an RGI is then assessed by summing up the above four values, which all vary in the range (0–1), and it is therefore represented by the total value (TotVal): the higher TotVal, the greater the suitability.

The suitability map representing the Sardinian RGI (Figure 2) is next overlaid with the zoning schemes of the MMPs provided in Figure 3. Through a spatial intersection between the two layers, for each resulting polygon a vector having components (Zone, NatVal, ConVal, RecVal, LandVal, TotVal) is produced, where "Zone" represents the zone type assigned by the MMP and can take one of the ten values listed in Table 1.

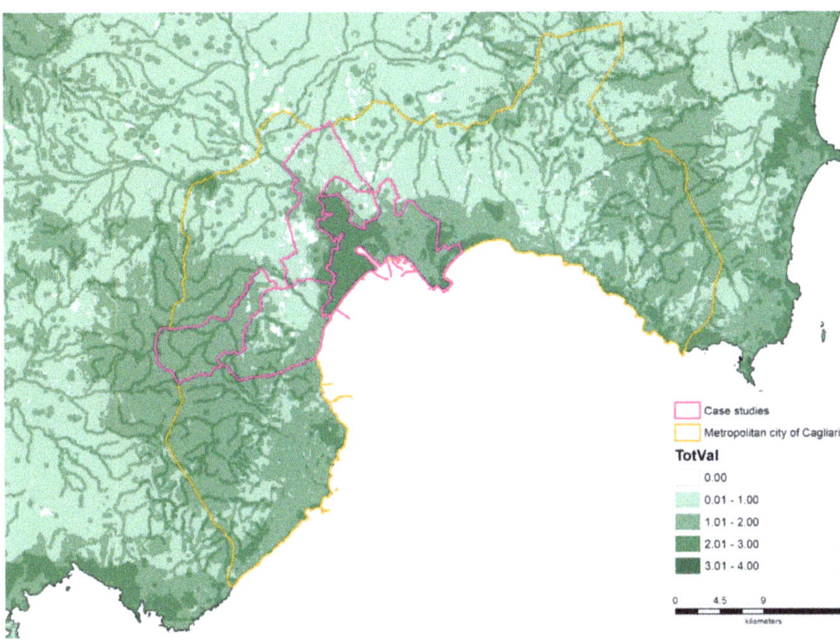

Figure 2. Map of the total value, which identifies the eligibility of patches to be included in the Regional green infrastructure.

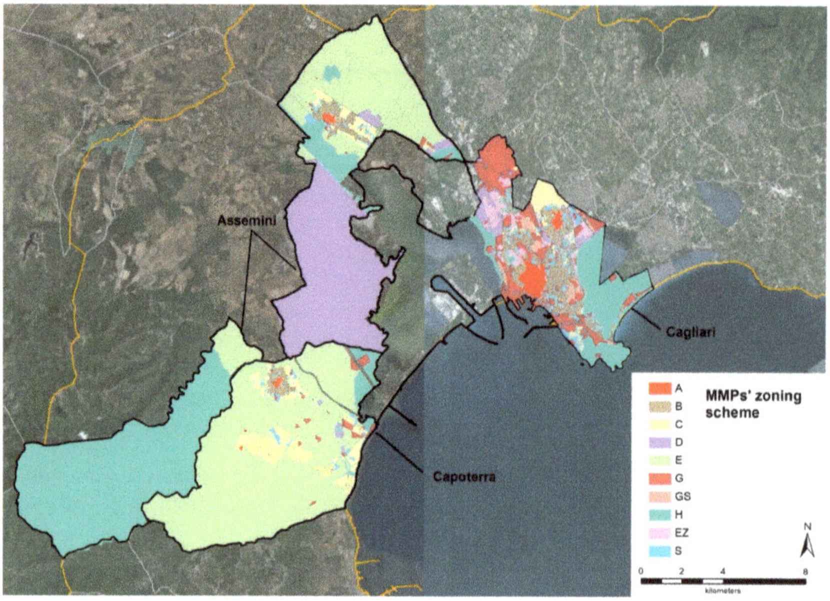

Figure 3. The zoning layout of the MMPs of Assemini, Cagliari and Capoterra.

Next, for each of the three municipalities here taken as case studies a multiple linear regression is performed:

$$TotVal_k = \beta_{0,k} + \beta_{1,k}A + \beta_{2,k}B + \beta_{3,k}C + \beta_{4,k}D + \beta_{5,k}E + \beta_{6,k}G + \beta_{7,k}GS + \beta_{8,k}H + \beta_{9,k}EZ + \beta_{10,k}Area, \quad (1)$$

where

- "k" is the municipality;
- explanatory variables representing the zoning scheme ("A" to "EZ", see Table 1) are dichotomous, or Boolean, variables; each dichotomous variable can take only two values, 1 or 0, according to the following rule: if a patch is classed under the A zone type, the variable A equals 1, otherwise it equals 0; if a patch is classed under the B zone type, the variable B equals 1, otherwise it equals 0, and so on; each coefficient estimated by regression (1), β_i, i = 1, … , 9, identifies the change in TotVal related to a patch in case it is classed under the zone type identified by the variable associated to the coefficient β_i (i.e., A, B, etc.) with respect to the basic condition that the parcel of land under consideration was classed as "S" zone; the coefficients estimated by regression (1), β_i, i = 1, … , 9, define a taxonomy of the zone types based on the quantitative contribution to TotVal expressed by the values of β_i, i = 1, … , 9;
- "Area" is the size of the parcel of land under consideration, resulting from the spatial intersection between the zoning map and the RGI suitability map;
- results from the multiple linear regression are finally used to develop, for each municipality, an ordered list of the planning zones; for each municipality, the order depends on the value of the coefficients β_i, i = 1, … , 9, of regression (1).

As in several studies concerning urban and regional phenomena characterized by multiple factors, the use of multiple linear regression models is based on the grounds of the lack of identified predetermined relationships between the variables representing the factors (among many, [43–46]). Since no a priori interpretive hypothesis is available, the unknown surface, located in an n-dimensional space, which represents the functional relationship between n variables that characterize an urban phenomenon, can be approximated, point by point, by the small region identified by the tangential plane. The small region shared by the (unknown) surface and the (known) tangential plane is identified by a local linear relationship between the variables, which is a linear local approximation of the unknown general relationship between the variables. The tangential plane represents the trace of the local linear relationship between the variables over the unknown surface in the n-dimensional space. The equation of this plane can be estimated through a multiple linear regression like (1), where the trace of TotVal over the unknown surface is a ten-dimensional plane [47,48].

The variable Area is used to control the possible effect of the size of land parcels; if the estimated coefficient β_{10} were significant in terms of *p*-value hypothesis testing, this would imply that "TotVal," that is the eligibility of a parcel to be included in the RGI, is influenced not only by the zone type of a patch, but also by its size, or, that either an agglomeration ($\beta_{10} > 0$) or a dispersal effect ($\beta_{10} < 0$) is detected. A two-tailed significance test based on t-statistic is implemented in the estimated regressions in order to check if the estimated coefficient of Area is significantly different from zero (p-value < 5%). If the *p*-value is greater than 5%, this would imply that the correlation between TotVal and Area can be confidently excluded, or, that no agglomeration or dispersal effect does show up.

3. Results

The estimates of the regressions related to Cagliari, Assemini and Capoterra define the features of the effects of a zone type on the eligibility of a patch to be included in the RGI. Indeed, each coefficient of the dichotomous variables estimated in the regressions identifies the effect on the eligibility of a patch to be included in the RGI as a consequence of it being classified as a homogeneous zone type from "A" to "H," or as "EZ" or "GS" (only for the MMP of Cagliari, which does not show any "E" type zone) types, with respect to the basic situation of a patch being classified as "S" homogeneous zone type. This estimated effect equals the difference in TotVal, everything else being equal. Accordingly, a ranking of the homogeneous zone types can be defined, on the basis of the estimated effects, from the highest to the lowest.

Tables 2–4 show the results of the regression estimates related to Cagliari, Assemini and Capoterra.

Table 2. Regression results related to Cagliari.

Explanatory Variable	Coefficient	Standard Deviation	t-Statistic	*p*-Values	Mean of the Explanatory Variable
A	0.693	0.068	10.123	0.000	0.0096
B	−0.035	0.020	−1.731	0.083	0.2637
C	0.014	0.030	0.486	0.627	0.0660
D	−0.100	0.038	−2.624	0.009	0.0346
G	0.021	0.023	0.897	0.370	0.1481
GS	0.568	0.036	15.779	0.000	0.0400
H	0.750	0.022	34.883	0.000	0.2031
EZ	0.003	0.029	0.116	0.907	0.0706
AREA	0.000	0.000	−1.583	0.114	7925.00

Dependent variable: **TotVal:** Mean: 1.498; Standard deviation: 0.549; Adjusted R-squared: 0.337.

Table 3. Regression results related to Assemini.

Explanatory Variable	Coefficient	Standard Deviation	t-Statistic	*p*-Values	Mean of the Explanatory Variable
A	0.206	0.118	1.746	0.081	0.0125
B	−0.300	0.065	−4.649	0.000	0.1043
C	0.029	0.068	0.419	0.675	0.0771
D	0.851	0.057	14.914	0.000	0.3393
E	0.268	0.060	4.497	0.000	0.1960
G	0.095	0.094	1.009	0.313	0.0229
H	0.988	0.060	16.585	0.000	0.1991
AREA	0.000	0.000	−1.197	0.231	35505.19

Dependent variable: **TotVal:** Mean: 1.115; Standard deviation: 0.808; Adjusted R-squared: 0.304.

Table 4. Regression results related to Capoterra.

Explanatory Variable	Coefficient	Standard Deviation	t-Statistic	*p*-Values	Mean of the Explanatory Variable
A	−0.072	0.119	−0.601	0.548	0.0106
B	−0.713	0.055	−13.023	0.000	0.0769
C	0.007	0.042	0.169	0.866	0.2652
D	−0.086	0.111	−0.778	0.436	0.0124
E	0.263	0.040	6.587	0.000	0.3672
G	0.306	0.062	4.951	0.000	0.0525
H	1.159	0.051	22.939	0.000	0.1020
AREA	0.000	0.000	1.056	0.291	24266.91

Dependent variable: **TotVal:** Mean: 1.034; Standard deviation: 0.743; Adjusted R-squared: 0.317.

The estimates of the coefficients of the area of the patch are not significant in terms of *p*-value hypothesis testing. As a consequence, as put in evidence in Section 3, there is no sign of a marginal effect of the highly volatile size of patches on their eligibility to be included in the RGI. In other words, no agglomeration or dispersal effect is detected, which makes inferences and implications based on the estimates of the three regressions straightforwardly connected to the variables representing the zoning type (from "A" to "EZ").

In the rest of this study, we adopt the same test described in Section 3 as regards the significance of the estimated coefficient of AREA in order to assess the significance of the estimated coefficients related to the variables representing the zoning type (from "A" to "EZ"). By so doing, if the *p*-value is greater than 5%, this would imply that correlation between TotVal and the zoning type-related variable at stake can be confidently excluded. If this is the case, the variable is labeled as not significant (NS) in Table 5, based on the results shown in Tables 2–4, columns "*p*-values."

The regression estimates put in evidence that the "H" zone type always shows the highest effect on patch eligibility, whereas the "C" zone type, namely the residential expansion type, does not influence eligibility in any of the three regression estimates. The effect of the "EZ" zone, only identified in the zoning layout of the MMP of Cagliari, also related to residential expansion with an additional endowment of public services integrated into the residential areas, is not significant as well. In two out of three cases (Cagliari and Assemini), the effect on the eligibility of the "G" zone type, that is, general public services, is not significant, whereas its influence is positive as regards Capoterra, where the impact of the "D" zone type (industrial, craftsmanship and hypermarket areas) is not significant.

On the contrary, the impact of the "D" zone type is significant as regards Cagliari and Assemini, even though it is much more important, as regards its size, in the case of Assemini. The negative estimate of the coefficient of "D" (−0.100) indicates that, in the case of Cagliari, the effect of this zone type is lower than the effect of the "S" zone type (publicly-owned public service areas) by about one-tenth. The marginal effect of the "B" zone type is weakly significant in all of the three cases as well. This finding stresses that the areas showing large and pervasive land-taking processes, such as the "B", "C", "G" with the exception of Capoterra, "D" with the exception of Assemini, and "EZ" zone types, are almost totally inadequate to be included in the RGI. The highly urbanized "S" zones, which are targeted as already, or bound-to-become, publicly-owned public service areas, show very low influence as well.

The zone types which mainly help to characterize a patch as being eligible to be part of the RGI are (1) the "A" type, that is, historic and artistic center, featured by environmental values related to the built environment, with the exception of Capoterra, whose "A" zone is, by the way, less attractive and valuable than Cagliari's and Assemini's corresponding areas; (2) the "E" type, which identifies rural and agricultural areas, characterized by the lowest levels of soil sealing and land take (this type of zone is not present in the zoning layout of the MMP of Cagliari); and, above all, (3) the "H" type, which is characterized by patches which the MMPs identify as worth protecting because of their environmental and landscape-related features. The effect on the eligibility of the "GS" zone type, which identifies open spaces and recreational areas, that is, almost-totally unbuilt areas, and which is only included in the zoning layout of the MMP of Cagliari, is consistent with the effect of the "H" zone type as well.

Table 5 highlights the ranking of the zone types as regards their influence on the eligibility of patches to be included in the RGI, and the corresponding means of NatVal, ConVal, LandVal and RecVal, in order to identify the factors' influence in a comparative way.

Table 5. Ranking of the homogenous zones based on the contribution to TotVal implied by the regression results, and average values of the four factors which determine TotVal, related to each homogeneous zone (NP: the homogeneous zone is not present in the MMP's zoning rules; NS: the regression *p*-value entails that the coefficient is non-significant).

Zone Type	Cagliari					Assemini					Capoterra				
	Rank	Average Values: NatVal, ConVal, LandVal, RecVal				Rank	Average Values: NatVal, ConVal, LandVal, RecVal				Rank	Average Values: NatVal, ConVal, LandVal, RecVal			
A	2	0.432	0.000	1.000	0.573	4	0.000	0.000	0.722	0.087	NS				
B	6	0.030	0.000	1.000	0.261	6	0.038	0.000	0.200	0.065	5	0.049	0.000	0.027	0.068
C	NS					NS					NS				
D	7	0.233	0.000	0.952	0.039	2	0.644	0.162	0.636	0.010	NS	NS	NS	NS	NS
E	NP					3	0.482	0.028	0.352	0.006	3	0.529	0.061	0.523	0.010
G	NS					NS					2	0.448	0.057	0.639	0.019
GS	3	0.607	0.024	1.000	0.262	NP					NP				
H	1	0.675	0.204	1.000	0.195	1	0.748	0.187	0.647	0.005	1	0.696	0.282	1.000	0.038
EZ	4					NP					NP				
S	5	0.101	0.001	1.000	0.225	5	0.316	0.000	0.258	0.030	4	0.288	0.022	0.513	0.034

With reference to the "A" and "H" (and "GS," in the case of Cagliari) zone types, the average values of LandVal are comparatively high, since they are always higher than 0.6. The average values of NatVal of the "E" zones are lower than the "A" and the "H" zones' values, even though they are higher than the remaining zones. Moreover, the "H" zones show the highest average values of ConVal in all of the three cases, although there is room for improvement, since they are never higher than 0.3.

On the other hand, the conservation value on average equals zero as regards patches located in the "A" zones, whereas it is very close or equal to zero in already-urbanized areas or in areas characterized by ongoing advanced urbanization processes, such as "B," "C," "D," "G," "EZ" and "S", which is consistent with expectations, since it is very unlikely that habitats protected under the provisions of European Union rules can be found in these areas.

The results of the regressions show that the "A," "E" and "H" zone types are the most important in terms of impact on the eligibility of patches to be part of the RGI. Moreover, Table 5 stresses that there is still large room for improvement as regards all the zone types. For example, the almost-totally urbanized areas classed as "B,", "C," "D," "G, "EZ" and "S" zone types show non-null NatVal and RecVal, and often comparatively not so low, in each of the three MMPs, especially with reference to the recreational profile (RecVal), which gives credit to possible scope for improving RGI-related features of areas located in the three towns of the MCC.

Particularly relevant is the improvement margin related to agricultural areas ("E" zone type) and to the protection areas ("H" zone type) as regards all of the four values. This implies that the ruling framework related to these zone types would be worth exporting to other parts of the municipal land in order to increase the eligibility of patches to be included in the RGI.

4. Discussion

The study analyzes the relations between the land uses, defined in the MMPs of three local municipalities included in the MCC, and the RGI whose identification is based on the methodology proposed by Lai and Leone [37].

According to the results presented in Section 3, the "H" zones are the areas that mainly positively affect the eligibility of patches to be part of the RGI in the three study areas. In particular, in relation to

"H" zones, the average values of the four factors show the following similar trends (1) NatVal is higher than 0.5; (2) ConsVal and RecVal are lower than 0.5; and (3) LandVal equals 1 (maximum value) in the case of Cagliari and Capoterra and is lower than 0.7 in the case of Assemini. As a consequence, there is plenty of room for improving two out of the four factors (ConVal and RecVal).

ConVal is mainly influenced by the presence of habitat of community interest. "H" zones are conceived as areas of particular environmental and natural interest; thus, they may represent buffer zones to protect high-quality sites, such as Natura 2000 sites, or stepping stones along migration routes. A possible policy recommendation aims at extending the environmental protection regimes related to habitats and species beyond the boundaries of protected areas by identifying those patches that, in relation to their characteristics, could be suitable for species and habitats. Therefore, advancements of scientific knowledge related to habitats and species within "H" zones and awareness-raising activities are preliminary necessary steps in order to increase the size of protected areas. In line with this recommendation, Maiorano et al. [49] suggest that integrated management of Natura 2000 sites and of their neighboring areas may improve the effectiveness of conservation measures within protected areas due to control over human-induced activities in the surrounding areas.

Acting on elements that influence RecVal shows more room for improvement than ConVal due to its lower values in relation to "H" zones in the three study areas. RecVal is calculated on the basis of geotagged information retrieved from the social media Flickr, representing the attractiveness of a certain area to visitors in a defined time period. Several studies [50–52] show that recreational attractiveness of an area, conceived of as the demand for recreational activities, is influenced by different factors, such as accessibility and accommodation availability. Therefore, a possible recommendation concerns making these areas more accessible through infrastructures that, on the one hand, support slow mobility (such as cycle and pedestrian paths) and, on the other hand, do not increase habitat fragmentation. In fact, increased fragmentation of habitats is likely to result in decreasing values of ConVal and NatVal.

In relation to NatVal, although its average values are quite high (between 0.67 and 0.75) in all of the three case studies, there is still some room for improvement. NatVal is mainly influenced by land uses and threats to habitats, identified through standard data forms of regional Natura 2000 sites. From this standpoint, two types of policy actions should be taken into account as particularly effective: reduction of threat and mitigation of land-taking processes. Both these actions can include measures aiming at restoring ecosystems, also through the use of nature-based solutions (NBSs). The concept of NBSs was coined by the European Commission [53] to define techniques and solutions based on the use of nature in urban areas. NBSs are designed to address effectively several social challenges in terms of effective resources management, and, at the same time, to provide economic, social and environmental benefits. NBSs are more efficient and cost-effective solutions than traditional approaches [54]. The European Commission [53] identifies a series of NBSs to make cities more livable and sustainable, such as the restoration of abandoned and degraded areas, the use of permeable surfaces and of rain gardens to manage and control rainwaters within urban settlements. For example, in the city of Cagliari a significant and troubling phenomenon, represented by agricultural uses and informal settlements, characterizes a particular "H" zone, called "AR—Is Arenas" within the regional "Molentargius-Saline" park. In these areas, specific measures to mitigate threats caused by urban settlements are necessary.

Moreover, due to the positive influence of "H" zones on the eligibility of patches to be part of the RGI, both the increase of the existing "H" zones and the definition of new "H" zones at the expense of other zones could represent a possible policy action.

In relation to Capoterra and Assemini, "E" zones also influence positively the eligibility of patches to be part of the RGI. The average values of NatVal, ConVal, LandVal and RecVal are lower than those that can be found in "H" zones and, for this reason, there might be more room for improvement, in particular in relation to NatVal and ConVal. Natural value is mainly influenced by the quality of land covers, frequently threatened by intensive agricultural use and by habitat fragmentation due to rural settlements and infrastructure. He et al. [55] in a recent work, where they study the impacts of land covers on habitat quality, suggest improving habitat quality through agricultural policies that promote a more sustainable use of land, with particular attention to isolated rural settlements. In relation to ConVal, as promoted by the 2014–2020 Sardinian regional Rural Development Program, a possible policy could include sustaining agri-environment-climate commitments, comprising, among others, incentives to support those farmers who allocate part of their farmland for wildlife (e.g., establishing grass swards along wetlands, keeping unharvested conservation lands for wildlife, or maintaining hedgerows and drywalls for small vertebrates).

5. Conclusions

This study aims at identifying a methodological approach to include and implement GI within spatial planning at the city level. For this purpose, relations between the land uses defined by MMPs and the factors related to the spatial identification of the RGI are analyzed and assessed.

The study highlights two important aspects. First, as suggested by Mell [56,57], GI implementation is strongly influenced by the sometimes lacking continuity between what is proposed by the national and regional administrations and what is implemented by the municipalities. The lack of models that assess synergies and tradeoffs between different functions and benefits that GI can provide is a significant aspect [23,24]. According to Davies et al. [20], the social role of GI is strongly related to three issues: "where," "which interventions" and "in what circumstances." These three aspects require, on the one hand, accurate knowledge of the involved urban contexts, and, on the other hand, an adequate understanding of the available planning tools [18]. In relation to the implementation of the methodology proposed in this study, regarding three towns of the Metropolitan City of Cagliari, MMPs establish spatial taxonomies of the municipal lands based on land cover categories. In Italy, and in the regional context of Sardinia, the integration of GI within planning practices is strongly influenced by the expertise of the technical staff working in local administrations, and by political pressures from representatives of the municipal councils, which should be based on local communities' expectations and perceived needs related to land uses. However, under the provisions of the laws in force, direct participation of local communities in decision-making processes concerning the approval procedure of MMPs is limited and rather ineffective.

The second aspect focuses on the implementation of spatial transformations. In urban contexts, characterized by a weak ecosystem structure and by pressures and impacts generated by high-building-density settlements, ecosystem restoration measures play an important role in integrating GI within local planning [58]. Therefore, as highlighted by the findings of this study, consistently with the provisions of Action 6 of the Target 2 of the Convention on Biological Diversity 2020 Strategic Plan, GI should represent a significant tool to promote ecosystem restoration in urban and rural areas. Moreover, GI plays an important role in economic terms as well, since it increases the attractiveness of urban environments [20]. On the other hand, a collaborative approach to policy-making concerning local spatial planning procedures, which would imply an effective cooperation effort involving technical structures and representatives of the local administrations, and direct participation of the local communities, is likely to produce regulations which may eventually identify the most adequate decisions concerning the spatial taxonomy of land uses on a non-deterministic, multi-functional basis. Furthermore, other studies [17,57] analyze how GIs are implemented in spatial planning. For example, Mell [57] assesses the implementation of GI in the United Kingdom and in the United States through a comprehensive review of the existing literature. Di Marino et al. [17] investigate this issue through the analysis of policy and planning documents and by interviewing regional and city planners. This study goes beyond the traditional analysis of planning documents to understand, as suggested by Davies et al. [20], where, what type of interventions and in what circumstances the implementation of GI is likely to impact local (municipal) planning processes.

In conclusion, the proposed methodology can be regarded as a tool in support of decision-makers that can be exported to other European contexts, where Natura 2000 Network is established in compliance with the Habitat Directive. The main advantage of the proposed methodology is its flexibility, which makes it possible to add new values in order to include normative, social and economic aspects that characterize other European contexts. A first most significant limitation concerns the assessment of place attractiveness (RecVal) based on social media only, although some research has argued that social-media retrieved information can be used as a reliable proxy for visitation data (see for instance Wood et al. [59], Sessions et al. [60], Heikinheimo et al. [61]. A second limitation stems from the fact that the methodology for assessing natural value (NatVal) does not take NBSs (such as green roofs or green walls) into account, hence possibly underestimating the natural value in built-up areas. These limits could be addressed in future research.

Author Contributions: S.L., F.L., and C.Z. collaboratively designed the research and jointly wrote Sections 1 and 5. Individual contributions are as follows: S.L. wrote Section 2; C.Z. wrote Section 3; F.L. wrote Section 4.

Funding: This study was supported by the Research Program "Natura 2000: Assessment of management plans and definition of ecological corridors as a complex network", funded by the Autonomous Region of Sardinia for the period 2015–2018, under the Call for "Projects related to fundamental or basic research" of the year 2013, implemented at the Department of Civil and Environmental Engineering and Architecture (DICAAR) of the University of Cagliari, Italy.

Acknowledgments: The study was implemented within the Research project based on the Agreement between Dipartimento di Ingegneria Civile, Ambientale e Architettura (Department of Civil, Environmental Engineering and Architecture, DICAAR) of the University of Cagliari, Italy, and the Autonomous Region of Sardinia, Departmental Office of Environment Protection, finalized to the objectives of the Project "GIREPAM–Integrated Management of Ecological Networks through Parks and Marine Areas" (Programme INTERREG Marittimo Italy-France Maritime 2014–2020, Axis 2).

Conflicts of Interest: The authors declare no conflict of interest. The supporting sponsor had no role in the design of the study; in the collection, analyses, or interpretation of data; in the writing of the manuscript, and in the decision to publish the results.

Glossary

ConsVal	Conservation value
GI	Green infrastructure
GIS	Geographic information system
LandVal	Landscape value
NatVal	Natural value
MCC	Metropolitan City of Cagliari
MMP	Municipal masterplan
NBS	Nature-based solution
NP	Not present
NS	Not significant
RecrVal	Recreation value
RGI	Regional green infrastructure
RLP	Regional landscape plan

References

1. Mell, I.C. *Global Green Infrastructure: Lessons for Successful Policy-Making, Investment and Management*; Routledge: Abingdon, UK, 2016; ISBN 9781138854642.
2. Jongman, R.; Pungetti, G. *Ecological Networks and Greenways: Concept, Design and Implementation*; Cambridge University Press: Cambridge, UK, 2004; ISBN 978-0521827768.
3. Fábos, J.G. Greenway planning in the United States: Its origins and recent case studies. *Landsc. Urban Plan.* **2004**, *68*, 321–342. [CrossRef]
4. Ahern, J. Planning and design for sustainable and resilient cities: Theories, strategies and best practice for green infrastructure. In *Water-Centric Sustainable Communities*; Novotny, V., Ahern, J., Brown, P., Eds.; Wiley-Blackwell: Hoboken, NJ, USA, 2007; pp. 135–176. ISBN 978-0-470-47608-6.
5. Benedict, M.A.; McMahon, E. *Green Infrastructure: Linking Landscapes and Communities*; Island Press: Washington, DC, USA, 2006; ISBN 978-1559635585.

6. Kambites, C.; Owen, S. Renewed prospects for green infrastructure planning in the UK. *Plan. Pract. Research* **2006**, *21*, 483–496. [CrossRef]
7. Wright, H. Understanding green infrastructure: The development of a contested concept in England. *Local Environ.* **2011**, *16*, 1003–1019. [CrossRef]
8. Weber, T.; Sloan, A.; Wolf, J. Maryland's green Infrastructure assessment: Development of a comprehensive approach to land conservation. *Landsc. Urban Plan.* **2006**, *77*, 94–110. [CrossRef]
9. European Commission. *Communication from the Commission to the European Parliament, the Council, the European Economic and Social Committee and the Committee of the Regions. Green Infrastructure (GI)—Enhancing Europe's Natural Capital*; European Commission: Brussels, Belgium, 2013. Available online: https://eur-lex.europa.eu/resource.html?uri=cellar:d41348f2-01d5-4abe-b817-4c73e6f1b2df.0014.03/DOC_1&format=PDF (accessed on 27 December 2018).
10. Elbakidze, M.; Angelstam, P.; Yamelynets, T.; Dawson, L.; Gebrehiwot, M.; Stryamets, N.; Johansson, K.E.; Garrido, P.; Naumov, V.; Manton, M. A Bottom-up Approach to Map Land Covers as Potential Green Infrastructure Hubs for Human Well-Being in Rural Settings: A Case Study from Sweden. *Landsc. Urban Plan.* **2017**, *168*, 72–83. [CrossRef]
11. Millennium Ecosystem Assessment. *Ecosystems and Human Well-Being: A Framework for Assessment*; Island Press: Washington, DC, USA, 2003; ISBN 1-55963-403-0.
12. Vallecillo, S.; Polce, C.; Barbosa, A.; Castillo, C.P.; Vandecasteele, I.; Rusch, G.M.; Maes, J. Spatial Alternatives for Green Infrastructure Planning across the EU: An ecosystem service perspective. *Landsc. Urban Plan.* **2018**, *174*, 41–54. [CrossRef]
13. European Commission. *Communication from the Commission to the European Parliament, the Council, the Economic and Social Committee and the Committee of the Regions. Our Life Insurance, Our Natural Capital: An EU Biodiversity Strategy to 2020*; European Commission: Brussels, Belgium, 2011. Available online: https://eur-lex.europa.eu/legal-content/EN/TXT/PDF/?uri=CELEX:52011DC0244 (accessed on 27 December 2018).
14. Wilson, E.; Piper, J. *Spatial Planning and Climate Change*; Routledge: New York, NY, USA, 2010; ISBN 978-0415495912.
15. Lennon, M.; Scott, M. Delivering Ecosystems Services via Spatial Planning: Reviewing the Possibilities and Implications of a Green Infrastructure Approach. *Town Plann. Rev.* **2014**, *85*, 563–587. [CrossRef]
16. Votsis, A. Planning for Green Infrastructure: The Spatial Effects of Parks, Forests, and Fields on Helsinki's Apartment Price. *Ecol. Econ.* **2017**, *132*, 279–289. [CrossRef]
17. Di Marino, M.; Tiitu, M.; Lapintie, K.; Viinikka, A.; Kopperoinen, L. Integrating Green Infrastructure and Ecosystem Services in Land Use Planning. Results from Two Finnish Case Studies. *Land Use Policy* **2019**, *82*, 643–656. [CrossRef]
18. Albert, C.; Aronson, J.; Fürst, C.; Opdam, P. Integrating Ecosystem Services in Landscape Planning: Requirements, Approaches, and Impacts. *Landsc. Ecol.* **2014**, *29*, 1277–1285. [CrossRef]
19. Lennon, M.; Scott, M.; Collier, M.; Foley, K. The Emergence of Green Infrastructure as Promoting the Centralisation of a Landscape Perspective in Spatial Planning—The Case of Ireland. *Landsc. Res.* **2017**, *42*, 146–163. [CrossRef]
20. Davies, C.; Hansen, R.; Rall, E.; Pauleit, S.; Lafortezza, R.; De Bellis, Y.; Santos, A.; Tosics, I. Green Infrastructure Planning and Implementation. The Status of European Green Space Planning and Implementation Based on an Analysis of Selected European City-Regions. 2015. Available online: https://greensurge.eu/working-packages/wp5/files/D_5.1_Davies_et_al_2015_Green_Infrastructure_Planning_and_Implementation_v2.pdf_ (accessed on 1 March 2019).
21. Artmann, M.; Bastian, O.; Grunewald, K. Using the Concepts of Green Infrastructure and Ecosystem Services to Specify Leitbilder for Compact and Green Cities–the Example of the Landscape Plan of Dresden (Germany). *Sustainability* **2017**, *9*, 198. [CrossRef]
22. Grădinaru, S.R.; Hersperger, A.M. Green Infrastructure in Strategic Spatial Plans: Evidence from European Urban Regions. *Urban For. Urban Green.* 2018. [CrossRef]
23. Meerow, S.; Newell, J.P. Urban Resilience for Whom, What, When, Where, and Why? *Urban Geogr.* **2016**. [CrossRef]
24. Meerow, S.; Newell, J.P. Spatial Planning for Multifunctional Green Infrastructure: Growing Resilience in Detroit. *Landsc. Urban Plan.* **2017**, *159*, 62–75. [CrossRef]

25. Cannas, I.; Lai, S.; Leone, F.; Zoppi, C. Green Infrastructure and Ecological Corridors: A Regional Study Concerning Sardinia. *Sustainability* **2018**, *10*, 1265. [CrossRef]
26. Lai, S.; Leone, F.; Zoppi, C. Implementing Green Infrastructures beyond Protected Areas. *Sustainability* **2018**, *10*, 3544. [CrossRef]
27. Commission of the European Communities (CEC). *The EU Compendium of Spatial Planning Systems and Policies: Italy*; Office for Official Publications of the European Communities: Luxembourg, 2000; ISBN 92-828-2689-9.
28. ISTAT Censimento Popolazione Abitazioni [National Census]. Available online: http://dati-censimentopopolazione.istat.it/Index.aspx# (accessed on 1 March 2019).
29. Zoppi, C.; Lai, S. Assessment of the Regional Landscape Plan of Sardinia (Italy): A Participatory-action-research Case Study Type. *Land Use Policy* **2010**, *27*, 690–705. [CrossRef]
30. PUC Vigente Dal 27.8.2015 [MMP in Force Since 27 August 2018]. Available online: https://comune.assemini.ca.it/amministrazione/amministrazione-trasparente/pianificazione-governo-del-territorio/piani-programmi-16 (accessed on 27 December 2018).
31. Piano Urbanistico Comunale Pubblicato sul BURAS il 26/05/2016 [MMP Published on the Official Journal of the Regional Administration of Sardinia on 26 May 2016]. Available online: https://www.comune.capoterra.ca.it/index.php?option=com_content&view=article&id=2817&Itemid=532 (accessed on 27 December 2018).
32. Piano Urbanistico Comunale [Municipal Masterplan]. Available online: https://www.comune.cagliari.it/portale/it/at18_puc.page (accessed on 27 December 2018).
33. Cagliari SIT–Sistema Informativo Territoriale [Cagliari GIS–Geographic Information System]. Available online: https://sit.comune.cagliari.it/?filtro=puc#13/39.2238/9.0906 (accessed on 27 December 2018).
34. Billings, S.P. Do Enterprise Zones Work? An Analysis at the Borders. *Public Financ. Rev.* **2007**, *37*, 68–93. [CrossRef]
35. Wetherell, S. Freedom Planned: Enterprise Zones and Urban Non-Planning in Post-War Britain. *Twent. Century Br. Hist.* **2016**, *27*, 266–289. [CrossRef]
36. Arcidiacono, A.; Ronchi, S.; Salata, S. Managing Multiple Ecosystem Services for Landscape Conservation: A Green Infrastructure in Lombardy Region. *Procedia Eng.* **2016**, *161*, 2297–2303. [CrossRef]
37. Lai, S.; Leone, F. A Methodological Approach to Identify a Multifunctional Green Infrastructure at the Regional Scale. A Case Study from Sardinia, Italy. *Urban. Inf.* **2017**, *272*, 836–840.
38. InVEST (Integrated Valuation of Ecosystem Services and Tradeoffs) Is a Freely Available Software Program Comprising a Series of Tools to Assess and Map Various Ecosystem Services. Available online: https://naturalcapitalproject.stanford.edu/invest/ (accessed on 1 March 2019).
39. Carta dell'Uso del Suolo in Scala 1:25.000–2008 [1:25,000 Land Cover Map of Sardinia–2008 Edition]. Available online: http://www.sardegnageoportale.it/index.php?xsl=2420&s=40&v=9&c=14480&es=6603&na=1&n=100&esp=1&tb=14401 (accessed on 1 March 2019).
40. Natura 2000 Data–the European Network of Protected Sites. Available online: https://www.eea.europa.eu/data-and-maps/data/natura-9 (accessed on 1 March 2019).
41. European Landscape Convention. Available online: https://www.coe.int/en/web/conventions/full-list/-/conventions/rms/0900001680080621 (accessed on 1 March 2019).
42. Il Piano Paesaggistico Regionale [The Regional Landscape Plan]. Available online: http://www.sardegnageoportale.it/index.php?xsl=2420&s=40&v=9&c=14482&na=1&n=10&esp=1&tb=14401 (accessed on 1 March 2019).
43. Cheshire, P.; Sheppard, S. On the Price of Land and the Value of Amenities. *Econ. New Ser.* **1995**, *62*, 247–267. [CrossRef]
44. Stewart, P.A.; Libby, L.W. Determinants of Farmland Value: The Case of DeKalb County, Illinois. *Rev. Agric. Econ.* **1998**, 80–95. [CrossRef]
45. Sklenicka, P.; Molnarova, K.; Pixova, K.C.; Salek, M.E. Factors Affecting Farmlands in the Czech Republic. *Land Use Pol.* **2013**, *30*, 130–136. [CrossRef]
46. Zoppi, C.; Argiolas, M.; Lai, S. Factors Influencing the Value of Houses: Estimates for the City of Cagliari, Italy. *Land Use Pol.* **2015**, *42*, 367–380. [CrossRef]
47. Bera, A.K.; Byron, R.P. Linearised Estimation of Nonlinear Single Equation Functions. *Int. Econ. Rev.* **1983**, *24*, 237–248.
48. Wolman, A.L.; Couper, E. Potential Consequences of Linear Approximation in Economics. *Federal Reserve Bank Econ. Q.* **2003**, *11*, 17.

49. Maiorano, L.; Falcucci, A.; Garton, E.O.; Boitani, L. Contribution of the Natura 2000 Network to Biodiversity Conservation in Italy. *Conserv. Biol.* **2007**, *21*, 1433–1444. [CrossRef] [PubMed]
50. Amoako-Tuffour, J.; Martinez-Espineira, R. Leisure and the Net Opportunity Cost of Travel Time in Recreation Demand Analysis: An Application to Gros Morne National Park. *J. Appl. Econ.* **2012**, *15*, 25–49. [CrossRef]
51. Heagney, E.C.; Rose, J.M.; Ardeshiri, A.; Kovac, M. Optimising Recreation Services from Protected Areas—Understanding the Role of Natural Values, Built Infrastructure and Contextual Factors. *Ecosyst. Serv.* **2018**, *31*, 358–370. [CrossRef]
52. Font, A.R. Mass Tourism and the Demand for Protected Natural Areas: A Travel Cost Approach. *J. Environ. Econ. Manag.* **2000**, *39*, 97–116. [CrossRef]
53. European Commission. *Towards an EU Research and Innovation Policy Agenda for Nature-Based Solutions & Re-Naturing Cities. Final Report of the Horizon 2020 Expert Group on "Nature-Based Solutions and Renaturing Cities"*; Publications Office of the European Union: Luxembourg, 2015; ISBN 978-92-79-46051-7. Available online: https://ec.europa.eu/newsroom/horizon2020/document.cfm?doc_id=10195 (accessed on 27 December 2018).
54. Lafortezza, R.; Chen, J.; Konijnendijk van den Bosch, C.K.; Randrup, T.B. Nature-based Solutions for Resilient Landscapes and Cities. *Environ. Res.* **2018**, *165*, 431–441. [CrossRef] [PubMed]
55. He, J.; Huanga, J.; Li, C. The Evaluation for the Impact of Land Use Change on Habitat Quality: A Joint Contribution of Cellular Automata Scenario Simulation and Habitat Quality Assessment Model. *Ecol. Model.* **2017**, *366*, 58–67. [CrossRef]
56. Mell, I.C. Green Infrastructure: Concepts, Perceptions and Its Use in Spatial Planning. Ph.D. Thesis, University of Newcastle, Newcastle, UK, 2010. Available online: https://core.ac.uk/download/pdf/40036997.pdf (accessed on 1 March 2019).
57. Mell, I.C. Aligning Fragmented Planning Structures Through a Green Infrastructure Approach to Urban Development in the UK and USA. *Urban For. Urban Green.* **2014**, *13*, 612–620. [CrossRef]
58. Tucker, G.; Underwood, E.; Farmer, A.; Scalera, R.; Dickie, I.; McConville, A.; van Vliet, W. *Estimation of the Financing Needs to Implement Target 2 of the EU Biodiversity Strategy. Report to the European Commission*; Institute for European Environmental Policy: London, UK, 2013. Available online: http://ec.europa.eu/environment/nature/biodiversity/comm2006/pdf/2020/Fin%20Target%202.pdf (accessed on 1 March 2019).
59. Wood, S.A.; Guerry, A.D.; Silver, J.M.; Lacayo, M. Using Social Media to Quantify Nature-Based Tourism and Recreation. *Sci Rep.* **2013**, *3*, 2976. [CrossRef] [PubMed]
60. Sessions, C.; Wood, S.A.; Rabotyagov, S.; Fisher, D.M. Measuring Recreational Visitation at U.S. National Parks with Crowd-Sourced Photographs. *J. Environ. Manag.* **2016**, *183*, 703–711. [CrossRef] [PubMed]
61. Heikinheimo, V.; Di Minin, E.; Tenkanen, H.; Hausmann, A.; Erkkonen, J.; Toivonen, T. User-Generated Geographic Information for Visitor Monitoring in a National Park: A Comparison of Social Media Data and Visitor Survey. *ISPRS Int. J. Geo-Inf.* **2017**, *6*, 85. [CrossRef]

sustainability MDPI

Article

Natura 2000 Areas and Sites of National Interest (SNI): Measuring (un)Integration between Naturalness Preservation and Environmental Remediation Policies

Francesco Scorza, Angela Pilogallo *, Lucia Saganeiti and Beniamino Murgante

School of Engineering, University of Basilicata, 85100 Potenza, Italy; francesco.scorza@unibas.it (F.S.); lucia.saganeiti@unibas.it (L.S.); beniamino.murgante@unibas.it (B.M.)

* Correspondence: angela.pilogallo@unibas.it; Tel.: +39-320-421-0250

Received: 27 December 2019; Accepted: 2 April 2020; Published: 7 April 2020

Abstract: The Natura 2000 network was established as a tool to preserve the biological diversity of the European territory with particular regard to vulnerable habitats and species. According to recent studies, a relevant percentage of Natura 2000 sites are expected to be lost by the end of this century and there is widespread evidence that biodiversity conservation policies are not fully effective in relation to the management plans of the protected areas. This paper addresses the issue by analyzing a specific case in which there is a problem of integration between different competences and sectoral policies that leads to the lack of a monitoring system of territorial management performances. The study area, located in the Basilicata Region (Southern Italy), includes a Site of National Interest (SNI), for which several reclamation projects are still in the submission/approval phase, and a partially overlapping Natura 2000 network site. The tool used to monitor biodiversity in the study area is the degradation map obtained through the "habitat quality and degradation" InVEST tool which is used to assess the current trend and thus define a baseline for comparison with two medium and long-term scenarios applicable to the SNI's procedure of partial and total remediation. The proposed methodology is intended to be a part of a larger and more complex monitoring system that, developed within the framework of ecosystem services, allows for the overcoming of the limits related to fragmentation and contradictions that are present in land management by offering a valuable support to decision makers and the competent authorities in biodiversity conservation policy design.

Keywords: habitat degradation; ecosystem services; Natura 2000 network; Site of National Interest (SNI); spatial planning; Basilicata Region

1. Introduction

According to the Intergovernmental Science-Policy Platform on Biodiversity and Ecosystem Services (IPBES) [1] about 25 per cent of both animal and plant species are threatened and around 1 million species already face extinction, many within decades. Even if during the last twenty years remarkable progress has been made in deeply understanding how biodiversity impoverishment affects ecosystems' services (ES) and their relevant performance [2], and consequently the quality [3] of human life, the composition of the species' communities is rapidly decreasing with potentially unpleasant consequences for the resilience of the ecosystems [4].

If the current challenges included in the Sustainable Development Goals (SDGs) point toward balancing the often conflicting objectives of human development and biodiversity conservation (see SDGs 1, 2, and 8) [5], the aims of biological diversity preservation and the protection of vulnerable habitats and species have already been addressed by the European Union (EU) through the Birds and Habitats Directives and the subsequent establishment of the Natura 2000 network (N2K network).

Stretching over 18% of land areas and about 9.5% of marine territories across all 28 EU countries, N2K represents the largest international network of protected areas. Although the goal of constituting a vital backbone of the EU's green infrastructure has been reached, a big challenge still lies ahead: the appropriate management of sites [6]. As a matter of fact, protected areas should not only be considered as reservoirs of biological diversity but also as the sustainers of an ecosystem resilience and the fundamental providers of functions, goods, and services that are essential for human well-being and wealth [7,8]. Although the N2K network potential for the achievement of the conservation objectives is widely recognized [9], the results to date are not considered to be fully satisfactory [10,11], and a number of studies have been carried out in order to identify the main weaknesses [12–15].

As far as territorial governance is concerned, it emerges that the effectiveness of site designation and management depends on the decision-making and policy-design process [16], as the support of local stakeholders in their approval and participatory role is crucial for the long-term success of site management [17,18]. Another point considered to be very critical is that of the overlapping policies and responsibilities at different government levels [12] that are often reflected in cross-scale political contradictions [19], conflicts related to other sectoral policies [20], and a top-down governance gap [21].

The European Commission (EC) [22], in declaring the gap between spatial planning and its instruments for the implementation of the N2K network as one of the most significant causes for the lack of conservation objectives, points to territorial planning as the most appropriate framework for the creation of an improved synergy between different sectoral and environmental policies and for ensuring that developments comply with the EU sectoral and environmental legislation. Furthermore, it promotes the adoption of geo-spatial information technologies (GIS) and remote sensing as reliable information sources for decision-making processes and identifies Strategic Environmental Assessment (SEA) and Environmental Impact Assessment (EIA) as key instruments to guarantee knowledge-based prevention, mitigation, and compensation of sector-specific impacts on N2K sites.

In the same perspective, Leone et al. [23] assume that the consistency between N2K site management plans (MPs) and municipal masterplans (MMPs), in terms of sustainability objectives, should be guaranteed by the SEA of MMPs and/or MPs and considered to be a very effective technical procedure in order to support the implementation of ES into spatial planning.

This work examines a case study from the Basilicata Region where a Site of Community Interest (SCI)/ Special Protection Area (SPA) and a Site of National Interest (SNI) coexist and partially overlap. This area was developed as an industrial site between the post-World War II years and the 70s. Public-driven industrial policies stopped during the 80s and the area underwent a period of abandonment and decline with consequent environmental issues that derived from extensive pollution in abandoned industrial parcels. At the end of the 90s a part of the study area was proposed as a SCI/SPA site thanks to the variety of species and biodiversity richness present there. The procedure for the recognition of sites as nodes of the N2K of the Basilicata Region was completed in 2003, but it was only in 2017, following the drafting and approval of their management plans, that sites were designated as Special Areas of Conservation (SACs).

At the same time, due to the high levels of pollution and related effects on health, environment, and the local economy, Law 179/2002 established the "Val Basento" as a SNI and assigned its remediation responsibility to the Ministry of the Environment, Land, and Sea (MATTM).

The fragmentation of competences in territorial governance led to a lack of integration between those acts promoted by the different competent authorities (Basilicata Region for N2K sites and MATTM for SNI), who operated without any coordination.

The aim of this work is to describe and explain the conflict between SCI/SPAs and SNI sites in our case study via ES land-use analysis and produce an estimation of the positive effects that remediation actions produce on the environmental components in medium and long-term scenarios.

"Habitat quality", in referring to the recent debate on ES classification [24,25], is considered to be a provisioning ES as per the Millennium Ecosystem Services Assessment [26], but a supporting ES according to the most recent interpretation of The Economics of Ecosystems and Biodiversity

(TEEB) [27]. The last version of the Common International Classification of Ecosystem Services (CICES) [28], in reorganizing divisions and groups belonging to the "Regulating and Maintenance Services (biotic)", further explores the various aspects related to biodiversity. The closest reference to the meaning by which we have dealt with habitat quality is the class "maintaining nursery populations and habitats (including gene pool protection)" (code 2.2.2.3—CICES v.5), representing the provision of suitable habitats for wild plants and animals and the maintenance of the appropriate ecological conditions necessary for sustaining these populations.

In this work, we refer to habitat quality as a measure (or even as a proxy) for biodiversity by applying a tool that produces a spatial assessment of habitat quality and degradation based on land-use classes without any additional hypothesis on "genetic diversity" or "species richness" or any other specific biological/ecological meaning.

The analytical tool used is InVEST Habitat Quality, that proved to be effective for the assessment of how different change scenarios in land cover or, as in our case study, habitat threats might affect habitat quality and, consequently, biodiversity [29]. The results obtained in the study area highlight the potential of the proposed methodology to support the decision-making process, orienting reclamation procedures and improving management actions for both SCI/SPA and SNI sites within an integrated approach.

2. Study Area

The study area, located in the Basilicata Region (Southern Italy), extends for about 742.5 square kilometers and is located along the middle valley of the Basento River, partially including the municipalities of Grassano, Grottole, Miglionico, Pomarico, Montescaglioso, Bernalda, Pisticci, Ferrandina, Salandra, Garaguso, and Calciano, all belonging to the province of Matera.

The interest in this study area (see Figure 1) is based on the simultaneous presence of a large industrial area, a SNI, and areas of acknowledged naturalistic-environmental value.

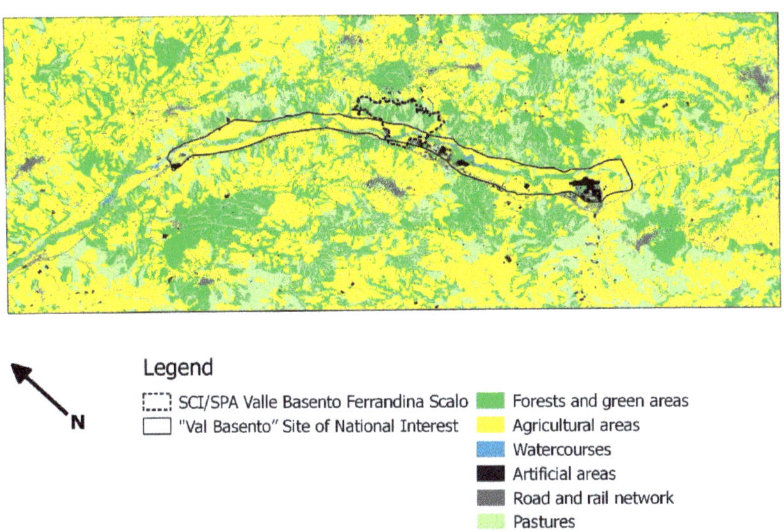

Figure 1. Represents the context of the study area from which emerges the dominance of agricultural land use linked to both cropland and extensive meadows. Along the Basento Valley there is one of the major industrial areas of the Basilicata Region, part of which was subsequently designated as a Site of National Interest (SNI). On the image is also visible the perimeter of the Natura 2000 (N2K) site and the overlapping area under investigation.

This area certainly falls within the Val Basento industrial agglomeration, one of the largest in Southern Italy, founded between the 50s and 60s subsequent to the discovery of a large methane deposit. After the starting phase of the construction works of the Ferrandina-Bari-Monopoli gas pipeline in 1961, many other industrial activities were established in the area, such as the petrochemical complex of the National Fuel Hydrogenation Company (ANIC). The international crisis that started in 1973 did not spare this industrial area and led to the shutdown of several establishments. A program agreement signed in 1987 gave the National Hydrocarbons Agency (Eni) full powers to relaunch the Val Basento industrial area, and the Matera Industrial Consortium was given the task of creating a technology park. In 1990 the so-called Tecnoparco Valbasento Spa was founded, which currently hosts production activities and companies involved in the environmental and energy sectors, providing services and infrastructures such as analysis laboratories; the production and distribution of electricity, nitrogen, and demineralized water production plants; and the collection, treatment, and disposal of liquid waste.

2.1. *"Val Basento" Site of National Interest (SNI)*

The Italian legislation defines SNIs as those areas in which the pollution of soil, subsoil, surface waters, and groundwater is so widespread and severe that they constitute a significant danger to public health and the natural environment.

Because of the seriousness of the contamination, with significant environmental, health, and socio-economic impacts, the administrative competence in remediation procedures is given to the MATTM, whereas local authorities are often involved in the role of operation-implementing bodies and are responsible for the development and protection of the territories affected by environmental contamination. The criteria for the identification of a SNI are defined by Law Number 134/2012, which, as a result, envisages the possibility that the SNI perimeter may change over time on the basis of new information on the potential and/or confirmed contamination of new areas or on the basis of a more accurate definition of the areas affected by potential contamination sources.

The Val Basento industrial area has been declared a SNI in accordance with Law Number 179/2002 "environmental provisions", while its perimeter is officially defined by the Ministerial Decree of 26.02.2003, which identifies an area of approximately 34 square kilometers. It is located along the lower altitudes of the Basento River middle valley, brushing three municipal territories.

Industrial settlements deemed responsible for a higher potential environmental impact include the Eni Desulphurization Plant (formerly AGIP) in the Salandra municipality; the chemical cluster of Ferrandina, including an asbestos treatment company (Materit), a plant for biodiesel production (Mythen), a chemical production plant (ex-Liquichimica, ex-Pozzi, now Syndial); the chemical-pharmaceutical pole of Pisticci with the presence of active ingredient production companies (Gnosis Biosearch), the manufacturing of plastics and chemical fibers (Dow, Nylstar, Politex, Equipolymers), and industrial wastewater treatment plants, such as that of the Tecnoparco Valbasento company and landfill areas (2C Dump, Enrico Mattei Airstrip).

On 31 December 2018 the report on the state of the remediation procedure issued by MATTM [30] declared that only 1% of the SNI's total area was under an approved project for the protection/remediation of both the soil and the aquifers.

2.2. *"Valle Basento–Ferrandina Scalo" Natura 2000 Site*

Since the 1980s, the struggle against the decline in biological diversity and habitat fragmentation has been extensively explored leading to remarkable progress toward understanding how biodiversity loss affects the functioning of ecosystems [3] and, consequently, human well-being. Aware that if this trend were to continue the sixth mass extinction would occur in less than 250 years [31], in 1998 the European Union approved a strategy to preserve biodiversity by actively implementing the Habitats Directive 92/43 and the Birds Directive 79/409 and proceeded with the establishment of the N2K network.

The N2K project assigns great importance to high-naturalness areas but also to semi-natural environments, which are essential for connecting areas that are spatially distant but close in ecological functionality [14]. This means that it assesses not only the current quality of a site but also the potential of the habitats to reach higher levels of complexity. The directive also takes into account those currently degraded sites where, nevertheless, habitats have retained their functional efficiency [32] and can, therefore, return to more evolved forms by reducing or eliminating degradation sources.

The habitats directive furthermore assigns the responsibility of ensuring the management of N2K sites, to member states. Although it is entirely left up to them to decide which option to follow among the management plans (either statutory or administrative measures), all have to take concrete actions to guarantee the conservation status of habitats and species. In Italy, the regions and autonomous provinces are responsible for the management of N2K sites. Most of them have delegated other administrations with this kind of responsibility.

The study area involves two sites of the N2K network: the SCI/SPA IT9220255 Valle Basento Ferrandina Scalo, which is entirely included, and part of the SCI/SPA IT9220260 Valle Basento Grassano Scalo. As it is assumed that the two sites are spatially connected within the ecological network of the Basilicata Region for the reasons mentioned above, the management plan approved by resolve of the Regional Council 1492/2015 is unique for both sites.

It is composed of two main parts: the first describes the characteristics of the two areas and illustrates the higher-ordered and sectoral regulatory framework in which the SCIs are located.

Although the management plan does not make explicit reference to the presence of the SNI, despite the spatial overlapping between these two areas, the planning tools to be considered include the regulatory plan of the industrial development area of Matera. It is prescribed that projects related to productive activities, in addition to their compliance with the obligations provided for by sectorial legislation, must be complemented by a specific report aimed at verifying the impacts and compatibility with neighboring activities and, more generally, with the territorial ecosystem and the urban settlements of the industrial and surrounding areas. Areas classified as SCIs and SPAs allow for the setup of craft activities and small and medium-sized industrial activities, characterized by low pollution levels or disturbances (gases, liquids, noise). Each plant, however, has to be subjected to ex ante Environment Impact Assessment (EIA).

As already mentioned, the interest in this case study arises from the overlay of a high naturalistic-environmental value area and a site considered to be particularly critical for the pollution level, especially with regards to the potential effects that this situation could cause from a social, health, and economic point of view. Equally critical is that no explicit reference to this conflicting situation is included in the management plan for the N2K site.

3. Methodology

The N2K network was created to meet the needs and wishes to preserve breeding and resting grounds for both rare and endangered species and certain endangered habitats. In other words, it was established to safeguard biodiversity [33].

In playing a key role at all levels of the ecosystem service hierarchy [34], biological diversity is the basis for the multifunctionality of N2K sites and is defined as a core planning principle [35] able to explicitly consider multiple and intertwining ecological, social, and economic functions. Sharing the positions of several authors [35–37], whereby an ES approach addresses crucial aspects relevant to multifunctionality planning, in this work we take a conservation perspective [34] by handling habitat quality as a proxy for biodiversity, i.e., as one of the ES provided by N2K sites.

As stated by Sallustio et al. [38], habitat quality assessment can, without a doubt, be considered an effective tool in order to both evaluate the effectiveness of tangible conservation policies and programs and orient management strategies toward improving biodiversity preservation. The tool is adopted in scientific literature to analyze how the intensity of human activity influences habitat quality [39,40] and determine the effects of ongoing threats [41] and expected reclamation actions on the current levels of biodiversity.

The habitat quality model, that belongs to the InVEST suite and is based on the hypothesis that a higher HQ corresponds to a higher abundance of species and vice versa [40,42–45], draws up two maps: habitat quality (Q_{xj}) and degradation (D_{xj}).

The HQ is directly related to the suitability of each land use/land cover (LULC) class to provide adequate conditions for the persistence of biodiversity (H_j) [46] and to (D_{xj}) according to the following formula:

$$Q_{xj} = H_j\left(1 - \left(\frac{D^z_{xj}}{D^z_{xj} + k^z}\right)\right) \quad (1)$$

where z is a constant equal to 2.5 and k is a scaling half-saturation parameter [42]. By default, $k = 0.5$ but users can modify its value in order to better highlight the spatial degradation on the landscape. In dealing with scenario analysis, users are recommended to set the k value equal to half the highest grid cell degradation and run all the subsequent elaborations with the same k value.

Hj represents habitat suitability, and its score, ranging from 0 to 1 and summarized in Table 1, was assigned considering the most suitable $(Hj = 1)$ woodland and freshwater LULC classes as they are considered the least modified habitats and consequently the most suitable for native species. On the other hand, the lowest values (Hj close to 0) were attributed to anthropic LULC classes, such as industrial and residential buildings, roads and railways, and landfills and mining areas. Intermediate values were finally given to the semi-natural land-use classes—grasslands, arable lands, and agricultural crops.

Habitat degradation (D_{xj}) is the function of the sensitivity of each LULC class to each threat (S_{jr}), of the relative weight of each threat (w_r), and of the impact i_{rxy} of the threat r in cell x originating in y and distant d_{xy}:

$$D_{xj} = \sum_r \sum_y \left(\frac{w_r}{\sum_r w_r}\right) r_y i_{rxy} \beta_x S_{jr} \quad (2)$$

where β_x, ranging from 0 to 1, represents the level of accessibility in grid cell x. In this work we always considered a complete accessibility; therefore $\beta_x = 1$.

With regards the assigned sensitivity values, it is possible to see from Table 1 that the highest values were allocated to woodlands, freshwaters, agricultural lands, and grasslands falling within the SCI/SPA area because they were considered to be of higher value and more vulnerable to the threats taken into consideration. In line with habitat suitability, zero sensitivity values were assigned to land-use classes with a strong anthropogenic component and, therefore, considered less vulnerable to threats. This process of value assignment comes from a qualitative expert interpretation of case-study-specific features in land use. In this specific case study, the assessment aimed to highlight, on the one hand, the greater suitability of the habitats included within the N2K site and, on the other hand, to amplify the value of the threat assigned to the industrial areas present within the SNI perimeter. The criterion was to transfer the overall approach of the research (comparing both the protected areas and the national interest sites remediation policies) to input values of the analytical model in order to verify the research hypothesis.

For each threat, users have to assign the maximum influence distance ($d_{r_{max}}$), which is to be considered as the threshold over which the threat r no longer has any impact, and the distance-decay function (linear or exponential):

$$i_{rxy} = 1 - \left(\frac{d_{xy}}{d_{r_{max}}}\right) \text{ if linear} \tag{3}$$

$$i_{rxy} = \exp\left(-\left(\frac{2.99}{d_{r_{max}}}\right)d_{xy}\right) \text{ if exponential.} \tag{4}$$

The advantage of using the InVEST HQ model stems not only from the possibility of representing the cumulative impact of multiple threats in a spatially explicit way but also in adequately considering the variability in the effects of each threat on different habitats (i.e., on each LULC) [40].

In general, a correct interpretation of the results should be achieved by following a joint reading of both habitat quality and degradation maps. As a matter of fact, while the first map depends on LULC and habitat characteristics and on the distribution and intensity of threats, the habitat degradation (HD) map is useful to emphasize areas where cumulative impacts, by different threats, influence HQ. With the same low HQ values, the HD map allows one to distinguish areas of poor naturalness (low H_j values) from those areas characterized by high habitat suitability and affected by a strong impact.

As this work aims to compare the positive effects of remediation actions, three scenarios were formulated, based on the current trend (I scenario), medium (II scenario), and long-term (III scenario) reclamation programs. According to MATTM, the polygons bounded and classified as sources of pollution included by the SNI perimeter and corresponding to abandoned or still-active industrial and production sites, are divided on the basis of the environmental remediation program progress. For some sites (polygons) the reclamation plan has already been approved, while for others it has been elaborated and formally submitted to the competent authority (MATTM) but is still not approved, so it is fair to assume that the reclamation times will be longer. For this reason, the mid-term scenario was created considering that only the sites (polygons) that correspond to approved remediation plans have been reclaimed. The long-term scenario, on the other hand, analyzed the effects of the completely reclaimed site. Both these scenarios (partial and total remediation) were compared to the current trend in which the weight of the threats corresponding to these polygons takes on its maximum value.

The weights have a value between 0 (less important) and 1 (very important) and were assigned on the basis of expert advice. The highest value was attributed to the industrial areas included within the perimeter of the SNI considering that the effects of remediation have a weight reduction from 1 to 0.6 (a reduction of 40%). The hypothesis is that the remediation process cannot recover a full degree of naturalness but has to render the environmental conditions of the sites comparable with other industrial ones. The minimum value of 0.2 was assigned to residential buildings, which in the study area correspond mainly to rural housing.

All the values used for threat definitions are summarized in the following table (Table 2). As can be seen, the assessment made by the expert did not lead to any distinction between the decay functions of the different threats. However, a relevant difference existed between the distances to which every threat exerts its impact. $d_{r_{max}}$ was in fact greater for the industrial areas included within the perimeter of the SNI.

Table 1. Sensitivity matrix in which the habitat suitability [0-1] for each land use/land cover (LULC) and the sensitivity of each habitat to the individual threat [0-1] is reported. Habitat suitability Site of Community Interest/Special Protection Area (SCI/SPA) regards the LULC inside the SCI/SPA.

		Sensitivity for each LULC											
		Arable Land	Woodlands	Freshwaters	Residential Buildings	Secondary Roads	Landfills and Mining Areas	Green Areas	Renewable Energy Sources (RES)	Primary Roads	Industrial Buildings	Railways	Grassland
	Habitat suitability	0,7	1	1	0,1	0,1	0,1	0,3	0,1	0,05	0,001	0,1	0,8
	Habitat suitability (SCI/SPA)	0,8	1	1	0,1	0,1	0,1	0,4	0,1	0,05	0,001	0,1	0,9
Threats	Agriculture	0	0,6	0,4	0	0	0	0	0	0	0	0	0
	Landfills and mining areas	0,6	1	1	0	0	0	0,3	0	0	0	0	0,6
	Industrial buildings	0,6	0,8	0,9	0	0	0	0,5	0	0	0	0	0,7
	RES fields	0,8	0,9	0,7	0	0	0	0,4	0	0	0	0	0,9
	Residential buildings	0,6	1	0,8	0	0	0	0,4	0	0	0	0	0,6
	Primary roads	0,7	0,9	0,7	0	0	0	0,3	0	0	0	0	0,7
	Secondary roads	0,6	0,7	0,5	0	0	0	0,4	0	0	0	0	0,6
	Railways	0,7	0,9	0,7	0	0	0	0,3	0	0	0	0	0,7
	Industrial areas with implemented prevention measures	0,6	0,7	0,7	0,6	0,4	0,4	0,5	0,1	0,4	0,2	0,4	0,9
	Industrial areas with reclamation plan presented but not approved	0,8	1	1	0,8	0,5	0,5	0,6	0,2	0,5	0,3	0,5	1
	Industrial areas with reclamation plan approved	0,8	1	1	0,8	0,5	0,5	0,6	0,2	0,5	0,3	0,5	1

Table 2. Summary table of values used for threat definition.

	Threat	Maximum Influence Distance ($d_{r_{max}}$) (Km)	Distance-Decay Function	Weight		
				I Scenario (Current Trend)	II Scenario (Mid-Term Scenario)	III Scenario (Long-Term Scenario)
	Agriculture	1.5	exponential	0.3	0.3	0.3
	Industrial Buildings	2	exponential	0.6	0.6	0.6
	Landfills and mining areas	2	exponential	0.8	0.8	0.8
	Primary roads	1	exponential	0.6	0.6	0.6
	Secondary roads	0.6	exponential	0.4	0.4	0.4
	Renewable energy sources (RES) farms	1	exponential	0.4	0.4	0.4
	Railways	0.8	exponential	0.5	0.5	0.5
	Residential buildings	1.5	exponential	0.2	0.2	0.2
Industrial areas included by the SNI perimeter	Industrial areas with implemented prevention measures	3	exponential	0.6	0.6	0.6
	Industrial areas with reclamation plan presented but not approved	3	exponential	1	1	0.6
	Industrial areas with reclamation plan approved	3	exponential	1	0.6	0.6

4. Results

The degradation maps obtained for the current trend and some detailed boxes showing partial (mid-term) and total (long-term) remediation scenarios, respectively, are shown below (Figure 2). Color nuances represent the areas where the reclamation actions have a weaker (light red) or a greater effect (dark red).

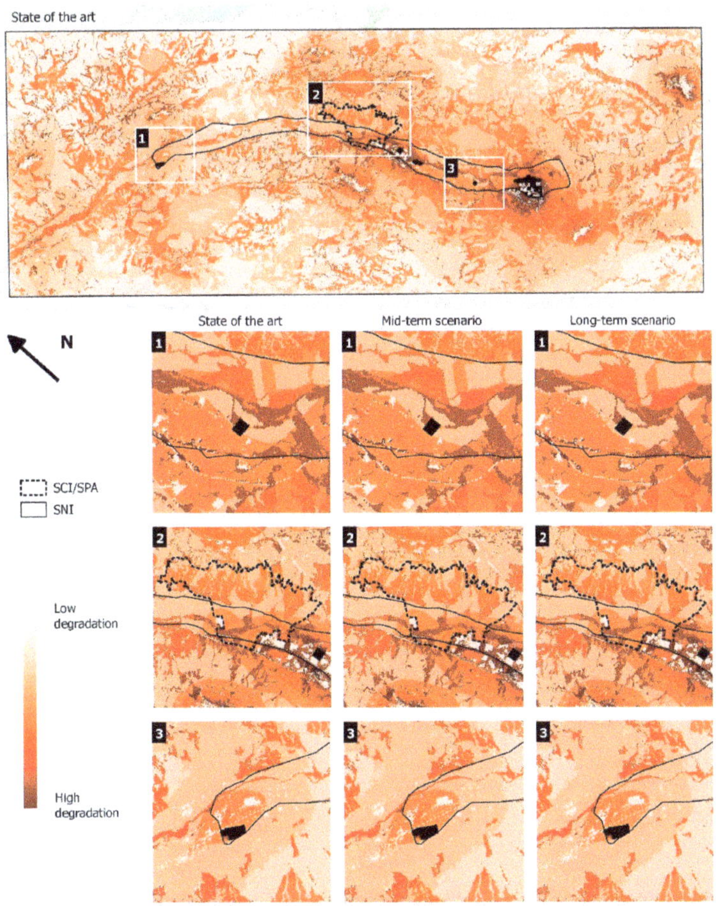

Figure 2. Degradation analysis: the large figure represents the entire study area in the current trend. The detail boxes display three areas where, thanks to the greater proximity to industrial sites that currently constitute a source of threat, the effects of remediation measures are more visible.

Out of a total of 21 polygons (i.e., industrial areas) identified as sources of impact according to the MATTM, the medium-term scenario showed the remediation effects of only one of these industrial areas accounting for 3.71 ha. The degradation decrease was, therefore, rather localized and was not extended to the entire study area

The effects of a complete remediation, which involves a total of three industrial areas, were more significant. One area was located in the northern part of the SNI zone and the other two were in the immediate downstream area with respect to the SCI/SPA perimeter. The surface area involved in this case was 30.06 ha, equal to 0.04% of the entire study area.

Figure 3, with respect to the difference in the degradation maps following partial and total remediation related to the current situation, shows the areas where the benefits of the interventions are felt.

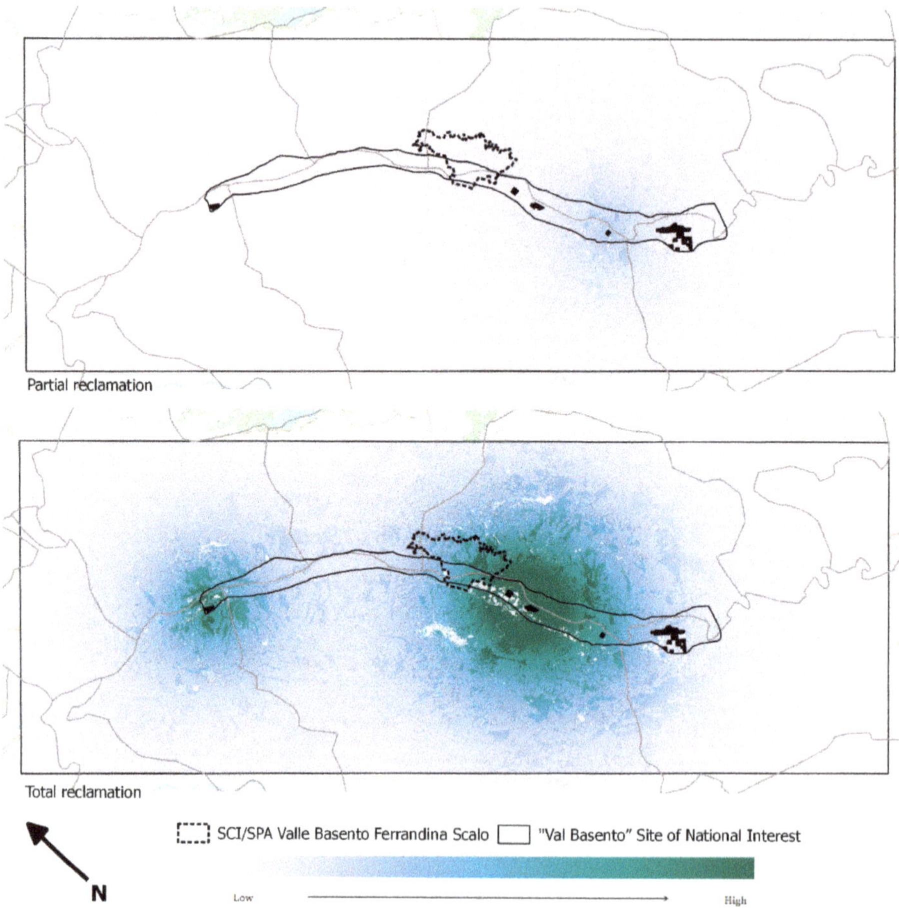

Figure 3. Effects of the reclamation procedure: decrease of the degradation degree in mid-term (partial reclamation) and long-term (total reclamation) scenarios. The figure shows the effects, obtained by difference, of the interventions foreseen in the medium and long-term scenario. While the area of influence for the partial remediation scenario is rather limited compared to the study area, total remediation has two much larger areas of influence.

In order to assess the variations following the implementation of a remediation program, the percentage changes were calculated using map algebra operations.

The maximum reduction in the degradation level (therefore corresponding to the long-term scenario) within the study area was 6%, whereas in the medium term, a decrease of less than 1% could be expected.

Within the SCI/SPA site, the partial reclamation (mid-term scenario) had no effect on changes in habitat degradation or quality. Differently, in the southern part of the area, the degradation reduction reached 3% as its maximum value (Figure 4).

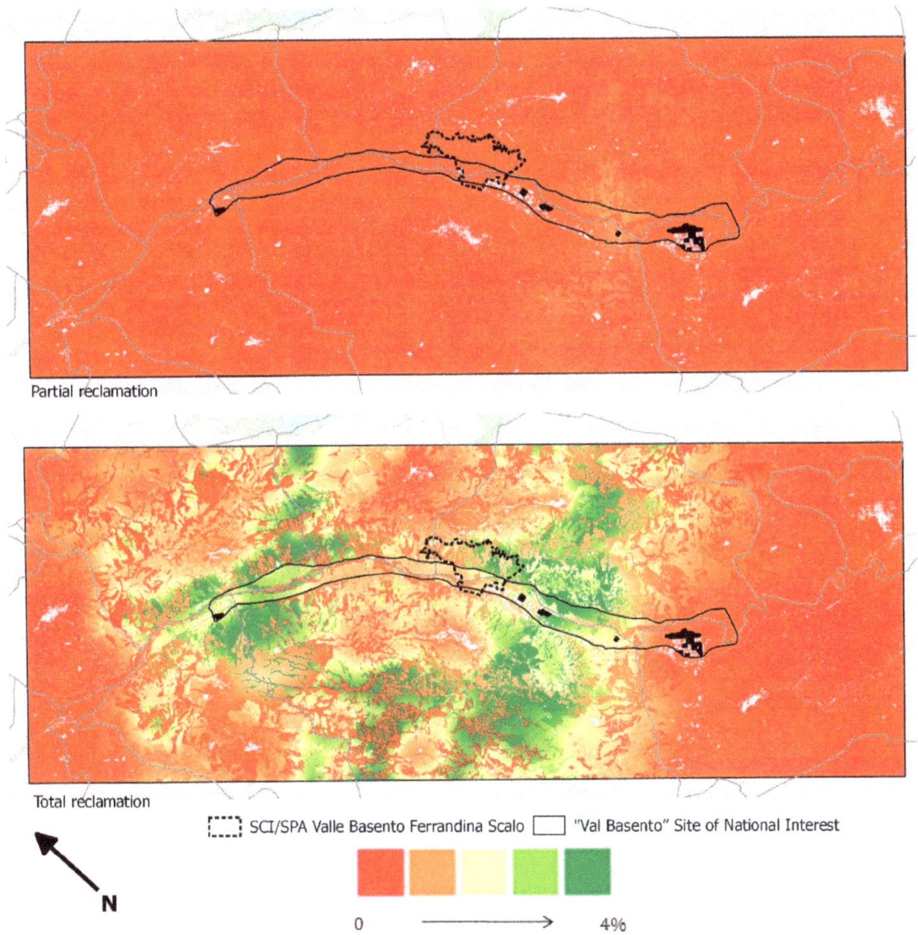

Figure 4. Percentage reduction in habitat degradation in mid-term (partial reclamation) and long-term (total reclamation) scenarios. It is possible to note that the effects of land reclamation were highly dependent on land use, with priority being given to meadows and grasslands. It is also possible to see that in areas where there are several threats that give rise to a cumulative impact, the effects of remediation were less significant. This was the case, for example, of the riparian areas close to the primary road and the railway line.

By overlapping the results obtained in terms of the percentage degradation decrease and land-use map, it appeared that the main beneficiaries of land reclamation interventions were mainly riparian areas close to creeks (reduction in degradation degree up to 4%) and grazing grounds. On the contrary, the wooded areas were affected to a lesser degree by the reclamation operations which, in these areas, led to a reduction of a maximum of 2%.

5. Discussions

An interpretation of the results must take into account that the model neglects the morphology of the territory and, therefore, the privileged directions of pollutant diffusion. The results are linked to land-use class characteristics in terms of habitat suitability and vulnerability to different threats. Because it is, in actual fact, a valley riverbed with two converging sides, the expected spatial distribution of reclamation effects—all other variables being equal—is not isotropic as, however, appears from the image. As a matter of fact, in the western part of the study area there is a sector along the slope where a relevant degradation reduction is recorded, especially with respect to meadows, grazing grounds, and agricultural crops. Moreover, no improvement in the wooded areas has been achieved, especially where they are surrounded by cultivated areas. This is due to the fact that agriculture is in itself a source of threat, and the relevant reclamation effects are clearly marginal.

However, in analyzing data distribution (Figure 5), it is evident that reclamation effects have a positive impact on different patches that, in the long-term scenario, are included in the lowest degradation class.

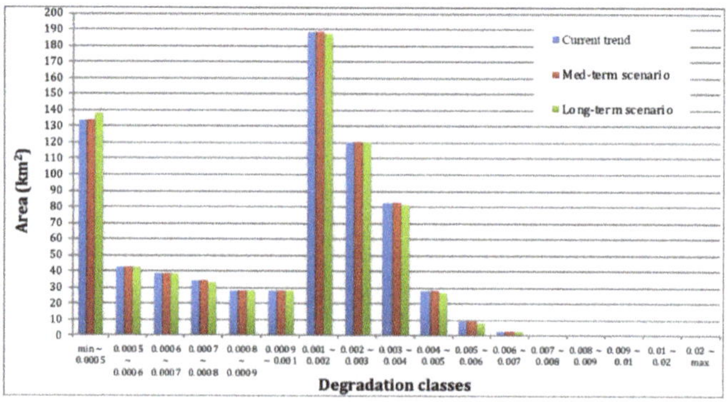

Figure 5. Distribution of degradation values in relation to the three scenarios.

In attentively analyzing the data, the partial reclamation (mid-term scenario) has an effect on a portion of territory that is equal to 0.56% of the total area. In particular, the classes that in the current situation are characterized by a high degradation, move to the low (0.43%) and medium (0.13%) degradation classes. The long-term remediation scenario, on the other hand, involves 3.79% of the total surface area which, from the higher degradation classes, shifts to the lowest degradation level.

The results show that habitat degradation in the study area is certainly due to the cumulative impact of multiple threats arising partially from the industrial area located along one of the main regional road infrastructures and partially from the road and rail network that from the river valley branches off along the slopes, also bordering or crossing areas of high naturalness.

6. Conclusions

Although the conservation of biodiversity is a priority on the European Union's agenda, the results reached so far are still lagging behind the objectives set [22]. The need to achieve a better integration between policies and stakeholders at all levels is highlighted with particular regard to the N2K network [16].

This work analyzes a case study where a strong contradiction in the management of the territory emerges. On the one hand, the presence of industrial areas with high pollution potential has led to the identification of a SNI whose remediation procedure is the direct responsibility of MATTM. On the other hand, the recognition of naturalistic and biodiversity conservation values has led to the identification of a N2K site, whose management plan, approved in 2015, completely neglects the presence of a SNI.

The aim of this work is to provide and test a methodology able to measure the effects of two overlapping and conflicting policy frameworks: the first is oriented toward naturalness preservation (N2K) and the second aims at solving environmental contamination issues (SNI). Both policies substantially ignore each other and demonstrate a fragmentation in the territorial governance system, where different authorities are responsible for specific fields of intervention.

As already pointed out by the authors in previous works [47–51], the ES framework allows for the integration and simultaneous consideration of multiple scales, multiple habitats, and multilevel environmental policies, thus offering the advantage of more holistic environmental management [52].

Therefore, the use of the InVEST Habitat Quality model allowed us to contribute to the general process for the provision of an effective territorial monitoring system, suitable to assess the effects of ongoing threats and environmental management actions on habitat quality. Although the model simplifies the complex reality linked to the phenomenon of pollutant diffusion, temporal and spatial variability, boundary conditions, and more generally the complex dynamics with which threats act to the detriment of habitat quality [46,53], it is useful to perform a scenario analysis in order to identify threats and habitats with respect to land use, especially in those area in which information on species abundance and composition, endemism, and functional significance is poor [40].

Considering the emerged strengths and weaknesses presented in the discussion section, future research should deepen the proposed methodology that proves to be efficient in approaching the issue of territorial governance by overcoming the limits of sectorial policy fragmentation. Therefore, further developments will be oriented toward testing alternative tools so as to better model the remediation processes in detail, including information on specific remediation actions (today not available) and integration input data with additional layers, such as morphology, water quality, air pollution, noise pollution, and evidence of climate-change effects.

The ES approach is considered adequate to deliver a common spatial evaluation framework in order to achieve a better integration of territorial governance, which has been fragmented into different decision centers (see also the work by Balletto et al. 2020 [54]). The proposed methodology should be useful for the construction of a cognitive framework that supports a regional landscape plan which we consider to be the appropriate planning level to manage such a contradictory case (SNI and SCI/SPA overlapping). This planning level is still lacking in the Basilicata Region even if the Regional Planning Law (23/99) foresees its development as a structural node in hierarchy planning.

Author Contributions: Conceptualization, A.P. and F.S.; methodology, F.S.; software, A.P.; validation, B.M., F.S. and A.P.; formal analysis, L.S.; investigation, A.P.; resources, A.P. and L.S.; data curation, L.S.; writing—original draft preparation, A.P. and L.S.; writing—review and editing, A.P. and F.S.; visualization, F.S.; supervision, F.S. and B.M.; project administration, B.M.; funding acquisition, F.S. All authors have read and agreed to the published version of the manuscript.

Funding: This research received no external funding.

Conflicts of Interest: Authors declare no conflict of interest. The funders had no role in the design of the study; in the collection, analyses, or interpretation of data; in the writing of the manuscript, or in the decision to publish the results.

References

1. IPBES. Summary for policymakers of the global assessment report on biodiversity and ecosystem services of the Intergovernmental Science-Policy Platform on Biodiversity and Ecosystem Services. 2019. Available online: https://ipbes.net/global-assessment. (accessed on 15 December 2019).
2. European Commission. Our life insurance, our natural capital: an EU biodiversity strategy to 2020. *Landsc. Ecol. Manag.* **2015**, *20*, 37–40. Available online: https://ec.europa.eu/environment/nature/biodiversity/comm2006/pdf/EP_resolution_april2012.pdf (accessed on 20 December 2019).
3. Cardinale, B.J.; Duffy, J.E.; Gonzalez, A.; Hooper, D.U.; Perrings, C.; Venail, P.; Narwani, A.; Mace, G.M.; Tilman, D.; Wardle, D.A.; et al. Biodiversity loss and its impact on humanity. *Nature* **2012**, *486*, 59–67. [CrossRef] [PubMed]
4. Oliver, T.H.; Isaac, N.J.B.; August, T.A.; Woodcock, B.A.; Roy, D.B.; Bullock, J.M. Declining resilience of ecosystem functions under biodiversity loss. *Nat. Commun.* **2015**, *6*, 10122. [CrossRef] [PubMed]
5. Waldron, A.; Miller, D.C.; Redding, D.; Mooers, A.; Kuhn, T.S.; Nibbelink, N.; Roberts, J.T.; Tobias, J.A.; Gittleman, J.L. Reductions in global biodiversity loss predicted from conservation spending. *Nature* **2017**, *551*, 364–367. [CrossRef]

6. Kati, V.; Hovardas, T.; Dieterich, M.; Ibisch, P.L.; Mihok, B.; Selva, N. The challenge of implementing the European network of protected areas Natura 2000. *Conserv. Biol.* **2015**, *29*, 260–270. [CrossRef]
7. Bastian, O. The role of biodiversity in supporting ecosystem services in Natura 2000 sites. *Ecol. Indic.* **2013**, *24*, 12–22. [CrossRef]
8. Doak, D.F.; Bakker, V.J.; Goldstein, B.E.; Hale, B. Tutorial reading 1—What is the future of conservation? *Trends Ecol. Evol.* **2014**, *29*, 77–81. [CrossRef]
9. Donald, P.F.; Sanderson, F.J.; Burfield, I.J.; Bierman, S.M.; Gregory, R.D.; Waliczky, Z. International conservation policy delivers benefits for birds in Europe. *Science* **2007**, *317*, 810–813. [CrossRef]
10. Jaeger, J.A.G.; Madrinán, L.F.; Soukup, T.; Schwick, C.; Kienast, F. Landscape fragmentation in Europe (EEA Report, Report No.: 2). Copenhagen, EEA: Bern: Federal Office for the Environment FOEN, European Environment Agency EEA. 2011. Available online: https://www.eea.europa.eu/publications/landscape-fragmentation-in-europe (accessed on 20 December 2019).
11. European Environment Agency. EU 2010 Biodiversity Baseline. Post-2010 EU biodiversity policy; Luxembourg. 2010. Available online: https://www.eea.europa.eu/publications/eu-2010-biodiversity-baseline (accessed on 20 December 2019).
12. Nolte, C.; Leverington, F.; Kettner, A. *Protected area management effectiveness assessments in Europe*; Bundesamt für Naturschutz (BfN) Federal Agency for Nature Conservation: Bonn, Germany, 2010; ISBN 978-3-89624-006-4.
13. Brambilla, M.; Bergero, V.; Bassi, E.; Falco, R. Current and future effectiveness of Natura 2000 network in the central Alps for the conservation of mountain forest owl species in a warming climate. *Eur. J. Wildl. Res.* **2015**, *61*, 35–44. [CrossRef]
14. Dimitrakopoulos, P.G.; Memtsas, D.; Troumbis, A.Y. Questioning the effectiveness of the Natura 2000 Special Areas of Conservation strategy: The case of Crete. *Glob. Ecol. Biogeogr.* **2004**, *13*, 199–207. [CrossRef]
15. Pellegrino, D.; Schirpke, U.; Marino, D. How to support the effective management of Natura 2000 sites? *J. Environ. Plan. Manag.* **2017**, *60*, 383–398. [CrossRef]
16. Beunen, R.; de Vries, J.R. The governance of Natura 2000 sites: The importance of initial choices in the organisation of planning processes. *J. Environ. Plan. Manag.* **2011**, *54*, 1041–1059. [CrossRef]
17. Dudley, N.; Jo Mulongoy, K.; Cohen, S.; Stolton, S.; Victor Barber, C.; Babu Gidda, S.; Nigel Dudley, C. *Towards Effective Protected Area Systems. An Action Guide to Implement the Convention on Biological Diversity Programme of Work on Protected Areas*; Secretariat of the Convention on Biological Diversity: Montreal, CA, USA, 2005; Volume 108.
18. Achim Steiner, J.M. The World's protected areas: Status, values and prospects in the 21st century. *Choice Rev. Online* **2009**, *46*, 3865.
19. Apostolopoulou, E.; Drakou, E.G.; Pediaditi, K. Participation in the management of Greek Natura 2000 sites: Evidence from a cross-level analysis. *J. Environ. Manag.* **2012**, *113*, 308–318. [CrossRef] [PubMed]
20. Winkel, G.; Blondet, M.; Borrass, L.; Frei, T.; Geitzenauer, M.; Gruppe, A.; Jump, A.; de Koning, J.; Sotirov, M.; Weiss, G.; et al. The implementation of Natura 2000 in forests: A trans- and interdisciplinary assessment of challenges and choices. *Environ. Sci. Policy* **2015**, *52*, 23–32. [CrossRef]
21. Stringer, L.C.; Paavola, J. Participation in environmental conservation and protected area management in Romania: A review of three case studies. *Environ. Conserv.* **2013**, *40*, 138–146. [CrossRef]
22. Natura 2000 and Spatial Planning. Final report for the European Commission (DG ENV) (Project 07.0202/2015/716477/ETU/ENV. B.3). Available online: https://www.wur.nl/upload_mm/9/5/2/7368439e-42ea-427a-8aec-f5d7d5827442_02635_Natura%202000_V5.pdf (accessed on 20 December 2019).
23. Leone, F.; Zoppi, C.; Leone, F.; Zoppi, C. Conservation Measures and Loss of Ecosystem Services: A Study Concerning the Sardinian Natura 2000 Network. *Sustainability* **2016**, *8*, 1061. [CrossRef]
24. La Notte, A.; D'Amato, D.; Mäkinen, H.; Paracchini, M.L.; Liquete, C.; Egoh, B.; Geneletti, D.; Crossman, N.D. Ecosystem services classification: A systems ecology perspective of the cascade framework. *Ecol. Indic.* **2017**, *74*, 392–402. [CrossRef]
25. Costanza, R.; de Groot, R.; Braat, L.; Kubiszewski, I.; Fioramonti, L.; Sutton, P.; Farber, S.; Grasso, M. Twenty years of ecosystem services: How far have we come and how far do we still need to go? *Ecosyst. Serv.* **2017**, *28*, 1–16. [CrossRef]
26. Alcamo, J.; Bennett, E.M. *Millennium Ecosystem Assessment (Program). Ecosystems and Human Well-Being: A Framework for Assessment*; Island Press: Washington, DC, USA, 2003; ISBN 9781559634038.

27. Ten Brink, P. *The Economics of Ecosystems and Biodiversity in National and International Policy Making*; Patrick ten Brink: Earthscan, London and Washingtonk, UK, 2012; ISBN 9781849775496.
28. Haines-Young, R.; Potschin, M. Common International Classification of Ecosystem Services (CICES) V5.1 Guidance on the Application of the Revised Structure; Barton in Fabis, Nottingham, NG11 0AE, U. 2018. Available online: https://cices.eu/content/uploads/sites/8/2018/01/Guidance-V51-01012018.pdf (accessed on 20 December 2019).
29. Arcidiacono, A.; Ronchi, S.; Salata, S. Ecosystem Services Assessment Using InVEST as a Tool to Support Decision Making Process: Critical Issues and Opportunities. *Lect. Notes Comput. Sci. (including Subser. Lect. Notes Artif. Intell. Lect. Notes Bioinformatics)* **2015**, *9158*, 35–49.
30. Documenti Sullo Stato di Avanzamento delle Procedure di Bonifica Ministero dell'Ambiente e della Tutela del Territorio e del Mare. Available online: https://www.minambiente.it/bonifiche/documenti-sullo-stato-di-avanzamento-delle-procedure-di-bonifica (accessed on 20 December 2019).
31. Hooper, D.U.; Adair, E.C.; Cardinale, B.J.; Byrnes, J.E.K.; Hungate, B.A.; Matulich, K.L.; Gonzalez, A.; Duffy, J.E.; Gamfeldt, L.; O'Connor, M.I. A global synthesis reveals biodiversity loss as a major driver of ecosystem change. *Nature* **2012**, *486*, 105–108. [CrossRef]
32. Cortina, C.; Boggia, A. Development of policies for Natura 2000 sites: A multi-criteria approach to support decision makers. *J. Environ. Manag.* **2014**, *141*, 138–145. [CrossRef] [PubMed]
33. Friedrichs, M.; Hermoso, V.; Bremerich, V.; Langhans, S.D. Evaluation of habitat protection under the European Natura 2000 conservation network—The example for Germany. *PLoS ONE* **2018**, *13*, e0208264. [CrossRef] [PubMed]
34. Mace, G.M.; Norris, K.; Fitter, A.H. Biodiversity and ecosystem services: A multilayered relationship. *Trends Ecol. Evol.* **2012**, *27*, 19–26. [CrossRef] [PubMed]
35. Hansen, R.; Pauleit, S. From multifunctionality to multiple ecosystem services? A conceptual framework for multifunctionality in green infrastructure planning for Urban Areas. *Ambio* **2014**, *43*, 516–529. [CrossRef] [PubMed]
36. Haase, D.; Schwarz, N.; Strohbach, M.; Kroll, F.; Seppelt, R. Synergies, Trade-offs, and Losses of Ecosystem Services in Urban Regions: An Integrated Multiscale Framework Applied to the Leipzig-Halle Region, Germany. *Ecol. Soc.* **2012**, *17*, art22. [CrossRef]
37. Chan, K.M.A.; Satterfield, T.; Goldstein, J. Rethinking ecosystem services to better address and navigate cultural values. *Ecol. Econ.* **2012**, *74*, 8–18. [CrossRef]
38. Sallustio, L.; De Toni, A.; Strollo, A.; Di Febbraro, M.; Gissi, E.; Casella, L.; Geneletti, D.; Munafò, M.; Vizzarri, M.; Marchetti, M. Assessing habitat quality in relation to the spatial distribution of protected areas in Italy. *J. Environ. Manag.* **2017**, *201*, 129–137. [CrossRef]
39. Gao, Y.; Ma, L.; Liu, J.; Zhuang, Z.; Huang, Q.; Li, M. Constructing Ecological Networks Based on Habitat Quality Assessment: A Case Study of Changzhou, China. *Sci. Rep.* **2017**, *7*, 46073. [CrossRef]
40. Terrado, M.; Sabater, S.; Chaplin-Kramer, B.; Mandle, L.; Ziv, G.; Acuña, V. Model development for the assessment of terrestrial and aquatic habitat quality in conservation planning. *Sci. Total Environ.* **2016**, *540*, 63–70. [CrossRef]
41. Scorza, F.; Pilogallo, A.; Saganeiti, L.; Murgante, B.; Pontrandolfi, P. Comparing the territorial performances of Renewable Energy Sources' plants with an integrated Ecosystem Services loss assessment: A case study from the Basilicata region (Italy). *Sustain. Cities Soc.* **2020**, *56*, 102082. [CrossRef]
42. The Natural Capital Project Habitat Quality—InVEST 3.6.0 Documentation. Available online: http://data.naturalcapitalproject.org/nightly-build/invest-users-guide/html/habitat_quality.html (accessed on 13 November 2019).
43. Aneseyee, A.B.; Noszczyk, T.; Soromessa, T.; Elias, E. The InVEST Habitat Quality Model Associated with Land Use/Cover Changes: A Qualitative Case Study of the Winike Watershed in the Omo-Gibe Basin, Southwest Ethiopia. *Remote Sens.* **2020**, *12*, 1103. [CrossRef]
44. Golden, H.E.; Lane, C.R.; Amatya, D.M.; Bandilla, K.W.; Raanan Kiperwas, H.; Knightes, C.D.; Ssegane, H.; Lüke, A.; Hack, J.; Brauman, K.A.; et al. The Use of Scenario Analysis to Assess Water Ecosystem Services in Response to Future Land Use Change in the Willamette River Basin, Oregon. *J. Environ. Manag.* **2014**, *53*, 4–14.

45. Gong, J.; Xie, Y.; Cao, E.; Huang, Q.; Li, H. Integration of InVEST-habitat quality model with landscape pattern indexes to assess mountain plant biodiversity change: A case study of Bailongjiang watershed in Gansu Province. *J. Geogr. Sci.* **2019**, *29*, 1193–1210. [CrossRef]
46. Leh, M.D.K.; Matlock, M.D.; Cummings, E.C.; Nalley, L.L. Quantifying and mapping multiple ecosystem services change in West Africa. *Agric. Ecosyst. Environ.* **2013**, *165*, 6–18. [CrossRef]
47. Scorza, F.; Murgante, B.; Las Casas, G.; Fortino, Y.; Pilogallo, A. *Investigating Territorial Specialization in Tourism Sector by Ecosystem Services Approach*; Springer: Berlin/Heidelberg, Germany, 2019; pp. 161–179.
48. Mazzariello, A.; Pilogallo, A.; Scorza, F.; Murgante, B.; Las Casas, G. Carbon Stock as an Indicator for the Estimation of Anthropic Pressure on Territorial Components. In *Lecture Notes in Computer Science (Including Subseries Lecture Notes in Artificial Intelligence and Lecture Notes in Bioinformatics)*; Springer Verlag: Berlin, Germany, 2018; Volume 10964, pp. 697–711. ISBN 9783319951737.
49. Pilogallo, A.; Saganeiti, L.; Scorza, F.; Las Casas, G. *Tourism Attractiveness: Main Components for a Spacial Appraisal of Major Destinations According with Ecosystem Services Approach*; Springer: Berlin/Heidelberg, Germany, 2018; pp. 712–724.
50. Pilogallo, A.; Saganeiti, L.; Scorza, F.; Murgante, B. *Ecosystem Services Approach to Evaluate Renewable Energy Plants Effects*; Springer: Berlin/Heidelberg, Germany, 2019; pp. 281–290.
51. Scorza, F. Improving EU Cohesion Policy: The Spatial Distribution Analysis of Regional Development Investments Funded by EU Structural Funds 2007/2013 in Italy. In *Lecture Notes in Computer Science (Including Subseries Lecture Notes in Artificial Intelligence and Lecture Notes in Bioinformatics)*; ICCSA: Ho Chi Minh City, Vietnam, 2013; Volume 7973 LNCS, pp. 582–593. ISBN 9783642396458.
52. Maltby, L.; van den Brink, P.J.; Faber, J.H.; Marshall, S. Advantages and challenges associated with implementing an ecosystem services approach to ecological risk assessment for chemicals. *Sci. Total Environ.* **2018**, *621*, 1342–1351. [CrossRef]
53. Tallis, H.; Mooney, H.; Andelman, S.; Balvanera, P.; Cramer, W.; Karp, D.; Polasky, S.; Reyers, B.; Ricketts, T.; Running, S.; et al. A Global System for Monitoring Ecosystem Service Change. *Bioscience* **2012**, *62*, 977–986. [CrossRef]
54. Balletto, G.; Milesi, A.; Fenu, N.; Borruso, G.; Mundula, L. Military Training Areas as Semicommons: The Territorial Valorization of Quirra (Sardinia) from Easements to Ecosystem Services. *Sustainability* **2020**, *12*, 622. [CrossRef]

sustainability
MDPI

Article

Landscape Features of Costal Waterfronts: Historical Aspects and Planning Issues

Donatella Cialdea

L.a.co.s.t.a. Laboratory, University of Molise, Via De Sanctis, 86100 Campobasso, Italy; cialdea@unimol.it; Tel.: +39-0874-404-970

Received: 8 January 2020; Accepted: 16 March 2020; Published: 18 March 2020

Abstract: This paper investigates the relationship between different factors that impose on the productive and settlement structures on coastal areas through an analysis carried out on the Italian Adriatic Sea coast. In the panorama of medium- and small-size cities, the relationship between the city, the territory, and the sea very often plays an important role. The main issue of this article is to expose a methodology developed for the definition of landscape quality objectives in the planning of the coast of a region in Southern Italy, Molise. Effort was concentrated on the creation of a territorial survey matrix that could be exploited by local authorities. In drawing up the criteria on which to base the New Regional Landscape Plan, this study provided for the recognition of the identifying matrices for landscape interpretation, creating a database organized in five resource systems. For each resource system, three basic grids were created: each of them collects and processes different information series. These three grids were useful for defining the new protection that is proposed for the sample area. Different conditions emerge in this area, in which two coastal strips have been identified, to the east and to the west of the historical centre.

Keywords: landscape regeneration; urban built environment; identity values; smart and resilient land; Natura 2000 Network

1. Introduction

The waterfront design and the coastal areas planning are topical issues. The international debate has shown that the potential urban quality of the waterfront in many cases is not in itself sufficient to guarantee an optimization of resources and the management of territorial transformations.

The remarkable interest in urban and territorial planning and regeneration has developed over time: many national and international researchers have defined numerous principles and guidelines on the theme of the coast and of resilient waterfronts in order to adapt them to ecological and sustainable growth and to new transformation models. Some general principles are evoked in order to underline that port cities everywhere are protagonists of major changes, both in terms of physical change and in terms of change in social tissues. Moreover, the specificity of European cities offers a panorama of similar solutions: the coexistence of strong historical references makes theories and application very comparable to each other, especially for the cities that are located in the Mediterranean basin.

Many regeneration interventions involving port areas have proved to have an important role in defining a smart sustainable development model. Starting from local cultural resources, they often manage to combine economic activities of the port with a regeneration of cultural heritage, also through the creativity of their inhabitants [1,2]. Many European cities have done so. Just think of some Spanish cities, like Barcelona or Valencia; some English cities, like Liverpool or Glasgow; some cities in the Netherlands, like Rotterdam or Amsterdam; some French cities, with the important case of Marseille; or even some German cities, with the well-known case of Hamburg. These examples concern cities in Europe. They highlight a sort of "European specificity" for which the new regained waterfronts tend to combine architectural boldness with territorial interventions, able to create a new face for the cities as a whole [3]. The policies of urban regeneration, in the final analysis, have taken on different faces over the last thirty years, but they shared the enhancement of the common heritage of the city.

In particular, the cases mentioned are apparently similar but instead represent very different urban planning solutions. The continuous interest of urban planning which, in recent decades, has thoroughly examined the issues pertaining to the "waterfront renaissance" [4–7] concerns every waterfront regeneration processes. In the regeneration interventions, the recovery of the built compared to land consumption not only is a physical and environmental recovery but also acts on the local economy; on the quality of living; on the social integration of its inhabitants; and, not less important, on financial instruments related to feasibility, costs, and benefits. The well-known transformations of Barcelona, for example, started as early as the end of the 1980s, were strongly characterized by the impulses linked to the great events [8].

Another example may be the city of Valencia, where the uses of the coastal strip historically had already undergone strong variations over time due to the construction of the port and later for its extension. The port area, in fact, has invaded valuable agricultural areas, with ancient orchards, and above all, it had never built a dialogue with the urban part of the city [9,10].

The case of the port city of Bilbao, where regeneration has involved the course of the Nerviòn river, especially in the part of its estuary, is also a very well-known example. Its transformation from a place of iron and steel business to a new urban reality is interesting from many points of view. It is particularly attractive due to the territorial dimension of urban regeneration policies, which are well integrated with the various intervention scales [11].

The Bilbao metropolitan strategic plan [12], definitively approved in 2003, includes the 34 municipalities of the metropolitan area, which develops along the course of the Nerviòn and includes a population of about one million inhabitants. The purpose of the strategic plan is to carry out a series of interventions, divided into "Aciones Estructurantes" and "Operaciones Estrategicas", which will transform the Bilbao area into a district of advanced services in a modern industrial region.

The case of Marseilles is also an opportunity for reflection: the work of waterfront redevelopment starts from the Euroméditerranée projects and the European Capital of Culture 2013, which aimed to reconnect the city to the sea and to heal a huge rift generated between these two elements when the port was moved to the northern part of the city. Historically, the city is divided into two parts because of the deep bay where the port was present. Marseille is a city that has never completely separated from the sea: it has rather incorporated it and tamed it, making it almost unrecognizable from the land in the cove of the Vieux Port.

Consequently, the need was felt to give continuity to the two shores of the Vieux Port area. A big part of the interventions, in fact, focused on the Vieux Port, classified as a UNESCO heritage site, and on the abandoned area of the J4 pier. Norman Foster won the international competition for the redevelopment and semi-customization masterplan of the Vieux Port in Marseille announced by Marseille Provence Métropole in 2010 [13,14].

Many local administrations prefer to rely on a model already experimented elsewhere and considered a winner: that of the great polarizing building of the ludic-cultural activity of the entire area concerned. This strategy, called the "Bilbao effect", involves assigning great names in architecture, able with their prestige to automatically confer a good degree of legitimacy on the regeneration operation. However, this operation is not always enough in itself to reestablish the authentic bond between the sea and the city. A reconstruction of the relationship between city and port must be concerted with all the actors involved such as local administrations; port authorities; and, above all, the local populations who live this relationship every day with its strengths and its criticalities.

This work analyses the context of major changes in the coastal territory and creates a system of territorial investigation that takes into account all transformations. Environmental, natural, and historical-cultural elements are threatened by growing anthropization, which accurate landscape planning could help to safeguard. Regarding the multiplicity of factors involved, the final goal is to give a contribution to the definition of how the productive economic system is able to construct new geographies in the panorama of national landscape features. The coastal systems are the ones that are most present in the Italian peninsula and that enrich the number of its geographies, also related to changes in land use over time.

The research background undoubtedly includes multiple activities, which naturally vary from context to context, especially in relation to the economic value of the coastal territories. What this research seeks to highlight, in order to offer new insights, is the desire to involve a framework, the one used in the landscape plan setting, to create a methodology based on territorial analysis, which also supports the choices for the waterfront. The identification of resource systems as an opportunity to compare the values of the area under study is considered the innovative element of the work. The final goal was the collection and subsequently the creation of a Geographical Information System of territorial governance tools that affect the analysed process with the help of GIS software. In this way, data processing and territorial analysis aimed at identifying critical issues and possible urban regeneration scenarios.

This is the result of previous and ongoing research, conducted in the L.a.co.s.t.a. Laboratory (**La**boratorio per le **A**ttività **CO**llegate allo **S**viluppo **T**erritoriale ed **A**mbientale) of the Molise University, which is at present responsible for the drawing up of the new landscape plan of the Molise Region due to the Agreement between University and Region, who funded this work.

Recognition of landscape protection interventions, oriented to preserve landscape values in harmony with planning tools, have been carried out. Moreover, different productive features, mainly in areas transformed from unproductive to agricultural use, through extensive reclamation operations were focused. The regional territories have been the object of settlement projects related to both tourism and production development. In continuation of what has already been elaborated over the last few years—by virtue of the financing of international projects using funds from Community Initiative Programs [15–17]—the work explores these issues in particular from the landscape point of view, following dictates of the Italian recent law, named the "Urbani Code" [18].

This study is structured in four sections. The introduction (Section 1) has described the main issues of the paper. Section 2 describes the methodology, which also includes a review of methodological approaches for landscape analyses. The next sections (Section 3 with the results, Section 4 with the discussion, and Section 5 with the conclusions) contain the description of the experimental results and comments, geared towards the extension of this work for future research.

2. Materials and Methods

The methodology implemented for the New Regional Landscape Plan preparation was oriented to searching the relationship between territory and production processes.

The figure below describes the methodology applied in the different phases of the work (Figure 1).

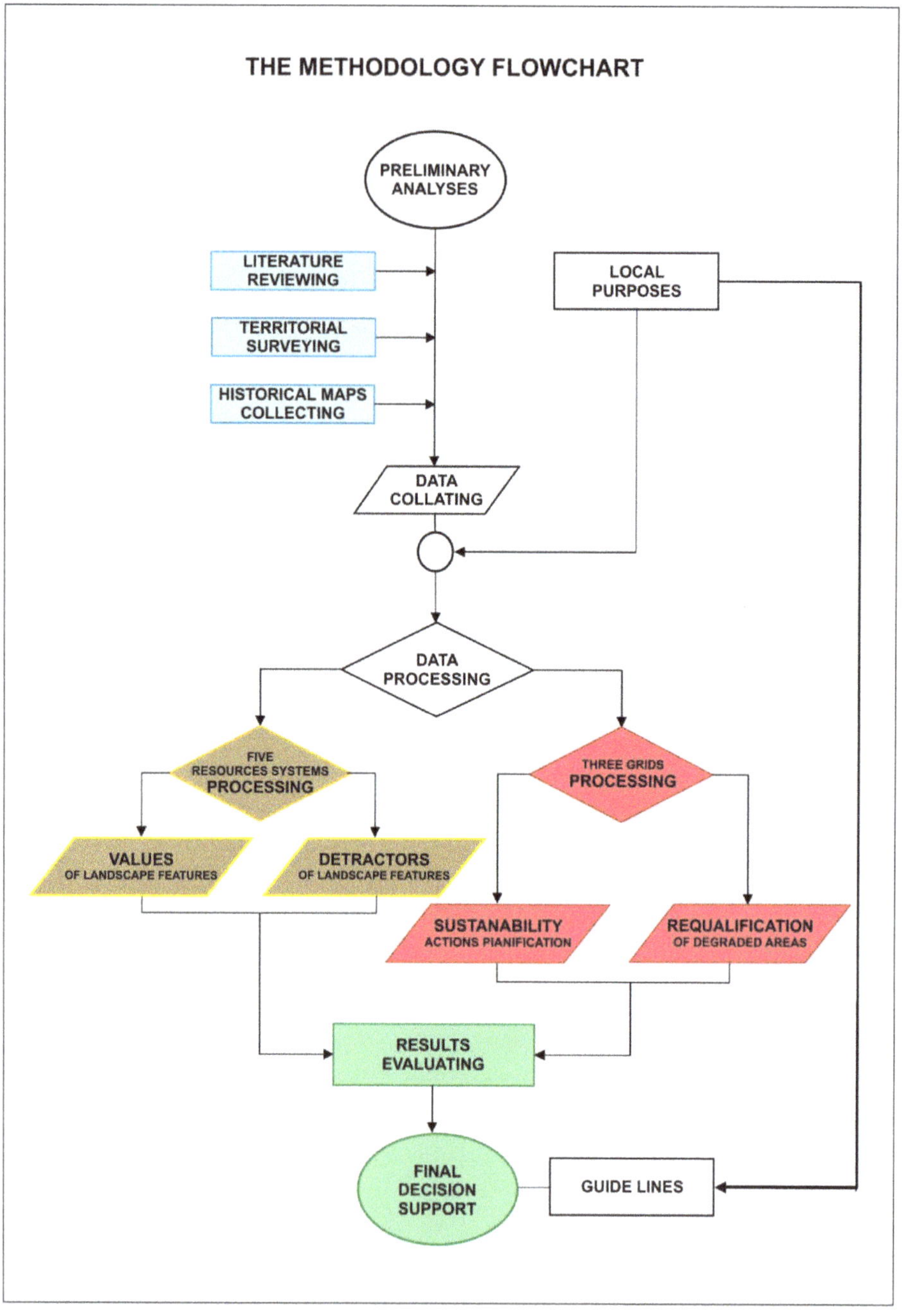

Figure 1. The flowchart illustrating the methodology (Source: l.a.co.s.t.a. Laboratory 2018).

2.1. Preliminary Analyses

The preliminary analyses, as highlighted in the figure, include the stages of literature review, territorial surveying, and the creation of thematic maps. The first phase includes the process of drawing up documents deriving from the analysis of the waterfront regeneration examples and from the European documents. In fact, over the years, European legislation has paid increasing attention to the city and to its need for renewal.

In 1975, the "European Charter of the Architectural Heritage" was drawn up, aimed at protecting the heritage through the cooperation of the States (members and nonmembers) of the European area [19]. The document also reflects the emerging interest in environmental issues, as it identifies protection as a tool that has the potential to combat soil waste and the occupation of new agricultural land. In the following years, attention was focused on the "rebirth of the city" and the themes supporting the initiatives to improve the environment were defined, also from the social point of view for the new quality of life. In the mid-eighties, guidelines were beginning to be defined for sustainable development, "a development that meets the needs of the present, without compromising the ability of future generations to meet their own needs" [20].

In 1990, the document known as "Green Paper on the Urban Environment" took shape. The document "promotes the arrest of urban expansion by rejecting the principles of functionalist urbanism, encouraging instead the multi-functionality and a development of densification of the existing" [21]. It also focuses on the problem of urban pollution and identifies actions to stem the phenomenon.

You must get to the mid-nineties with the Aalborg Charter [22] to see definitively "abandoned the functionalist approach proclaimed with the Charter of Athens (1933) from the Modern Movement. The new urban ideology replaces the concept of zoning with that of multi-functionality; the extension of the city with densification; the operations tabula rasa with rehabilitation operations or renewal; the technical and sectorial approach with the partnership approach" [23].

The policies for the city, born in the mid-seventies, received an ulterior impulse from the report on sustainable cities in 1996 [24]. This document promotes the sustainability theory by establishing a series of principles (urban management, political integration, principle of ecosystems, cooperation, and partnership) to pursue and evaluate sustainability strategies and policies in urban areas.

In the following years the debate deepened, up to the drafting of the framework of action for sustainable urban development in the European Union [25]. With this document, the European Commission defines the modalities for the allocation of structural funds for the "urban areas in difficulty" where there is a strong urban decay.

The Hanover conference of 2000 clearly introduces urban regeneration [26] of which the declared objectives are "social justice, reduction of poverty, and social exclusion and a more liveable urban environment" to be achieved through the construction of an economy that is more equitable socially and that is capable of protecting the environment.

Subsequently, in 2004, the document Aalborg + 10 was prepared, according to the guidelines dictated by the Conference, developing the well-known 10 topics [27]: the primary objectives were oriented to revitalize and requalify abandoned or disadvantaged areas, applying principles for sustainable design and construction.

In May 2007, the Leipzig Charter came to life. This document sanctions the need to resort to political strategies aimed at integrated urban development (as a global approach for suitable urban development). It defines concrete development objectives for the urban area and develops a "vision for the city" through action strategies for the "integrated urban development policy" [28].

2.2. Database Creation

The study, which provided for the recognition of the identifying matrices for landscape interpretation, led the definition of indicators related to different resource systems. Moreover, the criteria for the selection of the indicators—which will be useful for an evaluation of transformations through time of the territories being studied—has been defined. Five resource systems have been selected. They are physical-environmental; landscape-visual; historical-cultural; agricultural-productive; and demographic-tourism (Table 1).

Table 1. The five resource systems.

Resource Systems	Indicators	Sources
Physical-Environmental System	The first System includes the indicators relating to Climate and Atmospheric conditions, Water and waterways, and Marine and coastal environments.	The indicators selected derive from different sources: the Council Department for Public Works of the Molise Region, the Arpa Molise, and the Consortium for the Industrial Development of the Biferno Valley. The analysis of the data, which can be found in historical series, were compared with information obtained from soil science and geological maps as well as from geomorphologic maps. Maps of hydrological restrictions were consulted and a map of environmental risk related to landslide and hydrological risk based upon the most recent regional studies was produced.
Landscape-Visual System	The second System aims at defining the distinctiveness of the territory.	The morphology of the land was examined through an interpretation of the values already attributed to them in current landscape plans. A map was produced of the officially recognized natural ecosystems based upon the SCI sites (Sites of Community Importance) identified by the Natura 2000 Network. Further information was obtained from Regional Vegetation Maps and from maps drawn up by the Regional Administration of Corine Land Cover level IV soil use.
Historical-Cultural System	The third System was analysed through the investigation of elements and areas subject to historical restrictions, through an identification of building typology and through an analysis of landscape visibility.	The analysis of areas subject to restrictions was made by studying each protected historical building and archaeological site (National Cultural Heritage Ministry: historical building and archaeological site and update). The systems of buildings were analysed with an emphasis on the different typologies such as historical centres, rural buildings, towers and coastal defence systems, buildings that were a product of land reforms, large estates, post-earthquake reconstructions, monastic and religious buildings, buildings linked to cattle-tracks, and buildings linked to waterways.
Agricultural-Productive System	The fourth System, related to productive-agricultural resources, aims at defining the functions of agriculture. This entails an analysis of the land areas and the fruition of the land in agricultural terms (based upon local council indicators as well as indicators based upon the presence of farms).	All activities linked to agriculture were examined: the traditional farm type, the industrial type, and agricultural tourism. Particular attention was paid to irrigation, given that the coastal areas, as well as the pre-coastal strips, are major areas of irrigation. Information derives from the Historical National Agriculture Census and update, especially for irrigated areas, in the coastal and pre-coastal strips.
Demographic-Touristic System	The fifth System was analysed, following the subdivision of local township indicators. These indicate demographic changes, including changes in the farming population, which were then compared to specific indicators linked to industrial activity.	Data derive from the Historical National Population Census information and subdivision of local townships indicators; Local Council Urban Planning Tools and update. Verification of local council urban planning tools was included, paying attention to the large infrastructures foreseen, as these are responsible for major landscape variations, particularly those linked to the sea, ports and inter-ports as well as land communication systems, whether these include further development of pre-existing systems or the creation of new infrastructures.

2.3. Data Processing

The prevailing perspective of the new plan was to improve the protection of the territory's assets—refining the setting of the landscape value—and, above all, to enhance the potential of productive areas in the context of sustainable development of the Molise region, to which, moreover, give an innovative impulse. For this reason, the "values" have always been examined in relation to the presence on the territory of "detractors" who in fact change the value itself. These "detractors" are elements—infrastructural, industrial, or energy related—that invaded the territory over time, creating an uncontrolled development of the territory itself. The territorial analysis was carried out according to the criteria established in the protocol defined in agreement with the Molise Region [29–31]. The collected data was entered into a database, as shown in Table 2.

The sources of ancillary data are shown in the table. Obviously, these data came from heterogeneous databases concerning the various sectors of land use—from the urban sector to the agricultural sector and again to the industrial sector. They have been validated, updated, and subsequently processed to obtain useful information to be included in the Geographical Information System, exclusively created for the Molise Region. The cartographic basis for the territorial analyses is the Regional Technical Map in scale 1:5000, created in the nineties by the Cartographic Research Center of the Molise Region. It was updated for this work, having proceeded to purchase a series of photographic and satellite images (Landsat TM, Spot, and Quick Bird) from the mid 90s to present day; particular attention was paid to data by the MIVIS (Multispectral Infrared and Visible Imaging Spectrometer) sensor, appropriately corrected radio-metrically and, above all, geo-coded (orthorectified)) and georeferenced.

An innovative approach was set up too. For each resource system, three basic *grids* were created: each of them collects and processes different information series. In fact, in addition to the spatial dimension, a time horizon has also been introduced in territorial analyses. The data were grouped according to interpretation that take into account the following:

- for the first grid, the landscape assessments according to the criteria of the old landscape plans (it was called "grid" *A* actual state);
- for the second grid, the evolution of land use over time, from the mid 1950s (it was called *grid E* state evolved over time);
- the third grid analyzes the inconsistencies that derive from the dictates of the urban planning tools of municipalities in order to take into account homogeneous territorial units on a larger scale (it was called *grid P* previsional state).

The availability of all information, already accessible in a GIS environment, allowed further integration of the information in a much simpler and more effective way than would have been possible. It was useful to consider all the information available as a global mass of integrated ancillary data to the point of generating new knowledge. However, the use of the information contained in all these maps has been implemented through data verification and updating.

New information obtained by consideration and rations that emerged from the system of correlations and information queries in the new Geographical Information System were therefore classified.

In drawing up the criteria on which to base the New Regional Landscape Plan, the territorial elaborations carried out for the previous landscape plans of the region were also taken into account. They too were built in the late nineties and constitute an important starting point, even if not complete since the eight floors still in force do not cover the entire regional territory. Since the new legislative provision provides that, instead, the landscape plan involves the entire regional territory, they were found for the missing areas and, for the areas where the data had already been collected and the maps created, they were verified and update. A grids' comparison was made, and some images are shown in the following paragraph that will lead to the definitive conclusions.

Table 2. The data collection for the landscape analyses.

Resource Systems	Values	Detractors
PHYSICAL-ENVIRONMENTAL SYSTEM *1 ***1 Sources**: Molise Region, Department for Public Works; Arpa Molise; Consortium for the Industrial Development of the Biferno Valley, historical series and update.	**1 WATERBODIES** -River Basin -Artificial Lake -Natural Lake **2 NATURA 2000 NETWORK** -SCI (Sites of Community Importance) and SPA -(Special Protection Areas) -Management Plan **3 WOODLANDS** -Forests -State Forests -Nature Reserve -Oasis **4 GEO-SITES**	**1 SEISMIC CONSTRAINT** (zone 1, 2, 3, 4) **2 HYDROGEOLOGICAL CONSTRAINT** **3 HYDRAULIC SETUP** **4 SLOPE STRUCTURE** **5 L 445/1908** Inhabited areas to be consolidated and/or transferred **6 LACK OF MICROZONATION STUDIES**
LANDSCAPE-VISUAL SYSTEM *2 ***2 Sources**: Molise Region, current landscape plans; SCI sites (Sites of Community Importance) identified by the Natura 2000 Network; Regional Vegetation Maps updated to Corine Land Cover level IV soil use and update.	**1 RESTRICTED AREAS** -Under Law N°. 1497/39 -Ope_Legis -State property Areas **2 SIGNIFICANT ELEMENTS** **3 NATURALISTIC VALUES** -SCIs SPAs IBAs Areas -Established Protected Areas -Proposed Protected Areas **4 RESIDUAL AREAS** -Dunes -Civic Uses -Forests -Cattle-tracks Pathways -Pastures	**1 WIND-POWER PLANTS** -Carried out -Planned **2 GROUND-MOUNTED PHOTOVOLTAIC PLANTS** -Carried out -Planned **3 INFRASTRUCTURE NETWORK** -Roads -Railways -Power Lines **4 QUARRIES** **5 LANDFILLS** **6 TREATMENT PLANTS** **7 DEWATERING PUMPS** **8 INDUSTRIAL CENTRES** **9 OTHER INFRASTRUCTURES** -Airports -Ports -Interportsdata
HISTORICAL-CULTURAL SYSTEM *3 ***3 Sources**: National Cultural Heritage Ministry: historical building and archaeological site and update.	**1 ARCHAEOLOGICAL VALUE** -Restricted Areas under L 1089/39 -Areas certified by well-known researches -Areas certified by field surveys **2 ARCHITECTURAL VALUE** -Restricted Areas under L 1089/39 -Areas certified by well-known researches -Areas certified by field surveys **3 URBANISTIC VALUE** -Restricted Areas under L 1089/39 -Areas certified by well-known researches -Areas certified by field surveys **4 PATRIMONIAL VALUE** -Restricted Areas under L 1089/39 -Areas certified by well-known researches -Areas certified by field surveys	**1 ARCHAEOLOGICAL VALUE** -Land ownership -Open Sites **2 ARCHITECTURAL VALUE** -Conservation Status -Restoration Projects **3 URBANISTIC VALUE** -Recovery Plans -Regeneration Plans **4 ACCESSIBILITY TO SITES** -Presence of Highways -Presence of State Roads -Presence of Province Roads

Table 2. *Cont.*

Resource Systems	Values	Detractors
AGRICULTURAL-PRODUCTIVE SYSTEM*4 ***4 Sources:** Historical National Agriculture Census information and subdivision of local townships and update, especially for irrigated areas, in the coastal and pre-coastal strips.	**1 TERRITORIAL SURFACE (TS)** -Municipality TS (ha.) **2 FARMLAND SURFACE (FS)** -hectares -% ST/FS **3 UTILIZED AGRICULTURAL SURFACE (UAS)** -hectares -% UAS/FS **4 BIOLOGICAL CROPS SURFACE** -If > 10% UAS **5 FARMS** -Small < 10 ha. -Medium sized 10–50 ha. -Large > 50 ha. **6 AGRI-FOOD SPECIFICITIES** -D.O.C. (Quality Legislation for Controlled Origin) -D.O.P. (Protected Designation of Origin) -I.G.P. (Protected geographical indication) **7 RECLAMATION CONSORTIUM**	**1 INDUSTRIAL DEVELOPMENT CONSORTIUM** **2 INDUSTRIAL DISTRICT** **3 PRODUCTION PLANS (PIP)** **4 TOWARDS ALTERNATIVE ENERGY** Biogas Production -Forest Biomass -Wood-Chip -Wood chip heating systems **5 OIL MILLS** **6 SUNFLOWER DISTRICT**
DEMOGRAPHIC-TOURISTIC SYSTEM*5 ***5 Sources:** Historical National Population Census information and and subdivision of local townships indicators; Local Council Urban Planning Tools and update.	**1 PLANNING TOOLS** **2 DEMOGRAPHY** -Resident population (2011) -Demographic Trend 2010 (+ve/-ve) -Age classes -% 0–18 years -% 19–40 years -% 41–65 years -% over 65 years **3 ACCOMMODATIONS FACILITIES** -Hotels/Camping/Tourist Villages/Rental accommodation -Farmhouse accommodation/Country-Houses/ Youth hostels/ -Holiday homes/Bed & Breakfast	**1 PRESENCE OF ASBESTOS** -No sites -Class 1 -Class 2 -Class 3 -Class 4 -Class 5 **2 WASTES (QUANTITY IN TONNES)** -Waste Recycling -Co-mingled Waste Collection -Total -Non-hazardous waste recovery companies -Landfill site **3 SEWAGE TREATMENT PLANTS**

3. Results

On the specific case of the city of Termoli and, in particular, on its coastal territory, the overlapping of the three grids has made it possible to highlight the need to intervene to safeguard the landscape with differentiated suggestions in the different parts of the municipal territory. Its condition is comparable to other port cities along the Adriatic Sea and to numerous European waterfront situations.

Termoli is the biggest coastal town in Molise, on the Adriatic Sea coast. This case-study provides a starting point for the theoretical discussion. It was examined because it is of particular interest due to two factors: its dimensions and characteristics. The first condition is that it is a significant medium-sized city; the second is that its condition is flanked by the history of a troubled and undeveloped port. In the discussion, it was necessary to have a broader territorial outlook: the Molise coast, in fact, lends itself well to an analysis of production systems for many reasons because many similar phenomena have occurred over time. The most important variations concern the physical conditions due to erosion on a coast that is generally low-lying, with the exception of short stretches of the high coast of the town of Termoli, which stands on a promontory.

However, along the coast, there are some very important elements, such as the port facilities, including those of a tourist nature, the industrial core, and the large infrastructures, which are also the only significant ones in the whole region, and even several tourist settlements [32–34]. The city of Termoli has the oldest port of the region; recently, a further intervention was carried out on the near recent tourist port of Campomarino and a tourist port was also built in the municipality of Montenero di Bisaccia (although subject to controversy and contrary opinions). On the area immediately behind the coast, there is the Termoli Industrial Centre, undoubtedly the most significant in the region. Located close to the city in the Rivolta del Re district, it is home to the major industries in the area. Its presence since the early 1970s strongly determined a new layout of the territory with the increase in the demographic movement that involved also their surrounding area. Another significant phenomenon was that of land reclamation operations that had various evolutions over time, from the first interventions of the 1930s to the most significant ones of the 1950s by the Apulia, Lucania, and Molise Regions Reclamation Authority. Furthermore, there are a lot of tourist settlements, in building types such as small villas or small owned houses, with widespread phenomena in the 70s and 80s. Finally, behind the small strip of beach, there are the three main infrastructures of Molise, namely the A14 motorway, the State Road No. 16, and the railway. Going through three routes almost always parallel to one another, these infrastructures mark the entire coast.

The Molise Region is characterized by a low population density (about 70 inhabitants/km^2), a condition which, together with the difficult geomorphological features, prevented a strong urbanization. Only the coast (about 35 km) has population densities comparable to those of the neighbouring regions. Over the last few years, the city, albeit with a noticeable slowness due to the economic contingency, has undergone some transformations: the construction of a tourist port; the new planning of the port area; the integrated urban development projects involving the seafront and connections between the city and the port; and, finally, the provision of a commercial interport, a new connection between the city and the industrial district, located in the pre-coastal area.

The grid technique takes place by working on territorial analysis, conceived in vector format (which is well suited to the processing of discrete data, collected from the databases described above, very often available on a territorial basis), which was transformed in georeferred raster in order to ratify data deriving from unhomogeneous datasets [35–37].

The process led to defining the final synthetic value of orientation with respect to the objectives of landscape quality through a series of intermediate indicators obtained by processing the related data according to the hierarchical structure (system-elements-elementary data).

In this case, in particular, the accuracy level is that typical of 1:25,000 maps which conventionally equals the intrinsic error in graphics (such is considered the margin of error deriving from the pen mark equal to 0.35 mm, which is the equivalent of just under 10 metres in the nominal scale 1:25,000), which it is not possible to go below in traditional maps. However, it must be stressed that the choice of the 10-metre pixel does not degrade the information to a level that is any lower than the least accurate data and, as such, results in being widely usable for territorial analyses, where even 10 metres are below a significant threshold from the moment that they represent only 100 m^2.

The first *grid*, called A, that is the reference for the analysis of *the actual state*—as already declared—was developed from the present "Vast Area Landscape Environmental Physical Plan". In this sample is a plan covering the whole Molise's coastal area and describing the zones with elements of value recognised by the landscape plan itself. In fact, it describes the elements of historical-archaeological value, visual value, productive-agricultural value, natural value, and geological risk.

This plan undertook learning and assessment investigations through diversified maps. The map analysed in this sample area is the map S1 ("Map of the Territory's Quality"). It identifies the elements of which the importance means that they are localised and selected, characterised, and given a value.

The elements useful to the realisation of synthetic analysis were divided by category in accordance with the diverse categories of "interest":

- ✓ elements of historic, urban, archaeological, and architectural interest;
- ✓ agricultural-productive elements of interest for natural characteristics;
- ✓ naturalistic elements of interest for physical-biological characteristics;
- ✓ areal elements that are geologically unstable; and
- ✓ elements and environments of visual interest.

The grid relative to the "evolved state" (E) was created on the basis of land use: in this sample area, it highlights the main variations with particular attention to wooded areas (in particular, for areas corresponding to reforestation and deforestation), dunal areas (with the aim of defining the zones that have disappeared and those that remain), urban areas (and related progressive expansion), and areas under cultivation (with particular attention for the zones affected by the land reforms).

The categories attributed to the polygons for this map have been streamlined and simplified so as to obtain the following keys that are valid for both periods:

- ✓ urban areas;
- ✓ agricultural areas (including meadows and meadow-pastureland and arboriculture);
- ✓ grasslands and wastelands (shrub cover <40% and tree cover <20%);
- ✓ shrub and bushland (shrub cover >40%);
- ✓ chestnut plantations;
- ✓ broadleaf forests;
- ✓ coniferous forests;
- ✓ mixed forests of conifers and broadleaf;
- ✓ reforestation (forestry formation of conifers and broadleaf with $h < 5m$);
- ✓ bare ground areas (mountains, coasts, etc.); and
- ✓ water bodies and wetlands.

A comparative layer which analyses precisely the decrease or increase of the land with respect to the sea was introduced. Therefore, the key was structured in the following way:

- ✓ increase in urban areas;
- ✓ increase in agricultural areas;
- ✓ increases in woodland areas;
- ✓ increase in bare ground;
- ✓ increase in shoreline;
- ✓ decrease in agricultural areas;
- ✓ decrease in woodland areas;
- ✓ decrease in bare ground;
- ✓ decrease in shoreline; and
- ✓ no change.

Finally, the grid for the "previsional state" (P) has been derived from current urban planning tools and consists in the identification of the various designated zones with their relative attributes (indexes of suitability for building and of the various designated uses). In this sample area, it consists in identifying the various destination areas with their attributes. The degradation of the natural ecosystems characteristic of the coast has certainly begun in former times, but it has had a great increase in even more recent times, from the postwar period onwards. In fact, the spontaneous, chaotic, deregulated coastal development has pushed the anthropic presence ever closer to the shore. The Italian coast and, in particular, along the Adriatic Sea was affected above all by a strong increase in the construction of second homes right along the shoreline, also contravening the prohibition of construction for the coastal strip for its 300 m of depth, as was dictated since the aforementioned law on the landscape of 1939 (it is in fact a large part of houses illegally built but is subsequently remedied by building amnesty).

In the landscape analysis of the case study (Figure 2a), the most sensitive zone was deepened. They are, in particular, the MV1 areas (high level in visual values) and the MS areas (moderate visual value). For these areas, a medium sensibility to transformation was enshrined. In the MV1 area (in which the Industrial Area of Termoli is located), almost every use is allowed (under the conditioned transformation mode), except for settlements that are subject to the environmental assessment procedure. For the biggest MS area, with lots of residential settlements, intervention is possible, under the conditioned transformation mode.

This part of the regional coast also includes important areas from a naturalistic point of view: they are those designated as the A2N1 areas, which are strongly characterized by natural elements, and the A2N2 areas, where the natural vegetation is characterized by exceptional visual and naturalistic value (Figure 2b).

Their selection was implemented: in addition to sustainability indicators, other indexes have been added, which are considered important for the assessment of the sustainability of regeneration processes. Moreover, in addition to these indicators derived from the literature and from the regional databases, further specific indicators were identified as necessary to assess the landscape quality, and they have been added to our work. Consequently, elements related to the categories of natural, visual, perceptive, historical, and productive elements have been identified as landscape-related indicators. As regards the relationship between sea and city, studies carried out for the old town area underline some negative aspects, such as the isolation of the old town and the impossibility of enjoying the sea from the town centre. The presence of the sea is only partially perceived through a few viewpoints, and recent building has further contributed to the closure of the view from possible panoramic points.

The two zones of the old town and the 19th century town, which form the heart of Termoli, are in reality the hub around which the entire town's main economic, cultural, and tourist activities revolve. The first obstacle to an organic expansion of the town is formed by the railway, (inaugurated on 1864) which isolates the "19th century town" from the rest of the territory. The second barrier, built around 1960, is the motorway. The Adriatic State road No.16, which runs parallel to the motorway, adds to this series of infrastructural axes. In fact, the zone between the motorway and the Adriatic State road No.16 forms a kind of artificial island.

Therefore, specific analyses were carried out about the current urban plan (Figure 2c). The city of Termoli is the first of the four towns along the Molise coast to have a master plan. The first plan, adopted in 1971, was approved by the regional council on 1972. Three years later, the general variant to this master plan was adopted, and it was approved in 1977. Later, various partial variants have been proposed that have completely modified the plan's original design in order to adapt it to new needs. A general variant was adopted by the town council in 2003, but it was never approved by the region [38].

Really, the plan created a transversal link between the three homogeneous strips parallel to the coast and delimited by the great transport networks:

- the coastal strip (north coast, old town, Rio Vivo);
- the intermediate strip (delimited by the railway and motorway); and
- the external strip (beyond the Road Variant to the State Road No. 16)

However, the determinant factor in the original choices of the plan is the town's vocation to tourism activities and the will to adequately upgrade this aptitude that Termoli has gone.

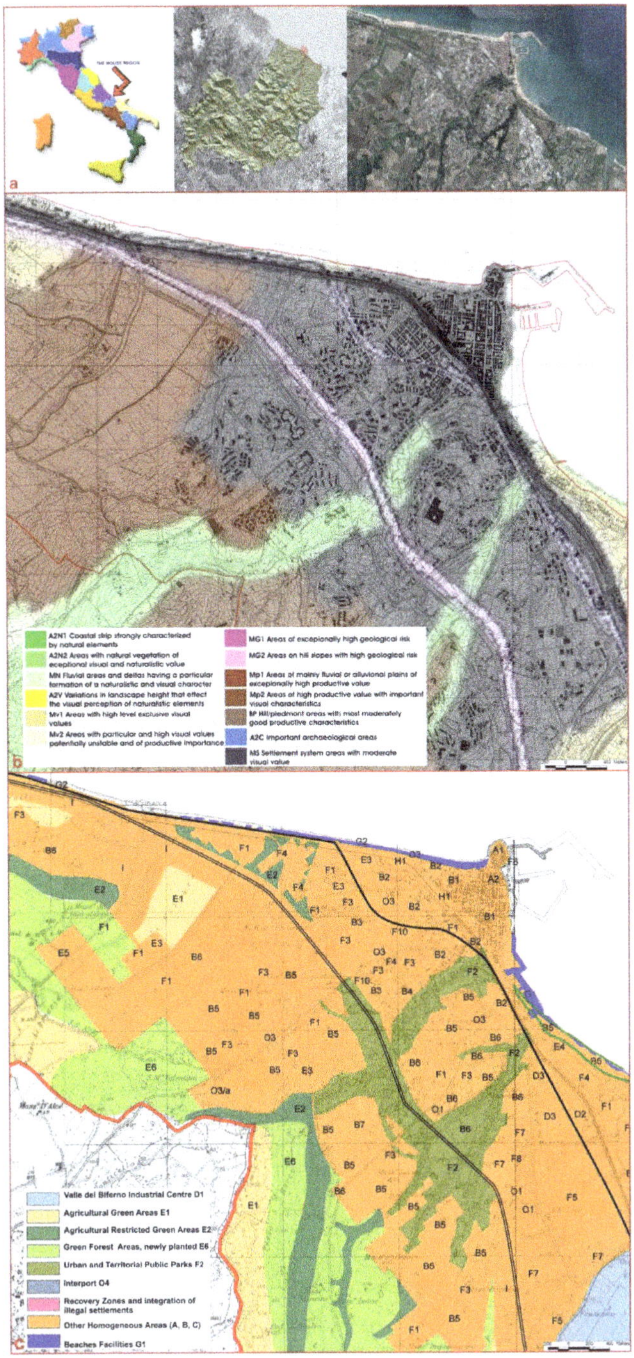

Figure 2. The sample area: (**a**) localization and orthophoto with the Molise Region and the Termoli Municipality; (**b**) transformation tools provided by the current landscape plan for the area surrounding Termoli; and (**c**) Termoli master plan and its zoning (Source: l.a.co.s.t.a. Laboratory 2019).

The local plan provides further interventions oriented to a) the valorisation of tourist accommodation and connected structures; b) the organization of artistic, archaeological, historical, and environmental resources; c) the identification of structures for leisure time use; and d) the construction of transport infrastructures and car parks. In the coastal strip, seasonal commercial structures providing services for beach activities are planned. The intermediate strip is planned in order to locate an exhibition-show area with ample spaces dedicated to permanent exhibits of local crafts; furthermore, an office district combining large infrastructures such as spaces for entertainments, offices, and city services is planned. The green area located just south of the park, which is unused and abandoned, in the plan will be turned into a theme park. In the external strip, green areas, services, and perimeters of the zoning are redrawn and new building areas are created.

In addition, some projects have recently been funded and carried out for the urban waterfront. The new waterfront of Termoli comprehends the new master plan of the Commercial and Tourist Port and the urban historical centre regeneration. The new distribution of functions and services improve the relationships between the activities of the harbour and the fisherman village. Moreover, in the Rio Vivo—Marinelle zone, there is the only urban park in Termoli, designed in 1962 and definitively approved in 1964. In the following years, some active sports facilities were added, including tennis and soccer fields, and finally a large swimming pool that is now in disuse. The current Termoli's master plan classified this area as a densely urbanized agricultural area (E1), although in the 1977 variant it was intended for the port equipment (F5). It is delimited to the south by the industrial zone and, more precisely, from the strip to the ever-built interchange to the southeast by the reclamation channel of contrada Marinelle; to the southwest from the road infrastructure (State Road No. 16) and the railway (the so-called Ferrovia Adriatica) and its coastal fragmentation barriers; and to the north by the Urban Beach Plan, which is a sandy beach subject to remarkable erosive phenomena attributable to the port of Termoli. The strategic environmental assessment for the master plan of the port provided for a development scenario until 2025, and it highlighted how the current configuration causes a significant backsliding. Moreover, the area under study is included in site of community importance named "Foce del Biferno—Campomarino Coast", in which there is a strong anthropization, considered the cause of the destruction of the dune, as well as detecting the danger to the pine wood now in contact with the marine waters due to the beach retreat.

4. Discussion

Territorial analyses have highlighted, within the study area, zones in which the greatest conflict occurs: for these areas, specific rules for the regeneration policies are necessary [39].

In particular, the municipality examined is located in a region with strong agricultural features: the excessive consumption of agricultural and natural soil not only deeply affects changes in the landscape but also involves problems in ecological assessment, especially inhibiting natural regeneration capacity of environmental resources. Moreover, this is well known to be a more and more of growing risk for urban areas, affected by poor planning choices and filled with many cultural elements [40]. Furthermore, the approach necessarily started from the current settlement assessment, with its economic and environmental conditions, looking for a new ecological situation based on the balance between the environmental resource availability and their uses. In this paper, therefore, the portion of the city of Termoli—which is located within a significant naturalistic and agricultural context—has been examined. In this land portion, it is however necessary to consider the emergent factors of the coastal landscape. Analyses were carried out, taking into account the resilience as a capacity of the system to be able to maintain stability compared to an initial state of equilibrium [41–43].

As well known, the most delicate situations in the analysis of resilience are both in the face of calamitous events, which produce instant change of state, and in the face of slower changes that occur through the growth of the city.

In this case-study, both conditions are true: in fact, significant flood events have occurred in recent years and continuous mutations are producing gradual changes over time [44]. Especially in cases characterized by prolonged crisis, it is important that systems were able to adapt to preserve their own identity. A major role should be attributed to "environmental infrastructure" and in particular to so-called "blue infrastructures", namely watercourses and coastal waters. Such elements should be attributed to new city development policies.

In order to prepare the new regional landscape plan, a table of the new landscape quality aims as listed in art. 135 of the National Code (L. 42/2004) was organized: it contains for each homogeneous area specific requirements and provisions, oriented to the conservation of the constituent elements; to the rehabilitation of compromised or degraded areas; to the protection of landscape features; and to the identification of the lines for the development, with particular attention to the preservation of rural landscapes and sites included in the UNESCO World Heritage List. Moreover, about the costal area in which our sample test is, the final targets are related to the conservation, protection, management, and planning of exceptional, ordinary, and degraded landscapes with particular reference to typical natural landscapes such as rivers, lakes, hills, mountains, coastal and rural landscapes, forestry, and agro-pastoral, not to mention historic, rural, urban, industrial, and infrastructure sectors (Table 3).

The objectives identified are also related to the government of the processes of urbanization and abandonment of the territory and to the preservation of material cultural values and intangible values such as the traditions and history of the region.

The general objective was subdivided into specific objectives, as shown in the table, in which these objectives were finally associated with landscape quality directions that indicate policies to adopt and those who have an interest in achieving these objectives as well as the measures required to adapt the urban planning instruments to the indications of the new regional landscape plan.

Therefore, the landscape quality targets in this area aim to safeguard the surviving heritage in the area, to recover and improve the landscapes altered and degraded by human activity and to define quality standards.

Subsequently, the three grids, set by this methodology, were useful for defining the new protection that is proposed for the coastal area in the municipality of Termoli (Figure 3). They are derived from the analysis of the tool dedicated to the protection of the landscape (starting from the definition of the current plan), from the city planning tool (master plan of the municipality of Termoli) and from the dictations of the other tools in force in this zone, which is the port master plan of the and the beaches and coasts plan. The diversity of the two areas of the coast clearly emerges, located around the historical urban centre. For these two areas, there is a strong concentration of the "built" but not predominantly "urban": the two coastal strips (zone A to the west and zone B to the east) that surround the promontory where the historic centre is located and includes the port area, which has taken on different and increasingly broad connotations over time (Figure 4).

Table 3. The landscape quality aims declined for the five resources systems in the sample area.

RESOURCESYSTEMS	GENERAL AIMS	SPECIFIC AIMS
PHYSICAL-ENVIRONMENTAL SYSTEM	1 PROMOTE THE PRESERVATION OF THE INTEGRITY OF AREAS OF HIGH NATURALNESS AND HIGH ECOSYSTEM VALUE	1.1 Safeguard geological-geomorphological systems with high integrity (geological formations, ravines, cliffs, crags) 1.2 Safeguard protected areas and areas of high environmental value such as those covered by the Nature 2000 Network 1.3 Safeguard and improve environmental functionality of river and lake systems of Molise 1.4 Safeguard and rebuild coastal marine habitats of Molise 1.5 Safeguard woods and forests of mountainous and hilly areas of Molise 1.6 Redevelop and redesign the coastal landscapes of Molise
LANDSCAPE-VISUAL SYSTEM	2 PROMOTE IMPROVED INTEGRATION OF LANDSCAPE AND THE QUALITY OF INFRASTRUCTURES	2.1 Define territorial and landscape quality standards in the settlement of new network infrastructure 2.2 Define territorial and landscape quality standards in the settlement of new energy infrastructure 2.3 Define territorial and landscape quality standards in the settlement of new productive activities
HISTORICAL-CULTURAL SYSTEM	3 PROMOTE THE PRESERVATION OF CULTURAL VALUES	3.1 Preserve cultural value and witnesses of settlements and historical manufacts 3.2 Preserve cultural value of traditional rural buildings 3.3 Preserve the visible cattle-tacks residual 3.4 Redevelop the historic rural landscapes
AGRICULTURAL-PRODUCTIVE SYSTEM	4 PROMOTE THE CONSERVATION OF AGRICULTURAL LANDSCAPES	4.1 Develop the agricultural landscape of Molise, recognize and promote its social functions 4.2 Preserve open landscapes of the reclamation as a characteristic aspect of identity of coastal landscape of Molise 4.3 Redevelop the agricultural landscape of Molise
DEMOGRAPHIC-TOURISTIC SYSTEM	5 PROMOTE THE IMPROVEMENT OF THE QUALITY OF THE SETTLEMENTS	5.1 Improve quality of urban settlements and their environmental performance, for greater well-being of the population 5.2 Redevelop degraded contemporary urbanization landscapes 5.3 Improve urban quality and touristic settlements 5.4 Improve urban quality of agricultural and productive settlements 5.5 Improve soft mobility quality (walking, cycling, trekking on horse) and its interconnection with traditional mobility

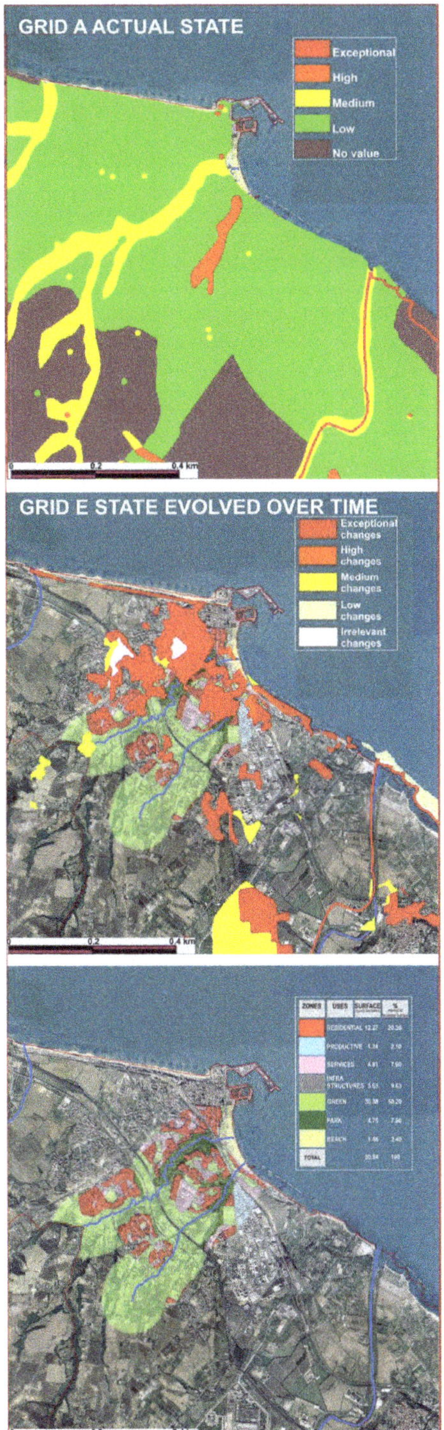

Figure 3. The three grids for the Molise coastal zone (Source: l.a.co.s.t.a. Laboratory 2019).

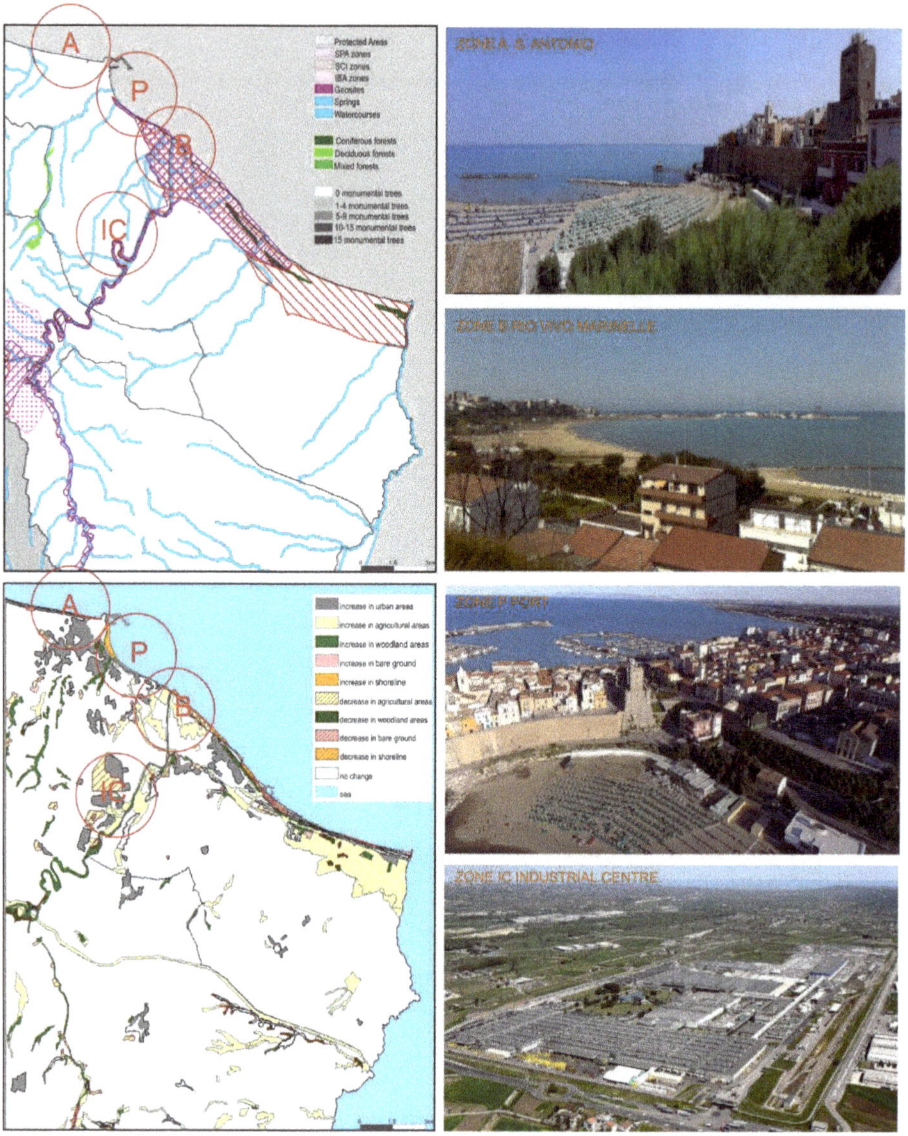

Figure 4. The two different coastal areas surrounding the urban centre (Source: l.a.co.s.t.a. Laboratory 2019).

These two different conditions were deepened using the matrices created with the data of the five systems and of the differences in land use in the respective pre-coastal areas.

5. Conclusions

The study clearly showed that anthropogenic pressures on the coast are increasingly evident. Even comparing this area to the rest of the regional territory, in fact, the work cannot help but highlight that the greatest transformations, with strong consequences on the landscape image of the territory, took place right on the coast. Obviously, for the constitution of the water basins (located in the pre-coastal zone), the urbanization occurred quickly and without controls (especially second homes) and new assets related to increasing infrastructures contributed to a strongly incoherent arrangement.

The nonurban spaces are the central focus of attention as they are no longer tied only to agricultural production (that pervaded in the past, especially the pre-coastal area as a result of reclamation operations) but increasingly to the creation of services to the population. In these areas, it is necessary to maintain the quality of the landscape with its identifying characteristics and with its differentiations related to the places' traditions.

On the whole, as far as the analyzes carried out are concerned, the clear agricultural and production vocation emerges: in particular, the lands in the valley and the irrigated ones in the hills are of exceptional value, in light of both the geopedological analyzes and cropland attitudes. No less important is the presence of SCI (Sites of Community Importance) areas that involve, as seen, the entire coastline of the Molise region. The element that has most changed the landscape is the increase in the built environment; furthermore, a substantial change was defined by the construction of the vast industrial area.

The final considerations, therefore, concern above all the destinies of the "built", now considered part of the well-established fabric of the urban areas, even if lacking in specific identifying characteristics.

The under-study areas, however, are clearly demarcated, on one side, by high infrastructure lines (railway, highway, and Adriatic road) and, on the other, by the sea line, as seen invested by sites of community importance areas. The opportunities that the new regional landscape plan can offer, also on the basis of the experiences of the other plans already approved or in the process of being approved in the other Italian regions, can be harbingers of a new philosophy of safeguarding the coastal areas, which does not destroy the existing but is able to restore a "landscape balance".

Coastal planning is a central aspect of this study. In fact, the most critical elements of the demographic and economic development of the regional realities concerned converge on the coast areas. The government of the territory can be implemented through an intervention perspective that concerns a plurality of characters inherent in natural and artificial resources to guarantee protection of environmental and territorial peculiarities. The analysis and comparison between objectives, plan strategies, environmental compatibility criteria, environmental reference indicators, and monitoring methods are all parameters that come into play, but they are not sufficient. Therefore, this vast study is intended to be a contribution to the deepening of the relationship between territorial planning and landscape planning, since the urbanization processes of the coastal areas inevitably influence even the innermost territories, also affected by related economic pressure tourism development that endangers the integrity of the most fragile environmental and landscape contexts. The analysis carried out demonstrates the need to set up a methodology for the landscape which, although starting from data collection based on a homogeneous approach, in reality must always be adapted to local conditions, even within the same region. The final goal is to create vital landscapes.

This study was performed in two different scales: at the territorial level and at the local level, that is, the municipal level. At the territorial level, the indications are intended to address and propose planning actions. At local level, the municipal plans should act in accordance with the landscape plan suggestion: within the areas mentioned above, different situations are identified (in relation to the features and values of the natural environment and human actions) that require specific norms and guidelines. The overlap of the current and forecast condition also allowed us to have an immediate (and quantifiable) perception of dangerous, intense, extensive, and persistent modifications, especially where there is neglect and poor preservation.

The future improvement of the research will be the suggestions for interventions that municipal administrations could promote, as is already happening in some other regions, as active tool for the protection of the landscape within their municipal planning tools, in compliance with what is defined by the regional landscape plan. This tool must naturally understand how the individual management tools for parts of their territories already foresee but further implement them in a more specific logic for urban planning.

Funding: This research was funded by the Molise Region, under Agreement with the l.a.co.s.t.a. Laboratory (Director prof. D. Cialdea) of the University of Molise for the realization of the "*New Regional Landscape Plan of Molise*", signed 11.02. 2011, REG. n. 11 del 28.02.2011.

Conflicts of Interest: The author declares no conflict of interest.

References

1. Rivas, M. From Creative Industries to the Creative place. Refreshing the Local Development Agenda in Small and Medium-Sized Towns, Urbact Creative Clusters Final Report, 2011. Available online: hup://urbact.eu/creative-clusters (accessed on 20 November 2019).
2. Martorell, J. *Transformación de un Frente Marítimo. Barcelona. La Villa Olímpica, 1992*; Editorial Gustavo Gili: Barcelona, Spain, 1988.
3. Savino, M. Waterfront d'Italia. Piani, politiche, progetti. In *Waterfront of Italy. Plans, Policies, Projects*; FrancoAngeli: Milano, Italy, 2010.
4. Grimaldi, M.; Pellecchia, V.; Fasolino, I. Urban plan and water infrastructures planning: A methodology based on spatial ANP. *Sustainability* **2017**, *9*, 771. [CrossRef]
5. Giovinazzi, O.; Moretti, M. Città portuali e waterfront urbani: Trasformazioni e opportunità. Port cities and urban waterfront: Transformations and opportunities. *TeMALab* **2009**, *2*, 59–66.
6. Desfor, G.; Laidley, J. *Transforming Urban Waterfronts*; Routledge: Abingdon, UK, 2010.
7. Sgobbo, A. Water sensitive urban planning. In *Approach and Opportunities in Mediterranean Metropolitan Areas*; INU Edizioni: Roma, Italy, 2018.
8. Busquets, J. Barcelona. In *La Construcción Urbanística de la Ciudad Compacta*; Ediciones del Serbal: Barcelona, Spain, 2004.
9. Pecourt, J.; Piñón, J.L. La Valencia Marítima of 2000, Colegio Oficial de Arquitectos de la Comunidad Valenciana. Available online: www.arquitectosdevalencia.es (accessed on 9 October 2019).
10. Martí, P.; García Mayor, C.; Melgarejo, A. Waterfront landscapes in Spanish cities: Regeneration and urban transformations. *WIT Trans. Built Environ.* **2018**, *179*, 45–56.
11. Quercio, N. Regeneration of Coastal Areas. Landscape and Design Aspects of the Waterfront. Ph.D. Thesis, University of Molise, Campobasso, Italy, 2014.
12. Leira, E. Bilbao Metropolitan, the Proposal: Plan Territorial Parcial of Bilbao Metropolitano, Gobierno Vasco. Available online: http://www.bilbaopedia.info/plan-territorial-parcial-bilbao-metropolitano (accessed on 4 January 2017).
13. D'Onofrio, P. Roadscape. The Project of Fast Roads in the Contemporary City and Landscape. Ph.D. Thesis, University of Naples Federico II, Napoli, Italy, 2006.
14. Buslacchi, M.E. Vista dal mare. La riqualificazione del waterfront di Marsiglia, tra edifici-icona e Mediterraneo. A view from the sea. The regeneration of Marseille waterfront: Iconic-buildings and Mediterranean sea. *TRIA* **2013**, *11*, 47–60.
15. Project "Sustainable Management of Coastal Areas", GES.S.TER. Funded by INTERREG III A, 2004–2007 (Director D. Cialdea University of Molise). Available online: http://europa.molisedati.it/molise (accessed on 2 December 2019).
16. Project "Model Implementation of Landscape Coastal Units" Funded by MIUR—Interlink, 2004–2006 (Director D. Cialdea University of Molise). Available online: https://interlink.miur.it/2004/selezionati.html (accessed on 2 December 2019).
17. Project "Archaeological and Environmental Sites", SEA-EAS Funded by INTERREG/CARDSPHARE, 2007–2008 (Director D. Cialdea University of Molise). Available online: http://www.regione.molise.it/nvi (accessed on 2 December 2019).
18. Repubblica Italiana, Decreto Legislativo 22 Gennaio 2004, n. 42 Codice dei Beni Culturali e del Paesaggio, ai Sensi Dell'articolo 10 Della Legge 6 Luglio 2002, n. 137. Available online: https://www.bosettiegatti.eu/info/norme/statali/2004_0042.htm (accessed on 10 December 2019).
19. Council of Europe. *European Charter of the Architectural Heritage—1975*; ICOMOS International Secretariat: Charenton-le-Pont, France, 1975.

20. United Nations General Assembly Brundtland Commission. Report of the Word Commission on Environment and Development: Our Common Future. 1987. Available online: https://sswm.info/sites/default/files/reference_attachments/UNWCED201987Brundtland/ (accessed on 10 December 2019).
21. Commission of the European Community. *Green Paper on the Urban Environment: Communication from the Commission to the Council and Parliament*; Office for Official Publications of the European Conununities: Luxembourg, 1990.
22. European Conference on Sustainable Cities & Towns, Charter of European Cities & Towns Towards Sustainability. Available online: https://www.eea.europa.eu/help/glossary/eea-glossary/charter-of-european-cities-and (accessed on 24 November 2019).
23. Aloni, R. The Narration of Urban Regeneration Processes in France. The Case of the PNRU. Ph.D. Thesis, University of Palermo, Sicily, Italy, 2017.
24. European Commission, Directorate General XI Environment, Nuclear Safety and Civil Protection, Expert Group on the Urban Environment. *European Sustainable Cities Report*; Office for Official Publications of the European Conununities: Luxembourg, 1996.
25. Commission of the European Communities. Sustainable Urban Development in the European Union: A Framework for Action. Available online: https://searchworks.stanford.edu/view/4373599 (accessed on 20 December 2019).
26. 3rd European Conference on Sustainable Cities & Towns. *The Hannover Conference 2000 Report*. 2000. Available online: https://sustainablecities.eu/conferences/hannover/ (accessed on 2 December 2019).
27. European Sustainable Cities and Towns Campaign. *Aalborg + 10—Inspiring Futures*. 1994. Available online: https://sustainablecities.eu/conferences/aalborgplus10/ (accessed on 2 November 2019).
28. European Union. *Leipzig Charter on Sustainable European Cities*; Final Draft: Leipzig, Germany, 2007.
29. Cialdea, D. (Ed.) *Environmental Assessment Interreg Reports*; Materials for Adriatic Cross Border Project GES.S.TER./Interreg IIIA, Report no. 3; Arti Grafiche La Regione: Campobasso, Italy, 2007.
30. Regione Molise. Laboratory (Director Prof. D. Cialdea) of the University of Molise for the Realization of the "New Regional Landscape Plan of Molise", Signed 11 February 2011. Available online: http://www3.regione.molise.it/flex/cm/pages/ServeBLOB.php/L/IT/IDPagina/1 (accessed on 24 November 2019).
31. Cialdea, D.; De Montis, A.; Vitiello, M.; Saio, G. The GIS architecture elements for the coastal areas along the Adriatic Sea. In Proceedings of the Ersa 2006, 46th Congress of the European Regional Science Association, Volos, Greece, 30 August–3 September 2006.
32. Cialdea, D. Valorisation strategies for Molise coastal area: Archaeological sites and settings of environmental value. In *Cultural Tourism and Sustainable Local Development*; Nijkamp, P., Fusco Girard, L., Eds.; Ashgate: Farnham, UK, 2009.
33. Cialdea, D. Il paesaggio tra infrastrutture e dissesti. Un'analisi territoriale applicata alla region. In *Edilizia Rurale e Paesaggio Agrario tra Passato e Future Rural Construction and Agricultural Landscape between Past and Future*; Alinea Editrice: Firenze, Italy, 2009.
34. Cialdea, D. *Ancient Territories and New Scenarios*, 2nd ed.; Arti Grafiche La Regione: Campobasso, Italy, 2019.
35. Cialdea, D.; Mastronardi, L. Integrated approach in the planning stage for landscape conservation. *WIT Trans. Ecol. Environ.* **2014**, *188*, 296–306.
36. Cialdea, D.A. Planning activities in coastal areas Italian and cross-border approaches along the Adriatic Sea. *Int. J. Sustain. Dev. Plan.* **2017**, *12*, 853–866. [CrossRef]
37. Cialdea, D. Smart land: Regeneration and sustainability in lost scenarios and new performances. In *Sustainability and Mobility in the Age of Change*; Springer: Berlin/Heidelberg, Germany, 2018.
38. Comune di Termoli. *Report of the General Modification to the Masteplan of the Municipality of Termoli*; Comune di Termoli: Termoli, Italy, 2002.
39. Cialdea, D.; Quercio, N. Dimensioni e contesti negli interventi di rigenerazione. Dimensions and contexts in regeneration interventions. In *Cambiamenti. Responsabilità e Strumenti per L'urbanistica al Servizio del Paese Changes. Responsibilities and Tools for Urban Planning to the Benefit of the Country*; Planum Publisher: Roma, Italy, 2017.
40. Mantini, P. Dall'urbanistica per piani e accordi al governo liberale del territorio per sussidiarietà e concorsualità. From town planning for plans and agreements to the liberal government of the territory for subsidiarity and competition. *Riv. Giur. Edilizia* **2013**, *3*, 141–151.

41. Gabellini, P. Fare urbanistica. Esperienze, comunicazione, memoria. In *Urban Planning. Experiences, Communication, Memory*; Carocci Editore: Roma, Italy, 2010.
42. Gasparini, C. Resilienza. *Urban. Inf.* **2015**, *263*, 261–262.
43. Lombardini, G. Resilienza e regione urbana: Sfide e opportunità per la pianificazione di scala intermedia. Resilience and urban region: Challenges and opportunities for mid-scale planning. *Urban. Doss.* **2013**, *4*, 151–153.
44. Cialdea, D.; Maccarone, A. Territorial diachronic maps for the Regional Landscape Plan. In *Planning Support Tools: Policy Analysis, Implementation and Evaluation*; Campagna, M., De Montis, A., Isola, F., Lai, S., Pira, C., Zoppi, C., Eds.; FrancoAngeli: Milano, Italy, 2012.

sustainability

MDPI

Article

The Contribution of Ecosystem Services in Developing Effective and Sustainable Management Practices in Marine Protected Areas. The Case Study of "Isola dell'Asinara"

Maddalena Floris [1], Vittorio Gazale [2], Federica Isola [1], Francesca Leccis [1,*], Salvatore Pinna [1] and Cheti Pira [1]

[1] Department of Civil and Environmental Engineering and Architecture, University of Cagliari, 09123 Cagliari, Italy; maddalenafloris@gmail.com (M.F.); federica.isola@unica.it (F.I.); pinnasalvador@gmail.com (S.P.); ing.cheti.pira@gmail.com (C.P.)

[2] "Isola dell'Asinara" Marine Protected Area, 07046 Porto Torres, Italy; gazale@asinara.org

* Correspondence: francescaleccis@gmail.com; Tel.: +39-347-356-6551

Received: 29 December 2019; Accepted: 31 January 2020; Published: 4 February 2020

Abstract: Ecosystem Services (ESs) are assuming a constantly increasing importance in management practices due to their key role in ensuring a sustainable future to fauna and flora on Earth. In addition, ES degradation and quality loss jeopardize current human activities. For this reason, it is essential to develop methodologies and practices able to efficiently assess environmental and socio-economic impacts in terms of ES deterioration, especially within protected areas. Norms and regulations have to be able to identify habitat and species categories to be preserved, and to determine the cost of their destruction and decline, according to a holistic vision, which includes social and economic impacts, besides the environmental ones. The paper illustrates the case study of the "Isola dell'Asinara" Marine Protected Area (MPA) in Sardinia, where an experimental methodology was developed with the aim to draw new regulations that integrate conservation measures of Natura 2000 sites included in its territory, provisions determined by the integrated coastal zone management (ICZM) protocol and the Standardized Actions for Effective Management of MPAs (ISEA) project. Subsequently, in order to assess the status of ESs and impacts on ESs located within the MPA territory, an ecosystem-based approach was implemented and applied to the actions defined for the new regulation proposal. Results show that regulations are in this way valuably enriched by environmental aspects of the MPA that would otherwise be overlooked.

Keywords: ecosystem services; millennium ecosystem assessment; marine protected areas; Natura 2000 network; integrated coastal zone management; standardized actions for effective management of marine protected areas; sustainable management; agenda 2030

1. Introduction

Commitment to biodiversity preservation has been increasing dramatically since the Convention on Biological Diversity in 1992. Thanks to international cooperation and stronger awareness of the close relationship between human needs and ESs, attention to ES preservation has grown exponentially over the years. Indeed, a constantly increasing number of scientific studies underline the fundamental role of marine and coastal ecosystems in providing important social and economic advantages [1–4], including food provisioning, nutrient cycling, biodiversity, climate regulation, cultural values, recreation and amenities [5] and the growing human dependence on marine ecosystems and their services [6].

However, overexploitation and mismanagement of marine and coastal resources, habitat destruction and water pollution jeopardize the wealth of ecosystems [7] and, consequently, the services they provide [2], thus threatening the well-being of local communities and of the global community as a whole in developing and industrial countries [8]. Nowadays, many coastal areas have been shown to be vulnerable due to ocean warming, sea level rise, flooding, storm surges, beach erosion, changes in ocean freshwater balance [9], overfishing, tourism and pollution [10,11]. Therefore, appropriate marine and coastal area management and governance are essential to limit environmental and ecosystem damages caused by human activities, to protect biodiversity [12–15] and to improve ecosystem resilience [2] and ecosystem services supply [16]. Scientific research shows that Ecosystem Services (ESs) have to be included in planning discourses [17], but this is still little done in practice [18]. Therefore, following the suggestions of the Strategic Plan for Biodiversity 2011–2020 and the related Aichi Targets, it is necessary to add a new conservation paradigm that integrates ESs in protected areas' planning rules in addition to the existing sets of biodiversity conservation measures [19]. The Strategic Plan for Biodiversity 2011–2020 was adopted in 2010 by the parties to the Convention on Biological Diversity, which has been ratified by 196 countries. Since the contracting countries are committed to the achievement of a number of targets by 2020, in 2011 the EU adopted the EU Biodiversity Strategy, which sets out six targets and 20 actions to halt the loss of biodiversity and ecosystem services in the EU by 2020. Among the objectives set by the strategy, this study focuses on those related to the preservation and protection of habitat, species and ESs located in Marine Protected Areas (MPAs) and Natura 2000 sites. In particular, regulation and planning aspects related to ESs are examined in the light of EU recommendations and communications in order to define research objectives and actions.

Marine Protected Areas (MPA) are globally recognized tools for managing marine ecosystems, specifically designed to safeguard biodiversity, to preserve marine ecosystem health, to maintain the supply of ESs [20], to prevent habitat loss [21] and to sustainably regulate human activities that affect the marine environment [22]. The aim is to reverse ecosystem and biodiversity loss, while sustaining local economy relying on sea and coastal resources [23]. The International Union for Conservation of Nature (IUCN) underlines that deep knowledge of the area is necessary in order to define ecological boundaries of the MPA and to set its objectives. It also calls attention to the vital support of the public and to established techniques for surveillance and monitoring of compliance with the provisions of the regulations. It plays an essential role in advocating the expansion of the MPA network through reliable science and by engaging with local stakeholders. In Europe, the State of the Environment Report (SOER) identifies three types of MPAs: Marine Natura 2000 sites, marine protected areas designed in the framework of regional marine agreements and the single national marine protected areas [24]. These three typologies of protected areas can overlap, that is, a specific zone can be disciplined according to more than one regulatory regime [25]. In Italy, MPAs are established through a ministerial order under the laws no. 979/1982 and no. 394/1991 and normally include three zones characterized by different levels of protection; namely, zone A, the "integral reserve", which is a no entry and no take zone; zone B, the "general reserve", which surrounds zone A where human activities and resource exploitation are severely restricted; and zone C, the "partial reserve", where human activities and resource exploitation are generally allowed but regulated [26]. It might occur that other regulatory regimes operate within the boundaries of an MPA, such as Natura 2000 network conservation measures for Special Protection Areas (SPAs), Special Areas of Conservation (SACs) and Sites of Community Importance (SCIs), the provisions determined by the integrated coastal zone management (ICZM) and by the Standardized Actions for Effective Management of MPAs (ISEA). SCIs, identified by European countries according to article 2 of European Directive 92/43/EEC on the conservation of natural habitats and of wild fauna and flora, subsequently labelled SACs, contribute towards ensuring bio-diversity through the conservation of natural habitats and of wild fauna and flora. SPAs, identified by European countries according to article 1 of European Directive 79/409/EEC on the conservation of wild birds, contribute to the conservation of all species of naturally occurring birds in the wild state in the European territory. In this way, fragmentation, spatial isolation and functional independence of protected areas often prevent a coordinated and integrated management of local activities, which might even present conflicts and contrasts that need to be handled. For example, the coexistence of tourist, commercial and fishery activities need multidisciplinary and integrated public policies [27]. In addition, managing protected areas as isolated reserves—without integrating them into wider spatial strategies—exposes them to the consequences of habitat alteration and destruction, pollution and overfishing that might occur outside their boundaries [28]. For these reasons, an integrated management tool able to adequately harmonize sector-specific policies according to a holistic vision of the territory that goes beyond fragmentation and sectoral policies is clearly crucial [29,30]. Researchers agree that its objectives must be clearly defined and compatible with one another [31–38] so that they need to be researched empirically through negotiation with stakeholders, balancing ecological concerns, economic interests, social issues and political power [39].

With the aim of developing the growth potential of marine and maritime economic activities in a sustainable way, the European Commission identified nine ongoing initiatives in the European Union, including the one on maritime spatial planning and integrated coastal zone management introduced by Directive 2014/89/EU and the one on the ecosystem-based approach introduced by the Marine Strategy Framework Directive (2008/56/EC) [40].

On the one hand, Directive 2014/89/EU defines maritime spatial planning as a cross-cutting policy tool enabling the application of a coordinated, integrated and trans-boundary ecosystem-based approach, which promotes smart, sustainable and inclusive growth and ensures sustainable use of marine and coastal resources [41]. On the other hand, Directive 2008/56/EC recommends ecosystem-based approaches for marine strategies, so that pressure of human activities does not compromise the capacity of marine ecosystems to respond to human-induced changes and the sustainable use of marine goods and services is ensured to present and future generations [42]. In Italy, according to article 2 subsection 3 and article 3 subsection 4 of the Decree of the Ministry of Environment, Land and Sea Protection (in Italian: Ministero dell'Ambiente e della Tutela del Territorio e del Mare, MATTM) published on 17th October 2007, the management of Natura 2000 sites located within the boundaries of an MPA is in charge of the MPA managing authority. Therefore, the MATTM demanded the integration of Natura 2000 conservation measures into MPA regulations in order to efficiently manage the area.

2. The Case Study

Among the tools identified by the European Union to promote trans-boundary cooperation there is the Interreg Maritime Italy–France Programme, which financed 87 projects [43] including the GIREPAM project (Integrated Management of Ecological Networks through Parks and Marine Areas; in Italian: Gestione Integrata delle Reti Ecologiche attraverso i Parchi e le Aree Marine), which involves 16 partners located in five Italian and French regions (Sardinia, Corsica, Provence-Alps-Côte d'Azur, Liguria and Tuscany) and pursues the following three objectives:

1. To improve conservation and promotion of marine and costal zones and to direct public accessibility to natural sites.
2. To improve the efficiency of planning and public governance of marine and coastal areas in the area of cooperation.
3. To raise awareness on the economic value of the natural capital and to promote green and blue growth [44].

The research group of the Department of Civil and Environmental Engineering and Architecture (DICAAR) of the University of Cagliari worked on the second objective by outlining an Experimental Procedure (in Italian: Protocollo Sperimentale, PS) aiming at formulating appropriate regulations for the management and control of the Marine Protected Areas (MPAs) named "Tavolara-Punta Coda Cavallo" and "Isola dell'Asinara" located in north-eastern and north-western Sardinia, respectively, as shown in Figures 1 and 2.

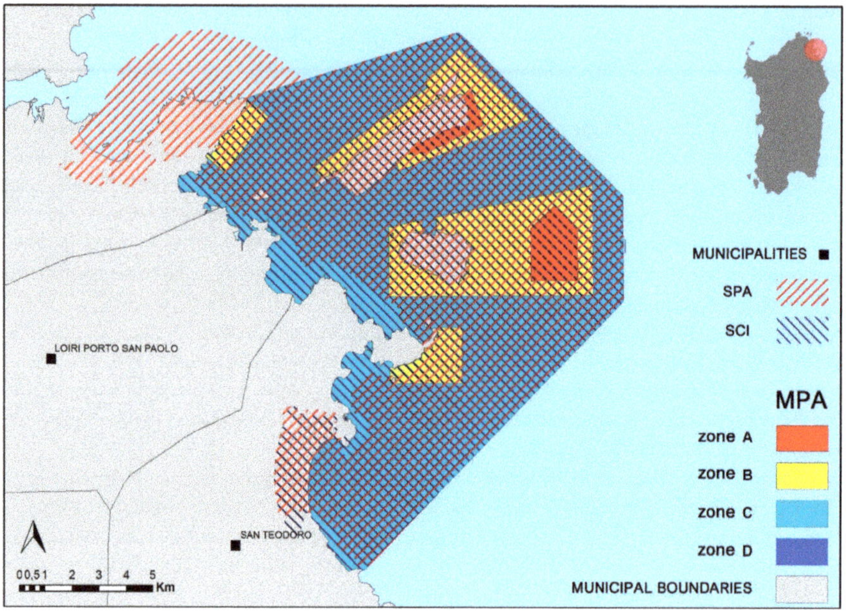

Figure 1. The territory of "Tavolara-Punta Coda Cavallo" Marine Protected Area (MPA) and of the Natura 2000 sites that overlap with it. Source: Authors' elaboration.

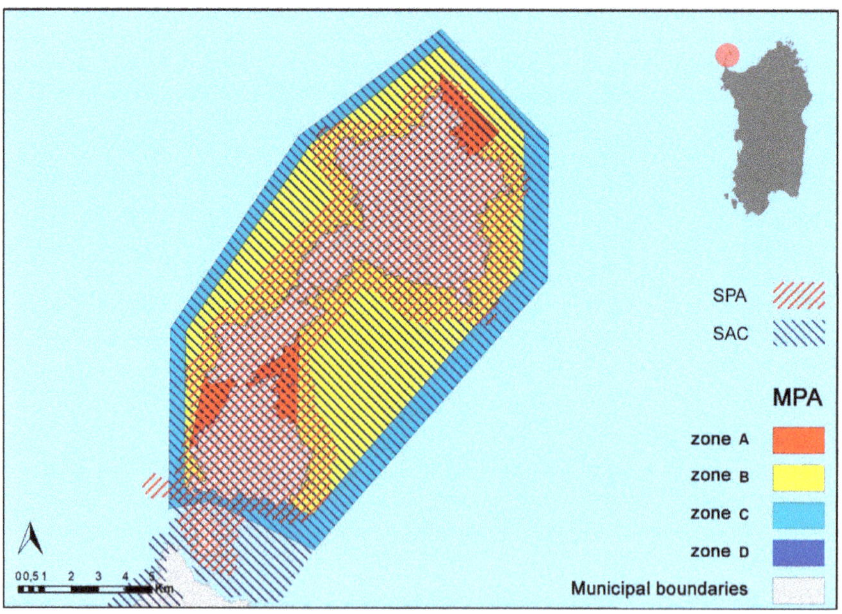

Figure 2. The territory of "Isola dell'Asinara" MPA and of the Natura 2000 sites that overlap with it.Source: Authors' elaboration.

Figures 1 and 2 show that SPAs, SACs and SCIs are located within the boundaries of the MPAs of "Tavolara-Punta Coda Cavallo" and "Isola dell'Asinara". Consequently, conservation measures identified by the Natura 2000 network for SPAs, SACs and SCIs are in force in the MPAs. In addition, the ICZM protocol and the ISEA project are adopted by the MPAs; therefore, the provisions they determine are in force in the MPAs as well. The PS aims at integrating these legislative measures in a single document, coherently with a holistic territory vision able to ensure favorable conservation status of habitats, thus preserving biodiversity.

Within the framework of the study carried out by the DICAAR research group, following the provisions provided by the European Directives 2014/89/EU and 2008/56/EC mentioned above, this paper focuses on the definition of a pioneering methodology to define "ecosystemic objectives", in accordance with the Sustainable Development Goals (SDGs) of Agenda 2030, to enrich the traditional assessment for sustainability provided by the Strategic Environmental Assessment (SEA). In this way, ecological and socio-economic disciplines are integrated to provide decision-makers with a synthesis of complex information to inform strategies definition. The aim is to provide a scientific framework, which integrates ecosystem services in environmental assessment procedures, in order to guide managers and legislators in making sound decisions.

For the purpose of this paper, only data concerning the MPA of "Isola dell'Asinara" are considered. The MPA was established in 2002 by a decree of the MATTM and its current regulations were approved by the MATTM through the decree of the 30th of July 2009 [45]. Its area covers approximately 108 km^2 in the sea and 79.64 km^2 in the mainland entirely comprised within the boundaries of the Municipality of Porto Torres, in the Province of Sassari [46]. The MPA territory overlaps with the following Natura 2000 sites: The SPA "ITB010001 Isola Asinara", the SPA "ITB013011 Isola Piana di Porto Torres" and the SAC "ITB010082 Isola dell'Asinara", as shown in Figure 2 [47]. It is characterized by an extensive coverage of various high-quality assets of natural capital as shown in Figure 3.

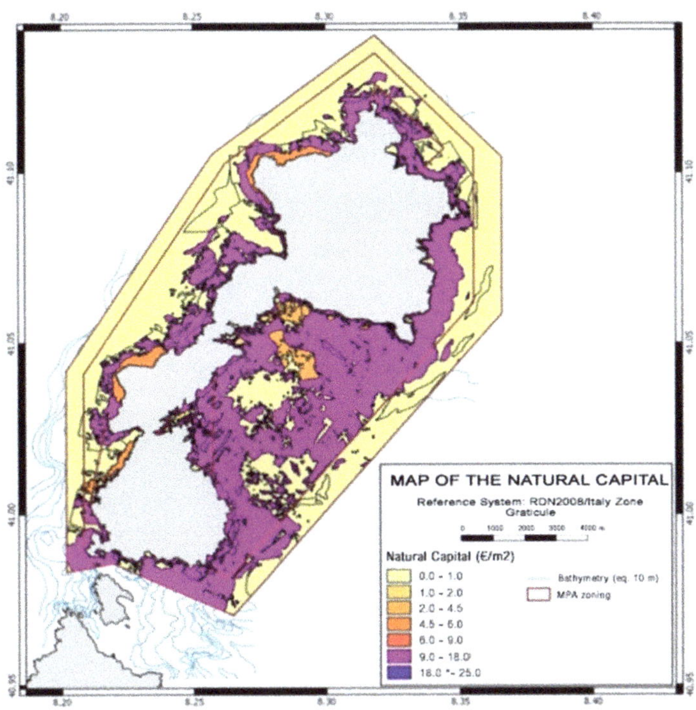

Figure 3. Map of the natural capital. Source: Povero et al., 2018 [48].

3. Methodology

The approach adopted for the definition of the new MPA regulation proposal is based on the outline of the PS. It draws inspiration from the Strategical Environmental Assessment (SEA), with particular reference to articles 1 and 6 of Directive 2001/42/CE, which concern decisional processes and public participation. In particular, the PS recalls the principle that the SEA is not a simple assessment of environmental impacts related to a decision, but it plays a strategic role in all the stages of the whole decisional process [49,50]. In this respect, the SEA is an effective political instrument for territorial governance, able to define strategical actions through the integration of diverse approaches and tools [49,51–55].

The PS applies these concepts with the aim of enhancing decisional processes and management practices. In particular, it promotes public participation in a holistic vision pointing at going beyond the fragmentation and sectorization that characterize the planning and management tools in force. It is structured into the sections illustrated in Figure 4.

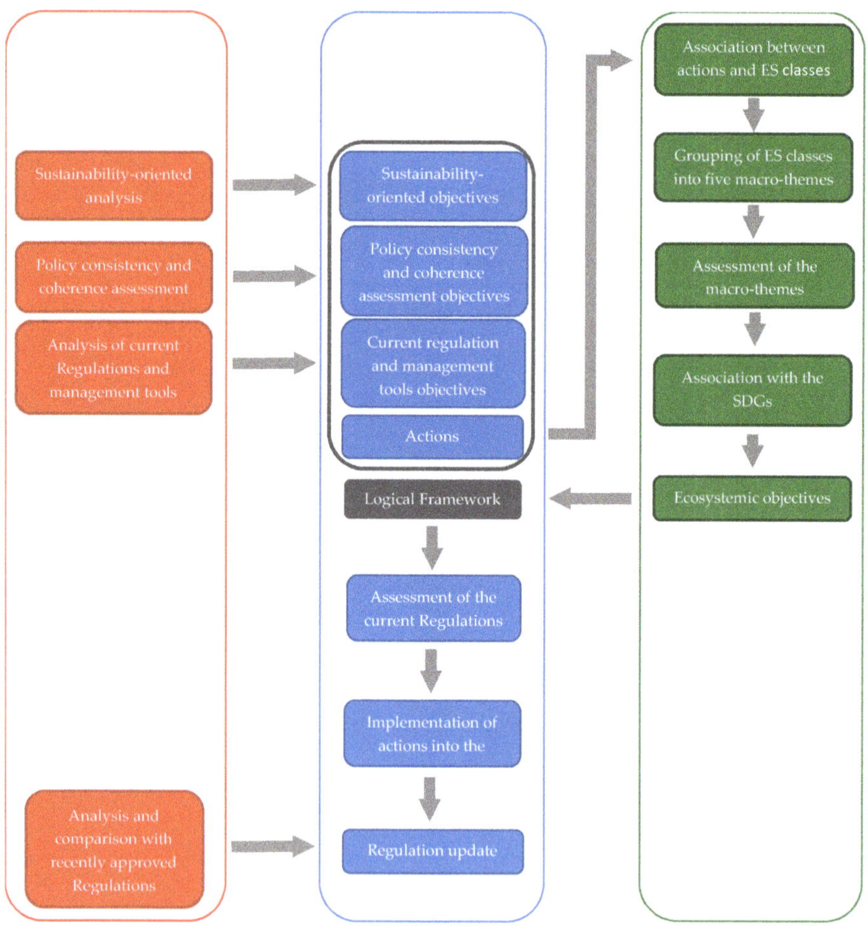

Figure 4. The Experimental Procedure (PS) scheme. Source: Authors' elaboration.

The core of the PS is constituted by the Logical Framework (LF), since it organizes all the objectives and actions identified in the previous sections, and connects them in a cascade relationship. The LF is organized according to a matrix structure, largely adopted in the programs promoted by the EU and other international entities. The LF is inspired to the so-called "programming by objectives", according to which the definition of the activities strictly follows the objective identification [56–58]. The LF structure ensures the identification of "conceptual connections between sustainability-oriented objectives related to the spatial contexts at stake and the operational planning actions concerning the integration of conservation measures related to" Natura 2000 sites into the new regulation proposal [59]. It is organized into four levels, which express the relationships among the various objectives drawn from the diverse protection instruments analyzed during the process of regulation elaboration (see Figure 4) and the actions defined to pursue the objectives selected for the prospective regulations.

The first level lists the sustainability-oriented objectives—formulated on the basis of the SWOT analysis—which depict the environmental status of the area through the analysis of each environmental component that characterizes the area [60]. The second level encompasses the objectives derived from the policy consistency and coherence assessment, which analyzes plans and programs in force at the regional, provincial and local level. The aim of this analysis is to identify the potential effects of plans and programs within the MPA territory and their interaction with the regulations. The third level enumerates specific objectives. They represent the regulation aims, as their codification is based on the contents of normative and management tools in force within the MPA. This section is the core of the PS, due to its importance within the process of upgrading the regulations, and its subsequent implementation.

The last level of the LF is the operative one, since it identifies the actions that address regulation objectives. In order to draw an efficient new regulation proposal, actions are formulated considering the planning and management tools in force (specific objectives) and the results of the SWOT analysis. These actions are then compared with the legislative instruments in force, which are in this way integrated in the new regulation proposal.

The Construction of the Logical Framework

As illustrated in Figure 4, the first level of the LF is constituted by the sustainability-oriented objectives. These objectives result from the analysis of various environmental and socio-economic aspects that describe the analyzed area. The environmental components are identified on the basis of the list provided by the guidelines issued by the Sardinian Region adequately adapted to the MPA studied [61]. In particular, the following ten "environmental components" are examined:

1. Air;
2. Water;
3. Waste treatment;
4. Soil and marine geo-morphology;
5. Flora, fauna and biodiversity;
6. Landscape;
7. Settlement layout and functions;
8. Tourism, recreation and outreach activities;
9. Fishing and other production activities;
10. Mobility and marine and terrestrial accessibility.

Characteristics of each component are reported in a summary sheet, whose scheme is illustrated in Table 1. The sheet contains the essential information assessing the status of natural resources and the pressures of anthropic and economic factors, which can be relevant in the phase of implementation of the regulations. Contextually, a SWOT analysis is conducted to analyze each environmental component, considering data drawn from the context analysis, in order to finally define a set of sustainability-oriented objectives. Subsequently, these objectives are better focused thanks to the ten sustainability criteria identified by the EC [48].

Table 1. Summary sheet for the analysis of the environmental components.

<table>
<tr><td colspan="2">Environmental Component K</td></tr>
<tr><td colspan="2">The introduction qualitatively describes the environmental component K in the analyzed territorial context (i.e., air).
The description is organized according to themes (i.e., air quality, meteorological conditions), which are in turn articulated in specific aspects (i.e., source of pollution, temperature, annual precipitation, etc.). The latter provide a quantitative analysis based on appropriate indicators.</td></tr>
<tr><td colspan="2">Theme n</td></tr>
<tr><td colspan="2">This section provides a brief qualitative description of the theme n, which refers to the environmental component K.</td></tr>
<tr><td colspan="2">Specific aspect t</td></tr>
<tr><td colspan="2">Brief description of the aspect through its distinguishing elements such as legislative and management tools, operational tools, thematic maps, infrastructure, etc.</td></tr>
<tr><td>Indicator 1 [unit]</td><td rowspan="3">Notes</td></tr>
<tr><td>. . .</td></tr>
<tr><td>Indicator n [unit]</td></tr>
<tr><td colspan="2">Source: . . .
Year: . . .</td></tr>
<tr><td colspan="2">SWOT Analysis</td></tr>
<tr><td>Strengths</td><td>Weaknesses</td></tr>
<tr><td>. . .</td><td>. . .</td></tr>
<tr><td>Opportunities</td><td>Threats</td></tr>
<tr><td>. . .</td><td>. . .</td></tr>
<tr><td>Sustainability-oriented Objectives</td><td>Environmental sustainability criteria</td></tr>
<tr><td>Ps_Ob_SA_n . . .</td><td>. . .</td></tr>
</table>

Table 2 shows the SWOT analysis conducted for the "air" component, in relation to the sustainability-oriented objective: "Increasing air quality by mitigating pollution impacts, in order to control negative effects caused by climate change". It highlights strengths, weaknesses, opportunities and threats related to the component "air". Subsequently the identified sustainability-oriented objective is compared to the appropriate environmental sustainability criteria, designated by the EC [42] (The ten sustainability criteria are the following: (i) Minimization of the use of non-renewable resources; (ii) use of renewable resources within limits of capacity for regeneration; (iii) environmentally sound use and management of hazardous/polluting substances and wastes; (iv) conservation and enhancement of the status of wildlife, habitats and landscapes; (v) maintenance and improvement of the quality of soils and water resources; (vi) maintenance and improvement of the quality of historic and cultural resources; (vi) maintenance and improvement of the local environmental quality; (viii) protection of the atmosphere (global warming); (ix) development of environmental awareness, education and training; (x) promotion of public participation in decisions involving sustainable development). In particular, the sustainability-oriented objective here identified is better focused by considering the environmental sustainability criteria numbers i, ii, iii, vii and viii.

Table 2. SWOT analysis.

SWOT Analysis	
Strengths	**Weaknesses**
— Absence of air pollution sources within the MPA.	— Absence of monitoring networks able to promptly provide information within the MPA territory. — Potential air pollution within the MPA boundaries, due to motor boats.
Opportunities	**Threats**
— Favorable meteorological conditions during the summer period, and mild weather during the winter.	— Presence of pollution sources near the MPA, namely the Porto Torres industrial area and Fiume Santo power station.
Sustainability-oriented Objective	**Environmental sustainability criteria**
— To enhance air quality by reducing the pollution consequences in order to mitigate negative effects caused by climate change.	i, ii, iii, vii, viii

The second level of the LF is represented by the policy consistency and coherence objectives.

The assessment of policy consistency and coherence considers all the plans and programs in force at any level in the Marine Protected Area by examining the economic, social and spatial aspects relevant to the analyzed context. All the objectives illustrated in each plan or program are analyzed, and those considered meaningful for the regulations are adopted, either identical to their original formulation or adequately rephrased in order to be specifically tailored to the context.

For example, the objectives identified in relation to the Regional Air Quality Plan are listed below:

- To pursue general environmental enhancement, in relation to other zones and types of polluting substances;
- to integrate environmental needs in sectorial policies (especially in the sectors of energy, industry and transportation) with the aim of ensuring sustainable socio-economic development;
- to increase citizen awareness in order to promote environmentally friendly behavior;
- to integrate authorizations and inspection and monitoring activities in order to best ensure the implementation of the planning and regulation actions.

The third level of the LF is constituted of the specific objectives drawn from the various regulatory regimes in force in the "Isola dell'Asinara" MPA, with the aim of grouping them into a single regulatory framework. In correspondence with each objective, one or more actions are identified in order to guarantee the implementation of the objectives of the new regulation proposal. Simultaneously, actions pursue the aim of eliminating—or at least reducing—pressures exerted on natural resources by anthropic factors. In Tables 3 and 4 excerpts of the LF referring to the "air" component are reported.

Table 3. LF excerpt referring to the "air" component.

Sustainability-Oriented Objectives	Policy Consistency and Coherence Objectives	Regulation Objectives (Specific Objectives)
Ps_Ob_SA_1 To enhance air quality by reducing pollution consequences, in order to mitigate the negative effects caused by climate change.	Ps_Ob_CE_4 To increase citizen awareness in order to promote environmentally friendly behaviors.	PS_Ob_REO_10 To organize information, promotional, awareness-raising and educational programs in order to guarantee sustainable fruition of the area and of its natural resources.
		PS_Ob_REO_11 To promote the development of economic, social, cultural and tourist activities, which are compatible with conservation needs.
		PS_Ob_REO_12 To promote programs and projects directed to the enhancement of marine ecosystems.
		PS_Ob_REO_16 To ensure efficiency in the integrated management of the MPAs.
	Ps_Ob_CE_5 To monitor natural phenomena and processes in order to support authorization and control procedures for a better implementation of the regulation actions.	PS_Ob_REO_18 To efficiently organize surveillance and assistance activities.
		PS_Ob_REO_19 To promote scientific investigation programs.
		PS_Ob_REO_20 To monitor the state of conservation of habitats and species.
	Ps_Ob_CE_11 To integrate sustainable development policies, in order to reduce and prevent climate change and desertification impacts, by reducing the emission of climate-changing gas.	PS_Ob_REO_8 To efficiently organize recreational boat activities, mooring and docking, in compliance with the rules in force in each zone of the MPA and with the related Code of Conduct.
		PS_Ob_REO_13 To reduce the main anthropic pressure factors.
		PS_Ob_REO_16 To ensure efficiency in the integrated management of the MPAs.
	Ps_Ob_CE_13 To increase knowledge, applied research and experimentation.	PS_Ob_REO_19 To promote scientific investigation programs.
	Ps_Ob_CE_16 To reduce energy consumption, pollutant emissions and their impacts on the MPA territory, landscape and environmental value.	PS_Ob_REO_8 To efficiently organize recreational boat navigation, mooring and docking, in compliance with the rules in force in each zone of the MPA and with the related code of conduct.
		PS_Ob_REO_13 To reduce the main anthropic pressure factors.

Table 4. LF excerpt referring to the "air" component.

Sustainability-Oriented Objectives	Regulation Objectives (Specific Objectives)	Regulation Actions
Ps_Ob_SA_1 To enhance air quality by reducing pollution consequences, in order to mitigate the negative effects caused by climate change.	Ps_Ob_REO_11, Ps_Ob_REO_13	Act_REO_3 Positioning of landing stages and buoys for the mooring of recreational boats in proximity of the Asinara island, in order to preserve *Posidonia oceanica* habitats.
	Ps_Ob_REO_11, Ps_Ob_REO_13	Act_REO_4 Realization of artificial submerged barriers in order to constrict illegal trawling and to protect *Posidonia oceanica* meadows.
	Ps_Ob_REO_12	Act_REO_6 To elaborate strategies and programs for the sustainable use of the MPA territory.

Table 4. *Cont.*

Sustainability-Oriented Objectives	Regulation Objectives (Specific Objectives)	Regulation Actions
	Ps_Ob_REO_18	Act_REO_7 To adopt measures to facilitate the rescue of animals in distress or disoriented.
	Ps_Ob_REO_10	Act_REO_8 To regulate photo shoots, filming and TV recordings.
	Ps_Ob_REO_8, Ps_Ob_REO_10, Ps_Ob_REO_13	Act_REO_11 To regulate activities allowed in each MPA zone.
	Ps_Ob_REO_10, Ps_Ob_REO_13	Act_REO_12 To organize the tourism fruition system within the MPA.
	Ps_Ob_REO_8	Act_REO_13 To control and limit ship and boat speed, in order to reduce underwater noise pollution and the risk of collision with cetaceans and turtles.
	Ps_Ob_REO_20	Act_REO_14 To regulate monitoring activity.
	Ps_Ob_REO_10, Ps_Ob_REO_13	Act_REO_18 To design and create submarine paths.
	Ps_Ob_REO_12	Act_REO_19 To launch seabed cleaning campaigns.
	Ps_Ob_REO_19	Act_REO_20 To regulate technical and scientific research activity.
	Ps_Ob_REO_13	Act_REO_21 To regulate bathing and swimming.
	Ps_Ob_REO_8	Act_REO_24 To regulate recreational boat activities.
	Ps_Ob_REO_8	Act_REO_25 To regulate docking of recreational boats.
	Ps_Ob_REO_8	Act_REO_26 To regulate mooring of recreational boats.
	Ps_Ob_REO_11	Act_REO_31 To regulate chartering and leasing—even temporary—of recreational boats.
	Ps_Ob_REO_11, Ps_Ob_REO_13	Act_REO_33 To define a set of indicators to asses carrying capacity of the MPA during the summer.
	Ps_Ob_REO_11	Act_REO_34 To regulate authorizations for the development of socio-economic, tourist, social and cultural activities within the MPA.
	Ps_Ob_REO_11	Act_REO_35 To adopt measures to preserve and protect the archaeologic, historic and cultural submarine heritage.
	Ps_Ob_REO_8	Act_REO_41 To systematize the mooring infrastructure.
	Ps_Ob_REO_20	Act_REO_42 To launch monitoring campaigns to collect data on air quality.

Once the LF is completely filled, actions are compared with the regulations in force in order to evaluate their mutual coherence. In the event that the regulations in force in the MPAs do not include—or only partially include—a particular action, the regulations are updated either by adding new articles and subsections, or by modifying the existing ones. In order to best pursue this upgrading, regulations in force are also compared with regulations recently approved in other Italian MPAs, which are considered as best practices from which inspiration is taken. Table 5 shows the methodological model adopted to revise the regulations in force and is based on a double level of assessment.

Table 5. Methodology adopted to revise the regulations in force.

Title							
Regulations in Force							**Regulation Update**
Article from Regulations in Force	**First Assessment Level: Comparison between Actions and Articles of Regulations in Force**			**Second Assessment Level: Comparison with Recently Approved Regulations**			**Article of the Updated Regulations**
Subsection	Action	Article/action assessment	Remarks about the article/action assessment	Recent approved regulations		Remarks about the comparison with recent approved regulations	New subsection proposal

As a result of this double comparison, the new regulation proposal includes the aspects that characterize the sectorial norms that coexist in the area and those drawn from the regulations recently approved in other Italian MPAs. This integration aims at reinforcing biodiversity and environmental conservation, while guaranteeing efficient management of all the socio-economic activities in the area.

4. Definition of the PS Actions

While the objectives constitute the structural dimension of the PS, the actions, in turn, embody the operational aspect, by providing detailed indications and restrictions for the organization and management of the MPA. Following the criteria adopted for the identification of the regulation objectives, actions are codified considering the current restrictions imposed by the legislation in force. These are the MPA establishment decree, the regulations, the management plans of the Natura 2000 sites, and the provisions of the ICZM protocol and of the ISEA project. In addition, the results of the environmental analysis—in particular, the SWOT analysis—for each environmental component have been considered. In this way, actions result strictly linked to the regulation objectives and well connected to the strategic themes raised during their definition.

Once defined, actions were grouped by the categories that characterize the regulations in force as illustrated below:

- Dispositions and regulatory norms;
- monitoring;
- promotional, awareness-raising, information and educational programs;
- security and surveillance;
- institutional coordination.

Attention has been focused on the dispositions and regulatory norms showing a strong link between the sustainable use of natural resources and the protection of marine habitats and species.

Table 6 reports the action set defined.

Table 6. Regulation actions. Source: PS of "Isola dell'Asinara" MPA.

Code	Action
Action category: DISPOSITIONS AND REGULATORY NORMS	
ACT_Reg_1	To adopt protection and conservation measures for habitats and species.
ACT_Reg_2	To regulate the management of stranded *Posidonia oceanica*.
ACT_Reg_3	Positioning of landing stages of buoys for the mooring of recreational boats in proximity of the Asinara island in order to preserve *Posidonia oceanica* habitats.
ACT_Reg_4	Realization of artificial submerged barriers in order to constrict illegal trawling and to protect *Posidonia oceanica* meadows.

Table 6. *Cont.*

Code	Action
ACT_Reg_5	To adopt measures to prevent coastal erosion.
ACT_Reg_6	To elaborate strategies and programs for the sustainable use of the MPA territory.
ACT_Reg_7	To adopt measures to facilitate the rescue of animals in distress or disoriented.
ACT_Reg_8	To regulate photo shoots, filming and TV recordings.
ACT_Reg_9	To launch recovery campaigns of abandoned waste and of the polluted areas of the MPA.
ACT_Reg_10	To launch eradication campaigns of all invasive or potential invasive alien plants.
ACT_Reg_11	To regulate activities allowed in each MPA zone.
ACT_Reg_12	To organize the tourism fruition system within the MPA.
ACT_Reg_13	To control and limit ship and boat speed, in order to reduce underwater noise pollution and the risk of collision with cetaceans and turtles.
ACT_Reg_14	To regulate monitoring activities.
ACT_Reg_15	To define identification criteria of fishing gear hallmarking.
ACT_Reg_16	To define a plan for the recovery of ghost fishing gear.
ACT_Reg_17	To adopt adequate measures to rehabilitate dunes and bars (protection, signage and maintenance).
ACT_Reg_18	To design and create submarine paths.
ACT_Reg_19	To launch seabed cleaning campaigns.
ACT_Reg_20	To regulate technical and scientific research activity.
ACT_Reg_21	To regulate bathing and swimming.
ACT_Reg_22	To regulate guided scuba diving tours.
ACT_Reg_23	To regulate sea-watching activities.
ACT_Reg_24	To regulate recreational boat activities.
ACT_Reg_25	To regulate docking of recreational boats.
ACT_Reg_26	To regulate mooring of recreational boats.
ACT_Reg_27	To regulate activities of public passenger transport services, sailboat charters and guided tours.
ACT_Reg_28	To regulate professional fishing activities.
ACT_Reg_29	To regulate activities of recreational fishing and ichthyic tourism.
ACT_Reg_30	To regulate whale-watching activities.
ACT_Reg_31	To regulate chartering and leasing, even temporary, of recreational boats.
ACT_Reg_32	To assess vulnerability and risk to accordingly adopt measures of prevention, mitigation and adaptation to face consequences of natural disasters and environmental emergences.
ACT_Reg_33	To define a set of indicators to asses carrying capacity of the MPA during the summer.
ACT_Reg_34	To regulate authorizations for the development of socio-economic, tourist, social and cultural activities within the MPA.
ACT_Reg_35	To adopt measures to preserve and protect the archaeologic, historic and cultural submarine heritage.

Table 6. *Cont.*

Code	Action
ACT_Reg_36	To adopt temporary protection measures in the areas and during the reproductive periods of vulnerable species that nest on cliffs or minor islands, such as ban on entry of boats, bathing prohibition and stop of recreational and commercial tourist activities.
ACT_Reg_37	To tighten controls on sea-water quality.
ACT_Reg_38	To establish rules and regulations concerning wastewater and hydrocarbons discharge into the sea.
ACT_Reg_39	To circumscribe areas at risk of landslide.
ACT_Reg_40	To define motor parameters of boats according to the Italian Legislative Decree No. 171/2005.
ACT_Reg_41	To systematize the mooring infrastructure.
Action category: MONITORING	
ACT_Reg_42	To launch monitoring campaigns to collect data on air quality.
ACT_Reg_43	To launch monitoring campaigns of habitats and species to guarantee biodiversity preservation.
ACT_Reg_44	To launch monitoring campaigns of marine and coastal invasive alien plants.
ACT_Reg_45	To launch monitoring campaigns of geomorphological dynamics of beach-dune systems and of sandy coastlines.
ACT_Reg_46	To launch census and monitoring campaigns of fish species.
ACT_Reg_47	To launch monitoring campaigns of tourist activity impacts.
ACT_Reg_48	To launch monitoring campaigns of the marine avifauna of conservation interest.
ACT_Reg_49	To launch monitoring campaigns of professional fishing and its impacts.
ACT_Reg_50	To launch census and monitoring campaigns for the protection of coralligenous biocenosis.
ACT_Reg_51	To launch census and monitoring campaigns for the protection of *Posidonia oceanica* meadows.
ACT_Reg_52	To launch monitoring campaigns of fishing and ichthyic tourism.
Action category: PROMOTION, SENSIBILIZATION, INFORMATION AND EDUCATION	
ACT_Reg_53	To launch awareness-raising campaigns to prevent alien species importation.
ACT_Reg_54	To launch awareness-raising campaigns for the installation of low-impact-lighting systems aimed at reducing animal disorientation.
ACT_Reg_55	To launch awareness-raising and information campaigns to protect the species of the marine and coastal environment.
ACT_Reg_56	To launch awareness-raising and information campaigns to reduce anthropic disturbance of the avifauna in the MPA.
ACT_Reg_57	To launch awareness-raising and information campaigns to reduce waste abandonment.
ACT_Reg_58	To launch environmental education programs and activities.
ACT_Reg_59	To promote the MPA territory.
ACT_Reg_60	To promote interdisciplinary scientific research projects.
ACT_Reg_61	To plan awareness-raising, education and training activities on integrated management of coastal areas.

Table 6. *Cont.*

Code	Action
ACT_Reg_62	To launch awareness-raising, education and control activities to prevent material removal from the coast and the sea, according to the Regional Law 16/2017.
ACT_Reg_63	To launch campaigns to promote environmental and landscape heritage.
ACT_Reg_64	To launch information and promotional campaigns on the economic activities of the MPA.
ACT_Reg_65	To launch information and promotional campaigns addressed to the economic actors who work in the MPA.
ACT_Reg_66	To launch awareness-raising campaigns addressed to the operators of the fishing sector against the abandonment of fishing equipment.
Action category: SURVEILLANCE	
ACT_Reg_67	To organize and intensify the surveillance in the most problematic zones, paying particular attention on marine and terrestrial areas with important tourism flows.
ACT_Reg_68	To implement the surveillance camera system.
ACT_Reg_69	To launch surveillance campaigns to protect habitats and species.
ACT_Reg_70	To organize and improve the surveillance of the fishing sector.
ACT_Reg_71	To promote poaching suppression.
Action category: INTERINSTITUTIONAL COOPERATION	
ACT_Reg_72	To encourage national, regional and local initiatives through coordinate actions of promotion, cooperation and partnership.
ACT_Reg_73	To adopt adequate measures to reinforce regional, national and international cooperation.
ACT_Reg_74	To coordinate institutional intersectoral activities promoted by administrative services and regional and local authorities in charge in the coastal zones.
ACT_Reg_75	To organize access and parking in the MPA according to the environmental carrying capacity.
ACT_Reg_76	To enhance and coordinate, in cooperation with the surrounding local entities, the public transportation system that connects the Asinara Island with the municipalities of Stintino and Porto Torres.
ACT_Reg_77	To coordinate the activities for cetacean and sea turtle rescue.

Selection of Ecosystem Services for MPAs

Given the need of guaranteeing coherency and compatibility among the diverse assessments, and the aim of properly integrating ES assessment into the PS in order to update the regulations in ecosystemic terms, this paper proposes a selection of ESs for the "Isola dell'Asinara" MPA based on the Common International Classification of Ecosystem Goods and Services (CICES). CICES consists of a hierarchical scheme for ES classification based on the three main categories of ecosystem services [62]:

1. Provisioning services: Provision of genuine goods, such as food, water, timber, fiber, fuel and other raw materials, but also genetic material and ornamental species;
2. regulation services: Regulation of climate, air quality and water, soil formation, pollination, waste assimilation, and mitigation of natural hazards such as erosion, weeds, etc.;
3. cultural services: Non-material benefits such as heritage and cultural identity, spiritual enrichment and intellectual, aesthetic and recreational values.

The work has been conducted on the basis of a study realized in the "Isola dell'Asinara" MPA, concerning environmental accounting in Italian marine protected areas (in Italian: "Contabilità ambientale nelle aree marine protette italiane") [63], which assigns an ecological value to environmental heritage and subsequently identifies the ecological functions and the relative ESs that characterize it.

On the basis of data provided by the MPA, a set of ESs has been identified and classified according to the CICES scheme (Table 7). The criterion adopted for ES identification was the influence wielded over them by the protection regime in force within the MPA, and their suitability to MPA management.

Table 7. Proposal of a common Ecosystem Service (ES) classification [64].

CICES for Ecosystem Accounting			
Section	**Division**	**Group**	**Class**
This column lists the three main categories of ES.	This column divides section categories into the main types of output or process.	The group level splits division categories by biological, physical or cultural type or process.	The class level provides a further sub-division of group categories into biological or material outputs and bio-physical and cultural processes that can be linked back to concrete identifiable service sources.

The three sections identified by the CICES scheme are: Regulation and maintenance, provisioning and cultural. It must be noticed that these categories are not designed for a specific spatial scale nor for a unique governance level. Thus, the ES assessment needs to be tailored for MPAs and Natura 2000 sites by considering that:

- Terrestrial ES (cultivated crops, surface water for drinking, etc.) are out of the scope of this study;
- the MPA objectives focus more on biodiversity protection rather than on recreational functions and food productions (i.e., fishing).

Tables 8–10 illustrate the match between ES classes and PS actions related to the ES sections "regulation and maintenance", "provisioning" and "cultural", respectively.

Table 8. Association between ES classes and PS actions related to the ES section "provisioning".

Section: Provisioning	
ES Class	**PS Action**
Cultivated crops	
Reared animals and their outputs	
Wild plants, algae and their outputs	
Wild animals and their outputs	
Plants and algae from *in-situ* aquaculture	
Animals from *in-situ* aquaculture	
Surface water for drinking	
Ground water for drinking	
Fibers and other materials from plants, algae and animals for direct use or processing	To adopt protection and conservation measures for habitats and species
	To regulate professional fishing activities
	To regulate activities of recreational fishing and ichthyic tourism
Materials from plants, algae and animals for agricultural use	
Genetic materials from all biota	
Surface water for non-drinking purposes	
Ground water for non-drinking purposes	
Plant-based resources	
Animal-based resources	

Table 9. Association between ES classes and PS actions related to the ES section "regulation and maintenance".

Section: Regulation and Maintenance	
ES Class	**PS Action**
Bio-remediation by micro-organisms, algae, plants, and animals	
Filtration/sequestration/storage/accumulation by ecosystems	To regulate the management of stranded *Posidonia oceanica*
	Positioning of landing stages of buoys for the mooring of recreational boats in proximity of the Asinara island, in order to preserve *Posidonia oceanica* habitats
	Realization of artificial submerged barriers in order to constrict illegal trawling and to protect *Posidonia oceanica* meadows
Dilution by atmosphere, freshwater and marine ecosystems	
Mediation of smell/noise/visual impacts	
Mass stabilization and control of erosion rates	To circumscribe areas at risk of landslide
Buffering and attenuation of mass flows	
Hydrological cycle and water flow maintenance	
Flood protection	To regulate the management of stranded *Posidonia oceanica*.
	To adopt measures to prevent coastal erosion.
	To adopt adequate measures to rehabilitate dunes and bars (protection, signage and maintenance)
	To assess vulnerability and risk to accordingly adopt measures of prevention, mitigation and adaptation to face consequences of natural disasters and environmental emergencies
	To launch monitoring campaigns of geomorphological dynamics of beach-dune systems and of sandy coastlines
Storm protection	To regulate the management of stranded *Posidonia oceanica*
	To adopt measures to prevent coastal erosion.
	To adopt adequate measures to rehabilitate dunes and bars (protection, signage and maintenance)
	To assess vulnerability and risk to accordingly adopt measures of prevention, mitigation and adaptation to face consequences of natural disasters and environmental emergencies
Ventilation and transpiration	
Pollination and seed dispersal	
Maintaining nursery populations and habitats	To adopt protection and conservation measures for habitats and species
	To elaborate strategies and programs for the sustainable use of the MPA territory
	To adopt measures to facilitate the rescue of animals in distress or disoriented
	To regulate photo shoots, filming and TV recordings
	To launch eradication campaigns of all invasive or potential invasive alien plants
	To regulate activities allowed in each MPA zone
	To organize the tourism fruition system within the MPA
	To control and limit ship and boat speed, in order to reduce underwater noise pollution and the risk of collision with cetaceans and turtles

Table 9. *Cont.*

Section: Regulation and Maintenance	
ES Class	**PS Action**
	To regulate monitoring activities
	To design and create submarine paths
	To launch seabed cleaning campaigns
	To regulate technical and scientific research activity
	To regulate bathing and swimming
	To regulate guided scuba diving tours
	To regulate sea-watching activities
	To regulate recreational boat activities
	To regulate docking of recreational boats
	To regulate mooring of recreational boats
	To regulate activities of public passenger transport services, sailboat charters and guided tours
	To regulate professional fishing activities
	To regulate activities of recreational fishing and ichthyic tourism
	To regulate whale-watching activities
	To regulate chartering and leasing, even temporary, of recreational boats
	To define a set of indicators to asses carrying capacity of the MPA during the summer
	To regulate authorizations for the development of socio-economic, tourist, social and cultural activities within the MPA
	To adopt temporary protection measures in the areas and during the reproductive periods of vulnerable species that nest on cliffs or minor islands, such as ban on entry of boats, bathing prohibition and stop of recreational and commercial tourist activities
	To systematize the mooring infrastructure
	To launch monitoring campaigns of habitats and species to guarantee biodiversity preservation
	To launch monitoring campaigns of marine and coastal invasive alien plants
	To launch census and monitoring campaigns of fish species
	To launch monitoring campaigns of tourist activity impacts
	To launch monitoring campaigns of the marine avifauna of conservation interest.
	To launch monitoring campaigns of professional fishing and its impacts.
	To launch census and monitoring campaigns for the protection of coralligenous biocenosis
	To launch monitoring campaigns of fishing and ichthyic tourism
	To launch awareness-raising campaigns to prevent alien species importation

Table 9. *Cont.*

Section: Regulation and Maintenance	
ES Class	**PS Action**
	To launch awareness-raising campaigns for the installation of low-impact-lighting systems aimed at reducing animal disorientation
	To launch awareness-raising and information campaigns to protect the species of the marine and coastal environment
	To launch awareness-raising and information campaigns to reduce anthropic disturbance of the avifauna in the MPA
	To launch awareness-raising and information campaigns to reduce waste abandonment
	To plan awareness-raising, education and training activities on integrated management of coastal areas
	To launch awareness-raising, education and control activities to prevent material removal from the coast and the sea, according to the Regional Law 16/2017
	To launch awareness-raising, education and control activities to prevent material removal from the coast and the sea, according to the Regional Law 16/2017
	To launch campaigns to promote environmental and landscape heritage
	To launch information and promotional campaigns on the economic activities of the MPA
	To launch information and promotional campaigns addressed to the economic actors who work in the MPA
	To launch awareness-raising campaigns addressed to the operators of the fishing sector against the abandonment of fishing equipment
	To promote poaching suppression
Pest control	
Disease control	
Weathering processes	
Decomposition and fixing processes	
Chemical condition of freshwaters	
Chemical condition of salt waters	To adopt protection and conservation measures for habitats and species
	Positioning of landing stages of buoys for the mooring of recreational boats in proximity of the Asinara island, in order to preserve *Posidonia oceanica* habitats
	Realization of artificial submerged barriers in order to constrict illegal trawling and to protect *Posidonia oceanica* meadows
	To elaborate strategies and programs for the sustainable use of the MPA territory
	To launch recovery campaigns of abandoned waste and of the polluted areas of the MPA
	To launch eradication campaigns of all invasive or potential invasive alien plants
	To regulate activities allowed in each MPA zone
	To organize the tourism fruition system within the MPA
	To regulate monitoring activities
	To launch seabed cleaning campaigns
	To regulate bathing and swimming

Table 9. *Cont.*

Section: Regulation and Maintenance	
	To regulate recreational boat activities
	To regulate activities of public passenger transport services, sailboat charters and guided tours
ES Class	**PS Action**
	To regulate chartering and leasing, even temporary, of recreational boats
	To define a set of indicators to asses carrying capacity of the MPA during the summer
	To regulate authorizations for the development of socio-economic, tourist, social and cultural activities within the MPA
	To tighten controls on sea-water quality.
	To establish rules and regulations concerning wastewater and hydrocarbons discharge into the sea.
	To define motor parameters of boats according to the Italian Legislative Decree No. 171/2005
	To launch monitoring campaigns of tourist activity impacts
	To launch census and monitoring campaigns for the protection of *Posidonia oceanica* meadows
	To launch awareness-raising and information campaigns to reduce waste abandonment
	To plan awareness-raising, education and training activities on integrated management of coastal areas
	To launch campaigns to promote environmental and landscape heritage
	To launch information and promotional campaigns on the economic activities of the MPA
	To launch information and promotional campaigns addressed to the economic actors who work in the MPA
Global climate regulation by reduction of greenhouse gas concentrations	Positioning of landing stages of buoys for the mooring of recreational boats in proximity of the Asinara island, in order to preserve *Posidonia oceanica* habitats.
	Realization of artificial submerged barriers in order to constrict illegal trawling and to protect *Posidonia oceanica* meadows
	To launch census and monitoring campaigns for the protection of *Posidonia oceanica* meadows
Micro and regional climate regulation	Positioning of landing stages of buoys for the mooring of recreational boats in proximity of the Asinara island, in order to preserve *Posidonia oceanica* habitats
	Realization of artificial submerged barriers in order to constrict illegal trawling and to protect *Posidonia oceanica* meadows
	To launch monitoring campaigns to collect data on air quality
	To launch census and monitoring campaigns for the protection of *Posidonia oceanica* meadows

Table 10. Association between ES classes and PS actions related to the ES section "cultural".

Section: Cultural	
ES Class	**PS Action**
Experiential use of plants, animals and land-/sea-scapes in different environmental settings	To adopt protection and conservation measures for habitats and species
Physical use of land-/sea-scapes in different environmental settings	To adopt protection and conservation measures for habitats and species
	To define identification criteria of fishing gear hallmarking.
	To define a plan for the recovery of ghost fishing gear
	To regulate bathing and swimming
	To regulate guided scuba diving tours
	To regulate sea-watching activities
	To regulate recreational boat activities
	To promote the MPA territory
	To launch campaigns to promote environmental and landscape heritage
	To launch awareness-raising campaigns addressed to the operators of the fishing sector against the abandonment of fishing equipment
Scientific	To promote interdisciplinary scientific research projects
Educational	To launch environmental education programs and activities
	To plan awareness-raising, education and training activities on integrated management of coastal areas
	To launch awareness-raising, education and control activities to prevent material removal from the coast and the sea, according to the Regional Law 16/2017
Heritage, cultural	To adopt measures to preserve and protect the archaeologic, historic and cultural submarine heritage
Entertainment	To regulate bathing and swimming
	To regulate guided scuba diving tours
	To regulate sea-watching activities
	To regulate recreational boat activities
	To regulate whale-watching activities
Aesthetic	To launch awareness-raising, education and control activities to prevent material removal from the coast and the sea, according to the Regional Law 16/2017
Symbolic	
Sacred and/or religious	
Existence	
Bequest	

Tables 8–10 show that the previously defined actions of the PS match all the three ES sections and that the majority of the actions refer to the "regulation and maintenance" and "cultural" ES sections. In particular, the highest number of matches is observable in the classes "maintaining nursery populations and habitats" and "chemical condition of salt waters" in the section "regulation and maintenance" and in the class "physical use of land/sea-scapes in different environmental settings" in the section "cultural". However, not all the ES classes are matched to PS actions. Nevertheless, some of them are relevant for defining an appropriate management of the MPA. For example, "wild animals

and their outputs", "bio-remediation by micro-organisms, algae, plants and animals", "dilution by atmosphere, freshwater and marine ecosystems", "symbolic", "existence" and "bequest".

Subsequently to this classification, in order to assess the management of the MPA, ES classes have been grouped into five macro-themes according to the proposal of the Millennium Ecosystem Assessment proposal [62] for urban, coastal and marine areas as follows:

- Tourism fruition and eco-tourism;
- fish biomass supply for business or entertainment purposes;
- climate regulation;
- water quality regulation;
- landscape and cultural heritage.

These macro-themes have been assessed through a methodological approach similar to the one that guided the environmental assessment of the PS. Indeed, they have been described through a proper summary sheet with the aim of identifying a set of objectives in addition to those already defined in the PS. In this way, it is possible to update the regulations according to an "ecosystemic" perspective. Table 11 provides an example of a summary sheet for the ES "k".

Table 11. Example of a summary sheet for the ES "k".

ES k		
The introductive section qualitatively describes the ecosystem Service K in the analyzed territorial context. Moreover, the activities that best represent the analyzed ES are identified and articulated in specific aspects. These provide a quantitative analysis based on appropriate indicators.		
Activity 1		
This section provides a brief qualitative description of the activity and its characterizing aspects.		
ES Class	*Indicator*	*Measure unit*
Activity n		
This section provides a brief qualitative description of the activity and its characterizing aspects.		
ES Class	*indicator*	*Measure unit*
Flow calculation		

Among the previously identified "ecosystemic macro-themes", for the purpose of this study, it is worth focusing on the following:

- Fish biomass supply for business or entertainment purposes;
- landscape fruition for recreational activities (bathing, diving and yachting) and scientific/educational purposes.

In the operational phase, after identifying and mapping where each ES fruition occurs in order to analyze each ES macro-theme, an interview sample was realized with the aim of monitoring anthropic pressures related to ES fruition. Data collection focused on bathing, scuba diving, professional and small-scale fishing and yachting and relied on fieldwork and interviews conducted in the MPA headquarters, especially through:

- Questionnaires to users (i.e., tourists and scuba divers);
- collection of information and documentation related to financial reports, studies and resource exploitation in the MPA headquarters. This part was conducted by the MPA staff in collaboration with the University of Genova;

- analysis of the documentation on the authorizations granted by the MPA in order to assess anthropic pressures and to gather information concerning the typology of the related operator or user;
- outline of a new authorization system for the simultaneous collection of information about the authorization-release procedure and of all the data needed for the application of the environmental accountability [63].

Human benefits have been assessed according to an anthropocentric approach. In particular, those that derive from ES fruition by operators, tourists and visitors of the MPA were considered as "environmental benefits" and thus evaluated in monetary terms. Differently, biophysical benefits (i.e., those related to the environment and not to humans) were assessed according to an eco-centric approach, which is currently being developed by the Department of Earth Sciences of Environment and Life of the University of Genova within a number of projects, such as GIREPAM. Environmental benefits deriving from the use of the MPA natural heritage have been re-classified on the base of the ES from which they originate [63].

Tables 12 and 13 show the summary sheets related to "tourism fruition" and "fish biomass supply for business or entertainment purposes", elaborated as an example of the previously illustrated analysis.

Table 12. Summary sheet related to "tourism fruition".

ES "tourism fruition and eco-tourism"
The "Isola dell'Asinara" MPA allows the following tourism activities: Bathing. According to article 11, bathing is forbidden in zone A, and allowed in zones B and C, according to the ordinances of the of the competent maritime authority. Diving. According to articles 12 and 13 of the regulations, non-organized scuba diving is forbidden. Scuba diving tours are allowed exclusively in zones B and C during the day and at night, prior authorization and according to the procedures provided in the same regulations. Recreational boats. According to article 14 of the regulations, navigation is allowed to recreational boats with inboard engine exclusively in zone C and in the entry corridors to the mooring of "La Reale" and "Cala d'Oliva", prior authorization and according to the procedure provided in the same regulations.
Bathing
Since in the "Isola dell'Asinara" seaside facilities are absent, users mainly bath in "Cala Sabina", "Cala d'Oliva" and "dell'Ossario" beaches, which are also included in guided tours. These beaches accommodate the majority of bathers, because access to other sites is more complicated. The yearly number of bathers, obtained through monitoring activity, is estimated to be approximately 44,450. During the summer of 2016, a survey was conducted on a sample of 343 bathers. They were all tourists, since the island has only one resident. 77% of the respondents were long-stay tourists, the average of their stay in Sardinia was 11 days, of which only one was spent in the MPA. Among them, 15% spent their holidays in a second home, 31% rented a house and 44% chose other options (i.e., hotel, bed and breakfast, etc.). The remaining 23% of the interviewees were one-day tourists. Figure 5 shows tourist distribution according to the following categories of tourists: (i) Usual stay; (ii) occasional stay; (iii) usual one-day stay; (iv) occasional one-day stay. Thus, among the visitors of the "Isola dell'Asinara" MPA every year, 76% are occasional tourists. With respect to the origin, 95% of bathers are residents in Italy (of them, 25% are from Sardinia and 38% from the north of Italy). All the interviewees consider the MPA important or even a priority.

Table 12. *Cont.*

Class type	*Indicator*	*Measure unit*
Landscape fruition for recreational and scientific activities.	No. of visits.	Average annual attendances (2014–2016 period).
Scuba Diving		
Diving centers must be authorized by the MPA managing entity. Authorization is subject to verification of a number of specific requirements for diving centers. In order to protect the seabed, the managing entity monitors underwater activities, identifies dive sites and mooring points, and establishes the maximum number of dives per day and of the marine vessels daily employable by each diving center. The number of dives per day is fixed at 24 (intended as single plunges), for each of the 25 sites identified by the managing entity. The maximum number of diving centers authorized every year is 5, and the maximum number of marine vessels for each center is 2. During the three-year-period 2014–2016, the annual average of dives was 2223. With the aim of deeply analyzing habits of scuba divers, and how they perceive the MPA, a sample of 61 scuba divers were interviewed; 67% of them were men, and 55% were aged between 30 and 50. The majority of divers (92%) were from Italy, and 87% were tourists who were staying in Sardinia; 62% of divers considered the institution of the MPA a priority, whereas the remaining 38% considered it important but not a priority.		
Class type	*Indicator*	*Measure unit*
Landscape fruition for recreational and scientific activities.	No. of visits.	Average annual attendances (2014-2016 period).
Flow calculation		
The flow of the "tourist fruition" ES is calculated through the numbers of users of each fruition category.		

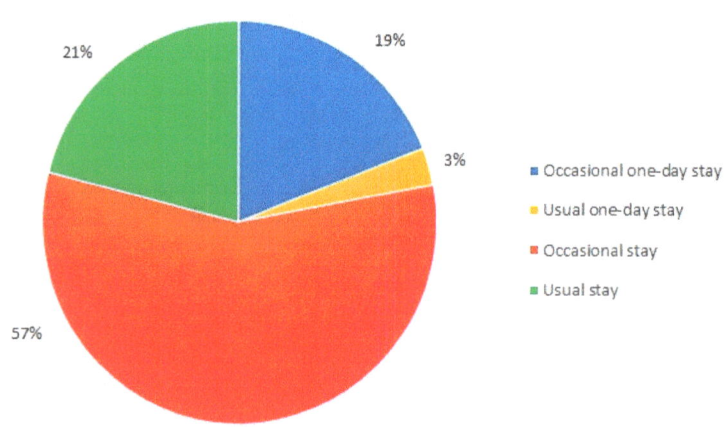

Figure 5. Tourists' stay distribution. Source: Authors' elaboration.

Table 13. Summary sheet related to "fish biomass supply for business or entertainment purposes".

ES: "Fish Biomass Supply for Business or Entertainment Purposes"		
Professional Small-Scale Fishing Related to the ES The number of fishing boats that work in the MPA is 26, of which 73% are wooden vessels and the remaining 27% are made of fiberglass. The average length is 8.76 m, and the propellant is diesel.		
The regulations of the "Isola dell'Asinara" MPA establishes that: — Professional small-scale fishing is forbidden in zone A; — trawling, purse-seine and fishing with light sources is forbidden; — aquaculture and active repopulation are forbidden; — professional small-scale fishing is allowed in zones B and C only to the fisheries registered in the municipality lists of the MPA, according to the procedures established by the regulations.		
With regard to the calculation of fish biomass, data are obtained through visual census campaigns (conducted by the MPA). In particular, the MPA signed an agreement with the Environmental Office of Corsica (OEC, Managing Entity of the Natural Reservation of the Strait of Bonifacio, France), to monitor the fish fauna of the SAC ITB010082 "Isola dell'Asinara" and the SPA ITB010001 "Isola Asinara".		
The methodology adopted was the direct count via the "visual census" technique, choosing fixed circular points and areas of 100 m^2, with a depth between 10 and 20 m. Each fixed point was determined by using a rope of 11.3 m length, corresponding to the diameter of a circle whose area is 100 m^2. The biomass was calculated on the basis of density data, conferring to each individual of each species an average weight according to its size class.		
By multiplying the biomass value (calculated for area units) by the areas, the total amount of fish biomass was obtained: 943 tons of carbon, corresponding to 3.63% of the benthonic biomass stored in the MPA.		
This amount is distributed mainly on the biocoenosis of *Posidonia oceanica*, on infralittoral algae and on coastal detrital.		
Calculation of the fish stock underestimates the real value, because the "visual census" technique cannot detect some species, as the pelagic ones. Thus, the species captured by professional, small-scale and leisure fishers should be added to the results obtained with this methodology. However, even in this case, the sum would underestimate the real stock, because these catches would not represent the entirety of the individuals that compose the stock.		
Class type	***Indicator***	***Measure unit***
Wild animals and their outputs	Quantity of fish caught	Kg/year
Flow calculation The flow of the "wild animals and their outputs" ES was calculated by previous studies conducted by the AMP (Vargiu, 2012, 2014, 2015), considering data on small-scale professional fishing in the three-year period between 2014 and 2016. The estimated quantity of fish caught was of 61,562 kg/year.		

Definition of "Ecosystemic objectives" was realized on the basis of the strategy of Agenda 2030 towards sustainable development (Agenda 2030 is an action program joined by 193 ONU countries in September 2015. Agenda 2030 strongly highlights the unsustainability of current economic, environmental and social development models. This vision shifts away from the common idea that sustainable development is just an environmental issue by simultaneously integrating the diverse development dimensions into one coherent vision. The agenda gathers 17 Sustainable Development Goals (SDGs) into an action program articulated in 169 "targets" or "goals". The objectives were launched in 2016, and the subscribing countries committed themselves to achieve them by 2030). In particular, with the aim of defining the "ecosystemic objectives" referring to the studied MPA, the Sustainable Development Goals of Agenda 2030 refer to the themes on which the case study focuses. They have been selected and linked to the macro-themes previously defined. The selected goals are the following:

- Goal 7: Ensure access to affordable, reliable, sustainable and modern energy for all;
- Goal 13: Take urgent action to combat climate change and its impacts;
- Goal 14: Conserve and sustainably use the oceans, seas and marine resources for sustainable development.

Subsequently, each SDG has been matched with the previously identified macro-themes. Table 14 shows the definition of the ecosystem objectives in relation to the ES chosen for the case study and in correspondence with the selected SDG. These objectives are drawn from the SDGs, but are carefully rephrased in order to be adapted to the context.

Table 14. Ecosystemic objectives.

ES macro-themes	SDG	Ecosystemic Objective
Touristic fruition	Goal 14 Conserve and sustainably use the oceans, seas and marine resources for sustainable development	— Significantly prevent and reduce marine pollution; — Efficiently regulate fishing activities; — Improve the economic benefits deriving from activities like fishing, aquaculture and tourism through the sustainable use of marine resources.
Fish biomass supply for business or entertainment purposes	Goal 14 Conserve and sustainably use the oceans, seas and marine resources for sustainable development	— Efficiently regulate fishing activities; — Significantly prevent and reduce marine pollution; — Ensure access to natural resources and marine markets to small-scale fishermen.
Climate regulation	Goal 13 Take urgent action to combat climate change and its impacts	— Integrate the regulations with measures to contrast climate change.
Energy production from renewable resources	Goal 7 Ensure access to affordable, reliable, sustainable and modern energy for all	— Promote the use of renewable energy in the regulations

5. Discussion

In this section, results previously obtained and presented are discussed, highlighting the main issues that characterized the two phases of the research. The results offer a way to cope with the critical issues related to the definition of MPA regulations, identified in the previous sections.

The research was articulated into two phases: In the first phase, the methodological approach to update the regulation in the light of Natura 2000 conservation measures was defined; in the second phase, the analysis of the constitutive elements of the PS and their assessment in light of the ESs was conducted.

Concerning the first phase of the research, it is worth noting the following two significant results:

- The integration of the conservation measures provided in Natura 2000 management plans, and in the ICZM protocol and ISEA project into the regulations;
- the involvement and active participation of competent authorities, public bodies, stakeholders and experienced actors—who contributed to regulation improvement by virtue of their diverse needs and visions—which reflect the contrast between environmental protection and socio-economic development. It is important to underline that this moment is not for actors to passively approve decisions already taken according to a top-down approach, but it constitutes a valuable opportunity to share knowledge among stakeholders characterized by diverse backgrounds. It is during this phase that the PS process swings into action, by actually designing the new regulation proposal. Participation is essential to effectively integrate conservation measures and sectoral policies because norms are, in this way, assessed from a legislative point of view and interpreted in the light of their impacts on MPA users' daily life. For this reason, participation is strongly encouraged from the very first steps of the PS definition process. Therefore, sharing is not limited only to results, but it also includes the working methodology.

The two results reported above allow the overcoming of some critical issues that arose during the construction of the PS:

- The overlapping of local, regional and national competencies. In particular, while the draft of the Natura 2000 management plans is produced by municipalities, their approval is entitled to regional administrations. Similarly, MPA regulations are proposed by the MPA authority, but they are approved by the MATTM. Interaction of diverse authorities complicates the process and entails numerous difficulties, such as the fragmentation and sectorization that characterize the planning and management tools in force;
- lack of operational objectives and actions in the regulations in force, which are able to adequately address a number of territorial problems and issues related to the allowed activities in the MPA. Response to this criticism is currently offered by the annual procedural guidelines, which establish detailed norms and operational conditions for the activities allowed in the MPA. In this sense, the regulation updating responds to the territorial problems and issues related to the allowed activities in the MPA by integrating the indications of the annual procedural guidelines into the new regulations.

Concerning the second phase of the research, it is worth noting the following three significant results:

- Acquisition of knowledge and awareness of ESs and of their potential contribution to human wellbeing. This is obtained through a set of objectives oriented to promote and protect the natural capital of the MPA by giving the regulation updating an ecosystem focus. In fact, it is fundamental to orient human actions to a sustainable use of ES in order to conserve their flows, especially considering that their economy relies on specific ESs (i.e., agriculture and fishing are strictly linked to provisioning ESs, tourism takes advantage of cultural ESs, etc.). With reference to this, "ecosystemic objectives" are defined on the basis of ES and are contextualized in the area of the case study coherently with the regulation actions defined in the PS. This represents the pivotal element for the regulation process to update coherently with the EU guidelines regarding ESs. In particular, with regard to the ES "climate regulation", it was necessary to define a corresponding objective aiming at integrating measures to combat climate change into the regulations. This aspect was totally absent in the existing regulations; thus, it represents an innovation with respect to the "traditional" issues that arose during the implementation of the PS;
- the formulation of a set of ecosystemic objectives and actions in coherence with Agenda 2030 objectives. This connection allowed addressing the existing objectives to specific future challenges (as in the case of objectives related to fishing in light of Objective 14 of Agenda 2030 and the fish biomass supply for business or entertainment purposes).

6. Conclusions and Suggestions for Further Research

In light of the previously discussed results, the main obstacles to efficient MPA management are the high variety of objectives regarding protection and development of MPAs and the need to provide users with information about norms and rules concerning the sustainable fruition of these areas. The need to overcome these discrepancies and to simultaneously facilitate users' fruition, by reducing information fragmentation, called for the construction of a dynamic and interdisciplinary instrument able to consider, in an integrated way, all the ecosystem fragilities, the coastal landscape and the human impacts on them.

The proposed conceptual approach represents the starting point for the definition of guidelines on the systematic application of the PS model to spatial planning and to the management of natural areas characterized by an overlapping of diverse normative levels.

Thus, governance and management of marine areas can be innovated according to a holistic territorial vision, able to efficiently contrast biodiversity loss and ecosystem degradation.

Possible further research projects can be developed according to this vision, as, for example:

- The definition of a monitoring system related to the regulation implementation, in order to assess the impacts of the coastal ecosystem management on ESs;
- ES measurement through adequate accountability and statistical methods to support the decisional process. Indeed, notwithstanding their fundamental role for socio-economic development, the importance of ecosystems for human wellbeing is currently not adequately evaluated, especially at planning and regulation levels [65]. Indeed, ecosystems provide vital support to life quality, and their functions and services can be assessed in monetary terms thus including the value of natural capital in resource management and spatial planning. ES assessment should take into account ecosystem mapping, anthropic pressures that threaten them and interactions between natural capital and its potential in terms of ES provision. This requires the definition of a set of adequate and representative indicators [66];
- ES monetary assessment. In this sense, it is necessary to re-consider the whole management process in order to identify and highlight ES value, and to understand the role of their safeguarding in influencing their quality and quantity in relation to an improvement of the natural capital [22], and also in economic terms;
- the introduction of an evaluation procedure for the MPA regulations. The proposed PS stresses the significative impacts of the regulations on the environment and formulates an innovative approach, inspired by SEA principles and methodology, which allows the inclusion of environmental objectives into the decisional process.

In conclusion, this study proposes an innovative ecosystem approach based on the interaction of diverse disciplines [67] that allows a draft of a new regulation proposal that is more effective with respect to the themes of environmental safeguards and biodiversity protection [68]. It demonstrates that it is possible to realize efficient ex-ante and ex-post assessment for public policies through the development of accounting systems, indicators and assessment methods related to the impact of these policies on the state of the natural capital [66]. This innovative approach can be replicated in other similar contexts for sustainable planning in coastal territories.

Author Contributions: Conceptualization, M.F., V.G., F.I., F.L., S.P. and C.P.; methodology, M.F., F.I., F.L., S.P. and C.P.; investigation, M.F., F.I., F.L., S.P. and C.P.; resources, M.F., V.G., F.I., F.L., S.P. and C.P.; writing—original draft preparation, M.F., F.I., and C.P.; writing—review and editing, F.L. and S.P.; visualization, F.L. and S.P. All authors have read and agreed to the published version of the manuscript.

Funding: This research was funded by the Research project based on the Agreement between Dipartimento di Ingegneria Civile, Ambientale e Architettura (Department of Civil, Environmental Engineering and Architecture, DICAAR) of the University of Cagliari, Italy, and the Autonomous Region of Sardinia, Departmental Office of Environment Protection, finalized to the objectives of the Project "GIREPAM–Integrated Management of Ecological Networks through Parks and Marine Areas" (Programme INTERREG Marittimo Italy-France Maritime 2014–2020, Axis 2).

Acknowledgments: Maddalena Floris gratefully acknowledges Sardinia Regional Government for the financial support of her Ph.D. scholarship. (P.O.R. Sardegna F.S.E. Operational Program of the Autonomous Region of Sardinia, European Social Fund 2014-2020 - Axis III Education and training, Thematic goal 10, Priority of investment 10ii.).

Conflicts of Interest: The authors declare no conflict of interest. The funders had no role in the design of the study; in the collection, analyses, or interpretation of data; in the writing of the manuscript, or in the decision to publish the results.

References

1. Agardy, T. Information needs for marine protected areas: Scientific and societal. *BMS* **2000**, *66*, 875–888. [CrossRef]
2. Agardy, T.; Alder, J.; Dayton, P.; Curran, S.; Kitchingman, A.; Wilson, M.; Catenazzi, A.; Restrepo, J.; Birkeland, C.; Blaber, S.; et al. Coastal System. In *Current State and Trends*; Global Assessment reports; Island Press: Washington, DC, USA, 2005; Chapter 19; Volume 1, pp. 513–549.

3. Arkema, K.K.; Guannel, G.; Verutes, G.; Wood, S.A.; Guerry, A.; Ruckelshaus, M.; Kareiva, P.; Lacayo, M.; Silver, J.M.; Lacayo-Emery, M. Coastal habitats shield people and property from sea-level rise and storms. *Nat. Clim. Chang.* **2013**, *3*, 913–918. [CrossRef]
4. Australian Government. *The Logical Framework Approach, AusGUIDElines, AusAID*; AusAID: Canberra, Australia, 2000.
5. Bonvoisin, N. Decision-Making for Sustainable Development: How Assessment Can Help. Available online: https://www.unece.org/fileadmin/DAM/oes/nutshell/2009/6_SusDev.EIA.pdf (accessed on 4 October 2019).
6. Borja, A. Grand challenges in marine ecosystems ecology. *Front. Mar. Sci.* **2014**, *1*, 1. [CrossRef]
7. Blackman, R. *Project Cycle Management*; Tearfund: Teddington, UK, 2003.
8. Brown, A.L.; Thérivel, R. Principles to Guide the Development of Strategic Environmental Assessment Methodology. *Impact Assess. Proj. Apprais.* **2012**, *18*, 183–189. [CrossRef]
9. Bussi, F. *Progettare in Partenariato. Guida Alla Conduzione di Gruppi di Lavoro Con il Metodo GOPP*; Franco Angeli: Milano, Italy, 2001.
10. Cattaneo Vietti, R.; Tunesi, L. *Le aree marine protette in Italia*; Problemi e prospettive: Aracne, Roma, 2007; Available online: http://www.aracneeditrice.it/pdf/9788854814363.pdf (accessed on 26 November 2019).
11. Cicin-Sain, B.; Belfiore, S. Linking marine protected areas to integrated coastal and ocean management: A review of theory and practice. *Ocean Coast. Manag.* **2005**, *48*, 847–868. [CrossRef]
12. Cicin-Sain, B.; Knecht, R.W.; Vallega, A.; Harakunarak, A. Education and training in integrated coastal management: Lessons from the international arena. *Ocean Coast. Manag.* **2000**, *43*, 291–330. [CrossRef]
13. Comitato per il Capitale Naturale, Secondo rapporto sullo stato del capitale naturale in Italia, 2018. Available online: https://www.minambiente.it/sites/default/files/archivio/allegati/sviluppo_sostenibile/II_Rapporto_Stato_CN_2018_2.pdf (accessed on 29 November 2019).
14. Devuyst, D.; Hens, L.; De Lannoy, W. Chapter 5: Linking Impact Assessment with Sustainable Development and the Introduction of Strategic Environmental Assessment. In *How Green Is the City?* Columbia University Press: New York, NY, USA, 2001.
15. European Commission; DG XI Environment, Nuclear Safety and Civil Protection. *A Handbook on Environmental Assessment of Regional Development Plans and EU Structural Funds Programmes*; Environmental Resources Management: London, UK, 1998; Available online: http://ec.europa.eu/environment/archives/eia/sea-guidelines/handbook.htm (accessed on 2 December 2019).
16. European Commission. Directive 2008/56/EC of the European Parliament and of the Council. Available online: https://eur-lex.europa.eu/legal-content/EN/TXT/PDF/?uri=CELEX:32008L0056&from=EN (accessed on 10 November 2019).
17. European Commission. Report from the Commission to the European Parliament and the Council on the progress in establishing marine protected areas (as required by Article 21 of the Marine Strategy Framework Directive 2008/56/EC). Brussels, 1/10/2015. Available online: https://ec.europa.eu/transparency/regdoc/rep/1/2015/EN/1-2015-481-EN-F1-1.PDF (accessed on 25 September 2019).
18. European Environmental Agency. Marine Protected Areas. Available online: https://www.eea.europa.eu/themes/water/europes-seas-and-coasts/assessments/marine-protected-areas (accessed on 25 September 2019).
19. European Environment Agency. Marine environment. Available online: https://www.eea.europa.eu/soer-2015/europe/marine-and-coastal#note21 (accessed on 25 September 2019).
20. European Union. 2012. Blue Growth. Opportunities for Marine and Maritime Sustainable Growth. Available online: https://ec.europa.eu/maritimeaffairs/sites/maritimeaffairs/files/docs/publications/blue-growth_en.pdf (accessed on 10 November 2019).
21. European Union. Directive 2014/89/Eu of the European Parliament and of the Council. Available online: https://eur-lex.europa.eu/legal-content/EN/TXT/PDF/?uri=CELEX:32014L0089&from=EN (accessed on 10 November 2019).
22. Federparchi, Contabilità ambientale nelle Aree Marine Protette Italiane. Idea progettuale, 2014. Available online: http://www.torredelcerrano.it/wp-content/uploads/2016/11/2_Documento-Contabilit%C3%A0_ambientale_FEDERPARCHI-Roma-15-aprile-2014.pdf (accessed on 23 November 2019).
23. Fernandes, L.; Day, J.; Lewis, A.; Slegers, S.; Kerrigan, B.; Breen, D.; Cameron, D.; Jago, B.; Hall, J.; Lowe, D.; et al. Establishing representative no-take areas in the Great Barrier Reef: Large-scale implementation of theory on Marine Protected Areas. *Conserv. Biol.* **2005**, *19*, 1733–1744. [CrossRef]

24. Fischer, T.B. Towards a more Systematic Approach to Policy, Plan and Programme Environmental Assessment—Some Evidence from Europe. In *SEA in Australasia*; Marsden, S., Dovers, S., Eds.; Sydney Place Federation Press: Sydney, Australia, 2002.
25. Gaston, K.J.; Pressey, R.L.; Margules, C.R. Persistence and vulnerability: Retaining biodiversity in the landscape and in protected areas. *J. Biosci.* **2002**, *27*, 361–384. [CrossRef] [PubMed]
26. GIREPAM, Gestione Integrata delle Reti Ecologiche attraverso i Parchi e le Aree Marine. Il progetto, n.d. Available online: http://interreg-maritime.eu/it/web/girepam/checosarealizza (accessed on 10 November 2019).
27. Gomes, I.; Pérez-Jorge, S.; Peteiro, L.; Andrade, J.; Bueno-Pardo, J.; Quintino, V.; Rodrigues, A.M.; Azevedo, M.; Vanreusel, A.; Queiroga, H.; et al. Marine biological value along the Portuguese continental shelf; insights into current conservation and management tools. *Ecol. Indic.* **2018**, *93*, 533–546. [CrossRef]
28. Groves, C.R. *Drafting a Conservation Blueprint: A Practitioner's Guide to Planning for Biodiversity*; Island Press: Washington, DC, USA, 2003.
29. Groom, M.J.; Vynne, C.H. Habitat Degradation and Loss. In *Principles of Conservation Biology*, 3rd ed.; Groom, M.J., Meffe, G.K., Carroll, C.R., Eds.; Sinauer Associates, Inc.: Sunderland, MA, USA, 2012; pp. 173–212.
30. Gurrutxaga San Vicente, M.; Lozano Valencia, P. *La Integración de la Conectividad Ecológica en los Instrumentos de Ordenación y Planificación Territorial: Una Revision*; Boletín de la A.G.E.: Madrid, Spain, 2019; n. 49; pp. 45–66.
31. Henson, S.A.; Beaulieu, C.; Ilyina, T.; John, J.G.; Long, M.; Seferian, R.; Tjiputra, J.; Sarmiento, J.L. Rapid emergence of climate change in environmental drivers of marine ecosystems. *Nat. Commun.* **2017**, *8*, 14682. [CrossRef]
32. Hilborn, R. The role of science in MPA establishment in California: A personal perspective. *Environ. Conserv.* **2012**, *39*, 195–198. [CrossRef]
33. Interreg. Il Programma Italia-Francia Marittimo. Available online: http://interreg-maritime.eu/it/programma (accessed on 10 November 2019).
34. IPBES (Intergovernmental Science-Policy Platform on Biodiversity and Ecosystem Services). *The IPBES Regional Assessment Report on Biodiversity and Ecosystem Services for the Americas*; Rice, J., Seixas, C.S., Zaccagnini, M.E., Bedoya-Gaitán, M., Valderrama, N., Eds.; IPBES: Bonn, Germany, 2018; 656p.
35. IPCC (Intergovernmental Panel on Climate Change). *Climate Change 2001: The Scientific Basis. Contribution of Working Group I to the Third Assessment Report*; Houghton, J.T., Ding, Y., Griggs, D.J., Noguer, M., Van der Linden, P.J., Dai, X., Maskell, K., Johnson, C.A., Eds.; Cambridge University Press: Cambridge, UK, 2003.
36. Italian Official Gazette. Regolamento di Esecuzione ed Organizzazione Dell'area Marina Protetta "Isola dell'Asinara. Available online: www.parcoasinara.org/it/contenuti/articoli/dettagli/553/ (accessed on 29 January 2020).
37. Jameson, S.C.; Tupper, M.H.; Ridley, J.M. The three screen doors: Can marine "protected" areas be effective? *Mar. Pollut. Bull.* **2002**, *44*, 1177–1183. [CrossRef]
38. Jentoft, S.; Chuenpagdee, R.; Pascual-Fernández, J.J. What are MPAs for: On goal formation and displacement. *Ocean Coast. Manag.* **2011**, *54*, 75–83. [CrossRef]
39. Jones, P.J. A review and analysis of the objectives of marine nature reserves. *Ocean Coast. Manag.* **1994**, *24*, 149–178. [CrossRef]
40. Kaiser, M.J. Are marine protected areas a red herring or fisheries panacea? *Can. J. Fish. Aquat. Sci.* **2005**, *62*, 1194–1199. [CrossRef]
41. Kajanus, M.; Leskinen, P.; Kurttila, M.; Kangas, J. Making use of MCDS methods in SWOT analysis—Lessons learnt in strategic natural resources management. *For. Policy Econ.* **2012**, *20*, 1–9. [CrossRef]
42. Kelleher, G. *Guidelines for Marine Protected Areas*; International Union for the Conservation of Nature: Gland, Switzerland, 1999.
43. Kørnøv, L.; Thissen, W. Rationality in Decision- and Policy-Making: Implications for Strategic Environmental Assessment. *Impact Assess. Proj. Apprais.* **2012**, *18*, 191–200. [CrossRef]
44. Leone, F.; Zoppi, C. Conflictual Issues Concerning Land Uses Related to Ecosystem Services under the Provisions of the Habitats and Birds Directives. Available online: http://www.input2016.it/conference_2016/conference_proceedings (accessed on 29 January 2020).

45. Linee Guida per la Valutazione ambientale strategica dei Piani urbanistici comunali [Guidelines Manual on Strategic Environmental Assessment of Municipal Masterplans], Annex to the Decision of the Regional Government of Sardinia n. 44/51 of 14 December 2010. Available online: http://www.sardegnaambiente.it/documenti/18_269_20110203150553.pdf (accessed on 2 December 2019).
46. Lundquist, C.J.; Granek, E.F. Strategies for Successful Marine Conservation: Integrating Socioeconomic, Political, and Scientific Factors. *Conserv. Boil.* **2005**, *19*, 1771–1778. [CrossRef]
47. Margules, C.R.; Pressey, R.L. Systematic conservation planning. *Nature* **2000**, *405*, 243–253. [CrossRef]
48. Povero, P.; Vassallo, P.; Paoli, C.; Dapueto, G.; Massa, F.; Pozzi, M.; Armenio, M.; Rigo, I.; Congiatu, P.; Gazale, V.; et al. *Contabilità Ambientale nelle Aree Marine Protette Italiane. Ecorendiconto dell'Isola dell'Asinara*; Report finale; Parco Nazionale dell'Asinara, CoNISMa: Genova, Italy, 2018.
49. Marino, D. *Le aree marine protette italiane. Stato, politiche, governance*; Franco Angeli: Milano, Itlay, 2011.
50. ME (Millennium Ecosystem Assessment), Ecosystems and Human Well-being. A Framework to Assessment, Island Press, 2003. Available online: http://pdf.wri.org/ecosystems_human_wellbeing.pdf (accessed on 13 November 2019).
51. Ministero dell'Ambiente, della Tutela del Territorio e del Mare (MATTM). Studio e analisi delle forme di coesistenza e criticità tra sviluppo economico- sociale e conservazione della natura. Il ruolo dei processi partecipati. Verso la strategia nazionale per la biodiversità, 2018. Available online: https://www.minambiente.it/sites/default/files/archivio/biblioteca/dpn_tavolo1_partecipazione.pdf (accessed on 29 November 2019).
52. Murray, S.N.; Ambrose, R.F.; Bohnsack, J.A.; Botsford, L.W.; Carr, M.H.; Davis, G.E.; Dayton, P.K.; Gotshall, D.; Gunderson, D.R.; Hixon, M.A.; et al. No-take Reserve Networks: Sustaining Fishery Populations and Marine Ecosystems. *Fisheries* **1999**, *24*, 11–25. [CrossRef]
53. Navarro Ortega, A. Las áreas marinas protegidas: ¿Un instrumento para la gestión eficiente del medio marino en España? In Proceedings of the III Congresso Internacional do Mar "A Proteção do Ambiente Costeiroe Marinho", Universidade Lusíada, Lisboa, Portugal, May 2014.
54. Nobre, A.M.; Ferreira, J.G. Integration of ecosystem-based tools to support coastal zone management. *JCR* **2009**, *SI 56*, 1676–1680.
55. Norse, E.A. *Global Marine Biological Diversity: A Strategy for Building Conservation into Decision Making*; Island Press: Washington, DC, USA, 1993; pp. 3–9.
56. Paoli, C.; Povero, P.; Burgos, E.; Dapueto, G.; Fanciulli, G.; Massa, F.; Scarpellini, P.; Vassallo, P. Natural capital and environmental flows assessment in marine protected areas: The case study of Liguria region (NW Mediterranean Sea). *Ecol. Model.* **2018**, *368*, 121–135. [CrossRef]
57. Parco Asinara. Aspetti Storico-Geografici. Available online: http://www.parcoasinara.org/it/contenuti/articoli/dettagli/544/ (accessed on 12 November 2019).
58. Parco Asinara. Contenuti. Available online: http://www.parcoasinara.org/it/contenuti/articoli/dettagli/512/ (accessed on 12 November 2019).
59. Parsons, T.R. Biological coastal communities: Productivity and impacts, in Coastal systems studies and sustainable development. *UNESCO TPMS* **1992**, *64*, 27–37.
60. Partidàrio, M.R. Strategic environmental assessment: Key issues emerging from recent practice. *Environ. Impact Assess. Rev.* **1996**, *16*, 31–55. [CrossRef]
61. Potts, T.; Burdon, D.; Jackson, E.; Atkins, J.; Saunders, J.; Hastings, E.; Langmead, O. Do marine protected areas deliver flows of ecosystem services to support human welfare? *Mar. Policy* **2014**, *44*, 139–148. [CrossRef]
62. Rogers, A.; Sumaila, U.; Hussain, S.; Baulcomb, C. *The High Seas and Us: Understanding the Value of High-Seas Ecosystems*; Report 2014; Global Ocean Commission: Oxford, UK, 2014.
63. Sadler, B.; Verheem, R. *Strategic Environmental Assessment—Status, Challenges and Future Directions*; Ministry of Housing, Spatial Planning and the Environment: The Hague, The Netherlands, 1996.
64. Salm, R.V.; Clark, J.R. *Marine and Coastal Protected Areas: A Guide for Planners and Managers*; IUCN: Gland, Switzerland, 2000.
65. Scott, J.M.; Davis, F.W.; McGhie, R.G.; Wright, R.G.; Groves, C.; Estes, J. Nature reserves: Do they capture the full range of America's biological diversity? *EA* **2001**, *11*, 999–1007. [CrossRef]
66. Sheate, W.R.; Dagg, S.; Richardson, J.; Aschemann, R.; Palerm, J.; Steen, U. Integrating the environment into strategic decision-making: Conceptualizing policy SEA. *Eur. Environ.* **2003**, *13*, 1–18. [CrossRef]

67. Smith, V.; Tilman, G.; Nekola, J. Eutrophication: Impacts of excess nutrient inputs on freshwater, marine, and terrestrial ecosystems. *Environ. Pollut.* **1999**, *100*, 179–196. [CrossRef]
68. Zoppi, C. Integration of Conservation Measures Concerning Natura 2000 Sites into Marine Protected Areas Regulations: A Study Related to Sardinia. *Sustainability* **2018**, *10*, 3460. [CrossRef]

sustainability

MDPI

Article

From Degradation to the Regeneration of Territorial Heritage. An Eco-Systemic Vision for the Promotion of the Natural, Urban and Landscape Capital of the Metropolitan City of Reggio Calabria

Concetta Fallanca [1,*], Antonio Taccone [1] and Chiara Corazziere [2]

[1] PAU Department, Mediterranean University of Reggio Calabria, Via dell'Università 25, 89124 Reggio Calabria, Italy; ataccone@unirc.it

[2] LASTRE Lab, PAU Department, Mediterranean University of Reggio Calabria, Via dell'Università 25, 89124 Reggio Calabria, Italy; chiaracorazziere@gmail.com

* Correspondence: cfallanca@unirc.it; Tel.: +39-335-406209

Received: 29 October 2019; Accepted: 26 November 2019; Published: 28 November 2019

Abstract: The results of the research conducted on the subject of regeneration of areas of land suffering degradation were presented, studied and analyzed to establish "families" of causes and effects, to forecast lines of action commensurate with the reversibility of the damage. The area in question is the Metropolitan City of Reggio Calabria covering the entire territory of the former province and including the Aspromonte National Park. The methods of direct observation and critical interpretation of the phenomena, in terms of extension and incidence, are aimed at advancing a protocol of interventions for the definition, evaluation and implementation of regeneration prospects through experimental pilot laboratories for University-Region coordination. Of particular interest was the role of prevention for areas that are used improperly with the consequent loss of habitat quality, limitations of the effectiveness of ecosystem services, for those environments that express values integrated from a cultural, naturalistic point of view, identity and are subject to risks of fragmentation that endanger ecological connectivity.

Keywords: ecosystem services; habitat quality; ecological connectivity; depopulation of inland areas; territorial heritage; reversibility of degradation

1. Introduction

The latest data on the soil erosion of the ISPRA report a "virtuous" Calabria with values that appear to show caution in transformations [1]. In reality the demographic decrease that affects the entire regional territory and the current economic stagnation due to severe forms of recession (maximum unemployment values for all ages and youth, aged population, migration, per capita income among the lowest) reduces any form of "anthropic pressure" to transformations and at the same time limits every action aimed at the maintenance, care and protection of the territory. So an indicator that may seem positive reveals a condition of progressive degradation of resources, identities, landscapes. The Metropolitan City of Reggio Calabria which, should be clarified immediately in order not to give rise to misunderstandings, concerns the entire territory of the 97 municipalities of the former province of 3210.37 km^2, with Reggio Calabria exceeding 180,000 inhabitants, Gioia Tauro with about 20,000 inhabitants and with the centers over 10 thousand of Palmi, Siderno, Taurianova, Rosarno, Villa San Giovanni, Locri, Melito di Porto Salvo, Polistena and Cittanova and with countless small towns under 1000 inhabitants for a total of 548,009 inhabitants (01/01/2019—ISTAT). A territory crossed by numerous rivers that has about 220 km of coastline from the Tyrrhenian to the Ionian and reaches peaks almost

2000 m high (Aspromonte 1956 m). Aptly this reality is self-preserved and envelops the entire National Park of Aspromonte, an immense core area with extraordinary but still unexpressed potential that presents, ever more exacerbated, all the decisiveness of an Apennine mountain area in a geographical reality that expresses the discomforts of a deep south area of Southern Italy. In this moment, the main risks for the territories of the Metropolitan City of Reggio Calabria come from a scarce anthropogenic presence in the territory, from a lack of defense, from a lack of interest in the values and resources that appear to be underused, if we exclude the coastal areas, where the pressure on transformation for second homes is still strong; we witness an immense urban heritage of abandoned villages where building units are put up for sale and abandoned fields and terraces, untreated woods (the surface of the forests of the Metropolitan city is about 102,000 Ha, equal to over 30% of the total area) are systematically set on fire to gain pastures invigorated through sprouting. Only in 2016, across the entire regional territory, there were 985 forest fires that covered a total area of 8037 hectares, of which 5302.3 wooded and 2734.8 not wooded [2].

Thus, soil erosion especially in inland mountain areas is limited to that related to improper and basically reversible uses, the need to provide effective forms of regeneration of correct land use for productive purposes and in full consistency with the presence of a national park and woods, river areas, coastal dune areas that represent a precious and still substantially preserved ecological network that structures the entire territory of the Metropolitan City.

Calabria, over time, has tried to stem the impact of settlements on natural and environmental systems, even introducing the concept of "physical integrity of the territories" and "promoting and protecting the landscape, the environment and agricultural activity, considering the soil as a common good and non-renewable resource that performs functions and produces ecosystem services, with the effects of prevention and mitigation of hydrogeological instability events and in line with mitigation strategies and adaptation to climate change" [3]. The legislation of some Regions has not, at times, overcome the concept of mere "containment of land degradation", at other times it has been relaunched by prescribing actions to regenerate existing urban fabrics in order to avoid new urbanisations. The ten-year commitment of urban planning to contain soil erosion has spread slowly into collective acquisitions of forms of awareness, when the scientific-planning debate has already been appropriately addressed, seeking methods and strategies for ecological compensation, regeneration of urban areas, re-naturalization of abandoned areas, recovery of agricultural heritage, with an aim towards environmental and functional qualification of the territory. Today we understand that the action of containment must be accompanied by the serious action of redesigning, rethinking the performance of the areas, to raise the quality of the areas in function of what they express with respect to the broad context of belonging [4]. The regional regulations offer exciting hints on the subject, firstly by expanding the concept of land depletion [5] with the inclusion of functions and uses such as mining, construction site areas, sports-recreational areas, which even if they do not imply waterproofing of the soil, induce the territory to lose its natural features (Piedmont Regional Law 3/2013). Widespread orientation can direct the transformations towards the already urbanised areas, with the identification by the municipalities of the settled area, outside which new recognition will have to be considered in exceptional cases defined by law (Bolzano Province). The common interest is to direct the transformations towards the degraded areas, underused or disused, to be redeveloped and regenerated starting from the recovery of the real estate and the existing urban fabric [6]. An urban regeneration that is sometimes explicitly oriented to reduce seismic vulnerability even with the completion and construction of new public works (Lazio LR 7/2017). Concepts of environmental ecological endowment are made up of all the spaces, works and interventions that contribute to combating climate change and its effects on human society and the environment and improving the quality of the urban environment, mitigating the effects of heating to counter the *heat island phenomenon* (Emilia Romagna LR 24/2017). The regeneration process is considered a strategy and tool that also favors a balanced distribution on the territory and the protection of the agricultural territory from transformations that would make it "non-agricultural", as in the case of Sardinia (LR 8/2015) which established measures for the protection of rural areas, such as the criteria for consolidating the volumes of agricultural holdings and attributes to municipal planning instruments for the purpose of safeguarding agricultural areas from improper exploitation, and Tuscany with Regional Law 65 of 2014, which sets the objectives of *"conservation and management of the territorial heritage, promoting its enhancement in function with sustainable and sustainable local development"* and contrasts the degradation of land by promoting *"the development of the multifunctional potential of agricultural and forest areas"*. The definitive efficacy of the concepts of territorial heritage are exemplary, understood as *"a common good constituting the regional collective identity"* and of *"quality of the rural territory"* expressed by the consolidation and strengthening of its multi-functionality, the reproduction of the landscape, the hydrogeological balance, also the economic well-being of the region", fundamental for a sustainable and lasting development, guaranteeing food quality and the environment.

In the overall objective of making cities and human settlements inclusive, safe, resistant and sustainable, the role that urban planning and spatial planning are required to play using new approaches and every available opportunity becomes fundamental. At this moment it seems possible to converge the objectives of governing transformations on the theme of metropolitan cities, integrated mobility, the revitalization of internal territories, pro-village initiatives, river and coastal contracts, towards the new development of the 2030 Agenda on the road to sustainability and resilience.

The reform that introduced metropolitan cities, which for now appears to be suspended pending further elucidation, presupposes a new way of understanding urban roles and the territory to which they belong; it is not by chance that words like competition, growth and words like cohesion, development have been replaced. The conceptual acquisitions on the theme of the enhancement of internal areas confirm that the metropolitan cities must assume an important role in favoring the vast territory and the suffering villages and Apennines towards the full attestation of a cultural, economic and social identity that contributes to the mass by valuing the whole system [7]. It means going beyond the dichotomous vision "pulp" and "bones" of the territory, which carries concepts of centrality and marginality, where the most frequent words are abandonment, old age, isolation, to recognise different ways of living and the richness of this great variety of living cultures that is connotative of our country [8].

In the particular case of the territory of the Metropolitan City of Reggio Calabria and even more so in the part of the Aspromonte National Park (which occupies almost two thirds of the territory of the former province) the territorial and landscape heritage has a cultural, naturalistic and environmental eco-systemic of immense richness [9], a value overshadowed by risky social undertone making the sense of belonging and the identity of the communities increasingly marginal.

2. Materials and Methods

The research took root in the activities carried out over more than three decades, documented by publications and research reports on closely related topics: the land consumption of the regional territory of Calabria between 1954 and 1984 and the map of environmental risk in Calabria [10,11], the naturalistic environmental framework plan of the Territorial Coordination Plan of Calabria [12], the studies for the settlement heritage of the Aspromonte National Park [13], the studies for the natural heritage of the Province of Reggio Calabria [14], the drafting of the Charter of Places of the Calabrian Region [15,16].

This integrated accumulation of knowledge has enabled a vision capable of going beyond the sectoral approach, which usually leads to a reading of the territory by "levels" and individual problems, to use the method of direct observation and interpretative identification of phenomena and Recurrent "degradation families" and then the precise quantification of the extension of the direct area and the causes of degradation for the development of an assessment of the possible effects on wide and derived areas. The aim is to raise these issues by putting forward hypotheses of causes and direct and induced effects to propose specific in-depth laboratories for each type of territorial degradation with the necessary skills (geologists, chemists, agronomists) to start interventions with a Regional and University direction that can constitute lines for pilot action towards the solution of the problems and for the development of rules of behaviour and periodic monitoring, which would also be useful for every single local territorial reality.

The Aspromonte National Park, Design Core of the Metropolitan City of Reggio Calabria

The Metropolitan City of Reggio Calabria [17–22] presents a complex territorial system with different areas and contexts from the morphological, settlement and economic development point of view [23]. It is the only Italian and European metropolitan city that includes a national park, the Aspromonte National Park, of over 65,000 hectares, which is one of the five homogeneous territorial areas identified by the Statute of the metropolitan city. The other homogeneous areas are the Plain of Gioia Tauro, the Strait, the Locride and the Grecanica areas (Figure 1). The metropolitan city of the Strait has an intimate link between Mediterranean coastal environments and mountain landscapes, with the Apennine ridge that continues into the Peloritani of the Sicilian side.

The presence of the Aspromonte National Park, included entirely in the metropolitan perimeter [24], represents a rare condition, characterised by the exceptional naturalistic, landscape and cultural heritage and by a rural and mountain settlement system that is overcoming the depopulation phase to experience events related to tourism fruition centered on cultural and landscape, historical and archaeological, enogastronomic and naturalistic aims. The internal centers of the park express a culture of places characterized by a rural world that is joined in a thousand different narratives, focusing on a common and, at the same time, peculiar common project. Its central position in the "vast" context of the metropolitan territory, which includes 37 municipalities, one third of those that make up the entire area, represents a key point in the metropolitan process of the city of Reggio Calabria.

The characteristics of the settlement of the territory of the Aspromonte Park allow the identification of macro cultural areas, based on the different and homologous matrices of identity and the functional relationships that are present in the territory of the park, which have created structured relationships between the centers and the areas themselves. The configuration of the metropolitan city territory is therefore closely linked to the presence of the Aspromonte massif. The rugged orography does not favor the settlement of inhabited centers and productive settlements and the centers in the most inaccessible and less accessible areas of the territory do not reach one thousand inhabitants.

The territorial system of the metropolitan city is characterized by the presence of a network of fragile infrastructure both in the connections with the vast territory, in the relations between the internal areas and the progressive abandonment of the internal centers with the typical settlement dynamics of the rural Apennine mountain world, to the advantage of coastal areas that are subject to strong settlement pressure. These considerations lead to a literary definition of the settlement system at the heart of the metropolitan city, belonging to the Park territory, functional to the project of the metropolitan city and decisive for the structuring of a polycentric territorial armor made up of bearing systems, supra-local systems and local systems that concern both the functional system of services and the network of accessibility.

The design themes that emerge clearly from the interpretations of the settlement system, show how it is also possible through a conscious and innovative use of resources, to enhance the prospects of functional integration, to overcome the current fragmentation of places and produce socio-economic-cultural benefits which are decidedly relevant for the metropolitan area in its entirety.

The methods and means for operating relate not only to the prospective of the metropolitan city, but also to the opportunities offered by the economies of the Aspromonte National Park. There are direct interventions by the Park Authority to improve the relationship between the settled environment and the natural territory that must guarantee the weight and role of pilot interventions to indicate design guidelines and generate reverberating effects in the sphere of the contiguous transformations of the vast metropolitan context. In this sense, the initiatives to enhance the ruins of ancient villages, geo-sites and architectural works scattered along the waterways must be interpreted, as must the direct promotion of civic museums, archaeological trails and nature trails.

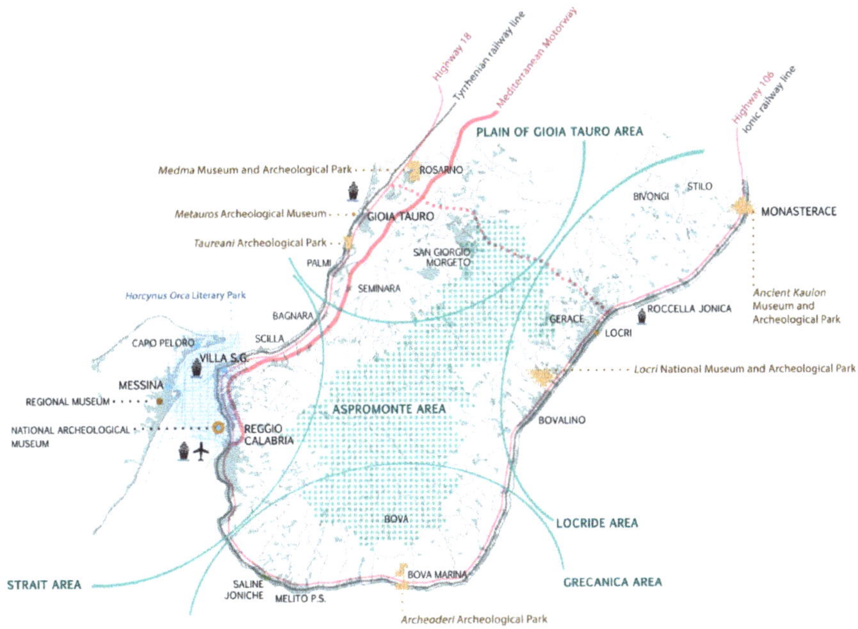

Figure 1. The homogeneous areas of Metropolitana City of Reggio Calabria. The central area, in green, is the National Park of Aspromonte.

A second modality is linked to the directing role that the Park Authority can play in guiding the transformation of governments towards the continuous improvement of the quality of settlements, centers, hamlets and nuclei, also through incentives for the progressive achievement of quality for the established environment.

A third project orientation concerns the activities that the Park Authority can express for the formation of a *new culture of residing and living* that is more adherent and responsive to the Park condition, conceived in a metropolitan context of reference.

The body of intervention methods referred to allows the first hypotheses to be calculated according to project themes: the *park's management areas*, for the promotion and enhancement of the park's centrality settlement system and the promotion and enhancement of the inhabited centers due to their great potential that can play complementary roles to the centrality system; *the material connection networks* for the valorization and multi-functional integration of the various existing types; *the network of receptivity and reception*, for the rationalization, strengthening and promotion of a widespread, qualified and diversified receptivity and for the promotion of local systems that offer tourism; *the network of cultural and recreational infrastructures*, for the promotion of projects for the enhancement of cultural structures and for the creation of didactic-scientific laboratories and of an integrated system offering leisure activities; *quality of the settlement environment and safety of park sites* for the promotion of projects aimed at progressively achieving quality for the settled environment, and integrated projects supporting high rural areas and the reevaluation of environmental interventions.

All this leads to transforming the settlement system into attractive poles providing economies that form new social and cultural aggregative models that contribute to raising the quality level of the entire metropolitan context [25].

It is here that the processes open up towards new geographies that challenge new issues: environmental, cultural-identity, strategic-infrastructural, which focus on the redevelopment of the territory and of the infrastructural network, on the reorganization of services and public space, on the need for physical and cultural accessibility of places and occasions, for reconnecting the different urban parts and the territory through an integrated project that cannot be separated from a serious reconnaissance, reading and regeneration of places that have lost meaning, functions and memory over time.

3. Results

3.1. Areas of Degradation Classified According to the Reversibility of Their Condition

The outcome of the research conducted on the regeneration of areas suffering land degradation, anticipated and analyzed by "families" of causes and effects, with a particular focus on the degradation of agricultural territory and the fragmentation of ecosystem services, providing for lines of action commensurate with the degrees of reversibility of the damage. The methods are based on an in-depth activity, carried out over decades of activity and intensified during the research, which can be considered ongoing, of direct observation of the phenomena experienced throughout the territory and facilitated by the aid of aerial photogrammetric survey systems employing drones for this purpose. The research is directly aimed at the quantitative, qualitative and critical understanding of the extent of the phenomena and its incidence in the environmental systems to which the interested areas belong in order to grasp the authentic sense of the questions for the definition and evaluation of the regeneration prospects. There is particular focus on the role of prevention aspects for those areas that are used improperly with consequent loss of quality of the habitats, with evident limitation of the effectiveness of ecosystem services, and to those contexts that express values integrated from a cultural, naturalistic, identity point of view and are subject to risks of fragmentation that endanger ecological connectivity.

3.1.1. Areas of Environmental and Social Degradation Affected by Productive Activities—Extractive, Manufacturing and Industrial—Inactive or Abandoned

Consider the different characteristics of land "waste" which apparently generate less serious consequences in the near and immediate context, not concerning the areas originally equipped to host productive activities and which, after a relatively short life cycle, have simply become inactive and therefore, due to a subsequent profound change in the economic and social fabric to which an inability to adapt and relaunch consistent with the abandoned vocations of the territory.

An emblematic case for the Metropolitan City of Reggio Calabria is the complex site of Saline Joniche, on the coast of the municipality of Montebello Jonico and articulated in the Liquichimica Biosintesi plant, in an industrial port and in the Great Repairs Workshop of the State Railways, today entirely in disuse.

The two ponds formed near the Liquichimica area due to the natural silting of the port are today a natural oasis used for resting by many migratory birds which, as a natural habitat guarantees the maintenance of biodiversity, a SIC area, whose protection lies not with the local council, but with the Metropolitan City (Figure 2).

This, like other production sites, now abandoned, do not have, officially, risks of contamination but would still require, in most cases, a priority reclamation of polluting substances and materials such as asbestos, which is still present as a building material.

A similar argument can be made for the areas of landfill or storage of processing waste, resulting from the continuation of important construction sites, or to demolition works by explosion such as those related to the modernisation of the Mediterranean Motorway section overlooking the Strait and disused tracts not yet regenerated.

These are large areas, therefore, that represent a factor of environmental and social degradation also because they interrupt the natural and urban scheme with surfaces and features that are physically inaccessible, as well as impeding potential processes of environmental protection and promotion and triggering new cultural strategies for and economic production and social innovation.

An analogous reasoning can be made for the areas and features affected by mining that for reasons similar to those of production or for simple reasons of exhaustion are now abandoned.

Figure 2. The inactive plant of Saline, on the jonic coast. On the right the port filled with sand, on the left the two ponds, protected as SIC area.

The Cave Report edited by Legambiente 2017 describes a drop in active sites and an increase in those abandoned, more than triple, according to an incomplete survey given that Calabria, along with Friuli Venezia Giulia and Lazio, is one of three regions not monitoring extraction activities, failing to draw up, as a consequence, a census of abusive or abandoned sites and able to count only 237 active and 49 abandoned quarries [26]. The Calabria Region, with the L.R. 40/2009, provides for a contextual recovery of the extractive areas with coordination between the phases of excavation, reorganisation and landscape and environmental recovery of sites. At the moment an excavation license is requested, in fact, it is obligatory to plan the restoration of sites for the reconstruction of the topographic, geomorphological, hydraulic and vegetation structure of the areas involved in cultivation activity, suitable for accommodating the uses and pre-existing destinations programmed by current planning. The interventions essentially favor the reconstitution of the functionality of the ecosystems. In this case, however, checks are often not carried out, due also to a still incomplete regulatory framework, given that Regional Law on the matter refers to the implementation rules of the Regional Mining Plan without the latter having been approved or definitively drafted. Therefore, the task of the Region is to ensure that the owners of disused quarries provide for their environmental landscape restoration as required by the aforementioned regional law.

The Legambiente Report also shows good practices and virtuous examples, such as the management of underground mining activities with the simultaneous recovery of the areas, the recovery of disused quarries to create parks and host tourist activities according to formulas of collective-private and temporary hybrid use, the reduction of the extraction from quarry through the recovery of inert material stemming from abandoned buildings.

3.1.2. Areas That Have Lost the Effectiveness of Ecosystem Services

The phenomenon of the illegal construction typical of the Metropolitan City of Reggio Calabria is manifested in a *extensive* form above all, because, as for the whole regional territory, about 4 rooms are registered per resident for a population in constant contraction. The need to rely on abusive building, therefore, even if practiced following the crisis at an alarming rate, in Calabria, 46.6% of illegal buildings compared to those authorised [27] have discontinuous sections, often with large impermeable surfaces, which occur most commonly along the coastal strips where they can even reach the sandy shoreline. It is estimated, in fact, that the consumption of land in the coastal strip has a 29.5% concentration within the 300 m distance of the coastline. Here, as in the more internal areas, where the phenomenon is more *diluted*, even if apparently less impacting on the balance of the territory, illegal building increases, in fact, the loss of effectiveness of ecosystem services due to the removal of natural areas and an increase in the fragmentation of land, also negatively influences the development prospects of sectors vital to the metropolitan economy, such as conscious tourism and quality agri-food production (Figure 3).

Figure 3. Greenhouses for intensive cultivation, on the jonic coast, increases the loss of effectiveness of ecosystem services and the fragmentation of land.

The correct functionality of the land [28] is also compromised by agro-silvo-pastoral activities inconsistent with the territory or whose nature is even illegal, such as, to mention the most striking case, the phenomenon of abusive pastures in some municipalities included in the homogeneous area of the Piana, whose destructive action was curbed by a complex and articulated plan, outlined and planned within the Inter-institutional Technical Table promoted and coordinated by the Prefecture of Reggio Calabria with the Aspromonte National Park that delivered an unprecedented result, despite its experimental character, when compared to a problem that has been manifesting for over forty years.

In addition to the phenomena of falling into the sphere of illegality, other activities, while not increasing the permeable surfaces, vary the coverage from natural or agricultural to artificial.

This is the case, for example, of construction site areas set up for large construction works—adjacent or near, at times, to productive agricultural areas and valuable landscapes—such as, for example, those preparatory works for the start of construction of the Bridge over the Strait, as the variant of the Cannitello-Villa San Giovanni railway line, or of the modernisation works of the Mediterranean Motorway or the Bovalino-Bagnara transversal artery connecting to the SS 106.

They are areas where for different reasons—works that have not achieved what was originally intended; a *connection* to the territory in the first case, or that the phase of environmental regeneration has not yet been completed in the second case, or, in the last case, *accompany* a construction site which has been interrupted for several years—they dominate the territory for long periods without having, at times, a forecast of dismantling and returning the land to its original destination.

The objective of integrating the protection of land functions into territorial planning could be pursued on two fronts; that of replenishing the land for the protection of agricultural and forest land that gives local authorities the possibility, as occurs in Poland, of demanding the removal of valuable arable land in the case of conversion of agricultural land so as to increase the fertility of other land or to reclaim degraded land elsewhere, and that of the identification of areas excluded from the construction of infrastructure in order to guarantee the subsistence of ecological networks, such as takes place in France and The Netherlands, with the delimitation of "blue and green landscape areas".

3.1.3. Areas with a Loss of Quality of Habitat, Ecosystem Services, with Erosion of Material or Immaterial Assets Linked to Incorrect Use or Accessibility

Of all forms, this is the type of degradation that, in the various forms, most characterises the territory of the metropolitan city.

The areas belonging to *sensitive* contexts belong to this family, preserving important naturalistic-environmental values, a cultural heritage with strong identity features due to secular processes of human settlement, territorial resources that can still be exploited for innovation and the strengthening of the economic fabric, as much as contrarily, they show a fragility linked to the morphological characteristics of the places, to the lack of infrastructures and basic services, to the problematic socio-economic conditions of the communities.

Parallel to the depopulation of inland areas, in fact, the metropolitan territory, like the regional one, is experiencing a gradual abandonment of agricultural and forestry activities. It is estimated, for example, that Calabria, between 2012 and 2018, lost crops capable of potentially producing about 40,000 quintals of fruit, mainly "antique varieties" (difficult to preserve and not very appreciated by the markets), with a serious loss of biodiversity, or that the production of woody raw materials—an ecosystem supply service guaranteed by natural forest surfaces—suffered, due to soil consumption, a 20% reduction in just six years, between 2012 and 2018.

Also the fragmentation and conformation of the agricultural soils that characterise the metropolitan territory and that do not allow an immediate productivity which is sufficiently profitable according to the traditional formula of mono-functional agriculture, has contributed, if only for economic reasons, to discouraging the formation of a new generation of entrepreneurs.

Although the phenomenon of abandonment may in some cases not determine serious consequences, if not those deriving from a negligence that can be circumscribed to the single areas, the slow but constant abandonment of agricultural and woodland soils has, however, seriously affected the action of systemic care of the territories that preserved the ecological and structural balance of the slopes, feeding the condition of hydrogeological risk and incidence of fires in the metropolitan area.

This specific issue was answered, in the past with good results, with a forest fire prevention plan implemented by the Aspromonte National Park Authority according to a "formula" developed for the defence of forest heritage and biodiversity of the Protected Area that entailed the involvement of the Volunteer and Civil Protection Associations, the Pastors, the Breeders and the direct Farmers, whose work complements the already operational inter-institutional system. The pastors in particular have been entrusted with the unpublished interpretation of the role of "eco-pastor" and of "guardians of aspromontan nature".

Both the internal areas of the metropolitan city, often invested by strategies and funds—such as, for example, the SNAI and the PSR—aimed at reducing the depopulation of the centers or preserving their natural and cultural characteristics, as well as the coastal strips, increased by infrastructure and services for the greater concentration of anthropic pressure, have experienced, at times, improper use of the building opportunities, often through *solutions* achieved over excessive amounts of time and achieved only once the initial needs had passed, and due to oversizing or incoherency with regards to the real demands of communities and territories.

This is also the case of the waterproofed surfaces of the numerous shopping centers scattered throughout the territory without any obvious relationship with the real anthropic presence which, in addition to *eroding* the adjacent commercial network to the detriment of local productivity, modifies the territory with large areas dedicated to buildings and parking areas characterised by short life-cycles, without being regenerated or used for new purposes, but simply re-proposed a few kilometres away, once the original function has been exhausted.

In addition, the structures supporting the Mediterranean motorway can be included, which has recently been modernised, moving to a different location, leaving behind uncontrolled decaying sites, such as service stations, already reduced, after only four years, to a state of ruin.

It is possible to refer then to all the widespread, disordered, spontaneous, sometimes abusive and rectified over the years, direct consequence of an uncontrolled and not always sustainable urban expansion that leads to occupation of the metropolitan land according to two main modalities. The first sees the exploitation of the spaces between the different territorial systems—river courses, coastlines, roads, etc.—which generates, in fact, a territorial fragmentation that reduces the surface of natural and semi-natural environments and an increase in their isolation, which is intensifying as we move from the heart towards the mid-coast and up to the coastal settlements. Therefore, the reduction cannot only refer to surfaces but also to ecological connectivity, resilience and the ability of habitats to provide ecosystem services without considering that fragmentation also damages agricultural activities, because it increases production costs and fuel consumption for processing.

The second follows a linear trend that follows the railway line and the two state roads, the 106 for the Ionian coast and the 18 for the Tyrrhenian one, and generates a continuous city [29] totally devoid of services in the points of greater distance from the major centers—so as to include coastal communities among the inland areas established by the SNAI—and which welcome a community of inhabitants forced into high commuting rates and dependence on private vehicles to reach workplaces, education and care, but also relationships and of cultural fruition.

In both cases these approaches of land use reflect negatively on the psycho-physical well-being of local communities and on the quality and value of the landscape, whose use, as well as cultural heritage, is often compromised in terms of accessibility and reachability as, at times, of correct interpretation.

We refer, for example, to the screens of second homes and tourist facilities that prevent, for long stretches not only the view, but physical access to the beach, or that inhibit the deposition of protected species' eggs, such as the caretta caretta turtle, or, the common beach structures built frequently on the coasts of the metropolitan city, or the generation of a dune landscape that would be so useful in countering coastal erosion (Figure 4).

It is already evident, with respect to the cases examined, how much more urgent and necessary it is to plan the *restoration and defragmentation of ecosystems and to favor ecological connections* [30] rather than giving priority to the management of new projects of large volume and building work. It is necessary, therefore, to plan a reconnection of the surviving spaces between them and the rivers, because they bring the ecological network back to the natural and functional radial pattern of the sea-mountain connection bands that have their core area in the Aspromonte National Park, due to new installations (public and collective permeable areas close to the riverbeds can also function as rolling tanks), the reconfiguration of intrusions (conversion of disused roads and railways) and the elimination of tampering (small demolitions or controlled deterioration actions) according to a process shared with communities that generally have more difficulty in accepting the conversion of the existing than the proposition of a new achievement [31].

Consider, once again, buildings, even legal ones, that intercept visions of particular value and invalidate the possibility of intercepting valuable landscapes or using architectural sites of documentary value, according to the way in which they were designed.

Figure 4. Groups of second houses built on the beach near the mouth of the river Mesima, on the Tyrrhenian coast, erode the dune landscape.

In the case of cultural assets as of any building that can foresee a re-use whose effects imply effects on the next context and community, it would be desirable not to rely on a traditional approach in which an expert chooses the solution between different alternatives, but to recognize that they can use the cognitive, economic and social resources distributed among more complementary actors. No longer an objectual, but relational approach, no longer a logic aimed at transforming the collective heritage merely into financial value, but in support of medium and long-term development, in the logic of investing in social capital and transforming problematic areas of the metropolitan city into opportunities for growth.

It is precisely in these cases where there is the possibility of affixing direct and indirect bonds to old sites and parts of the territory, endowing the superintendencies, if supported by the municipal administrations, with the ability to trigger correct regeneration processes, re-activation and involvement of the entire territory, starting from a new synthesis of cultural, social, economic-productive resources and offering the community a set of tools for using the territory.

Also the drafting of projects for participation in tenders aimed at the provision of quality public services and spaces, whether it involves a functional qualification of old containers within pre-existing materials, or regarding new projects, should aim to intercept real questions, propose new lifestyles, anticipate trends, feed the well-being of the citizens with public mobility or collective and sustainable health systems, accessibility to cultural, natural and landscape resources, to restore quality to habitats and increase ecosystem services, and vice-versa.

4. Discussion

To convert the conscience that invites us to rediscover a *common ethic* to regenerate the territories affected by improper uses and reshape the forms of the agrarian landscape into proactive energy, the research has assigned design directions to the three large families of problems identified, cataloged and measured the extent and severity of the degradation phenomena and described briefly in the previous paragraph.

We talk about realities ranging from abandonment and loss of territorial function and much more serious realities, to the limit of the recognition of environmental damage that are still far from being pursued and restored. The causes are of various types, and the research has classified them according to their "reversibility" in order to trace new directions towards a rehabilitation of the territories. These are conditions that escape control and rules, indiscriminate excavations to extract lithoide material with maximum profit and no obligation to remodel; of under-utilised or disused industrial-craft warehouses with an area of relevance frequently used as a landfill for industrial waste, special waste; of immense shopping centers with their disproportionate abandoned or underused parking lots. There is talk of the urban wreck of housing projects never completed, because for years, decades, had been seized due to business failures, for agonised inheritance procedures. Sometimes, there are also public housing districts in this condition. In light of the "families" of risk of land depletion, it seems useful to retrace the main lines of action commensurate with the reversibility of the damage (Figure 5).

Figure 5. (**a**,**b**) Variant of the Cannitello-Villa San Giovanni railway line, preparatory work for the start of construction of the Bridge over the Strait, never built; (**c**) modernisation works of Bovalino-Bagnara, a transversal artery connecting the Mediterranean Motorway to the SS 106, never completed; (**d**,**e**) abusive or abandoned building near the Tyrrhenian coast; (**f**) Aspromonte wood processing inactive plant of the XIX century. The buildings have industrial archaeology restrictions imposed by superintendence; (**g**,**h**) a collapsed terrace due to the abandonment of crops, then recovered and recultivated; (**i**) abandoned spinning mill of the XIX century.

4.1. Regeneration of Sites That Need Surface Remediation and Depth Reclamation, Areas of Production (Craft, Industrial) Abandoned or Underused

The research identified, mapped and quantified the sites that present these problems, in a general classification that requires specific surveys for the development of land remediation programs and projects with the verification and control for types of pollutants and of the risks of diffusion to the aquifers and/or of propagation in the surrounding lands and downstream with respect to the areas used for the burying of wastes from processing products or for the presence of ruins or skeletons or deposits of materials at risk.

For the sole purpose of reclamation for the areas with sites containing asbestos, the 2016 Arpacal Report states that one roof in ten has an eternit roof. The Environmental Protection Plan, decontamination, disposal and reclamation for the purpose of defence against the dangers deriving from asbestos prepared for the Regional Asbestos Plan for Calabria reports a contaminated area for the entire Region of approximately 65,000 m.

4.2. Re-Naturalisation of Sites that Have Lost Their Naturalness, Permeability and Biodiversity for Different Reasons

The research has identified innumerable areas, sometimes even of vast extensions, that present these conditions and that it is possible to trace back to codifiable and programmable types of intervention.

For these areas, the QTRP of Calabria provides for the "environmental and functional qualification of the territory through the enhancement of the territory's resources, protection, recovery, minor consumption of land, and therefore the recovery and enhancement of the landscape, of the environment and rural territory "as a productive component and at the same time as environmental protection as a prevention and overcoming of environmental risk situations, ensuring the coherence between landscape planning and territorial and urban planning strategies.

The lines of action promoted by the research are identifiable with interventions aimed at reducing the impermeable surfaces, with the demolition of the large concrete or paved squares with the restoration of the agrarian land and the remodelling of those places by creating terraces, compluvium lines and of displacement. Margins of innovation can be found in the possibility of experimentally creating water collection systems (ponds, reservoirs, cisterns) usable for irrigation purposes to extend the irrigated agricultural area.

4.3. Revitalisation, Enhancement, Reactivation of Abandoned Territories, Both Agricultural and Forestal, through Support Measures, Incentives and Facilities to Reduce the Phenomenon of Neglect and Degradation

For the purposes of the revitalisation of the agrarian territory specific support policies have now become urgent, even indirectly, towards replanting and reactivating crops. Other European countries implement measures to give new quality and productivity. Consider the rural landscape policy in France, which is achieving significant results with forms of progressive improvement linked to specific quality objectives to be achieved at specific times agreed between the local parties and departmental communities. Even Germany has activated a serious policy of limiting the consumption of land, with an indication of the objectives of reducing the new urbanisation of the land, to be achieved in specific time-frames, a subject that in Italy has been consummated without ever having produced an effective action or a measure of any kind.

In Calabria the value of agricultural soils is favored by the Regional Law 2017/31, with provisions to facilitate the access of young people to the primary sector and counter the abandonment and consumption of agricultural soils, which dictates the fundamental principles for the conservation of the soil as a common good and a non-renewable resource, decisive for the defence of the ecosystem and landscape characteristics, for the prevention of hydrogeological instability, for the enhancement of typical and quality agri-food products. In particular, provision is made for the granting in concession or lease of publicly owned land in favor of young farmers and social cooperatives with the restriction of agricultural use.

The incentives are foreseen by the various measures of the Calabria RDP 2014/2020 for the purpose of reducing the phenomenon of neglect and degradation, also pursuing transversal objectives of "environment and climate change", especially in the areas of the regional territory at risk of erosion.

The university, as part of the activities related to the Third Mission, is playing an advisory and guidance role for the support of local authorities called upon to respond to such serious problems also for the promotion of pacts linked to river and coast contracts, for the candidacy for funding programs for the recovery of the villages and for the fight against the phenomenon of abandonment and for study activities, knowledge, and intensifying the characteristics of the territory and its eco-systemic values. In particular, the Rural Development Plan (PSR) for the Calabria Region opportunities are being pursued in the context of measures for the reforestation of non-agricultural land, integrated management and environmental protection, prevention actions and restoration of potential forest damaged by fires and natural disasters and for integrated management and protection of the environment and forest ecosystems. The University of Reggio Calabria with its fields of Architecture, Engineering, Agriculture and Law has been taking care of the implementation of services related to education, health, mobility, affirmation of the culture of legality and the support of actions in favor of agriculture and agri-food production and of conscious and sustainable tourism.

5. Conclusions

The numerous research passages, performed over the years according to different paths and favoring the renewed governance forms of the territory, today metropolitan, and therefore of vast area, have allowed the extension and the vastness of the problem of "quality waste" of the territories, for the development of a range of site-specific solutions while at the same time a "pilot" for similar contexts in which to advance verified answers effectively.

The degradation conditions affecting the Metropolitan City of Reggio Calabria, which is a Vast Area Authority whose extension coincides with the territory of the former Province (3210,37 sq km, over 150 inhabited centers, for a total of 548,009 inhabitants of which 180,000 are residents in the city of Reggio Calabria, once the capital of the province), makes this extremely diverse territory an ideal laboratory to tackle the issue of the conceptual transition between contrast to consumption and land waste towards forms of authentic regeneration of naturalistic values and systemic reflection. In such contexts, which express naturalistic, environmental and cultural identity excellence of the territory, it is possible to map phenomena of environmental degradation that create relapses also from a social and economic point of view, if only for the occurrence of forms of collective addiction to degradation. It is therefore possible to recognise the ideal field for a theoretical and applied experimentation directed towards overcoming the concept of "consumption" towards the authentic identification of the causes, of the direct and indirect incidence of the areas, but above all of the possible measures of intervention, even of the human and social dimensions, as the vision of a desired integral ecology.

Precisely in this sense the objective of the proposed research is to stimulate the observation of degradation phenomena by advancing hypotheses of causes and direct and induced effects so that specific laboratories relative to each family of territorial degradation express the necessary competences, even in a way transversal to the different cases. Due to direct Region-University coordination, where the latter can provide the activities of a Third Mission; a real support to the role of government of local authorities, which identifies those interventions useful to constitute lines of action not only to provide solutions to problems, but above that guide towards a code of behaviour and periodic monitoring, applicable with equal effectiveness from a vast scale to the single local territorial reality.

Author Contributions: C.F., A.T., C.C. collaboratively designed the research and jointly wrote Sections 3–5; C.F., A.T. jointly wrote Section 1; C.C., A.T. jointly wrote Section 2.

Funding: This research received no external funding.

Conflicts of Interest: The authors declare no conflict of interest.

References

1. Munafò, M. *Consumo di Suolo, Dinamiche Territoriali e Servizi Ecosistemici*; Report SNPA 08/19; SNPA: Rome, Italy, 2019.
2. Regione Calabria. Delibera n. 159/2019 della Giunta Regionale. *Piano Regionale per la prevenzione e la lotta attiva agli incendi boschivi 2019*. Available online: https://portale.regione.calabria.it/website/portaltemplates/view/view_provvedimenti.cfm?16367 (accessed on 2 October 2019).
3. Regione Calabria. Circolare n. 170882 del 02 maggio 2019. *Disposizioni per azioni di supporto ed accompagnamento dei procedimenti istruttori finalizzati al rilascio dei pareri in materia edilizio-urbanistica e paesaggistico-ambientale*. Available online: https://www.regione.calabria.it/website/portaltemplates/view/view_norme.cfm?1332 (accessed on 25 October 2019).

4. Turri, E. *Il Paesaggio e Il Silenzio*; Biblioteca Marsilio: Venezia, Italy, 2004.
5. Senato della Repubblica. *Consumo di Suolo: Elementi di Legislazione Regionale*. Available online: http://www.senato.it/service/PDF/PDFServer/BGT/01105756.pdf (accessed on 7 September 2019).
6. Senato della Repubblica. *Contenimento del Consumo del Suolo e Riuso del Suolo Edificato*. Available online: https://www.senato.it/japp/bgt/showdoc/17/DOSSIER/0/982072/index.html?part=dossier_dossier1-sezione_sezione15 (accessed on 20 September 2019).
7. Schama, S. *Landscape and Memory*; Vintage Books: New York, NY, USA, 1996.
8. Zerbi, M.C. *Paesaggi Della Geografia*; Temi e discorsi; Giappichelli: Torino, Italy, 1993.
9. Taccone, A. La gestione dei paesaggi per il turismo di qualità. *ArcHistoR* **2019**, *4*, 256–265. [CrossRef]
10. Fallanca, C. *L' Ambiente nella pianificazione del territorio*; Collana del Dipartimento di Architettura e Analisi della Città Mediterranea Saggi; Gangemi: Roma, Italy, 1994.
11. Fallanca, C. La Calabria. La qualità ambientale come modello di sviluppo. In *Basilicata e Calabria due Sud tra Europa e Mediterraneo*; Caldaretti, S., Cellamare, C., Fallanca, C., Eds.; Jason: Reggo Calabria, Italy, 1999.
12. Brandmayr, P.; Fallanca, C. L'ambiente e il paesaggio. In *Antologia degli studi per il Piano Territoriale della Calabria*; Bios: Cosenza, Italy, 1998; Volume 4.
13. Fallanca, C. Master in Progettazione dei Parchi Naturali. Verifica degli obiettivi formativi e considerazioni sui temi di ricerca. In *Progettazione dei parchi naturali*; Fallanca, C., Ed.; Iiriti: Reggio Calabria, Italy, 2005.
14. Fallanca, C. Percosi ed esiti del Corso di Alta Formazione Pro. Mo. Ter. In *Reggio Calabria. Idee progettuali per la città e il territorio*; Fallanca, C., Ed.; Iiriti: Reggio Calabria, Italy, 2008.
15. Fallanca, C. Nuove Interazioni Tra Cultura e Sviluppo in Calabria. In *Il Territorio Speranza: Politiche Territoriali Possibili per il Mezzogiorno d'Italia*; Andriello, V., Belli, A., Eds.; Pianificazione territoriale urbanistica ed ambientale; Alinea: Firenze, Italy, 2002.
16. Fallanca, C. Il riverbero di una esperenza progettuale condivisa. In *Genius Loci: Governance e Territorio*; Fallanca, C., Ed.; Iiriti: Reggio Calabria, Italy, 2008.
17. Fallanca, C.; Taccone, A. La Città Metropolitana di Reggio Calabria. In *Pianificare le Città Metropolitane in Italia: Interpretazioni, Approcci, Prospettive*; De Luca, G., Moccia, F.D., Eds.; Accademia; INU Edizioni: Roma, Italy, 2007; p. 520.
18. Fallanca, C. Planning Metropolitan Scenarios for the Strait Area. *Adv. Eng. Forum* **2014**, *11*, 209–213. [CrossRef]
19. Fallanca, C. La forza delle idee nel progetto di Reggio Calabria Città Metropolitana. In *100 IDEE per Reggio Calabria Città Metropolitana*; Fallanca, C., Ed.; Iiriti Editore: Reggio Calabria, Italy, 2015.
20. Fallanca, C. The Cultural Project for the Inclusive Metropolitan City. *Procedia Soc. Behav. Sci.* **2016**, *223*, 608–613. [CrossRef]
21. Fallanca, C. The Urban Question in Seismic Risk Prevention. Priorities, Strategies, Lines of Action. In *New Metropolitan Perspectives*; Calabrò, F., Della Spina, L., Bevilacqua, C., Eds.; Springer International Publishing: Cham, Switzerland, 2019; Volume 101, pp. 681–690. [CrossRef]
22. Fallanca, C. Progresso, Libertà e Cultura Dell'abitare in Calabria. *ArcHistoR* **2019**, *4*, 220–229. [CrossRef]
23. Taccone, A. The Enhancement of the Urban Environment: Sustainable Mobility in Reggio Calabria. *Adv. Eng. Forum* **2014**, *11*, 214–219. [CrossRef]
24. Fallanca, C. Le Anime Urbane Dell'area Metropolitana Dello Stretto. Il Punto Di Vista Continentale. *Archivio Di Studi Urbani e Regionali* **2019**, *124*, 5–25. [CrossRef]
25. Taccone, A. The Metropolitan Plural City. *Procedia Soc. Behav. Sci.* **2016**, *223*, 238–243. [CrossRef]
26. Legambiente. *I numeri e gli impatti economici e ambientali delle attività estrattive nel territorio italiano*. Available online: https://www.legambiente.it/sites/default/files/docs/rapporto_cave_2017.pdf (accessed on 3 October 2019).
27. Legambiente. *Abbatti l'abuso. I numeri delle (mancate) demolizioni nei comuni italiani*. Available online: https://www.legambiente.it/wp-content/uploads/abbatti_labuso_dossier_2018.pdf (accessed on 12 September 2019).
28. Gioffrè, V. Surplus edilizio e paesaggi dell'abbandono. Reggio Calabria, Strada Statale 106 Jonica. In *Territori dell'abusivismo: Un progetto per uscire dall'italia dei condoni*; Curci, F., Formato, E., Zanfi, F., Eds.; Urbanistica che cambia; Donzelli Editore: Roma, Italy, 2017.
29. Minervino, M.F. *Statale 18*; Fandango Libri: Roma, Italy, 2010.

30. Ministero dell'Ambiente e della Tutela del Territorio e del Mare. *Strategia Nazionale per lo Sviluppo Sostenibile. Direzione Generale per lo Sviluppo Sostenibile, per il Danno Ambientale e per i Rapporti con l'Unione Europea e gli Organismi internazionali.* Available online: https://www.minambiente.it/sites/default/files/archivio_immagini/Galletti/Comunicati/snsvs_ottobre2017.pdf (accessed on 5 October 2019).
31. Attademo, A.; Formato, E. *Fringe Shift*; List: Trento, Italy, 2018.

sustainability

MDPI

Article

Local Development and Protection of Nature in Coastal Zones: A Planning Study for the Sulcis Area (Sardinia, Italy)

Federica Leone * and Corrado Zoppi

Department of Civil and Environmental Engineering and Architecture, University of Cagliari, Via Marengo 2, 09123 Cagliari, Italy; zoppi@unica.it

* Correspondence: federicaleone@unica.it; Tel.: +39-070-6755213

Received: 4 July 2019; Accepted: 12 September 2019; Published: 18 September 2019

check for updates

Abstract: In 2008, the Council of the European Union adopted the "Protocol on Integrated Coastal Zone Management" (ICZM Protocol), then ratified by Decision No. 2010/631/EU. The ICZM Protocol defines integrated coastal zone management as a dynamic and flexible process that accounts for the relations between coastal ecosystems and landscape as well as the activities and the uses that characterize coastal areas. Integrated management of coastal zones is still a critical process in terms of translating theory into practice. In this theoretical framework, strategic environmental assessment (SEA) helps to improve decision-making processes related to coastal spatial planning by integrating development goals and sustainability criteria. This study proposes a methodological approach concerning ICZM-based decision-making processes at the local level. The methodology is implemented in relation to three case studies concerning three towns located in southwest Sardinia. The results show a general consistency between the analyzed plans in terms of objectives and themes. Three specific issues are particularly relevant in terms of integration of economic and social objectives and sustainability goals, that is, relations between beach services and coastal ecosystems, protection of coastal ecosystems, and accessibility to the coastal zones.

Keywords: Natura 2000 Sites; coastal land use plans; management plans of Natura 2000 Sites; integrated coastal zone management

1. Introduction

The integrated approach to planning and management of coastal areas has been a key concept of the European spatial planning strategy since the early 1970s [1]. The Resolution of the Council of Europe No. (73) 29 on the protection of coastal areas, adopted by the Committee of Ministers on 26 October 1973, recommends considering " … the national coastal heritage as a whole … " (point 1).

At present, the integrated approach to coastal zone management has increasingly acquired importance within the international theoretical and technical debate, as it represents a key paradigm for the implementation of sustainable development-related policies concerning coastal areas [2]. In 2008, the Council of the European Union (EU) adopted a key document, the "Protocol on Integrated Coastal Zone Management" [3] (ICZM Protocol from now on), then ratified by Decision No. 2010/631/EU. The ICZM Protocol defines integrated coastal zone management as " … a dynamic process for the sustainable management and use of coastal zones, taking into account at the same time the fragility of coastal ecosystems and landscapes, the diversity of activities and uses, their interactions, the maritime orientation of certain activities and uses and their impact on both the marine and land parts" (article 2). Integrated coastal zone management should be related to the specificity of the local contexts because coastal and marine planning issues cannot be dealt with on the basis of a one-size-fits-all approach [4].

However, the integrated management of coastal zones is still a critical process in terms of translating theory into practice [5]. As suggested by Soriani et al. [4], two typologies of problems may occur. The first one is connected with strategies and policies, whereas the second is related to the definition and the implementation of spatial planning processes. Both issues depend on the concurrence of several factors such as: the increasing demand for new building sites in the coastal areas; the complexity of administrative and technical management of coastal and marine resources; the often difficult and conflicting relationships between academicians, politicians, and public officials at the various administrative and technical scales; problems concerning institutional governance; the complex implementation of the subsidiarity principle as regards the relationships among public administrations and the implementation of effective participatory processes; and the integration between objectives and policies aimed at protecting environment and natural resources within spatial planning processes [4,6,7].

In this theoretical and technical context, strategic environmental assessment (SEA) may increase the effectiveness of decision-making processes regarding coastal management [8]. Indeed, EU Directive No. 2001/42/EC (SEA Directive) indicates that "The objective of this Directive is to provide for a high level of protection of the environment and to contribute to the integration of environmental considerations into the preparation and adoption of plans and programmes with a view to promoting sustainable development, by ensuring that, in accordance with this Directive, an environmental assessment is carried out of certain plans and programmes which are likely to have significant effects on the environment" (article 1). In a nutshell, SEA-based procedures improve decision-making processes by integrating economic and social development goals and sustainability criteria [9]. Moreover, according to the United Nations Environment Programme (UNEP) [10], Harvey [11], and Partidário et al. [12], SEA is identified as a tool to support countries in the implementation of the ICZM Protocol in relation to national strategies and coastal management plans.

This study defines a methodological approach by building on SEA-based procedures in order to support decision-making processes in the implementation of the ICZM Protocol at the local (municipal) level. In particular, objectives deriving from different plans, such as coastal land use plans (CLUPs) and management plans (MPs) of Natura 2000 Sites [13], are analyzed and compared in terms of reciprocal consistency in order to integrate their planning strategies and to identify potentially-negative impacts of CLUPs on MPs. The proposed methodology is implemented in relation to three case studies concerning three towns located in southwest Sardinia.

The study develops through four sections. Following the Introduction (this section), the second section presents the methodological approach, identifies materials and documents necessary to carry out the analyses, and describes the three towns whose CLUPs and MPs are assessed through the proposed methodology. Results of the analysis are presented in the following section (third section), whereas the concluding section (fourth section) discusses implications, caveats, and directions for future research.

2. Materials and Methods

This study implements and discusses a methodology that aims at achieving consistency with the other MPs of Natura 2000 Sites and CLUPs through the implementation of a logical framework (LF) that integrates the conservation measures established by the MPs into the spatial strategies of the CLUPs. The LF reflects the concepts of sustainability, endoprocedimentality, and identification of alternatives that characterize SEA processes under the provisions of the Italian Law enacted by Decree No. 152 of 3 April 2006 that implemented the SEA Directive into the Italian legislation [9].

The proposed LF builds on previous studies [14,15], where municipal masterplans' objectives and MPs' objectives are compared in terms of mutual consistency. In this study, the LF aims at analyzing the relations between MPs and CLUPs in terms of sustainability-oriented objectives by identifying the potentially negative impacts of CLUPs' actions on MPs' objectives. The LF is reported in Table 1, which shows a diagram whose five columns represent: (i) sustainability-oriented objectives; (ii) spatial planning themes; (iii) specific objectives of CLUPs; (iv) specific objectives of MPs; and (v) actions planned by CLUPs that may generate negative impacts as regards specific conservation-related objectives of MPs.

Table 1. The structure of the logical framework (LF).

Sustainability-Oriented Objectives	Themes	CLUP's Specific Objectives	MP's Specific Objectives	Potentially Unfavorable CLUP's Actions
Sustainability-oriented objectives 1	Theme 1	Specific objective 1 of CLUP	Specific objective 1 of MP	Action 1 Action m
			Specific objective k of MP	Action 1 Action m
		Specific objective j of CLUP	Specific objective 1 of MP	Action 1 Action m
			Specific objective k of MP	Action 1 Action m
	Theme h	Specific objective 1 of CLUP	Specific objective 1 of MP	Action 1 Action m
			Specific objective k of MP	Action 1 Action m
		Specific objective j of CLUP	Specific objective 1 of MP	Action 1 Action m
			Specific objective k of MP	Action 1 Action m

CLUP: coastal land use plans; MP: management plans.

The proposed methodology is implemented in relation to three coastal municipalities—Calasetta, Carloforte, and Portoscuso—located in southwest Sardinia in the Sulcis sub-regional area (Figure 1). Calasetta, Carloforte, and Portoscuso are three small-medium sized towns with municipal areas of 31.06 km^2, 51.10 km^2, and 38.09 km^2, respectively, and population densities of 90.86 residents/km^2, 123.30 residents/km^2, and 137.46 residents/km^2 [16]. Moreover, Carloforte is located in a small island connected to the Sardinian island through the Port of Portovesme (in the municipality of Portoscuso) and through the Port of Calasetta. The choice of these three towns reflects the idea that they may be conceived as a homogeneous urban system where, although each municipality keeps its administrative autonomy, the coastal zone needs an integrated management approach.

Moreover, each town is characterized by the presence of a number of Natura 2000 Sites. In particular, three small-sized Special Areas of Conservation (SACs) (SAC ITB042210 "Punta Giunchera", SAC ITB042208 "Tra Poggio La Salina e Punta Maggiore", and SAC ITB042209 "A nord di Sa Salina") overlap the territory of Calasetta. As regards Carloforte, its entire territory is included within the SAC ITB040027 "Isola di San Pietro", and the Specially Protected Area (SPA) ITB043035 "Coste e Entroterra tra Punta Cannoni e Punta delle Oche—Isola di San Pietro" is located in the northwest part of the municipality. The territory of Portoscuso is partially overlapped by the SAC ITB040028 "Punta S'Aliga" and the SAC ITB040029 "Costa di Nebida".

The documental sources of the study are the following:

1. The CLUP of Calasetta, the MP of the SACs ITB042210 "Punta Giunchera," ITB042208 "Tra Poggio La Salina e Punta Maggiore", and ITB042209 "A nord di Sa Salina";
2. The CLUP of Carloforte, the MP of the SAC ITB040027 "Isola di San Pietro", and the MP of the SPA ITB043035 "Coste e Entroterra tra Punta Cannoni e Punta delle Oche—Isola di San Pietro";
3. The CLUP of Portoscuso, the MP of the SAC ITB040028 "Punta S'Aliga", and the MP of the SAC ITB040029 "Costa di Nebida".

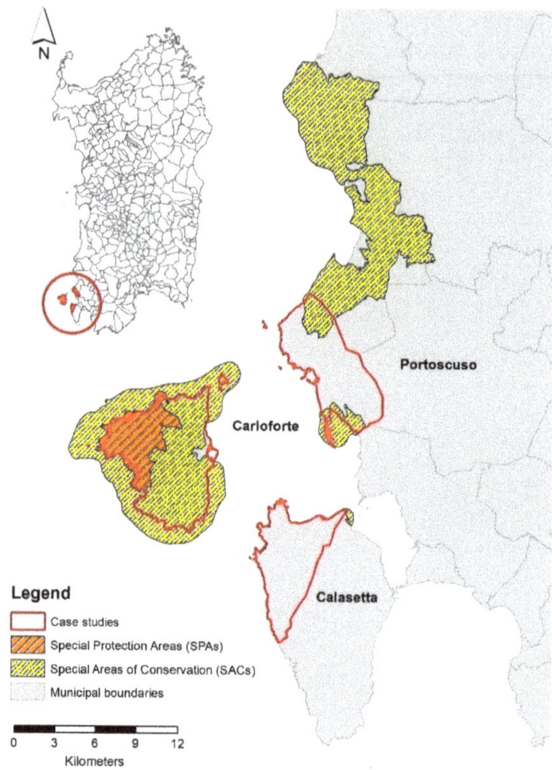

Figure 1. The study area.

3. Results

The proposed methodology and its implementation in relation to the CLUPs of the three towns of Sulcis identify coastal planning processes that integrate strategies at different scales, such as the local context, represented by the local municipalities that define and approve the CLUPs, and the regional and the national levels, which superintend planning policies oriented to environmental protection.

Tables 2–4 show the LFs related to the integration of CLUPs and the MPs related to the municipalities of Calasetta, Carloforte, and Portoscuso, respectively. In particular, taking account of the specific goals and contents of the MPs, each table focuses on sustainability-oriented objectives that account for protection of plants, animals, and biodiversity. The analyzed plans are consistent with each other in terms of objectives and themes. The three CLUPs address the following specific themes: (i) relations between beach services and coastal ecosystems (T_1); (ii) protection of coastal ecosystems (T_2); and (iii) accessibility to the coastal zones (T_3). For instance, in the case of Calasetta, the CLUP's goals are oriented towards the relations between beach services and coastal ecosystems, paying particular attention to ecosystem protection, the CLUP of Carloforte focuses on the relations between beach services and coastal ecosystem, considering accessibility as the core issue, and the CLUP of Portoscuso prioritizes ecosystem protection and accessibility. This is not surprising since: (i) the Carloforte's municipal land coincides with the SAC "Isola di San Pietro" and, that being so, each projected spatial transformation has to pass an Appropriate Assessment which, under the provisions of the Habitats Directive, is mandatory to evaluate if (and to what extent) it may possibly generate negative impacts on the SAC [17]; (ii) Portoscuso is characterized by the presence of an important industrial site featured by non-ferrous metals manufacturing plants and, as a consequence, the CLUP's objectives focus on protection of coastal ecosystems.

Table 2. LF of the integration of the CLUP and of the MP concerning the town of Calasetta [18].

Sustainability-Oriented Objectives	Themes	CLUP's Specific Objectives	MP's Specific Objectives	Potentially Unfavorable CLUP's Actions
Cal_SOO_1 Conservation and protection of biodiversity and coastal ecosystems	**T_1** Relations between beach services and coastal ecosystems	**Cal_CLUP_1** Planning beach services in relation to natural, rural, and urban contexts	**Cal_MP_1** Reduction or elimination of impacts induced by anthropic activities, animals and infrastructure on habitats and species	**Cal_CLUP_A_1** Authorization for pet-care services
				Cal_CLUP_A_2 Installation of pedestrian boardwalks
			Cal_MP_2 Restoration of the natural coastal morphology	
		Cal_CLUP_2 Definition of naturalistic measures in relation to reefs, in order to identify possible alternatives available to marine and coastal tourism	**Cal_MP_3** Mitigation of overuse of beaches, in particular during summer	**Cal_CLUP_A_3** Installation of dressing rooms and small cabanas
				Cal_CLUP_A_4 Placement of beach chairs and sun loungers
				Cal_CLUP_A_5 Installation of cabanas for the watchpersons
				Cal_CLUP_A_6 Installation of toilet and shower facilities
				Cal_CLUP_A_7 Installation of kiosks selling beverages and snacks
				Cal_CLUP_A_8 Installation of small stands in support of beach services and activities such as small boat charters, diving, and sailing schools

Table 2. *Cont.*

Sustainability-Oriented Objectives	Themes	CLUP's Specific Objectives	MP's Specific Objectives	Potentially Unfavorable CLUP's Actions
	T_2 Protection of coastal ecosystems	**Cal_CLUP_3** Prevention of coastal erosion processes and degradation	**Cal_MP_4** Mitigation of negative impacts on the natural status of dunes and prevention of their degradation	**Cal_CLUP_A_2** Installation of pedestrian boardwalks
			Cal_MP_5 Mitigation of coastal erosion effects and restoration of dunal systems	
			Cal_MP_6 Integration of measures aiming at removing *Posidonia oceanica* deposits from the beaches and at protecting coastal and marine habitats	
		Cal_CLUP_4 Promotion of environmental rehabilitation	**Cal_MP_1** Reduction or elimination of impacts determined by anthropic activities, animals, and infrastructure on habitats and species	**Cal_CLUP_A_1** Authorization for pet-care services
			Cal_MP_2 Restoration of the natural coastal morphology	
			Cal_MP_4 Mitigation of damages to the natural status of dunes e prevention of their degradation	**Cal_CLUP_A_2** Installation of pedestrian boardwalks
			Cal_MP_7 Protection and restoration of *Posidonia oceanica* meadows in the mooring areas	
		Cal_CLUP_5 Conservation of the salt pan	**Cal_MP_6** Integration of measures aiming at removing *Posidonia oceanica* deposits from the beaches and at protecting coastal and marine habitats	

Table 3. LF of the integration of the CLUP and of the MP concerning the town of Carloforte [18].

Sustainability-Oriented Objectives	Themes	CLUP's Specific Objectives	MP's Specific Objectives	Potentially Unfavorable CLUP's Actions
Car_SOO_1 Conservation and restoration of coastal ecosystems, paying particular attention to habitats and species of community interests	**T_1** Relations between beach services and coastal eco-systems	**Car_CLUP_1** Planning beach-related services and activities consistently with landscape and environmental protection goals	**Car_MP_1** Protection of marine waters	**Car_CLUP_A_1** Authorization released to small boat charters
			Car_MP_2 Conservation of the reef habitats	**Car_CLUP_A_2** Provision of the minimum service level in support of tourism in the most popular sandy and rocky beaches
			Car_MP_3 Conservation of dunal habitats	
			Car_MP_4 Promotion of sustainable uses of sites and related environmental resources	
			Car_MP_5 Conservation of arborescent matorral, thickets, and phrygana habitats	
			Car_MP_6 Conservation of important botanical species, such as *Astragalus maritimus* and *Rouya polygama*	
			Car_MP_7 Protection of the most important bird species living in the Natura 2000 Sites located in Carloforte	
			Car_MP_8 Protection of the local faunistic resources	
			Car_MP_9 Protection of *Caretta caretta*, a species of community interest	**Car_CLUP_A_1** Authorization released to small boat charters
	T_3 Accessibility	**Car_CLUP_2** Organization of an access point system and of the parking sites in order to regulate public access to beaches and coastal areas, minimizing environmental impacts	**Car_MP_2** Conservation of the reef habitats	
			Car_MP_3 Conservation of dunal habitats	**Car_CLUP_A_3** Installation of boardwalk access to beaches
			Car_MP_4 Promotion of sustainable uses of sites and related environmental resources	
			Car_MP_5 Conservation of arborescent matorral, thickets, and phrygana habitats	**Car_CLUP_A_4** Identification of parking sites in proximity of arborescent matorral, thickets, and phrygana habitats

Table 4. LF of the integration of the CLUP and of the MP concerning the town of Portoscuso [18].

Sustainability-Oriented Objectives	Themes	CLUP's Specific Objectives	MP's Specific Objectives	Potentially Unfavorable CLUP's Actions
Por_SOO_1 Biodiversity conservation	**T_2** Protection of coastal ecosystems	**Por_CLUP_1** Conservation and protection of coastal ecosystems	**Por_MP_1** Conservation of processes concerning ecological relations between biotic and abiotic components	
			Por_MP_2 Conservation and enhancement of physical and biological processes consistent with the ecosystems and their livability and development	
			Por_MP_3 Prevention of negative impacts on habitats and species protected under the Habitats Directive	**Por_CLUP_A_1** Installation of two portable floating docks for leisure fishing and nautical tourism
			Por_MP_4 Control and continuous assessment of ecological functioning and processes of habitats, floristic, and faunistic resources	
			Por_MP_5 Control and assessment of negative impacts on habitats and species protected under the Habitats Directive	
			Por_MP_6 Mitigation or prevention of negative impacts on habitats and species of community interest in terms of future expansion potential and conservation status	**Por_CLUP_A_2** Realization of new pedestrian and vehicle-accessible paths
			Por_MP_7 Increase in scientific knowledge and expertise on habitats and species of community interest in order to: (i) monitor the evolutionary dynamics and the population trends, (ii) assess the implementation of the MP; and (iii) fill the knowledge gap associated with the presence of new taxa	

Table 4. *Cont.*

Sustainability-Oriented Objectives	Themes	CLUP's Specific Objectives	MP's Specific Objectives	Potentially Unfavorable CLUP's Actions
		Por_CLUP_2 Promotion and enhancement of environmental restoration	**Por_MP_8** Mitigation of processes that cause quantitative and qualitative degradation of habitats, plants, and animals	**Por_CLUP_A_2** Realization of new pedestrian and vehicle-accessible paths
			Por_MP_9 Mitigation or prevention of pollution-related impacts generated by industrial activities	
			Por_MP_10 Reinstatement of the ecosystems as regards habitats and species protected under the Habitats Directive and their development potential	
			Por_MP_11 Reinstatement of degraded ecosystems damaged by industrial activities	
			Por_MP_12 Improvement or maintenance or reinstatement of habitats and habitats of Community interest through interventions aimed at preventing negative impacts generated by anthropic activities	**Por_CLUP_A_2** Realization of new pedestrian and vehicle-accessible paths
	T_3 Accessibility	**Por_CLUP_3** Improving the accessibility and the use of coastal areas to prevent erosion and degradation processes	**Por_MP_13** Optimization and management of site accessibility and internal mobility of vehicles and walkers	
			Por_MP_14 Promotion of tourist activities compatible with environmental conservation	
			Por_MP_15 Promotion of tourism and recreational activities in order to increase the economic attractiveness of the Special Areas of Conservation (SACs) through information and awareness-building campaigns	

In general, despite the apparent consistency between the CLUPs' and the MPs' specific objectives in terms of sustainability-oriented goals, the actions planned in the CLUPs may conflict with the MPs.

As regards the municipality of Calasetta (Table 2), the main potential conflicts concern the coastal ecosystems conceived as environmental assets that need protection and as sources of economic development based on tourism and recreational activities. Indeed, the CLUP aims at defining alternative planning options to marine and coastal tourism (Goal Cal_CLUP_2) and at preventing coastal erosion processes and degradation (Goal Cal_CLUP_3). On the other hand, CLUP's spatial transformations are oriented towards the development of coastal and marine tourist activities and towards making beaches more attractive for local and external visitors (Actions Cal_CLUP_A_2, Cal_CLUP_A_3, Cal_CLUP_A_4, Cal_CLUP_A_5, Cal_CLUP_A_6, Cal_CLUP_A_7, and Cal_CLUP_A_8). Moreover, the MP's objectives aim at discouraging tourism (Goal Cal_MP_3), at mitigating negative impacts on the natural status of dunes (Goal Cal_MP_4), and at reducing the impacts generated by anthropic activities, animals, and infrastructure on habitats and species (Goal Cal_MP_1).

Carloforte (Table 3) shows two inconsistencies. First, the protection of marine waters (Goal Car_MP_1) and of the brown turtle *Caretta caretta*, a species of community interest (Goal Car_MP_9), contrasts with the authorization released to small boat charters (Action Car_CLUP_A_1). Generally, small boats do not require licenses or certified skills. As a consequence, no specific technical knowledge is needed, which relates to marine ecosystems such as seabed, particular habitats such as *Posidonia oceanica*, or protected species such as *Caretta caretta*. The second inconsistency regards the accessibility of the beaches through the installation of boardwalks (Action Car_CLUP_A_3) and the location of parking sites in proximity of protected habitats and plants (Action Car_CLUP_A_4) that may generate potential conflicts in terms of conservation of dunal habitats (Goal Car_MP_3) and of arborescent matorral, thickets, and phrygana habitats (Goal Car_MP_5).

Portoscuso (Table 4) shows the same inconsistencies as Carloforte. In fact, the realization of new pedestrian and vehicular paths (Action Por_CLUP_A_2) is likely to generate negative impacts on habitat and species (Goal Por_MP_12) as regards future expansions and conservation status (Goal Por_MP_6). Under this perspective, the construction of pedestrian and vehicular paths may entail an increase in fragmentation of habitats that implies a loss of biodiversity. The second inconsistency concerns the installation of two floating docks for leisure fishing and nautical tourism (Action Por_CLUP_A_1) and: (i) the measures that aim at protecting habitats and species (Goal Por_MP_3); and (ii) the mitigation of degradation processes of habitats and species, both in quantitative and qualitative terms (Goal Por_MP_8). The presence of floating docks may possibly increase the number of boats and, as a consequence, the negative impacts on habitats and species. Moreover, floating docks are sometimes anchored to the seabed through concrete blocks that may damage the seabed and the marine species thereof.

The quantitative relevance of the potential conflicts between the CLUPs' actions and the MPs' objectives is identified by an overlay area that corresponds to around 10% of the SAC ITB040027 "Isola di San Pietro" located in the municipality of Carloforte [19] and to around 30% of the SAC ITB040028 "Punta S'Aliga" located in the municipality of Portoscuso [20]. Table 5 shows the habitat areas and the corresponding percentage shares as regards the overlay areas. In the case of the SAC ITB040028 "Punta S'Aliga" in the municipality of Portoscuso, a share of about 86% of the overlay area is covered by habitats classified under the provisions of the Habitats Directive (from now on indicated as "classified habitats"). The largest habitat is "Posidonia beds" that covers 73.81% of the overlay area. In relation to the SAC ITB040027 "Isola di San Pietro" in the municipality of Carloforte, the overlay area covered by classified habitats is around 41%. The largest overlay area shows the simultaneous presence of two classified habitats ("vegetated sea cliffs of the Mediterranean coasts with endemic *Limonium* spp." and "low formations of *Euphorbia* close to cliffs") that cover 18.31% of the overlay area.

Table 5. Classified habitats and percentage shares related to the overlay areas of MPs and CLUPs.

Typology of Classified Habitat	Carloforte		Portoscuso	
	Habitat Area Within the Overlay Area [m²]	Percentage of Habitat Surface in Relation to Overlay Area [%]	Habitat Surface Within the Overlay Area [m²]	Percentage of Habitat Surface in Relation to Overlay Area [%]
1110 "Sandbanks which are slightly covered by sea water all the time"	436.13	0.004	/	/
1120 * "Posidonia beds"	116.34	0.001	/	/
1150 * "Coastal lagoons"	/	/	1,549,216.87	73.810
1210 "Annual vegetation of drift lines"	4240.29	0.044	/	/
1210, 2110 "Annual vegetation of drift lines", "Embryonic shifting dunes"	79.95	0.001	/	/
1240 "Vegetated sea cliffs of the Mediterranean coasts with endemic *Limonium* spp."	611,432.23	6.373	/	/
1240, 5320 "Vegetated sea cliffs of the Mediterranean coasts with endemic *Limonium* spp.", "Low formations of *Euphorbia* close to cliffs"	1,757,256.61	18.310	/	/
1310 "Salicornia and other annuals colonising mud and sand"	/	/	180.82	0.009
1310, 1420 "Salicornia and other annuals colonising mud and sand", "Mediterranean and thermo-Atlantic halophilous scrubs"	/	/	304.55	0.014
1410 "Mediterranean salt meadows"	6802.84	0.071	5344.58	0.255
1410, 1420 "Mediterranean salt meadows", "Mediterranean and thermo-Atlantic halophilous scrubs"	229,628.30	2.393	5349.29	0.255
1420 "Mediterranean and thermo-Atlantic halophilous scrubs"	222,802.86	2.322	62,631.38	2.984
1420, 1410 "Mediterranean and thermo-Atlantic halophilous scrubs", "Mediterranean salt meadows"	/	/	25,092.77	1.195
2110 "Embryonic shifting dunes"	431.86	0.004	1566.87	0.074
2110, 2120, 2210 "Embryonic shifting dunes", "Shifting dunes along the shoreline with *Ammophila arenaria*", "*Crucianellion maritimae* fixed beach dunes"	10,021.93	0.104	/	/
2110, 1410. 2210 "Embryonic shifting dunes", "Mediterranean salt meadows", "*Crucianellion maritimae* fixed beach dunes"	/	/	59,689.98	2.844
2110, 2210 "Embryonic shifting dunes", "*Crucianellion maritimae* fixed beach dunes"	/	/	29,644.01	1.412
2120 "Shifting dunes along the shoreline with *Ammophila arenaria*"	/	/	25.59	0.001
2120, 2210 "Shifting dunes along the shoreline with *Ammophila arenaria*", "*Crucianellion maritimae* fixed beach dunes"	/	/	72.90	0.003
2210 "*Crucianellion maritimae* fixed beach dunes"	/	/	9007.92	0.429

Table 5. *Cont.*

Typology of Classified Habitat	Carloforte		Portoscuso	
	Habitat Area Within the Overlay Area [m^2]	Percentage of Habitat Surface in Relation to Overlay Area [%]	Habitat Surface Within the Overlay Area [m^2]	Percentage of Habitat Surface in Relation to Overlay Area [%]
2210, 2250* "*Crucianellion maritimae* fixed beach dunes", Coastal dunes with *Juniperus* spp.	/	/	5325.44	0.250
2230, 2250* "Malcolmietalia dune grasslands", "Coastal dunes with *Juniperus* spp."	1096.62	0.011	/	/
2230, 2250*, 2270* "Malcolmietalia dune grasslands", "Coastal dunes with *Juniperus* spp.", "Wooded dunes with *Pinus pinea* and/or *Pinus pinaster*"	/	/	2991.49	0.142
2250 * "Coastal dunes with Juniperus spp."	6699.08	0.070	/	/
2250 *, 2260 "Coastal dunes with *Juniperus* spp.", "Cisto-Lavenduletalia dune sclerophyllous scrubs"	/	/	27,431.58	1.307
2250 *, 2260, 2270 * "Coastal dunes with *Juniperus* spp.", "Cisto-Lavenduletalia dune sclerophyllous scrubs", "Wooded dunes with *Pinus pinea* and/or *Pinus pinaster*"	/	/	18,023.35	0.859
2250 *, 2270 * "Coastal dunes with *Juniperus* spp.", "Wooded dunes with *Pinus pinea* and/or *Pinus pinaster*"	15,448.55	0.161	/	/
2260 "Cisto-Lavenduletalia dune sclerophyllous scrubs"	1843.92	0.019	609.63	0.029
2270 * "Wooded dunes with *Pinus pinea* and/or *Pinus pinaster*"	/	/	597.93	0.028
2270 *, 2250 * "Wooded dunes with *Pinus pinea* and/or *Pinus pinaster*", "Coastal dunes with *Juniperus* spp."	/	/	2012.27	0.096
5210 "Arborescent matorral with *Juniperus* spp."	376,258.66	3.922	/	/
5330, 5430 "Thermo-Mediterranean and pre-desert scrub", "Endemic phryganas of the Euphorbio-Verbascion"	422,122.57	4.400	/	/
5430 "Endemic phryganas of the Euphorbio-Verbascion"	14.15	0.0001	/	/
6220 * "Pseudo-steppe with grasses and annuals of the Thero-Brachypodietea"	12,450.94	0.130	/	/
92D0 * "Southern riparian galleries and thickets"	323.39	0.003	/	/
9540 * "Mediterranean pine forests with endemic Mesogean Pines"	267,262.78	2.786	/	/

The star (*) stands for the EU standard notation for priority habitats.

Moreover, as explained in the examples below, the conflicts between CLUPs' actions and MPs' objectives are not only related to the overlay areas but also to impacts on classified habitats generated by CLUPs' actions that take place outside the overlay areas.

In the case of Calasetta, beach chairs, sun loungers, toilets, and shower facilities planned by the CLUP in order to make the beaches more attractive are likely to generate negative impacts on classified coastal habitats by increasing the number of tourists, even though they are located outside the MP's area. The overlay area just concerns the access to the beaches through pedestrian paths, as shown in Figure 2.

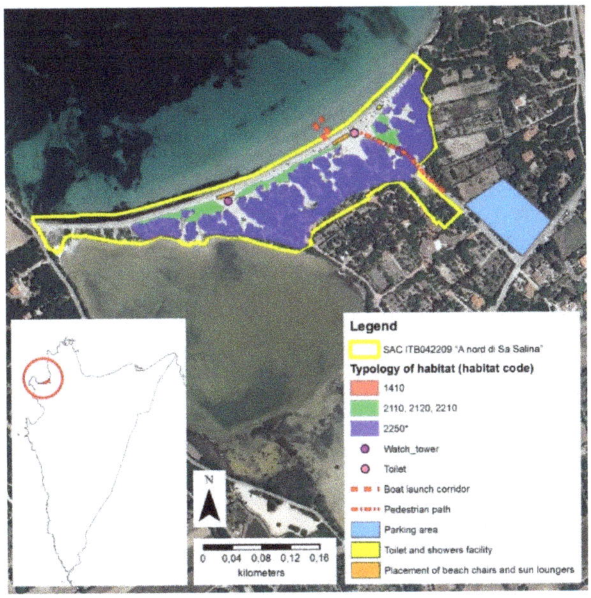

Figure 2. Overlay area related to the CLUP's actions and the SAC ITB042209 "A nord di Sa Salina" in the municipality of Calasetta.

Similarly, in the case of Carloforte, some conflicts are not specifically related to overlay areas. For instance, the conflict of Goals Car_MP_1 and Car_MP_9 with respect to Action Car_CLUP_A_1 is due to the potential negative impacts of small boat charters on classified marine habitats, which are not related to overlay areas. Furthermore, the conflicts of Action Car_CLUP_A_3 with respect to Goal Car_MP_3 and of Action Car_CLUP_A_4 with respect to Goal Car_MP_5 are due to the newly-planned beach accesses, whose negative impacts are not related to the very small size of the overlay areas. As shown in Figure 3, the total parking area of 490 square meters (the blue area within the green coastal patch) is about 5% of the area covered by habitats 1410 and 1420 ("Mediterranean salt meadows" and "Mediterranean and thermo-Atlantic halophilous scrubs") that are in the green coastal patch. The increase in parking areas, vehicles, and tourist presences will eventually generate relevant negative impacts on classified coastal and marine habitats that are not connected to their areal size.

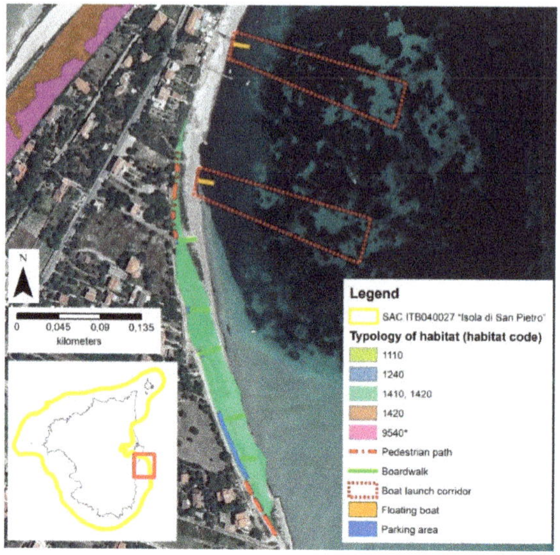

Figure 3. Overlay area related to the CLUP's actions and the SAC ITB040027 "Isola di San Pietro" in the municipality of Carloforte.

As regards the municipality of Portoscuso, no overlay area is detected with reference to the CLUP's actions and the classified habitat areas. For example, as shown in Figure 4, although the newly-planned floating docks for leisure fishing and nautical tourism (Action Por_CLUP_A_1) do not overlay any habitat area, the presence of these facilities may very possibly generate negative impacts on classified habitats and species.

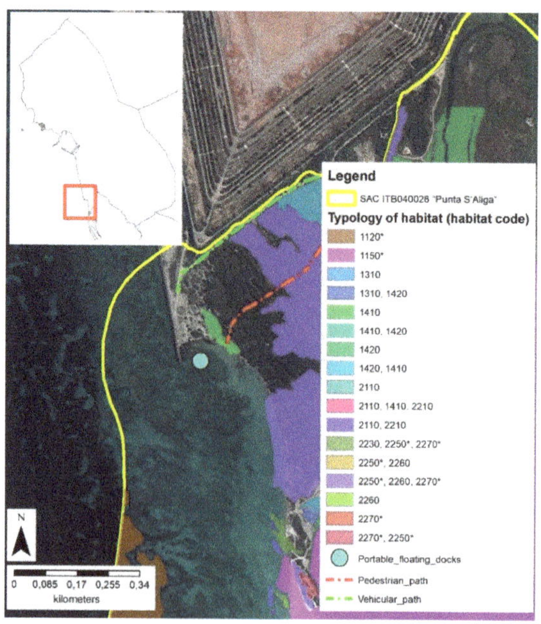

Figure 4. Overlay area between CLUP's actions and the SAC ITB040028 "Punta S'Aliga" in the municipality of Portoscuso.

4. Discussion and Conclusions

The three LFs show potential negative impacts that may be generated by the actions planned by the CLUPs. The assessment of the three cases concerning the towns of Calasetta, Carloforte, and Portoscuso highlights that the CLUPs and the MPs are characterized by a marked inconsistency since they were defined and approved through autonomous planning approaches, which were implemented by different public administrations, namely, the region as regards the MPs and the municipalities with reference to the CLUPs. Moreover, the two types of plan have different scopes since MPs state conservation measures related to Natura 2000 Sites, that is, protection of habitats and species, whereas CLUPs concern sustainable land-use approaches to coastal zone management.

Under this perspective, this study proposes and implements (as regards the coastal zones of the three municipalities of the Sulcis area) a methodology that aims at achieving consistency between the CLUPs and the MPs on the basis of the definition of the LFs.

The proposed detailed comparative assessment of the MPs and the CLUPs related to the three Sulcis municipalities implies a relevant contribution in this direction and, moreover, it leads the question of protection and improvement of species and habitats beyond the limits of sectoral plans regarding the Natura 2000 Sites. The proposed LF makes the issue of protection of habitats and species a central question in spatial policies as regards the coastal zones of the municipal areas regulated by the CLUPs. Supporting ecosystem services (ESs) provided by habitats and species [21] are the founding elements of the LFs since sustainability objectives, which are the LF's backbone, are largely based on the need to protect Natura 2000 Sites and to extend this conservationist approach outside the boundaries of the Natura 2000 Network [15].

Consistency of CLUPs and MPs entails overcoming the conflicts generated by different and sometimes conflicting objectives, which are mirrored by the planned land uses and spatial policies. These are the conflicts between the conservation of supporting ecosystems (habitats and species) and the increase in the supply of tourism-related ESs concerning the increase in the attractiveness of the coastal areas for tourists and local visitors, which is the main reason for the definition and the approval of CLUPs [22].

A central issue to be addressed and discussed as regards this study is how its implications can be generalized in order to resolve or at least mitigate the conflicts arising from the inconsistencies between the different priorities of coastal land use plans, defined and implemented by the local administrations, and of conservation measures established on Natura 2000 Sites, defined by the national governmental administrations of the EU countries under the provisions of the Habitats and Birds Directives.

From this point of view, it is evident that the quantitative spatial relevance of the conflicts is an important question since, if different priorities between local plans and Natural 2000 Sites are identified in large parts of the coastal areas, the effectiveness of public spatial policies will be heavily dependent on how these conflicts are addressed. This is the case of the municipalities of Carloforte and Portoscuso, where the conflicting areas cover about 10% of the SAC ITB040027 "Isola di San Pietro" in the municipality of Carloforte and about 30% of the SAC ITB040028 "Punta S'Aliga" in the municipality of Portoscuso. On the other hand, potential conflicts are not only ascribable to the overlay areas but also to the impacts of CLUPs' actions, though they are irrelevant with reference to their areal size, which may generate relevant negative effects on classified habitats. Consistent with what is observed in the planning processes of several coastal areas of the EU [23–25], the conflicts are essentially due to the fact that coastal land use plans aim at making coastal areas easily accessible for car, bike, and pedestrian traffic and attractive in terms of beach tourist services (see the potentially unfavorable CLUP's actions reported in the last column of Table 2), whereas the conservation measures related to Natura 2000 Sites aim at preventing negative impacts on coastal and marine habitats and species, which are likely to be generated by the implementation of the provisions of local coastal plans.

This premise entails that the nature of the conflicts and their relevant spatial size make the case of the Sulcis municipalities an important and significant paragon to address the general question of resolving the conflicts between local plans and conservation measures established by national plans.

This is even more the case since the conservation measures of the Natura 2000 Sites are based on the Standard Data Form (SDF) [26] established by the EU Decision taken on 11 July 2011. That being so, management and conservation measures derived from the SDFs are consistent with each other as regards all the countries of the Union. As a consequence, the implications coming from the outcomes of this study can be taken as general recommendations related to the solution of conflicts as regards all the countries of the EU and all the spatial contexts where ecological networks are established whose institutional and technical frameworks are similar to the Natura 2000 Network's.

Under this perspective, an outstanding implication can be derived by combining the results presented in the previous section with the conclusions of an important study of Kovács et al. on identifying and resolving conflicts related to trade-offs between protection of nature and natural resources and expectations regarding local development based on the exploitation of the provisions generated by the environment and its services [27]. The main point of the Kovács et al.'s study [27] is that the scale issues play a fundamental role in generating the conflicts. As a consequence, the solution of the conflicts has to be identified by dealing with the scale of the planning process. In Kovács et al.'s study [27], the scale issue was related to the fact that conservation measures related to three Natura 2000 Sites established by the Hungarian government in the Great Hungarian Plain were in conflict with the traditional agricultural land uses defined by the local rural plans. The argument of Kovács et al. [27] is that a fine-tuning of the spatial and the temporal layout of the conservation measures based on close cooperation between the national and the local authorities would have been the most effective approach to mitigating, if not avoiding, the negative impact of the Natura 2000 Sites-related spatial policies on the local economy mainly based on small farmers' production. The scale issue is quite similar with respect to the case of the Sulcis municipalities and of the other EU Natura 2000 Sites-related spatial contexts.

Thus, the integration of Kovács et al.'s article [27] and the findings of this study entails that a unique planning authority should be established, which should plan and implement both conservation measures related to the Natura 2000 Sites and regulations concerning coastal land uses. Representatives of the national Ministry of the Environment and of the local governments should participate in political and technical boards of this authority so as to make all priorities and expectations equitably visible and expressed in a decision-making process, which should result in a unique planning and regulatory instrument. Moreover, an important task of the authority should be the development of appropriate awareness-raising campaigns aimed at increasing the attitude of the local communities towards the relevant role of conservation of nature and natural resources in promoting and catalyzing the local social and economic development so as to bridge the obstacles generated by the under-evaluation of the temporal layout problems related to planning policies' implementation [27,28].

At present, a very heterogeneous situation characterizes the Italian approach to the definition and the implementation of conservation measures and MPs of the Natura 2000 Sites. While conservation measures are always approved by the national Ministry of the Environment, the definition of the measures can be either part of the MPs of the Natura 2000 Sites or established through a planning process not related to an MP. The definition of conservation measures and MPs can be under the responsibility of a number of public bodies, such as municipalities, regional administrations, regional parks and natural reserves, and regional environmental protection agencies. Public participation and public awareness-raising processes are not mandatory and are not frequently implemented. Finally, only three out of 21 regional administrations implement conservation measures and MPs into their spatial planning regulations [29]. The Italian situation, which is quite consistent with what occurs in the other EU countries, gives support to the implication discussed above, which integrates the findings of this study into Kovács et al.'s argument. This implication entails the need to establish a new authority responsible for the definition and the implementation of spatial policies that implement MPs and conservation measures into local planning regulations. In order to be effective and operational, the establishment of the new authority should be based on the provisions of a new regulation or directive approved by the European Parliament and the Council.

There are three promising future directions coming from this study. First, the restrictions on land uses imposed by the MPs and evidenced in the last column of the LFs of the three towns of the Sulcis area, which are likely to generate a decrease in the attractiveness of the coastal areas for tourists and local visitors, entail the need to assess the trade-off between the decreased provision of tourism-related ESs and supporting ESs provided by habitats and species [15]. Several studies in the current literature deal with the economic evaluation of tourism-related ESs, which are based on estimates of the market values of these services [30–33].

Furthermore, assessments of the economic value made available from species and habitats protected according to the Habitats and Birds Directives are rare, and, that being so, there is plenty of room for future research on the evaluation of the economic value of supporting ESs. The Millennium Ecosystem Assessment [21] and Busch et al. [34] estimate the economic value of habitats- and species-related ESs in terms of indirect or direct willingness to pay on the basis of the public-good nature, which implies the impossibility of identifying the value of these supporting ESs through their market price, which cannot be observed since they are non-excludable and non-rivalrous [15]. Estimates implemented on the basis of the hedonic approach in terms of direct (observed) willingness to pay [35–37] or of the contingent valuation methodologies in terms of indirect willingness to pay [34,38,39] are solid references for the assessment of the economic value of supporting ESs.

Research aimed at identifying the value of tourism-related and supporting ESs will make adequate estimates available that will make it possible to implement effective decision-making processes concerning the optimal mix between provisions of different types of ESs and, as a consequence, between CLUPs' and MPs' planning policies and related conflicting actions.

Secondly, the following relevant question should be carefully taken into account as regards the trade-offs between tourism-related and supporting ESs. The assessment of the economic value of the trade-offs between recreational and supporting ESs should entail the comparison between the additional demand for recreational ESs, which would eventually be unsatisfied as a consequence of the implementation of conservation measures aimed at preserving habitats and species and the increase in the demand for supporting ESs satisfied through these measures, since a balance of the gained and the lost economic value can only be associated with truly demanded ESs. This question is treated by Bastian et al. [40] from a theoretical standpoint on the basis of ecosystem potentials, properties, and services of ESs (EPPS). The integration of Bastian et al.'s findings [40] and the outcomes of this study could be very helpful in order to deal with the issue of the demand side as regards the trade-off between tourism-related and supporting ESs.

Thirdly, following Bastian [41] among many, it has to be highlighted that the exploration of trade-offs concerning the supply of other ESs—namely, provisioning, regulating, and cultural ESs—is quite promising as a future research direction.

Finally, addressing the conflicts that may arise between planning measures and actions just on the basis of the spatial zoning regulations of CLUPs and MPs can possibly result in scientifically-weak implications. An important complementary spatial analysis would entail the assessment of losses and gains in the supply of ESs on the basis of land cover taxonomies made available by CORINE Land Cover-based detailed datasets [42], such as the Copernicus Project Database [43]. In future research, the trade-offs between planning policies and actions identified through differences in the provision of categories of ESs should be qualified in terms of losses and gains in the supply of land uses rather than of land covers, since the intensity of management and the production of wealth can only be identified through land uses, which CORINE-like spatial taxonomies do not capture. Another relevant issue that needs to be addressed is the relationship between planning policies and measures and the resulting visual landscape, which plays an important role in shaping the attractiveness of spatial contexts [44–47].

Author Contributions: F.L. and C.Z. have made substantial contributions to the study's conception, background, and design remarks of Section 1 and to discussion and concluding remarks of Section 4. The methodological discussion proposed in Section 2 is by Federica Leone. The results presented in Section 3 are by C.Z.

Funding: This research received no external funding.

Acknowledgments: The study was implemented within the Research Program "TSulki: Tourism and Sustainability in the Sulcis sub-regional area", funded by the Autonomous Region of Sardinia for the period 2018–2019, under the provisions of the Call for the presentation of "Projects related to fundamental or basic research for the implementation of interventions in relation to the research context of the Sulcis Plan" of the year 2015, implemented at the Department of Civil and Environmental Engineering and Architecture (DICAAR) of the University of Cagliari, Italy.

Conflicts of Interest: The authors declare no conflict of interest. The supporting sponsor had no role in the design of the study; in the collection, analyses, or interpretation of data; in the writing of the manuscript, and in the decision to publish the results.

References and Notes

1. Saffache, P.; Angelelli, P. Integrated coastal zone management in small islands: A comparative outline of some islands of the Lesser Antilles. *J. Integr. Coast. Zone Manag.* **2010**, *10*, 255–279. [CrossRef]
2. Billé, R. Integrated coastal zone management: Four entrenched illusions. *Surv. Perspect. Integr. Environ. Soc.* **2008**, *1*, 12. [CrossRef]
3. The Text of the ICZM Protocol. Available online: https://eur-lex.europa.eu/legal-content/EN/TXT/PDF/?uri=CELEX:22009A0204(01)&from=EN (accessed on 10 May 2019).
4. Soriani, S.; Buono, F.; Camuffo, M. Problems and pitfalls in ICZM implementation: Lessons from some selected Mediterranean and Black Sea cases. *J. Integr. Coast. Zone Manag.* **2015**, *18*, S1-002. [CrossRef]
5. Burbridge, B.; Humphrey, S. Introduction to the special issue on the European demonstration program on integrated coastal zone management. *Coast. Manag.* **2003**, *31*, 121–126. [CrossRef]
6. Suárez de Vivero, J.L.; Rodríguez Mateos, J.C. Coastal crisis: The failure of coastal management in the Spanish Mediterranean Region. *Coast. Manag.* **2007**, *33*, 197–214. [CrossRef]
7. Shipman, B.; Stojanovic, T. Facts, fictions, and failures of integrated coastal zone management in Europe. *Coast. Manag.* **2007**, *35*, 375–398. [CrossRef]
8. Rochette, J.; Billé, R. (Eds.) Analysis of the Mediterranean ICZM Protocol: At the Crossroads Between the Rationality of Provisions and the Logic of Negotiations. Institute for Sustainable Development and International Relations (IDDRI), 2010. Available online: http://www.cirspe.it/gizc/Pubblicazioni/Pubblicazioni%202/4-%20ICZM_Med_IDDRI.pdf (accessed on 10 May 2019).
9. Leone, F.; Zoppi, C. The delicate relationship between capitalization and impoverishment of cultural and landscape resources in the context of Strategic Environmental Assessment of municipal master plans: A case study concerning Tertenia, Sardinia. In *Atti della XVIII Conferenza Nazionale SIU. Italia '45-'45. Radici, Condizioni, Prospettive. Venezia, 11–13 Giugno 2015 [Proceedings of the XVIII National Conference of SIU [The Italian Society of Urban and Regional Planners]. Italy '45-'45. Roots, Conditions, Perspectives. Venice, Italy, 11–13 June 2015]*; Various Authors, Ed.; Planum Publisher: Rome-Milan, Italy, 2015; pp. 1458–1467.
10. UNEP (United Nations Environment Programme); MAP (Mediterranean Action Plan). Priority Actions Program. Action Plan. for the Implementation of the ICZM Protocol for the Mediterranean—2012–2019. Draft (September 2011). Available online: https://www.pap-thecoastcentre.org/razno/Action%20Plan%20clear%20version%2013%20Sept%2011%20BS-2.pdf (accessed on 10 May 2019).
11. Harvey, N. Strategic environmental assessment in coastal zones, especially Australia's. *Impact Assess. Proj. Apprais.* **2000**, *18*, 225–232. [CrossRef]
12. Partidário, M.R.; Vicente, G.; Lobos, V. Strategic environmental assessment of the national strategy for integrated coastal zone management in Portugal. *J. Coast. Res.* **2009**, *2*, 1271–1275.
13. Natura 2000 sites includes three typologies of areas: Special protection areas (SPAs), established under EU Directive 2009/147/EC (the Birds Directive), and Sites of community importance (SCIs) and Special areas of conservation (SACs), established under EU Directive 92/43/EEC (the Habitats Directive).
14. Leone, F.; Zoppi, C. Ecosystem services as external drivers in the Strategic environmental assessment of management plans of the sites of the Natura 2000 network. *Urban. Inf.* **2015**, *42*, 34–38.
15. Leone, F.; Zoppi, C. Conservation measures and loss of ecosystem services: A study concerning the Sardinian Natura 2000 Network. *Sustainability* **2016**, *8*, 1061. [CrossRef]

16. Istat [The Italian National Institute of Statistics] Superfici delle Unità Amministrative. Superficie di Comuni, Province e Regioni Italiane al 9 Ottobre 2011 [Surfaces of Administrative Units. Surfaces of Italian Local Municipalities, Provinces and Regions on 9 October 2011]. 2011. Available online: https://www.istat.it/it/archivio/156224 (accessed on 10 May 2019).
17. "Any plan or project not directly connected with or necessary to the management of the site but likely to have a significant effect thereon, either individually or in combination with other plans or projects, shall be subject to appropriate assessment of its implications for the site in view of the site's conservation objectives [...] [T]he competent national authorities shall agree to the plan or project only after having ascertained that it will not adversely affect the integrity of the site concerned and, if appropriate, after having obtained the opinion of the general public." (Habitats Directive, art. 6, paragraph 3).
18. The information sources concerning the items of the five columns of Tables 2–4 are the following. (i) "Sustainability-oriented objectives," "Themes." "CLUP's specific objectives" and "Potentially unfavorable CLUP's actions." Calasetta: CLUP of Calasetta. Available online: http://albopretorio.comune.it/calasetta/AmministrazioneTrasparente/Pianificazionegovernoterritorio.aspx (accessed on 10 May 2019). Carloforte: CLUP of Carloforte. Available online: http://ww2.gazzettaamministrativa.it/opencms/opencms/_gazzetta_amministrativa/amministrazione_trasparente/_sardegna/_carloforte/190_pia_gov_ter/2015/Documenti_1424947761165/ (accessed on 10 May 2019). Portoscuso: CLUP of Portoscuso. Available online: https://drive.google.com/drive/folders/1LVQLSZlgYVgpdNyoHt-KknlYOOzzDjQi (accessed on 10 May 2019). (ii) "Sustainability-oriented objectives" and "MP's specific objectives." Calasetta: The MP of the SACs ITB042210 "Punta Giunchera," ITB042208 "Tra Poggio La Salina e Punta Maggiore" and ITB042209 "A nord di Sa Salina." Available online: http://www.sardegnaambiente.it/documenti/18_183_20160421121516.zip. Carloforte: The MP of the SAC ITB040027 "Isola di San Pietro" (Available online: http://www.sardegnaambiente.it/documenti/18_183_20170830162930.zip, accessed on 10 May 2019), and the MP of the SPA ITB043035 "Coste e Entroterra tra Punta Cannoni e Punta delle Oche—Isola di San Pietro" (Available online: http://www.sardegnaambiente.it/documenti/18_183_20170830163754.zip, accessed on 10 May 2019). Portoscuso: The MP of the SAC ITB040028 "Punta S'Aliga" (Available online: http://www.sardegnaambiente.it/documenti/18_183_20160418181506.zip, accessed on 10 May 2019), and the MP of the SAC ITB040029 "Costa di Nebida" (Available online: http://www.sardegnaambiente.it/documenti/18_183_20160419103004.zip, accessed on 10 May 2019). The "Sustainability-oriented objectives" are defined through the integration of the CLUPs' and the MPs' sustainability objectives, which are substantially consistent with each other.
19. The SPA ITB043035 "Coste e Entroterra tra Punta Cannoni e Punta delle Oche—Isola di San Pietro" is totally included within the SAC ITB040027 "Isola di San Pietro". That being so, the calculation of the overlay area is redundant.
20. In the case of the SAC ITB040029 "Costa di Nebida," located in the municipality of Portoscuso, and of all the SACs located in the municipality of Calasetta is not possible to calculate the the overlay share due to lack of reliable geo-referenced data.
21. Millennium Ecosystem Assessment. *Ecosystems and Human Well-Being: A Framework for Assessment*; Island Press: Washington, DC, USA, 2003; ISBN 1559634030.
22. Lai, S.; Zoppi, C. The influence of Natura 2000 Sites on land-taking processes at the regional level: An empirical analysis concerning Sardinia (Italy). *Sustainability* **2017**, *9*, 259. [CrossRef]
23. Jones, P.J.; Lieberknecht, L.M.; Qiu, W. Marine spatial planning in reality: Introduction to case studies and discussion of findings. *Mar. Policy* **2016**, *71*, 256–264. [CrossRef]
24. Piwowarczyk, J.; Wróbel, B. Determinants of legitimate governance of marine Natura 2000 Sites in a post-transition European Union country: A case study of Puck Bay, Poland. *Mar. Policy* **2016**, *71*, 310–317. [CrossRef]
25. Fraschetti, S.; Pipitone, C.; Mazaris, A.D.; Rilov, G.; Badalamenti, F.; Bevilacqua, S.; Claudet, I.; Carić, H.; Dahl, K.; D'Anna, G.; et al. Light and shade in marine conservation across European and contiguous seas. *Front. Mar. Sci.* **2018**, *5*, 420. [CrossRef]
26. The Standard Data Forms of all Natura 2000 Sites. Available online: http://natura2000.eea.europa.eu/ (accessed on 10 May 2019).

27. Kovács, E.; Kelemen, K.; Kalóczkai, A.; Margóczi, K.; Pataki, G.; Gébert, J.; Málovics, G.; Balázs, B.; Roboz, A.; Krasznai Kovács, E.; et al. Understanding the links between ecosystem service trade-offs and conflicts in protected areas. *Ecosyst. Serv.* **2015**, *12*, 117–127. [CrossRef]
28. As regards the three Sulcis municipalities, at present, there is no evidence of any attempt to deal with the spatial scale issue, that is to develop a virtuous cooperative planning process involving the national and local governments in order to bridge the gap between national and local planning processes, and related spatial regulations. Moreover, there is no evidence of any attempt to consider the temporal scale, since no time was dedicated by the national authorities to increase the local authorities' and communities' awareness concerning the importanve, and the implied value-added, for the local economies of protection and improvement of the quality of nature and natural resources.
29. The information presented in this paragraph is drawn from the analytical assessment of the process of definition and implementation of MPs and conservation measures concerning the Italian Natura 2000 Sites proposed in: Lai, S. Tra tecnocrazia e inclusione nella pianificazione per la tutela delle risorse naturali: un'analisi dei processi nei siti Natura 2000 in Italia [Between technocracy and inclusion in spatial planning for conservation of natural resources: An analysis of processes related to Natura 2000 sites in Italy]. In *Ricerca in Vetrina 2018. Ricerca è democrazia. Il Ruolo dell'Attività Scientifica nella Costruzione di un Futuro Equo e Solidale [Research in a Shop Window. Research is Democracy. The Role of Scientific Activity in the Building Process of a Fair and Equitable Future]*; Saiu, V., Mastinu, M., Angius, F., Leccis, F., Mei, G., Mura, E., Lai, L., Mais, S., Pinna, A., Cabras, R., et al., Eds.; FrancoAngeli: Milan, Italy, 2018; pp. 124–135. ISBN 9788891783806.
30. Campagne, C.S.; Salles, J.M.; Boissery, P.; Deter, J. The seagrass Posidonia oceanica: Ecosystem services identification and economic evaluation of goods and benefits. *Mar. Pollut. Bull.* **2015**, *97*, 391–400. [CrossRef] [PubMed]
31. Barbier, E.B.; Hacker, S.D.; Kennedy, C.; Koch, E.W.; Stier, A.C.; Silliman, B.R. The value of estuarine and coastal ecosystem services. *Ecol. Monogr.* **2011**, *81*, 169–193. [CrossRef]
32. Zander, K.K.; Straton, A. An economic assessment of the value of tropical river ecosystem services: Heterogeneous preferences among Aboriginal and non-Aboriginal Australians. *Ecol. Econ.* **2010**, *69*, 2417–2426. [CrossRef]
33. Lange, G.M.; Jiddawi, N. Economic value of marine ecosystem services in Zanzibar: Implications for marine conservation and sustainable development. *Ocean Coast. Manag.* **2009**, *52*, 521–532. [CrossRef]
34. Busch, M.; La Notte, A.; Laporte, V.; Erhard, M. Potentials of quantitative and qualitative approaches to assessing ecosystem services. *Ecol. Indic.* **2012**, *21*, 89–103. [CrossRef]
35. Zoppi, C.; Argiolas, M.; Lai, S. Factors influencing the value of houses: Estimates for the city of Cagliari, Italy. *Land Use Policy* **2015**, *42*, 367–380. [CrossRef]
36. Cheshire, P.; Sheppard, S. On the price of land and the value of amenities. *Economica* **1995**, *62*, 247–267. [CrossRef]
37. Palmquist, R.B. Estimating the demand for the characteristic of housing. *Rev. Econ. Stat.* **1984**, *66*, 394–404. [CrossRef]
38. Zoppi, C.; Lai, S. Assessment of the Regional Landscape Plan of Sardinia (Italy): A participatory-action-research case study type. *Land Use Policy* **2010**, *27*, 690–705. [CrossRef]
39. León, C.J. El metodo dicotomico de valoracion contingente: Una aplicacion a los espacios naturales en Gran Canaria [The dichotomous method of Contingent Valuation: An application to the natural resources of Great Canaria]. *Investig. Econ.* **1995**, *19*, 83–106.
40. Bastian, O.; Haaseb, D.; Grunewald, K. Ecosystem properties, potentials and services—The EPPS conceptual framework and an urban application example. *Ecol. Indic.* **2012**, *21*, 7–16. [CrossRef]
41. Bastian, O. The role of biodiversity in supporting ecosystem services in Natura 2000 Sites. *Ecol. Indic.* **2013**, *24*, 12–22. [CrossRef]
42. European Environment Agency. CORINE Land Cover. 2013. Available online: http://www.eea.europa.eu/publications/COR0-landcover (accessed on 10 May 2019).
43. Copernicus is the European Union's Earth Observation Programme, Looking at our Planet and Its Environment for the Ultimate Benefit of All European Citizens. It Offers Information Services Based on Satellite Earth Observation and in situ (non-space) data (text drawn from): The Copernicus Project Database Europe's eyes on Earth. Available online: https://www.copernicus.eu/en/about-copernicus;https://www.copernicus.eu/en/research/project-database (accessed on 10 May 2019).

44. Heagney, E.C.; Rose, J.M.; Ardeshiri, A.; Kovac, M. Optimising recreation services from protected areas—Understanding the role of natural values, built infrastructure and contextual factors. *Ecosyst. Serv.* **2018**, *31*, 358–370. [CrossRef]
45. Lai, S.; Leone, F.; Zoppi, C. Implementing green infrastructures beyond protected areas. *Sustainability* **2018**, *10*, 3544. [CrossRef]
46. Amoako-Tuffour, J.; Martinez-Espineira, R. Leisure and the net opportunity cost of travel time in recreation demand analysis: An application to Gros Morne national park. *J. Appl. Econom.* **2012**, *15*, 25–49. [CrossRef]
47. Font, A.R. Mass tourism and the demand for protected natural areas: A travel cost approach. *J. Environ. Econom. Manag.* **2000**, *39*, 97–116. [CrossRef]

sustainability MDPI

Article

A Dashboard for Supporting Slow Tourism in Green Infrastructures. A Methodological Proposal in Sardinia (Italy)

Ginevra Balletto [1], Alessandra Milesi [1], Mara Ladu [1] and Giuseppe Borruso [2,*]

[1] Department of Civil and Environmental Engineering and Architecture, University of Cagliari, Via Marengo 2, 09123 Cagliari, Italy; balletto@unica.it or ballettoginevra@gmail.com (G.B.); alessandra.milesi@gmail.com (A.M.); maraladu@hotmail.it (M.L.)

[2] Department of Economics, Business, Mathematics and Statistics "Bruno de Finetti", University of Trieste, Via Tigor 22, 34127 Trieste, Italy

* Correspondence: giuseppe.borruso@deams.units.it; Tel.: +39-040-558-7008

Received: 22 January 2020; Accepted: 20 April 2020; Published: 28 April 2020

Abstract: Spatial planning and territorial promotion can benefit largely from the application of the Information Communication Technology (ICT) at different scales. From knowledge acquisition to management and planning, their role in building an image of the territory, and constantly updating it to the benefit of users and planners, is of paramount importance. Institutional channels, together with social networks, are the means by which both a local community and a wider community of users share experiences and perceptions. ICTs are therefore strategic in supporting and promoting a sustainable tourism development of territories. Data and information aggregators as dashboards represent examples of decision support systems where digital data are organized and processed to produce an information output. The present paper is part of a wider research, related to the valorization of a former mining area in the Sulcis-Iglesiente area (Sardinia, Italy), where the extraction activity has left the place to abandonment, and only recently to tourism, stressing the concept of slow tourism. Such new opportunity has been launched with the Santa Barbara Walk (SBW), an ancient mining route currently trying to consolidate as a tourism attraction area. Such a territory is in constant transition with unique characters of anthropic and naturalistic characterization, setting itself as a green infrastructure, capable also of attracting a wide community of regional and extra-regional users. However, its digital network—consisting of intangible infrastructure and flows—is fragmented in terms of policies and contents. Additionally, a state of disorganization in slow tourism promotion activities can be observed. To implement the SBW capabilities, the present paper aims to develop a proposal for the framework of a circular dashboard applied to the SBW. In particular, we implement a set of indicators of performance of the SBW for the organization of information on the walk's main characters, to facilitate a shared governance and an effective tourism promotion. The SBW is recognized as a network connecting the main points of interest preferred by the slow tourism (This paper is based on the Research project TSULKI—Tourism and Sustainability in the Sulcis (Sardinia-Italy)—and on the agreement protocol between DICAAR Department of Cagliari University and Foundation of the Santa Barbara Walk, signed in December 2018).

Keywords: green infrastructure; landscape; smart dashboard; smart governance; slow tourism; Santa Barbara Walk; Sulcis-Iglesiente; Sardinia; Italy

---

1. Introduction

Information and Communication Technologies (ICTs) play an increasingly important role in the process of acquiring awareness and knowledge from a territory, as well as of its management, enhancement, and promotion at different scales [1–4]. Information is no longer managed only by a few institutional subjects through official portals, but built to a significant extent by smart communities that populate social networks and media, sharing experiences and perceptions. New ICTs are tools to support the promotion of sustainable regeneration and development of the territories, including tourism, especially if organized within a smart dashboard [5–13].

Sustainable development, from its roots, is thought to be based on the promotion and activation of local resources and communities. Putting it into the more specific case of a landscape in transition (Sulcis-Iglesiente), local sustainable development is thought of in terms of natural resources, social ones, as well as the historical and cultural heritage. In the information age, the Smart Community [14] represents a significant part of our contemporary society that makes an extensive use of the available technological tools, services, and devices to promote organization and efficiency, to improve the quality of life, but also to share personal knowledge, opinions, and interests [15]. For these reasons, it gradually became an extraordinary source of data for policy makers [16]. The 'heritage and new technologies' binomial contains an immense material and immaterial heritage of historical, cultural, and artistic values enhanced through the most recent forms of fruition of the Tourist Experience Design [17], according to which it becomes fundamental to offer each user personalized information and unique emotions, influencing both the cognitive and the emotional sphere.

The area of Sulcis-Iglesiente, located in the South-Western part of Sardinia (Italy), went through a phase of economic conversion based on sustainable slow tourism development [18–21] referred to abandoned mines [22,23]. This is the part of cultural tourism that arises from the reinterpretation, conservation, and enhancement of the material and immaterial mining heritage through the green infrastructures from the ancient mining tracks [24].

Such a heritage represents a peculiar character of the area, a hard one and apparently in contrast with a beautiful and still little-known landscape and endowment of natural resources. Among its peculiarities, there are those of having had a strong environmental impact on the area, still present in terms of pollutant, but that now represent also a character of the area, integrated and mitigated by the same nature, and being, itself, an element of attraction per se, a sign of past human action and of the transformation of the natural environment. It is a landscape in transition, given the fact that the relics of the human actions cannot be completely restored and secured, and therefore subject to the transformation of time and nature. In such terms, the same industrial archeological heritage can be expressed as a cultural heritage and the outcome of different stages of human-nature interaction. In the Sulcis-Iglesiente region, this type of tourism based on the use of ecosystem services is becoming a driving force for economic development [24], providing answers to the recent trend of preferring tourism activities for recreational and spiritual purposes [25], training or work, and slow and sustainable forms of mobility [26]. This is in line with the objectives of the Historical Environmental Geo-mining Park of Sardinia, established in October 2001 to safeguard and enhance the architectural and landscape heritage. Furthermore, it is also in line with the subsequent establishment (2017) of the SBW [27] and the birth of the homonymous Foundation, that allowed starting a new development phase based on the concept of slow tourism [28]. The risks and opportunities of this process of economic and social transformation, both in terms of policy and governance, led us to develop a specific research activity [28–31], the aim of which is putting together management, public administrations, and private players in a single network system, capable of fostering sustainable local development. In the area of the Sulcis-Iglesiente, in fact, the presence of points of historical-cultural and naturalistic interest, mining sites, reception, and refreshment activities that are not managed within a single network were observed. Many sites are equipped with portals and dedicated web pages, but the promotion is not performing well because of the presence of disorganized materials and poor management of the information. The website of the SBW Foundation is an interesting starting point, as it represents a first attempt of realizing an institutional network dedicated to slow tourism [32].

This paper's aim is twofold. On one side it is aimed at systematizing the information coming from institutions and from the users' communities to strengthen smart governance for sustainable development; on the other side, and as a successive part, its goal is to develop a proposal for the framework and design of a circular dashboard applied to the SBW. The dynamism of the data of the proposed dashboard allows elaborating coherent outputs with the institutional objectives and the comments and notes provided by the main social networks. The rest of the paper is organized as follows.

- The first section—Introduction and Literature review—focuses on the overview of the recent literature of new technologies for slow and smart tourism development and identification of the area of interest.
- The second section—Materials—discusses the topic of green infrastructures and slow tourism in the Sulcis-Iglesiente region and presents the case study of the SBW.
- The third section—Methods—develops the methodology to elaborate a proposal for the SBW Dashboard.
- The fourth section—Results—reports and discusses the main research results carried out.
- Conclusions are drawn in the fifth section, where the major findings are presented, together with the future developments of the research.

1.1. Literature Review

The paper is focused on a set of concepts apparently separated from each other, but actually linked and interrelated: Tourism, cultural heritage, ecosystem services, and smart dashboard for gathering data and information are the main elements tackled and revised here in terms of references and literature. In particular, tourism in this moment of health emergency—Covid-19—is, like all the others, a vulnerable sector, but also one of the engines of the economy that usually reacts first, if well managed. Nonetheless, the effects on it will depend on the duration of the crisis [33]. Events of this magnitude encourage us to plan the entire tourism system better, in a more sustainable way: The lesson is teaching us to be consistent in managing flows, in quantity and quality, proportional and related to the destination territories' capacities.

However, what emerges is that climate, polluted air, and infections travel together [34–36]. To highlight the link, for example, is the "Lancet Countdown Report 2019", which associates climate change with an increased spread of infectious diseases [37]. All anthropic activities are responsible for the emission of greenhouse gases and particulate pollutants that modify the composition of the atmosphere. Air quality and climate change are therefore two closely related environmental issues [37]. In other words, climate changes, on the one hand, affect the atmospheric processes (transport, reactivity, etc.), and on the other, cause changes in the functioning of terrestrial and marine ecosystems which can, in turn, affect the atmospheric processes [38]. However, these two environmental emergencies are still considered separately, both at the level of the scientific community and those responsible for environmental policies, as in the case of the recent Covid-19 emergency [39]. As a possible solution, there is the complete rethinking of our relationship with nature: Protecting biodiversity, stopping the climate crisis, curbing the destruction of forests [40–42], and reducing the consumption of resources. Although it is not yet possible to evaluate the scenarios, it is instead possible to believe that slow tourism, as it is based on the paradigm of sustainability, in contrast to over tourism, will not undergo profound changes. The Sulcis-Iglesiente does not have the same level of over tourism that places in Europe are suffering (Venice, Paris, Rome, etc.) [43]. In this sense, the medium and long-term sustainability of tourism depends on ensuring visitors do not wear out their welcome. Reorienting tourism to enhance local well-being is the way forward [44]. As strategic elements for the promotion of territories, the 'authenticity related to well-being' [45]. In the Sulcis-Iglesiente region, moreover, the peculiar characteristics linked to a "hard" and impactful activity such as mining are also included, which combined with the presence of natural and cultural resources characterize the context of the SBW [46].

In addition, Web and ICT instruments allow tourists and users to enjoy the territory for accommodation, tourist information, and for sharing information and experiences via social networks and apps for smartphones [47], that implying also operators and destination managers are prepared to interact with the same systems [48], realizing de-facto smart tourism destinations, appreciated both by locals and tourists [49]. In such sense, cultural heritage is the main concept around which the paper is realized, and is represented by the human-environment relationship and its outcomes, both nature-based—as the green infrastructure, and rural productions—and industry-based—as the resource extraction for industrial production and manufacturing [48,50,51].

These elements dealing with tourism and cultural heritage can, forcing the definition, find a synthesis into the idea of Ecosystem Services. Going to the definition provided by Costanza [52]—the benefits human populations derive, directly or indirectly, from ecosystems [53] and green infrastructures [51,54,55]—to be adapted to the case study hereby analyzed. The Sulcis-Iglesiente area is, in fact, a territory characterized by a coexistence of natural and cultural resources and human-made relics of mining extraction and industrial activities, that can represent elements to be protected and exploited, being inserted into spatial planning and territorial promotion policies in development paradigm, and also strengthened by the ongoing crisis [53–56].

In this sense, the dashboards demonstrate themselves as relevant instruments for supporting both users and decision makers in the understanding of a territory, and in allowing the managing of its assets. Some scholars [6] define the circular dashboard as "the process of data gathering, processing and organization of decision makers and users for planning purposes. In this domain the information obtained from the dashboard is used to evaluate urban performances and calibrate further and future city actions". The SBW dashboard is based on the simultaneous management of data from different sources—e.g., meteorological services, social network streams, geographic data, etc.—which makes it a tool, gathering data and information to be used and managed for making decisions, and therefore influencing input data for further steps [57].

2. Materials

2.1. Green Infrastructures and Slow Tourism in the Sulcis-Iglesiente (Sardinia, Italy)

Green Infrastructures (GI) are defined as a network of natural and semi-natural areas strategically planned with other environmental elements, designed and managed in such a way to offer ecosystem services [58]. In particular, according to the "Review of the progress made in the implementation of the EU strategy for green infrastructures (2019)" [59], the connection of natural capital in Europe is to be strengthened, through innovative technological solutions. Furthermore, with the report "Strategic GI and ecosystem restoration; geospatial methods, data and tools (2019)" [60] it is confirmed that GI also assumes a strategic role for risk management in which the present case study is fully inserted: The SBW, in the Sulcis-Iglesiente region (Sardinia, Italy) (Figure 1).

"Slow tourism" is one of the new trends in contemporary tourism. This trend emerged from the wider trend of the so-called "Slow Movement," which is characterized by the philosophy of "slow food" and the general philosophy of the "slow city" [61,62]. Slow tourism represents an "antidote" to mass tourism. In such a sense, a feverishly organized trip "to see as much as possible" is replaced by a slow journey of pleasure and relaxation. One of the key elements of slow tourism is the possibility for the traveler to become an active member of the local community and connect with the place and its people, creating stronger memories of the journey and the destination [63]. In other words, the ecological and cultural connectivity and the multifunctionality of ecosystems through slow tourism is consistent both with EU guidelines and with the local context of Sulcis-Iglesiente, characterized by a polycentric landscape, between abandoned mines and sites of environmental interest [46]. In this sense, a reference can be the "European Spatial Planning Observation Network" (ESPON) policy (2016), where the territorial condition requires 'polycentric development' as fundamental for sustainable development. Research done by "ESPON"—mapping the 'polycentricity' in Europe—illustrates the

Iglesiente area with a positive urban structure, composed of small and medium-sized towns, and policy recommendations conclude that the lower density implies more cooperation [64]. This aspect of the network is crucial for the enhancement and use through slow tourism of places and contexts with low urban density.

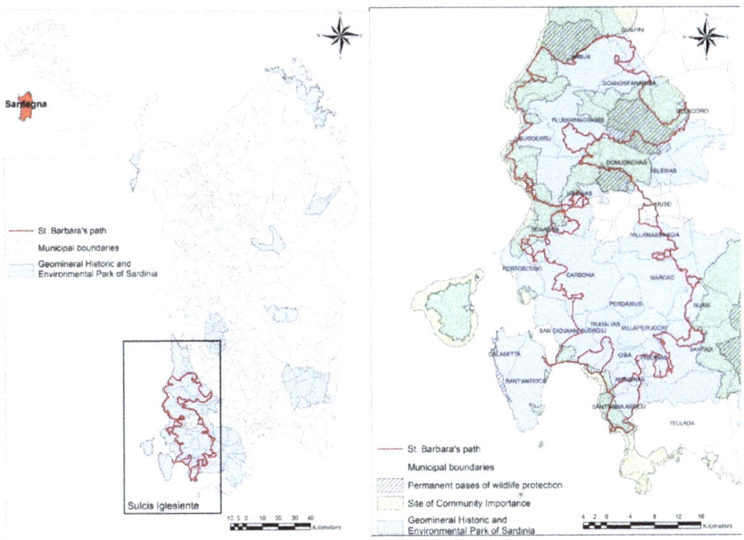

Figure 1. Santa Barbara Walk (SBW) territorial framework within the Geomineral Historic and Environmental Park of Sardinia and with natural sites.

The official tourism data provided by the Sardinia Region website reveal arrivals and presences in the entire region from 2017 to 2019.The data relating to 2019 was calculated by the authors in accordance with the percentage change estimated by the Sardinia Region compared to the previous year. The comparison shows that the flow of tourism in the Sulcis-Iglesiente confirms the growing regional trend, although it contributes modestly (Figure 2).

	2017	2018	2019 provisional
Sardinia Arrivals	3.136.422	3.299.951	3.431.949
Sulcis Iglesiente Region Arrivals	119.477	125.304	131.569
Sardinia Presences	14.385.685	15.052.251	15.654.341
Sulcis Iglesiente Region Presences	371.562	385.185	404.444
SBW presences	280	700	1895

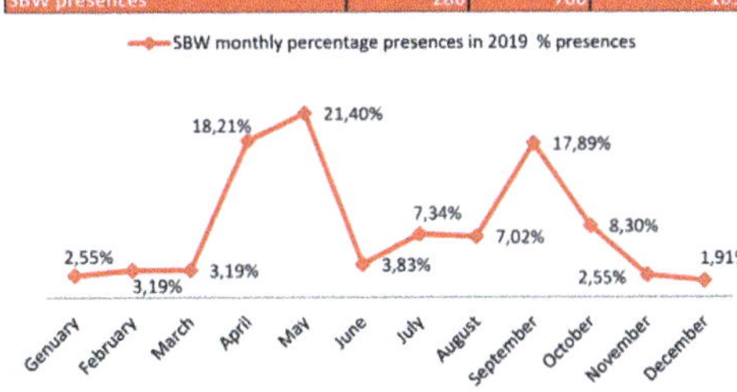

Figure 2. Comparison between tourist movements in Sulcis-Iglesiente region and in Sardinia in 2017–2018–2019 (provisional). Source Sardinia Region website [65] and Trend of SBW presences (source: SBW Foundation, 2020); percentage trend of monthly presences in the SBW in 2019 (source: SBW Foundation, 2020).

Within this context, the number of SBW users has also grown since 2017, the year of its establishment, affirming this infrastructure as a significant component of the slow tourism offer in Sardinia. Furthermore, the trend of monthly presences shows that this infrastructure promotes the extension of seasonal tourism. In fact, the greatest presence in 2019 was registered in April, May, and September, and not during the months most affected by the typical seaside tourism of Sardinia.

According to these analyses, the strategic role of historical paths and, therefore, also the SBW, should be considered as a driving force in defining a system of green infrastructures at the service of a slow tourist demand, based on knowledge and consistent with the objectives of the 2030 Agenda.

2.2. The Santa Barbara Walk and Its Social Network

The Santa Barbara Walk is an itinerary drawn in the Sulcis-Iglesiente area, in South-Western Sardinia. It follows the traces drawn by mining activities in the past decades in a unique environment as a mining area, located in a beautiful natural landscape. The SBW was brought to new life in sustainable slow tourism through a green infrastructure, characterized by the co-existence of newly drawn trunks of a Walk, the presence of a huge and massive industrial heritage, and the natural characteristics of a territory of particular beauty.

This historical region has, for several decades, been the most important center for international mining, as the main Sardinian economic source. The following crisis hitting the sector, followed by the mines dismantling in the 1990s, required rethinking the territory also through a form of tourism linked to industrial archeology in a particularly beautiful coastal landscape.

Within this context, in October 2001, the Geomineral Historic and Environmental park of Sardinia was established [66], divided into 8 areas covering 81 Municipalities of Sardinia, for a total area of approximately 3800 Km^2, which since 2007 is part of the "Global Geoparks Network" of UNESCO [67].The SBW is located within the "area 8" of the Park of Sulcis–Iglesiente which, other than being the largest, is the one most characterized by the presence of ancient mining activities.

In particular, the SBW spans over a complex territory characterized by a complex mining basin, which constitutes the The Geomineral Historic and Environmental park of Sardinia. This territory was the most important district for national and international mining due to its large production of lead and zinc. The crisis in the mining sector and the subsequent closure of the mines in the 1990s left a rich heritage of industrial archeology and infrastructure, as well as a unique landscape. The landscapes of Sulcis-Iglesiente are in fact deeply marked by the consequences of mining activities, with the presence of large open-air and underground excavations, mine adits, tunnels, and numerous mine wastes. The latter are constituted by accumulations of different types of waste rocks and tailings from mines and processing/metallurgical plants. All these elements highlight the vastity of mining operations carried out in the main mining places of the district, such as the great mines of Monteponi, San Giovanni, and Masua, and their related processing plants and handling systems, as the historical Laveria Lamarmora and Porto Flavia plants. The SBW then crosses a landscape rich in natural and anthropogenic elements—landfills, mine muds, and abandoned buildings—but at the same time mutable, because of its vulnerability. This condition of a continuously changing landscape—or landscape in progress—through slow tourism "provides economic, cultural, relational, and social benefits for both visitors and host communities".

Today, the SBW is included in the regional register of historical-religious paths of Sardinia, and in the year 2017, it was included in the Atlas walk of Italy of the Ministry of Cultural Heritage and Activities [68]. The walk extends along a 400 km ring, organized in 24 ways. It can be traveled walking, cycling, and on horseback [27,32]. The landscape of the SBW is characterized by the presence of numerous points of historical, cultural, and natural interest, enriching the already evocative mining landscape [28,30]. Put in different terms, it can be thought of as a green infrastructure both for the intrinsic aspects—recovery of the old mine tracks—and for the way of travel—walk, bike, and by horse.

However, to date the challenges are still many and constantly call for new policies to manage and enhance the SBW and cultural, natural, and landscape resources.

The main purpose of the SBW is inscribed within the recent slow tourism, characterized by the presence of innovative users, which we consider as a relevant part of the so called "Smart Community". With the term "community", we imply that providing user-generated contents, as, in this case, tourists or hikers actually enjoying the Walk and its points of interests, sharing their impressions and activities on social networks. In this particular context, the smart community corresponds to those actors which contribute to the awareness of the SBW by means of their activity on social networks and media and, therefore, they are potential players for its development and growth. The set of social network and media users, from Wikiloc paths collectors, to social networks and media users—both tourists and operators—becomes a pioneer of this SBW smart community. As a matter of fact, a true smart community does not yet exist, but a network of users willing to share their experiences is present. The data coming from this community is a significant input in this work. In the concept of the proposed dashboard, there is the primary object of helping the integration of the top-down management and the bottom up approach of the users. We assume that, using the data and information this community provides, we can build a bottom up participation process, in which the dashboard will represent the extent in which a true smart community is set up, linking together users and decision makers.

3. Methods

After describing the case study of the SBW and the role of its Smart Community, we decided to plan the realization of a dashboard to gather information and indicators on the walk. However, the conceptualization, design, and implementation of the dashboard represents, more importantly, an opportunity for putting all the different elements related to the SBW within a common system and framework, particularly related to the knowledge acquisition of the territory, to be used for planning future actions of territorial promotion. This required developing a methodology for designing the dashboard, articulated in the following phases, consisting of:

- Definition of the conceptual framework to organize the contents of the dashboard, consisting in the collection, analysis, and classification of the intrinsic and extrinsic characteristics of the Walk (Section 3.1), and of the user-generated ones from the smart community—these latter aimed at the production of internal reports for SBW managers (Section 3.2);
- Definition of the layout to organize the contents of the SWB circular dashboard (Section 3.3).

This methodology reflects the conceptual framework of the dashboard we proposed (Figure 3), which is thought of as a collector capable of organizing a comprehensive data set—characteristics and user generated—respectively, we elaborated and collected by the main social networks, as discussed below.

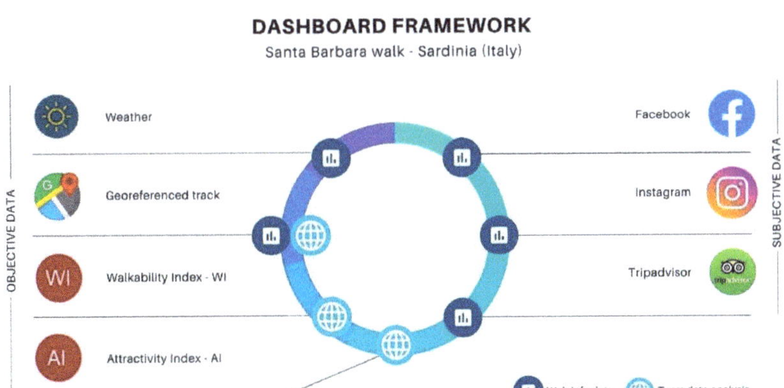

Figure 3. The Dashboard conceptual framework.

3.1. Survey and Analysis of the Intrinsic and Extrinsic Features of the SBW

Such a methodological phase regards the recognition of the elements that characterize the walk. These are the features of the Walk—intrinsic and extrinsic—but also a specific set of indicators that represent the 'attributes' of the same walk—the elements that allow the managers and planners on one side, and users on the other side, to express judgements on the walk's components.

The data concerning the intrinsic characteristics of the route, which describe the physical conformation of each of the 24 ways, were obtained from the SBW Foundation official website: Length (km), "difficulty", and duration (hours) (Figure 4).

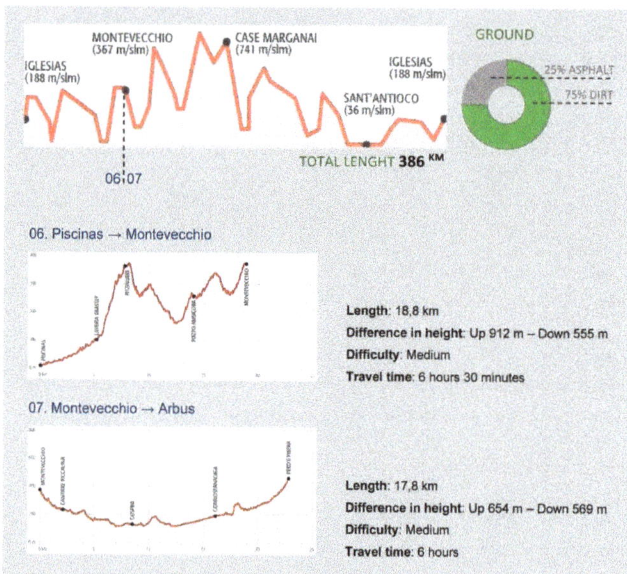

Figure 4. Intrinsic features of each of the 24 ways: Length (km), "difficulty" and duration (hours). Source: [24,31].

The extrinsic characteristics refer to the points of interest included in a 1 km buffer from the SBW's ways. These contribute to increasing the attractiveness of the SBW, and, among them, we consider also urban areas as qualifying factors for tourism activities, as well as dynamically linked to the context. The points of interest are classified according to the following categories (Figure 5):

(1). Abandoned mining sites (distribution areas and nodes): Villages and mining plants, industrial archeology facilities;

(2). Sites of historical and cultural interest (areas and nodes): Rural and archaeological villages, churches, museums, and historic buildings;

(3). Sites of naturalistic interest (areas and nodes): Natural parks, salt marshes, waterfalls and springs, caves, lakes, beaches, natural monuments;

(4). Urban areas.

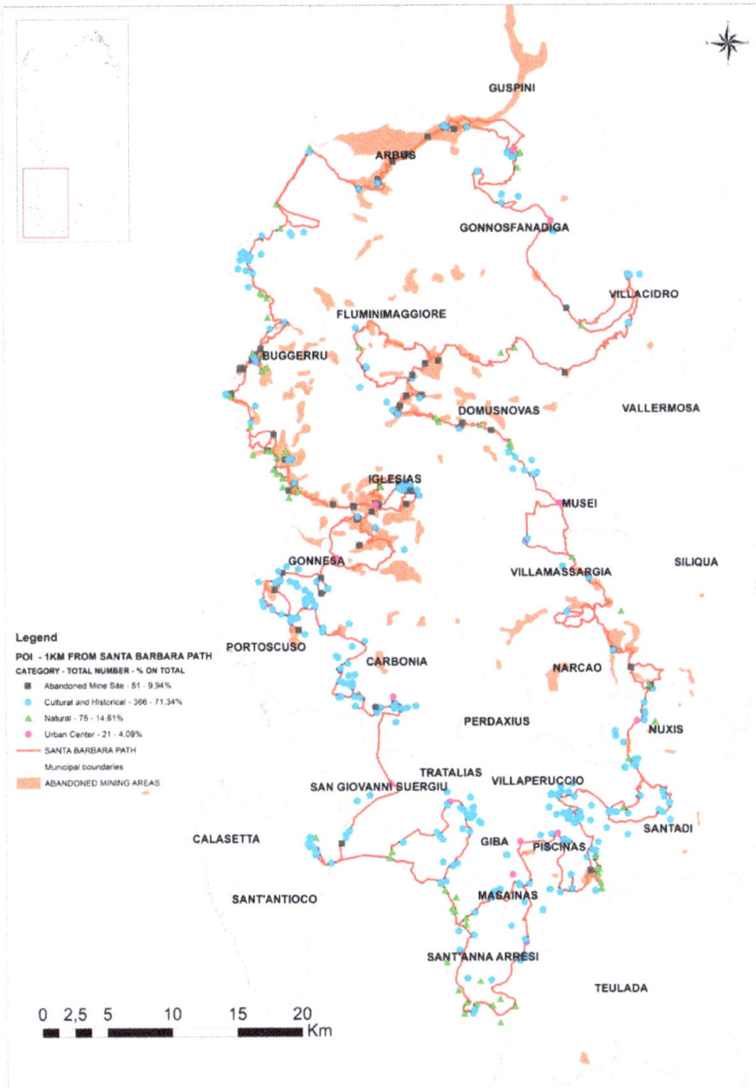

Figure 5. Points of interest (areas and nodes) within the 1 km buffer from the walk.

In organizing the extrinsic features, referred to the points of interest, we used the open source data from the Sardinia Region in shapefile format [69,70]. Then, we proceeded to classify the intrinsic and extrinsic characteristics, assigning specific score indexes referring to an average evaluation of difficulties:

(1). Intrinsic characteristics—a maximum weight has been assigned for the stages of shorter length, difficulty, and travel time.
(2). Extrinsic characteristics—a maximum weight has been assigned for the stages with the greatest number of points of proximity interest, including accommodation facilities and refreshment points.

Ranges and relative weights have been assigned, taking a non-expert user as a reference, also in consideration of the fact that the SBW is a religious path. Santa Barbara is the patron saint of miners, and veneration has always been very strong since ancient times. In general, the Way of Santa Barbara has been organized for a pilgrim who walks between 5 and 20 km per day. The pilgrim along the paths finds, every 5 km, services to eat or sleep (source: SBW Foundation, 2020).

As the Way was born and is addressed to pilgrims, therefore the weight attributed to the intrinsic characteristics increases with decreasing length, difficulty, and travel time. If for some reason you want to propose to expert walkers the weight attributed to the intrinsic characteristics, it will increase with increasing length, difficulty, and travel time. While the weight of extrinsic characteristics will coincide for both categories of walkers.

Below is reported the classification of intrinsic and extrinsic characteristics (Table 1).

Table 1. Score index—Intrinsic and Extrinsic characteristics of each SBW way.

Features of the Ways	Type	Range	Weight
Intrinsic [1]	length (Km)	5–10	3
		10–20	2
		20–30	1
	difficulty	Low	3
		Medium	2
		High	1
	travel time (h)	0–5	2
		5–10	1
Extrinsic [2]	n. of points of interest (buffer 1 km)	0–20	1
		21–40	2
		41–60	3
		61–80	4
	n. of tourist accommodation facilities (buffer 1 km)	0–10	1
		11–20	2
		21–30	3
		31–40	4
	n. of food and beverage points	0–10	1
		11–20	2
		21–30	3
		31–40	4

[1] Data source: https://www.camminominerariodisantabarbara.org/; [2] Data source: http://webgis2.regione.sardegna.it/catalogodati/card.jsp?uuid=ae5a776b-ef08-417b-acf3-752478f0d0b1.

This classification then allowed us to elaborate two indexes for each stage: The walkability index (WI), and the attractiveness index (AI). Within the debate on urban walkability [22,71], according to Blečić et al. (2015) [72], we have rethought the definition of walkability, adapting it to the peculiarities of the SBW, inserted in an extra-urban environment, marked by elements from the former industrial processes of considerable landscape value, together with the physical properties of the path and the reference context. As a prerequisite for the realization of the dashboard, and therefore as a knowledge instrument for managers and planners, we hereby propose a walkability index (WI). It is obtained by summing the weights attributed to the individual intrinsic characteristics (Wi), which can vary from a minimum of 3 to a maximum of 8 (1); and the attractiveness index (AI), understood as the sum of the weights attributed to the individual extrinsic characteristics (We), which can vary from a minimum of 3 to a maximum of 12 (2). To standardize the indexes to the more known and used TripAdvisor classification, WI is given by the product between the summation of Wi and the ratio of 5/8; while AI is given by the product between the summation of Wi and the ratio of 5/12.

$$WI = \frac{5}{8} \sum_{W=3}^{8} Wi \quad (1)$$

$$AI = \frac{5}{12} \sum_{W=3}^{12} We \tag{2}$$

The following is an example of the ways No. 02, 06, 07, and 21, selected for the application of the methodology because they are characterized by a significant landscape heritage and industrial archeology that makes them among the most attractive of the walk (Figure 6).

Track 06 - Piscinas-Montevecchio

	OBJECTIVE DATA TRACK 06	value	weight
INTRINSIC*	lenght (km)	18,8	2
	difficulty	medium	2
	travel time (hours)	6,3	1
	WI =3,1/5 medium		
ESTRINSIC*	point of interest (no.)	16	2
	tourism accomodation facilities (no.)	8	1
	food and beverage points (no)	4	1
	AI = 2/5 medium		

Track 07 - Montevecchio-Arbus

	OBJECTIVE DATA TRACK 07	value	weight
INTRINSIC*	lenght (km)	17,8	2
	difficulty	medium	2
	travel time (hours)	6	1
	WI =3,1/5 medium		
ESTRINSIC*	point of interest (no.)	23	3
	tourism accomodation facilities (no.)	20	2
	food and beverage points (no)	19	2
	AI = 2,9/5 medium		

Track 02 - Nebida-Masua

	OBJECTIVE DATA TRACK 02	value	weight
INTRINSIC*	lenght (km)	10,1	2
	difficulty	medium	2
	travel time (hours)	3,3	2
	WI =3,7/5 high		
ESTRINSIC*	point of interest (no.)	20	1
	tourism accomodation facilities (no.)	3	1
	food and beverage points (no)	1	1
	AI = 1,2/5 low		

Range WI: 0-1,6 = low; 1,6-3,3 = medium; 3,3-5 = high

Track 21 - S. Antioco-Carbonia

	OBJECTIVE DATA TRACK 21	value	weight
INTRINSIC*	lenght (km)	22,5	1
	difficulty	medium	2
	travel time (hours)	6	1
	WI =2,5/5 medium		
ESTRINSIC*	point of interest (no.)	45	3
	tourism accomodation facilities (no.)	40	4
	food and beverage points (no)	34	4
	AI = 4,6/5 high		

Range AI: 0-1,6 = low; 1,6-3,3 = medium; 3,3-5 = high

Figure 6. Walkability index (WI) and Attractiveness index (AI) of the No. 02, 06, 07, and 21 SBW ways.

3.2. *Survey and Analysis of the Data Provided by Social Network*

The second component of data necessary for the development of the dashboard concerns the user-generated characteristics of the Walk, provided by the smart community. They were found through the main social websites such as Facebook, Instagram, and TripAdvisor, for the main point of interest of the 24 ways, with reference to the abandoned mining sites. In fact, from the analysis of GPS ways related to walking and/or cycling routes downloaded from the Wikiloc platform (SBW's ways have been downloaded between 20 and 29 January 2018; bike tracks have been downloaded between 21 January and 3 February 2019) [28–30], these sites were found to be the most frequented by the smart community [31], as is shown from Figure 7. Furthermore, the disused mining sites were selected as a survey category to evaluate the approval rating of the smart community, which required the following analysis:

(1). Active Websites for each abandoned mining site (Figure 8);

(2). Profiles in the main social web (Facebook, Instagram, and TripAdvisor) (Figures 9–11).

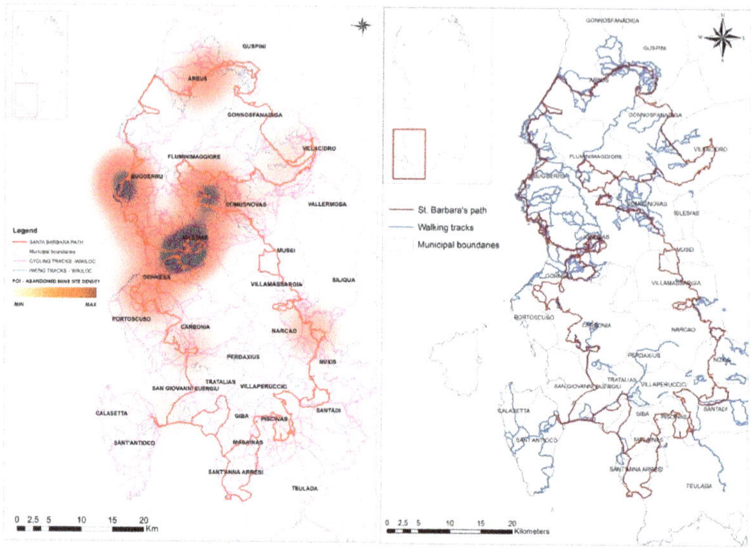

Figure 7. Comparison between the concentration of disused mining sites and the location of digital tracks (source Wikiloc) from the smart community.

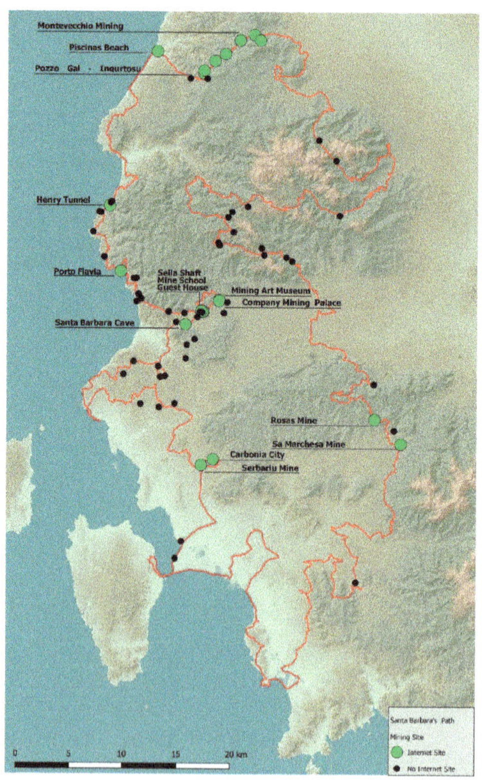

Figure 8. Web Site Map of the mining sites in the SBW. Period of investigation: 1–31 July 2019. Author: Michele Pinna.

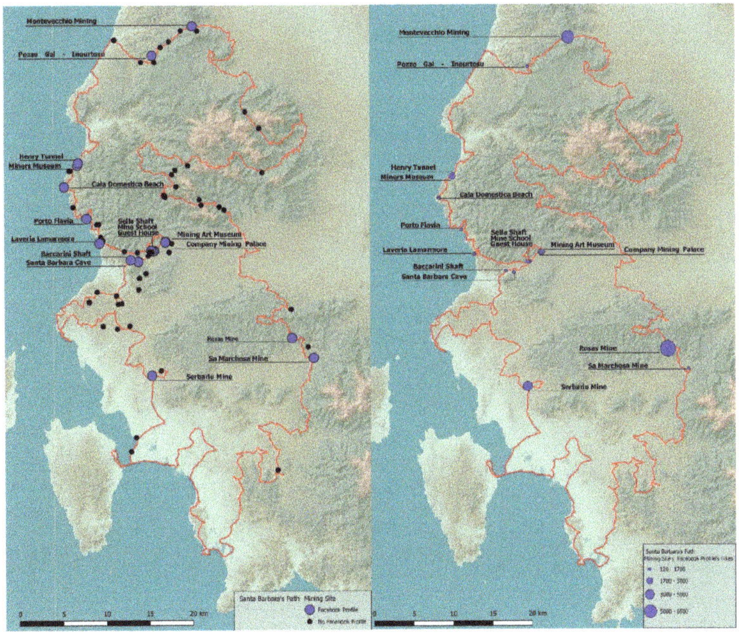

Figure 9. Mining sites with Facebook profile (**left**) and Facebook Social Rating (**right**). Period of investigation: 1–31 July 2019. Author: Michele Pinna

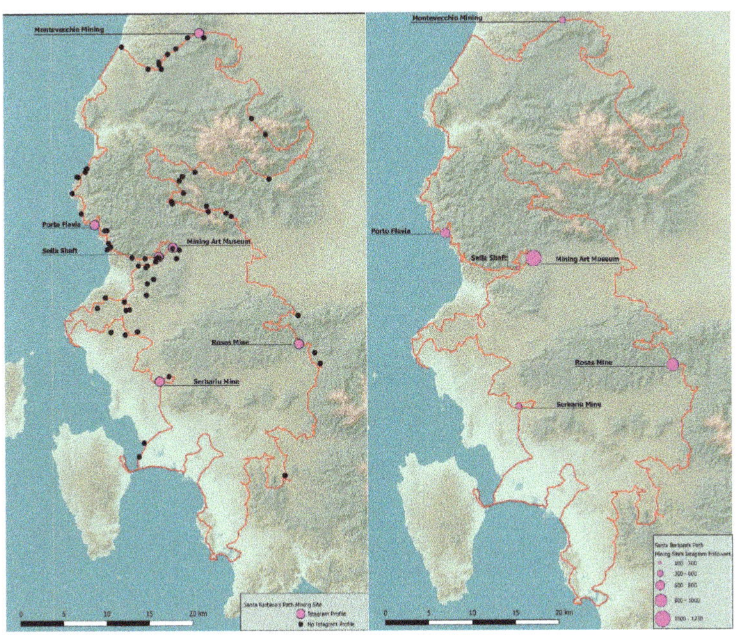

Figure 10. Mining sites with Instagram profile (**left**) and Instagram followers (**right**). Period of investigation: 1–31 July 2019. Author: Michele Pinna.

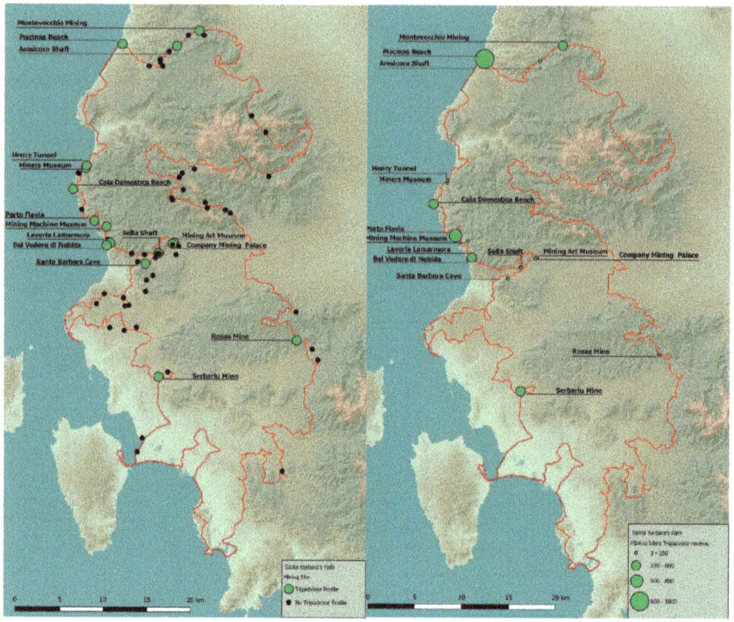

Figure 11. Mining sites with a TripAdvisor profile (**left**) and TripAdvisor Social Rating (**right**). Period of investigation: 1–31 July 2019. Author: Michele Pinna.

From the analysis, it emerged that only a minimal part of the abandoned mining sites is equipped with an official web page and a Facebook and/or Instagram profile to support the management and promotion activity (Figures 8–10).

In particular, it can be stated that the most important mining sites have a web page and social profiles, while in some cases the social profile is preferred to the website. The analysis confirms that the most significant mining complexes, such as Montevecchio and Serbariu, collect the greatest number of likes on Facebook (Figure 9, right), while the Museum of Mining Art in Iglesias reaches the greatest number of followers on Instagram, followed by the Mine Rosas (Figure 10, right).

The TripAdvisor platform is the main one currently in use at international level. It also provides the satisfaction index associated with the type of user according to its own classification (Figure 11 and Table 2). In this case, Piscinas and Porto Flavia are the sites that have collected the highest number of reviews, and therefore are also the most visited ones.

An analysis of the data provided implied the organization of the data from social web platforms in a matrix for each individual abandoned mining site. However, here we report only the matrix of social media analysis of the Montevecchio Mine Complex, as it is one of the most appreciated and best managed (Table 2). In fact, the Complex, in addition to having a well-structured and content-rich website, also has Facebook and Instagram profiles, as well as reviews on TripAdvisor. Montevecchio has a high rating on both Facebook (4.6/5) and TripAdvisor (4.5/5). The data provided by TripAdvisor shows that the site is visited mainly during the summer (June-August), and mainly by couples.

All these data reveal the state-of-the-art in terms of tourist promotion activities carried out by each mining site. At the same time, they are useful to inform future management and promote policies of the several points of interest. In particular, while Instagram informs about the number of followers, Facebook also provides the rating manifested by the smart community, along with comments and opinions, which reveal strengths and weaknesses of each point of interest, and therefore guide specific actions. Going beyond the data provided by social networks, Tripadvisor describes each site in terms of community satisfaction, type of visitors, and the visiting period. These data inform, in different cases, about the types of visitors attracted, as well as encourage their presence throughout the year in order to guarantee an extension of the seasonal tourism [29].

Table 2. Matrix of the data provided by the social network (Facebook, Instagram, TripAdvisor) for the Montevecchio Mine Complex. Period of investigation: 1–31 May 2019. Author: Michele Pinna.

MONTEVECCHIO www.minieramontevecchio.it				
Social Web	**Type of Data**		**No.**	**Download Date**
Facebook		Like	5891	May 2019
		Follower	6013	
		Review	200	
		Rating	4.6/5	
Instagram		Post	56	May 2019
		Follower	423	
		No. profile	49	
TripAdvisor (Period 2014–May 2019)	Valuation	No. review	360	May 2019
		Rating	4.5/5	
	Type of visitor	Family	64	
		Couple	135	
		Group	66	
		Single	5	
		Businessman	1	
	No. visitors/trimester	March-May	81	
		June-August	149	
		September-November	74	
		December-February	16	

3.3. The SBW Dashboard Layout

The classification of the intrinsic and extrinsic characteristics and user generated data of the SBW, as well as the definition of the walkability index (WI) and attractiveness index (AI) for each way, allowed us to develop the design of the SBW circular dashboard. After having framed the recent literature about data, techniques, and platforms of existing Dashboards [73] and Big data dashboards as decision support tools [74], we decided to use the open source Google Sites tools as a framework for developing the dashboard as compatible with the SBW Foundation website, so that the project presented here can be easily implemented and developed in the future. Moreover, we have considered the digital platform dedicated to the Regional Bicycle Mobility Plan of Sardinia [75], from which we extracted the tracks of the walk organized in biking tracks; furthermore, we selected the SBW Foundation website [32] for the walk and horse-riding tracks. The dashboard framework proposed (Figure 3) [76] is thought of as a collector capable of organizing a comprehensive data-set we elaborated and collected through the main social networks, as discussed in this study. It is not yet fully integrated and linked to the existing Foundation website. In the present stage of the research, we did focus on the design of a dashboard, that will support both potential tourists and planners in future decisions for the area and for the Region of Sulcis-Iglesiente.

So, from the design and style point of view, we decided to maintain the general layout of the existing website, in order to allow, once the dashboard is implemented, an easy integration with the existing material. In our vision, the dashboard consists of a homepage of general information of the SBW, and of several specific web pages dedicated to the characteristics of each of the 24 ways. In the figures, we present the dashboard's proposed structure and a mock-up of the set of data and information for one selected segment of the walk as a demo of the possible contents to be collected, gathered, and shared. The homepage (Figure 12) allows users to select each track and go to the next page.

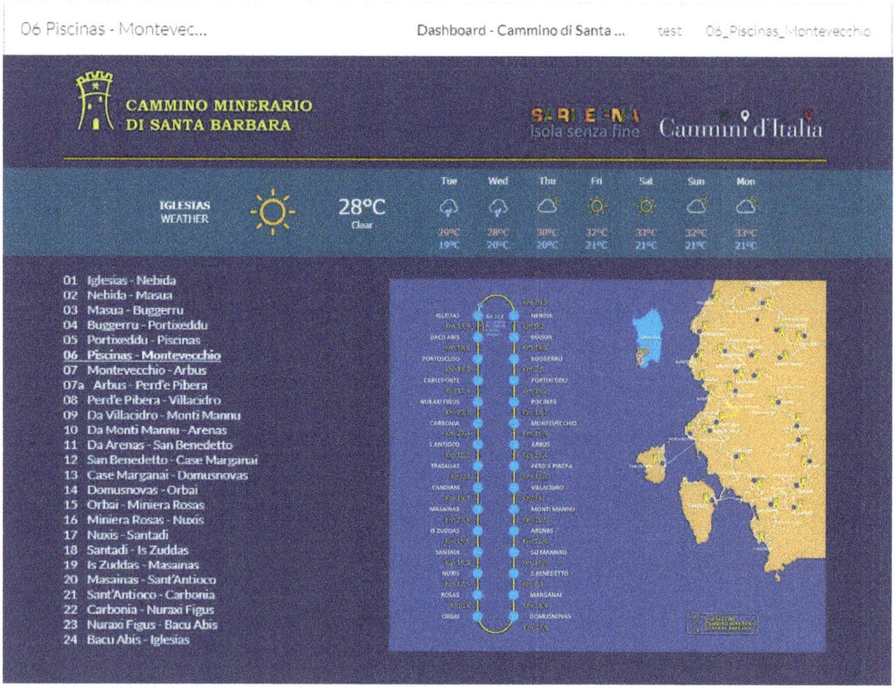

Figure 12. The SBW Dashboard layout.

The web page dedicated to the track No. 06 Piscinas-Montevecchio (Figure 13), provides four types of geospatial information:

(1). Ways and points of interest georeferenced in Google Maps (down on the left);
(2). Physical profile of the way, and its Walkability Index (WI) and Attractiveness Index (AI) (on the top);
(3). Rating provided by the main social networks analyzed (down on the right).

This will be implemented by means of RESTful API's as made available by the main social media portals:
Facebook—https://developers.facebook.com/docs
Instagram—https://www.instagram.com/developer/
TripAdvisor—https://developer-tripadvisor.com/content-api/documentation/
Instagram does not have ratings, but it has Likes and Followers. Real time photos can be shared on the profile.
The construction of the dashboard was supported by Iulia Tatomir, Sr. Analyst at Avanade Cagliari, Italy—https://www.avanade.com/en-gb

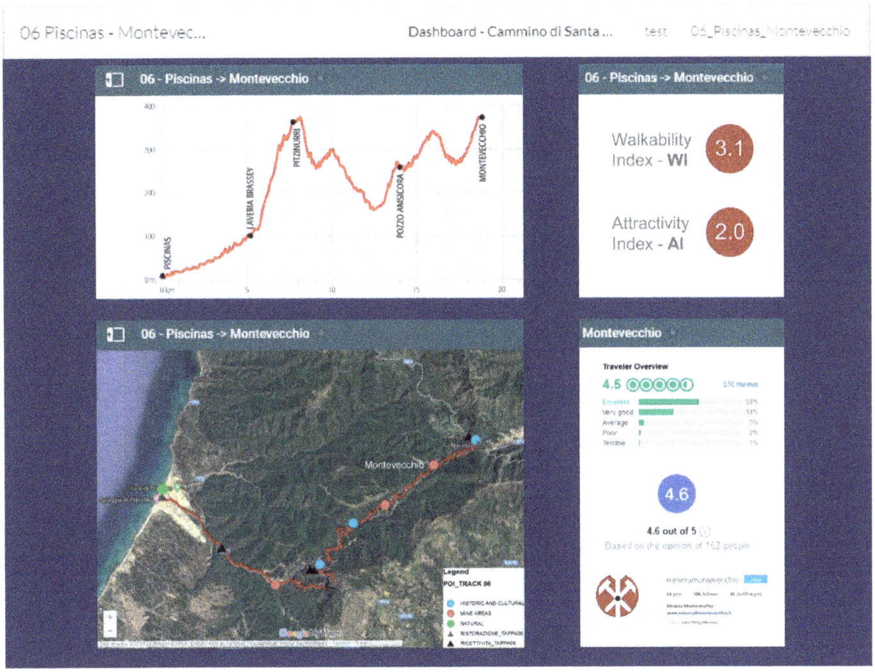

Figure 13. The web page of the way No. 06 Piscinas–Montevecchio.

The system here proposed can be defined as a smart dashboard because it is structured so to provide a constant update on performance indicators, with reference to dynamic information such as the weather, the Attractiveness Index (AI), and the rating provided by the main social networks. This is also a circular dashboard, as it can provide support to public institutions to observe and monitor what happens along the Walk, and the data provided are fundamental to inform future policies.

4. Results and Discussion

The main result of the present paper is the set-up of the organization of information contents on the Santa Barbara Walk—green infrastructure in the Sulcis–Iglesiente (Sardinia, Italy). Such a set of information and data are relevant for allowing tourists and decision makers to gain knowledge on the area and its characteristics: For the former to understand and discover the potentials of such a particular area; for the latter to plan actions for the management of the Walk and targeting initiatives. The idea of proposing a tool for sharing such an amount and kind of information led us to designing a dashboard for management and planning of the tourist offer of the area. Such research is inserted into the "TSulki" project (Tourism and Sustainability in Sulcis-Iglesiente area), currently in progress, regarding the sustainable tourism of the Sulcis-Iglesiente area.

A first result obtained deals with the survey and summary of the characteristics of the area, also in terms of slow tourism and smart tourism. Another result is in terms of gaining a wider awareness and knowledge on some of the characteristics of the SBW and its routes. We developed indexes to better evaluate trunks of the Walk, both for the planner's benefit and for users, these indexes providing qualitative and quantitative attributes for the same Walk.

The indexes deal with a WI—Walkability Index, based on intrinsic characteristics (as length, difficulty, travel time)—and AI—Attractiveness Index, based on extrinsic characteristics (points of interest, tourism facilities, food and beverage points)—that represent the context of SBW. In addition, the degree of interest was based on the comments and opinions from the community of users for each track coming from the main social networks—Instagram's followers, and Facebook's and TripAdvisor's ratings. Finally, the design of the SBW Dashboard was organized, with both users and decision makers in mind. The results obtained are in terms of a wider knowledge of the SBW, in terms of its intrinsic and extrinsic characteristics, as well as in terms of the perception of a particular subset of the users, as those leaving traces over the social networks and media. The analytical synthesis of these elements represents valuable contents for decision makers to target policies of slow tourism. These elements represent the main data to be summarized and used into the dashboard.

From the design point of view, as Table 3 summarizes, the dashboard is structured to be fed by a combination of data and information. Some main components deal with the fundamental and characterizing elements of the dashboard itself. As suggested above, the WI and AI indexes represent one of the important sources of added value of the project. Their presence at the present stage of development is thought of as a periodical update from the analysts and scholars involved in the SBW project, in the process of constant observation, maintenance, and update of the SBW. The proposed dashboard provides innovative contents and services which cannot be found in existing websites for the promotion of Paths through Italy, also in considering the digital Atlas of Paths developed by the Ministry of Cultural Heritage and Activities (MIBACT). The SBW dashboard has been conceived as a tool capable of collecting, monitoring, and sharing dynamic data such as Walkability Index (WI), Attractiveness Index (AI), and Rating, by direct connections with the main social networks.

Table 3. SBW Dashboard frame components (authors' elaboration).

Contents	Connectors	Main	Extendable
Analyzed	Periodical expert user update	WI; AI	Periodical Reports on SBW
User generated	API	Social media and networks ratings	Bot; social walls; etc.
Web generated	API & Frames	SBW Website	Weather forecast Civil protection information; tourist information

As a matter of fact, further data after a synthetic extraction through API from social network and media contents, can be assigned to the different ways of the SBW. A further level of data input comes from different websites linkable to the dashboard itself: The SBW websites; weather forecast widgets, etc. An extendable set of sources can be thought of in the further evolutions of the dashboards: Further reports generated by experts; other social networks, and media contents and synthesis; links to external websites for information on tourism.

5. Conclusions and Future Development

This paper had, as a main purpose, that of designing a dashboard for the promotion and management of slow tourism, enhancing the potentials of smart tourism in the Santa Barbara Walk (SBW) in the Sulcis–Iglesiente area (Sardinia, Italy).

In particular, we have systematized intrinsic and extrinsic information of the Santa Barbara Walk and developed a series of indices (walkability index and attractiveness index) aimed at increasing awareness and knowledge useful for slow tourism referring to a particular area.

We consider the unique combination of natural and past anthropogenic features related to mining activities as a unique set of elements, making the area an interesting example of landscape in transition, where the combination and interaction of both elements represent a strong element of characterization and attraction of the area for already existing and potential, future tourists. In such a sense, the area is considered as interesting in terms of slow tourism characterization and exploitable by means of smart tourism, whose embryonal elements are already present, in terms of social networks and sharing of information and knowledge among users by means of ICT instruments. That allowed us to obtain an image of the SBW as a green infrastructure, perfectly integrated into the surrounding area. It emerged that it is characterized not only by its physical features, but also by a set of several points of interest in its proximity—those located within a 1 km buffer from the different segments of the walk. Such first results have been expressed in terms of indexes and maps, helping a description and understanding the new slow tourism and the potentials of further development in the area. Such preliminary phase represented a starting point for developing the research methodology, consisting of three main phases.

The first phase was dedicated to the collection, analysis, and classification of the intrinsic and extrinsic characteristics of the walk that allowed developing a Walkability Index (WI) and an Attractiveness Index (AI) for each of the 24 ways. Both of these two indexes are considered useful to describe the different parts of the walk and, therefore, to guide the users' choice.

The second phase was dedicated to the collection, analysis, and classification of the user-generated characteristics of the points of interest located along the SBW. These data, provided by the SBW's users through the main social networks (Facebook, Instagram, and TripAdvisor), reveal the state-of-the-art in terms of tourist promotion activities.

The third phase appeared as that dedicated to the systematization of the web info data and the team data analysis to develop a design and a user-friendly graphical interface for the SBW circular smart dashboard. The analyses carried out revealed that a smart community is rising, from the SBW's users, contributing significantly to the affirmation of a new model of sustainable development based on slow tourism. Awareness and knowledge of history are fundamental to the re-construction of the past. Slow tourism is softly included in the Sulcis-Iglesiente area, free from compulsive mass tourism. The future developments of this study will cover the implementation of the dashboard, also for aspects linked to the safety of the place and the active role of the community of users, to be involved more and more in a bottom up approach in a smart community.

Such initial setup of the characters of the area, of its material and immaterial networks, as well as the designing of the dashboard, put the basis for the future research and actual development of the dashboard itself. This implies the validation of the data and information to be put into the same dashboard. In particular, we will submit the dashboard to a selected group of users, expert and non-expert users, looking forward to its inclusion on the SBW Foundation website and, therefore, usable for the public consultation. We will then proceed with the physical realization of the fully operational dashboard, taking care of the automation of the process of data collection and gathering, particularly of those coming from the social networks and media platforms involved, as well as of those feeding the indicators considered in the dashboard itself. In this sense, the proposed dashboard, including subsequent advances and possible implementations, represents a strategic tool to support the governance and the policies to promote slow tourism in the territory, in order to overcome the deep crisis left by the end of mining activities. The dashboard designed and presented here aims at representing a data, information, and report integrator, facilitating the decision process for both users as tourists and sport practitioners, as they can cross data concerning the quality and characteristics of the ways, the environmental conditions, as well as the comments from the other members of the community of users. From the policy maker's point of view, the dashboard is useful in providing such information in a dynamic point of view. The frequent reports on the AI and WI indexes, as well as the constant comments and evaluations through social networks and media, will help acquiring a major awareness of the area and of its preferred parts, as well as to highlights those parts where an intervention—of maintenance, major coordination between operators, municipality agreements, etc.—is needed.

The main challenge of this research work lies in the effective collaboration between the different stakeholders involved in the dashboard construction process. Institutions, associations, cooperatives, economic operators, and users are the main players involved in the collection of data from the territory and, consequently, in the definition of the development policies, in line with a sustainable strategic vision of the territory and the local communities' needs. In fact, as asserted in the introduction, in the area of the Sulcis-Iglesiente, the points of historical-cultural and naturalistic interest, mining sites, reception, and refreshment activities are not managed within a single network. The first attempt to address this condition has been the creation of the website of the SBW Foundation.

Author Contributions: The paper derives from the joint reflections of authors. However, Sections 1 and 1.1 can be attributed to G.B. (Ginevra Balletto); Section 2.1 to G.B. (Ginevra Balletto); Section 2.2 to A.M.; Section 3 to G.B. (Ginevra Balletto); Section 3.1 to M.L.; Section 3.2 to A.M. and M.L.; Sections 3.3 and 4 to G.B. (Ginevra Balletto), and Section 5 to G.B. (Giuseppe Borruso). All authors have read and agreed to the published version of the manuscript.

Funding: This research was funded by Foundation of Sardinia (Grant CUP F72F16003160002)—Research project RE-MINE—Restoration and rehabilitation of abandoned mining sites, and by Region of Sardinia, Fundamental or basic research projects for implementation of interventions in the field of research for the 'Sulcis Plan' (SULCIS-821319)—Research project TSULKI—Tourism and Sustainability in the Sulcis (Sardinia- Italy).

Conflicts of Interest: The authors declare no conflict of interest.

References

1. Kothari, T.; Xiang, Z.; Fesenmaier, D.R. The impact of a marketing information system: A case study of Smart-Baltimore. *Inf. Technol. Tour.* **2008**, *10*, 59–74. [CrossRef]
2. Battino, S.; Balletto, G.; Borruso, G.; Donato, C. Internal Areas and Smart Tourism. Promoting Territories in Sardinia Island. In Proceedings of the International Conference on Computational Science and Its Applications, Melbourne, Australia, 2–5 July 2018; Springer: Berlin/Heidelberg, Germany, 2018; pp. 44–57.
3. Buhalis, D. Technology in tourism-from information communication technologies to eTourism and smart tourism towards ambient intelligence tourism: A perspective article. *Tour. Rev.* **2019**, *75*, 267–272. [CrossRef]
4. Wan, C.B. Flourishing through smart tourism: Experience patterns for co-designing technology-mediated traveller experiences. *Des. J.* **2018**, *21*, 163–172. [CrossRef]
5. De Lange, M. From real-time city to asynchronicity: Exploring the real-time smart city dashboard. In *Time for Mapping*; Manchester University Press: Manchester, UK, 2018; pp. 238–255.
6. Balletto, G.; Borruso, G.; Donato, C. City Dashboards and the Achilles' Heel of Smart Cities: Putting Governance in Action and in Space. In *Computational Science and Its Applications–ICCSA 2018*; Gervasi, O., Murgante, B., Misra, S., Stankova, E.N., Torre, C.M., Rocha, A.M.A.C., Taniar, D., Apduhan, B.O., Tarantino, E., Ryu, Y., Eds.; Springer: Berlin/Heidelberg, Germany, 2018; pp. 654–668.
7. Portney, K.E. *Taking Sustainable Cities Seriously: Economic Development, the Environment, and Quality of Life in American Cities*; MIT Press: Cambridge, MA, USA, 2013.
8. Gullino, S. Urban Regeneration and Democratization of Information Access: CitiStat Experience in Baltimore. *J. Environ. Manag.* **2009**, *90*, 2012–2019. [CrossRef]
9. Kitchin, R.; Lauriault, T.P.; McArdle, G. Knowing and governing cities through urban indicators, city benchmarking and real-time dashboards. *Reg. Stud. Reg. Sci.* **2015**, *2*, 6–28. [CrossRef]
10. Batty, M. A perspective on city dashboards. *Reg. Stud. Reg. Sci.* **2015**, *2*, 29–32. [CrossRef]
11. Dameri, R.P. Urban smart dashboard. Measuring smart city performance. In *Smart City Implementation*; Dameri, R.P., Ed.; Springer: Berlin/Heidelberg, Germany, 2017; pp. 67–84.
12. Albusaidi, H.S.; Udupi, P.K.; Dattana, V. Integrated data analytic tourism dashboard (IDATD). In Proceedings of the 2016 5th International Conference on Reliability, Infocom Technologies and Optimization(Trends and Future Directions) (ICRITO), Noida, India, 7–9 September 2016; pp. 497–500.
13. Becken, S.; Miller, G. *Global Sustainable Tourism Dashboard*; Griffith University: Mount Gravatt, Australia, 2016.
14. California Institute for Smart Communities. *Smart Communities Guide Book*; California Institute for Smart Communities: San Diego, CA, USA, 2001.
15. Nisar, T.M.; Prabhakar, G.; Strakova, L. Social media information benefits, knowledge management and smart organizations. *J. Bus. Res.* **2019**, *94*, 264–272. [CrossRef]
16. Ladu, M.; Balletto, G.; Borruso, G. Sport and Smart Communities. Assessing the Sporting Attractiveness and Community Perceptions of Cagliari (Sardinia, Italy). In Proceedings of the International Conference on Computational Science and Its Applications, Saint Petersburg, Russi, 1–4 July 2019; Springer: Berlin/Heidelberg, Germany, 2019; pp. 200–215.
17. Barnes, S.J.; Mattsson, J.; Sørensen, F.; Friis Jensen, J. The Mediating Effect of Experiential Value on Tourist Outcomes from Encounter-Based Experiences. *J. Travel Res.* **2019**. [CrossRef]

18. Romano, F.; La Rocca, R.A. Slow Mobility and Cultural Tourism. Walking on Historical Paths. In *Smart Planning: Sustainability and Mobility in the Age of Change*; Springer: Berlin/Heidelberg, Germany, 2018; pp. 301–322.
19. Calzati, V. *Nuove Pratiche Turistiche e Slow Tourism. Il Caso Della Valnerina in Umbria*; Franco Angeli: Milan, Italy, 2016.
20. Agarwal, S.; Busby, G.; Huang, R. Slow tourism. In *Special Interest Tourism: Concepts, Contexts and Cases*; CABI: Wallingford, UK, 2017; pp. 183–195.
21. Tomić, S.; Leković, K.; Stoiljković, A. Impact of motives on outcomes of the travel: Slow tourism concept. *Školabiznisa* **2018**, *2*, 68–82. [CrossRef]
22. Beretić, N.; Plaisant, A. Setting the Methodological Framework for Accessibility in Geo-Mining Heritage Settings—An Ongoing Study of Iglesiente Area (Sardinia, Italy). *Sustainability* **2019**, *11*, 3556.
23. Beretić, N.; Cecchini, A.; Đukanović, Z. Geotourism as a Development Tool of the Geo-mining Park in Sardinia. *Geoheritage* **2019**, *11*, 1689–1704. [CrossRef]
24. Jelen, J. Mining Heritage and Mining Tourism. *Czech J. Tour.* **2018**, *7*, 93–105. [CrossRef]
25. Menne, A. *La Via Francigena–Tanta Vita–A piedi da Milano a Roma*; Edizioni Menne Antonino: Milan, Italy, 2019.
26. Perelli, G.; Sistu, G. Abitare il tempo... libero. Il sogno probabile del turismo minerario in Sardegna. In *Dove Finisceil Mare*; Iorio, M., Sistu, G., Eds.; Sandhi Editore: Cagliari, Italy, 2010; pp. 361–374.
27. Pinna, G. *Il Cammino Minerario di Santa Barbara.A Piedi in Sardegna tra Storia e Natura*; Terre di Mezzo: Milan, Italy, 2017.
28. Balletto, G.; Milesi, A.; Battino, S.; Borruso, G.; Mundula, L. Slow tourism and smart community. The case of Sulcis -Iglesiente (Sardinia -Italy). In *Computational Science and Its Applications-ICCSA 2019*; Misra, S., Gervasi, O., Murgante, B., Stankova, E., Korkhov, V., Torre, C., Rocha, A.M., Taniar, D., Apduhan, B., Tarantino, E., Eds.; Springer: Berlin/Heidelberg, Germany, 2019; pp. 184–199. ISBN 978-3-030-24310-4.
29. Balletto, G.; Milesi, A.; Mundula, L.; Borruso, G. Wave, walk and bike tourism. The case of Sulcis (Sardinia-Italy). In *Planning, Nature and Ecosystem Services*; Gargiulo, C., Zoppi, C., Eds.; FedOAPress: Naples, Italy, 2019; pp. 881–892.
30. Balletto, G.; Milesi, A.; Naitza, S.; Mundula, L.; Borruso, G. Smart community and landscape in progress. The case of the Santa Barbara walk (Sulcis, Sardinia). In *Planning, Nature and Ecosystem Services*; Gargiulo, C., Zoppi, C., Eds.; FedOAPress: Naples, Italy, 2019; pp. 893–903.
31. Dickinson, J.; Lumsdon, L. *Slow Travel and Tourism*; Earthscan: London, UK, 2010.
32. Cammino Minerario di Santa Barbara Web Site. Available online: https://www.camminominerariodisantabarbara.org/ (accessed on 15 September 2019).
33. WTO COVID-19: PUTTING PEOPLE FIRST, World Tourism Organization. Available online: https://www.unwto.org/tourism-covid-19-coronavirus (accessed on 25 March 2020).
34. Burke, M.; Auffhammer, M.; Burney, J.; Hsiang, S.; Lobell, D.; Roberts, M.; Schlenker, W. COVID-19 Reduces Economic Activity, Which Reduces Pollution, Which Saves Lives. G-FEED (Global Food, Environment and Economic Dynamics). Available online: http://www.g-feed.com/2020/03/covid-19-reduces-economic-activity.html (accessed on 25 March 2020).
35. Li, R.; Pei, S.; Chen, B.; Song, Y.; Zhang, T.; Yang, W.; Shaman, J. Substantial Undocumented Infection Facilitates the Rapid Dissemination of Novel Coronavirus (SARS-CoV2). *Science* **2020**. [CrossRef] [PubMed]
36. Setti, L. SIMA, Università di Bologna e Bari. *Relazione Circa l'effetto Dell'inquinamento da Particolato Atmosferico e la Diffusione di Virus Nella Popolazione.* Available online: http://www.simaonlus.it/wpsima/wp-content/uploads/2020/03/COVID19_Position-Paper_Relazione-circa-l%E2%80%99effetto-dell%E2%80%99inquinamento-da-particolato-atmosferico-e-la-diffusione-di-virus-nella-popolazione.pdf (accessed on 25 March 2020).
37. Roussel, I. Extreme Weather Events and Air Pollution Peaks in the Light of Climate Change: The Limits of the Notion of Risk. In *Extreme Weather Events and Human Health*; Springer: Berlin/Heidelberg, Germany, 2020; pp. 59–78.
38. Jacob, D.J.; Winner, D.A. Effect of climate change on air quality. *Atmos. Environ.* **2009**, *43*, 51–63. [CrossRef]
39. Setti, A.F.F. Climate Change, Human Health, and Sustainable Development. *Clim. Action* **2019**, *2020*, 253–263.

40. Kienast, F.; Helfenstein, J.; Grêt-Regamey, A.; Haines-Young, R.; Potschin, M. Ecosystem Services under Pressure. In *Landscape Planning with Ecosystem Services*; Springer: Berlin/Heidelberg, Germany, 2019; pp. 91–101.
41. Blum, J. Contribution of ecosystem services to air quality and climate change mitigation policies: The case of urban forests in Barcelona, Spain. In *Urban Forests*; Apple Academic Press: Palm Bay, FL, USA, 2017; pp. 21–54.
42. Bolund, P.; Hunhammar, S. Ecosystem services in urban areas. *Ecol. Econ.* **1999**, *29*, 293–301. [CrossRef]
43. Saarinen, J. Communities and sustainable tourism development: Community impacts and local benefit creation in tourism. In *A Research Agenda for Sustainable Tourism*; Edward Elgar Publishing: Cheltenham, UK, 2019.
44. Wu, D.; Shen, C.; Wang, E.; Hou, Y.; Yang, J. Impact of the Perceived Authenticity of Heritage Sites on Subjective Well-Being: A Study of the Mediating Role of Place Attachment and Satisfaction. *Sustainability* **2019**, *11*, 6148. [CrossRef]
45. De Salvo, P.; Calzati, V.; Soglia, S. Value for Time: Slowness, a Positive Way of Performing Tourism. In *Best Practices in Hospitality and Tourism Marketing and Management*; Springer: Cham, Switzerland, 2019; pp. 315–336.
46. Shang, W.; Qiao, G.; Chen, N. Tourist experience of slow tourism: From authenticity to place attachment–a mixed-method study based on the case of slow city in China. *Asia Pac. J. Tour. Res.* **2020**, *25*, 170–188. [CrossRef]
47. Flemming, S. The Geographies of Social Networks and Innovation in Tourism. *Tour. Geogr.* **2017**, *9*, 22–48.
48. Battino, S.; Lampreu, S. The Role of the Sharing Economy for a Sustainable and Innovative Development of Rural Areas: A Case Study in Sardinia (Italy). *Sustainability* **2019**, *11*, 3004. [CrossRef]
49. Buhalis, D.; Amaranggana, A. Smart Tourism Destination. In *Information and Communication Technologies in Tourism 2014*; Xiang, Z., Tussyadiah, I., Eds.; Springer: Berlin/Heidelberg, Germany, 2013; pp. 553–564.
50. Cohen, E. Authenticity and commoditization in tourism. *Ann. Tour. Res.* **1988**, *15*, 371–386. [CrossRef]
51. Weidinger, T. Encountering local inhabitants' perspectives in terms of authenticity: The example of rural tourism in Southern Germany. *DosAlgarves* **2015**, *25*, 4–24.
52. Costanza, R.; D'Arge, R.; De Groot, R.; Farber, S.; Grasso, M.; Hannon, B.; Limburg, K.; Naeem, S.; O'Neill, R.V.; Paruelo, J.; et al. The value of the world's ecosystem services and natural capital. *Nature* **1997**, *387*, 253–260. [CrossRef]
53. Leone, F.; Zoppi, C. Ecosystem services as external drivers in the Strategic environmental assessment of management plans of the sites of the Natura 2000 network. *Urban. Inf.* **2015**, *42*, 34–38.
54. Lai, S.; Leone, F.; Zoppi, C. Implementing Green Infrastructures beyond Protected Areas. *Sustainability* **2018**, *10*, 3544. [CrossRef]
55. Leone, F.; Zoppi, C. Local Development and Protection of Nature in Coastal Zones: A Planning Study for the Sulcis Area (Sardinia, Italy). *Sustainability* **2019**, *11*, 5095. [CrossRef]
56. Fetzer, T.; Hensel, L.; Hermle, J.; Roth, C. Coronavirus Perceptions and Economic Anxiety. Available online: https://arxiv.org/pdf/2003.03848.pdf (accessed on 25 April 2020).
57. Murgante, B.; Borruso, G. Smart cities or dumb cities? Smart communities, city dashboard e social network analysis. *GEOmedia* **2015**, *18*, 994.
58. Cannas, I.; Lai, S.; Leone, F.; Zoppi, C. Green infrastructure and ecological corridors: A regional study concerning Sardinia. *Sustainability* **2018**, *10*, 1265. [CrossRef]
59. European Union. Available online: https://eur-lex.europa.eu/legal-content/IT/TXT/PDF/?uri=CELEX:52019DC0236&qid=1562053537296- (accessed on 25 March 2020).
60. Estreguil, C.; Dige, G.; Kleeschulte, S.; Carrao, H.; Raynal, J.; Teller, A. *Strategic Green Infrastructure and Ecosystem Restoration: Geospatial Methods, Data and Tools*; EUR 29449 EN; Publications Office of the European Union: Brussels, Belgium, 2019.
61. Turrell, E.C. Taking Time: Craft and the Slow Revolution. Available online: http://takingtime.org/ (accessed on 25 March 2020).
62. Çiçek, M.; Ulu, S.; Uslay, C. The Impact of the Slow City Movement on Place Authenticity, Entrepreneurial Opportunity, and Economic Development. *J. Macromark.* **2019**, *39*, 400–414.
63. Serdane, Z.; Maccarrone-Eaglen, A.; Sharifi, S. Conceptualising slow tourism: A perspective from Latvia. *Tour. Recreat. Res.* **2020**, 1–14. [CrossRef]

64. European Route of Industrial Heritage. Available online: https://www.erih.net/fileadmin/Mediendatenbank/Downloads/Membership-Forms_en/ERIH_Membership_Brochure.pdf (accessed on 26 February 2020).
65. Sardinia Region Web Site. Available online: http://dati.regione.sardegna.it/dataset/movimenti-turistici-in-sardegna-2018-per-comune (accessed on 25 March 2020).
66. Parco Geominerario Web Site. Available online: http://www.parcogeominerario.eu/ (accessed on 15 September 2019).
67. UNESCO Web Site. Available online: http://www.unesco.org/new/en/natural-sciences/environment/earth-sciences/ (accessed on 15 September 2019).
68. Ministero Delle Politiche Agricole Alimentari Forestali e del Turismo Web Site. Available online: https://www.turismo.politicheagricole.it/home-cammini-ditalia/ (accessed on 15 September 2019).
69. Sardinia Region Web Site. Available online: http://webgis2.regione.sardegna.it (accessed on 15 September 2019).
70. Sardinia Region Web Site. Available online: http://dati.regione.sardegna.it/dataset (accessed on 15 September 2019).
71. Congiu, T.; Blečić, I.; Cecchini, A.; Fancello, G.; Talu, V.; Trunfio, G.A. Prospettive di sviluppo nella ricerca sulla camminabilità. *Urban. Inf.* **2015**, *263*, 9–12.
72. Blečić, I.; Cecchini, A.; Fancello, G.; Talu, V.; Trunfio, G.A. Walkability and urban capabilities: Evaluation and planning decision support//Camminabilità e capacità urbane: Valutazione e supporto alla decisione e alla pianificazione urbanistica. *Territ. Italia* **2015**, *1*, 51–66. [CrossRef]
73. Mauricio, R.; Guachi, T. *Dashboard Design to Assess the Impact of Distinct data Visualization Techniques 852 in the Dynamic Analysis of Survey's Results*; Polytechnic Institute of Leiria: Leiria, Portugal, 2018.
74. Kourtit, K.; Nijkamp, P. Big data dashboards as smart decision support tools for i-cities–An experiment 843 on Stockholm. *Land Use Policy* **2018**, *71*, 24–35. [CrossRef]
75. SardegnaCiclabile Web Site. Available online: http://www.sardegnaciclabile.it (accessed on 15 September 2019).
76. Balletto, G.; Borruso, G.; Milesi, A.; Ladu, M. Turismo lento nel Sulcis Iglesiente. Mappe di comunità per le sfide dello sviluppo turistico locale. In Proceedings of the XXIII National Conference ASITA, Trieste, Italy, 12–14 November 2019; pp. 595–602.

sustainability

MDPI

Article

'Greening' Green Infrastructure. Good Italian Practices for Enhancing Green Infrastructure through the Common Agricultural Policy

Stefano Magaudda [1,*], Romina D'Ascanio [2], Serena Muccitelli [3] and Anna Laura Palazzo [2]

1 DICEA, Department of Civil and Environmental Engineering, Sapienza University of Rome, 00184 Rome, Italy

2 Department of Architecture, Roma Tre University, 00184 Rome, Italy; romina.dascanio@uniroma3.it (R.D.A.); annalaura.palazzo@uniroma3.it (A.L.P.)

3 Department of City and Regional Planning, Cornell University in Rome, 00184 Rome, Italy; sm2522@cornell.edu

* Correspondence: stefano.magaudda@uniroma1.it

Received: 12 January 2020; Accepted: 11 March 2020; Published: 16 March 2020

Abstract: The Common Agricultural Policy (CAP) was established by the European Community in the 1950s to provide financial support to farmers in member states, increase agricultural productivity by promoting technical progress, and ensure a fair standard of living for farmers. Over time; awareness about the externalities of intensive farming would prompt environmentally friendly practices. These include, in the current programming period 2014–2020, the so-called "greening", which consists of: (i) crop diversification; (ii) the maintenance of permanent grassland surfaces; and (iii) the availability of 5% of arable land for ecological focus areas devoted to agricultural practices beneficial for the climate and the environment. These provisions, spurred by a decades-long debate that also stresses the importance of creating/restoring ecological connectivity on different scales to counter land fragmentation, are in tune with spatial planning initiatives throughout Europe. Here the point is how to combine these directions with either "ecological networks" (EN), designed as physical corridors to be preserved and enhanced for plants and animals' mobility needs; or "green infrastructure" (GI), defined on the European level as a "strategically planned network of natural and semi-natural areas with other environmental features designed and managed to deliver a wide range of ecosystem services" (European Commission; 2013). While in several European countries environmental measures targeting farmers and ecological networks directed at specific areas have been merged in a place-based approach, Italy is lagging behind. In general, no guidelines have been provided on the national level to support regional paths, while regions and municipalities lack the resources to implement GI. Conversely, while greening policies in the framework of the CAP are properly funded, they lack directions to be efficiently allocated. Against the backdrop of such concerns, this paper frames and reflects upon ongoing practices in three pilot areas in different Italian regions, selected based on desk analysis, in-depth interviews, and direct knowledge. Here, despite or thanks to the legislative framework, experimental approaches have been adopted to harness performance issues in targeted areas through broad participation by public and private stakeholders and multilevel governance schemes, opening possible pathways in view of the forthcoming programming period.

Keywords: common agricultural policy; green infrastructure; greening; ecological networks; regional planning; multi-level governance

1. Introduction

On the European level, there is broad agreement about the need to integrate environmental issues related to biodiversity and ecological continuity with other sectoral policies within a territorial framework in order to strengthen their effectiveness. Until recently, agriculture and environmental policies have followed separate, often conflicting paths [1–4].

From the turn of the century onward, increasingly demanding agri-environmental measures and greening obligations under the Common Agricultural Policy (CAP) have been conveyed into different national legal frameworks. Several experiences in various European countries have been harnessing the "greening" tool provided by the 2014–2020 programming period in areas devoted to ecological networks and more generally to environmental continuity. Regional planning tools have been supportive in addressing a wide range of "open space" features and types ranging from ecological networks (EN) to green infrastructure (GI). GI has recently been defined by the homonymous European strategy as "strategically planned network of natural and semi-natural areas with other environmental features designed and managed to deliver a wide range of ecosystem services," which are in turn defined "as the direct and indirect contributions of ecosystems to human well-being" [5–7]. This statement, encompassing the production of food and water, the control of climate and disease, nutrient cycles and oxygen production, and spiritual and recreational benefits among ecosystem services [8], turns to GI for sustainable development issues on various scales, allowing multi-level governance models while accommodating land-use provisions and requirements [9].

This article investigates possible correlations between greening measures and GI in some Italian contexts, mainly within regional legal frameworks and bottom-up governance schemes tied to agri-environmental policies [10]. In fact, the Italian case stands out for its delay in embedding such principles within a place-based strategy. In terms of multilevel governance, the problem is twofold. On the one hand, regional and local planning tools entrusted with environmental measures and lately with green infrastructure display critical overlap; on the other hand, according to long lasting tradition, sectoral approaches falling under different administrative jurisdictions prevail over comprehensiveness [11]. Other crucial issues relate to poor commitment to joint initiatives; farmers, local communities and elected officials do not engage in paths of cooperation, apart from a few instances [12].

The paper is structured as follows: Section 2—General overview—compares the particular point of view of European institutions after a decades-long debate fed by fundamental contributions from research and civil society. The current "state of the art" in the case of Italy fails to fulfill specific targets either within the CAP or within environmental policies.

Section 3—Materials and methods—relies on desk analysis, in-depth interviews, and direct participation in ongoing European projects to identify three study cases in Italy that differ in location, size, and type of agreements, yet align with the need to encompass environmental issues in the CAP 2014–2020 programming period.

Section 4—Results—investigates several outcomes, assessing and synthesizing mutual interactions, awareness raising, and capacity building in the selected areas to overcome rigid sectoral schemes among local authorities, practitioners, and local communities.

Section 5—Discussion and Conclusions—highlights crucial issues, both in terms of current vertical subsidiarity with top-down practices and horizontal subsidiarity by means of public-private partnership arrangements. A major critical aspect in these teams concerns their sustainability over time for the implementation and management of green infrastructure. Some recurring approaches have also been detected, allowing for innovative institutional, social, and financial perspectives with an eye on future CAP reform.

Ultimately, the essential issues for reshaping integrated sustainability agendas, coupling functional performance related to ecological networks and ecosystem services with design requirements and layouts for their best allocation (conformance criteria), draw upon area-based strategies and strong coordination among the different stakeholders.

2. General Overview

This article begins with a study of policy measures and tools promoted on the European level related to agriculture and the environment and eventually transcribed as commitments within the legislative and programming framework of member states.

The EU is currently encouraging integration between agriculture and the environment. On the one hand, the CAP has been enlarging its scope to also encompass environmental and biodiversity objectives, from boosting farming practices to maintaining grassland habitats in permanent pastures, protecting water and habitats through ecological focus areas (EFA), and diversifying crops. On the other hand, the European Biodiversity Strategy awaits implementation through national regulatory frameworks and planning initiatives that are capable of programming and managing ecological continuity and green infrastructure [13].

With reference to the Italian case, this section summarizes the evolution of a decades-long debate by European institutions regarding the treatment of environmental issues within the framework of the CAP (Section 2.1) and environmental issues, up to the formulation of the Green Infrastructure Strategy, which makes it possible to bridge the gap between measures addressing similar purposes (Section 2.2).

In fact, despite both environmental conditionality, which supports farmers in exchange for environmentally friendly practices, and the mandatory implementation of ecological continuity, practical applications are failing to frame greening policies within ecological connectivity.

2.1. Enhancing Conditionality within the Common Agricultural Policy

In Italy, an impassable barrier has long existed between environmental and agricultural issues. Some important driving forces in this regard have undoubtedly been the CAP's slow, steady, and self-reliant path, strong resistance from the agri-food sector to sustainable innovations in production and productivity horizons, and, last but not least, difficulties in connecting spatial planning—conferred on regions and municipalities—to beneficial environmental practices, bypassing local authorities [14].

For over 60 years, the CAP has played a crucial role in building Europe. Launched in 1958 at the Stresa Conference, it long remained unique within the portfolio of EU policies [15,16]. A series of adjustments and variations to ensure environmental values for agricultural activities gave rise to an autonomous reflection within the European institutions amid the backdrop of tangible concerns about the need to politically justify continuity in public support of a sector that was losing workers in both percentages and real terms.

CAP was originally intended to support agriculture income according to strong market protectionism and price support measures rather than to assist local communities with new facilities. The scope was to provide general funding for modernization in the six founding members (France, Germany, Italy, Belgium, The Netherlands, and Luxembourg). In this framework, food production was secured against the risk of supply shortage, even acting as a remarkable factor of cohesion [17].

In the mid-eighties, once the objective of food safety was achieved, agricultural policy would disclose a deep conflict between two opposing sides: supporters of a radical reform for the development of rural areas on the one hand, and on the other, defenders of the status quo intolerant of any innovation. In 1987, the Single European Act was enforced, stating that "action by the Community relating to the environment shall have the following objectives: (i) to preserve, protect, and improve the quality of the environment; (ii) to contribute towards protecting human health; and (iii) to ensure a prudent and rational utilization of natural resources" [18].

The Delors Plan, "Bringing the Single Act to success: a new frontier for Europe", at the launch of the structural policy, paved the way for relevant changes, since it argued that "rural development must not only be a collateral problem for agricultural market policy, but a legitimate goal with full rights" [19,20]. This statement was coupled with policies addressing the production and reproduction of environmental values regardless of the beneficiaries' condition as "farmers practicing farming as their main occupation". Since then, agri-environmental measures have been designed to encourage farmers to protect and enhance the environment on their farmland by providing payments in return for services such as good farming practices, encompassing forestry, tourism, craft activities, environmental protection, and conservation of natural areas.

At the turn of the century, the CAP would address increasingly ambitious issues such as: less intensive production methods in order to reduce environmental impacts; chemical emission reductions; organic farming; carbon retention; landscape care and the protection of natural resources; and public access to natural areas for recreational purposes [21].

Along with current direct support for farmers (Pillar I), rural development would be fostered by a specific budget heading (Pillar II) allowing for spatial allocation of expenditures within Rural Development Programmes (RDPs) designed to promote innovative paths after long-standing experience carried out with the special programs leader on a local basis [22,23] (Table 1).

Table 1. Main steps of the CAP from 2000 onwards (prepared by the authors).

<table>
<tr><th>Programming Period</th><th>Reform</th><th>Main Contents Related to Environmental Issues</th></tr>
<tr><td>2000–2004</td><td>Agenda 2000</td><td>CAP reforms broke the link between direct payments and production.
More emphasis was placed on the environment and animal welfare.
• "Decoupling": breaking the link between direct payments to farmers and the type (and amount) of produce.
• "Cross-compliance": farmers receive subsidies in the form of a single payment in return for keeping the land in good agricultural and ecological conditions, and fulfilling food safety, environmental, animal health, and welfare standards.</td></tr>
<tr><td>2005–2009
2010–2013</td><td>Fischler Reform
Health Check</td><td>Rural development plays an increasing role in setting CAP strategies, aligning with the EU's structural policy.</td></tr>
<tr><td>2014–2020</td><td>Europa 2020</td><td>The CAP budget is split into two headings (Pillar I and Pillar II)
Pillar I
Under the direct payment scheme, the greening payment is introduced as the second most important component after the basic payment, for an amount corresponding to 30% of the national ceiling, equal for all Member States. It is paid annually per eligible hectare.
Farmers should comply with the following agricultural practices beneficial for the climate and the environment:
• crop diversification (2 arable crops for holdings between 10 and 30 hectares, 3 arable crops for holdings exceeding 30 hectares);
• maintaining permanent grassland surfaces;
• the availability of 5% of the arable land for ecological focus areas (commonly referred to as EFAs).
Pillar II
This accounts for 28% of the CAP budget, allowing for spatial allocation of expenditure within Rural Development Programmes (RDPs). The measures common to all regional programs are:
• facilitating generational turnover;
• promoting agri-environmental and climate-change-related measures;
• managing Natura 2000 sites;
• supporting organic farming;
• supporting areas with natural constraints.</td></tr>
</table>

In the current programming period (2014–2020), the CAP accounted for 38% of the entire EU financial statement. In Italy, the budget devoted to RDPs, deemed to better support sustainable farming and innovation as well as boosting employment, increased by 18.5% at current prices compared to the programming period 2007–2013, while Pillar I decreased by 8.3% [24].

Within Pillar I, the greening, which is the "green" component of the basic income support, was introduced in order to make the EU direct payment system more environmentally friendly and thus to obtain a "basic level of environmental management" from all beneficiaries.

Despite the expectations raised by greening as a tool to produce environmental public goods, their territorial allocation is a key problem in Italy. Greening measures have not significantly shaped the Italian context due to land ownership fragmentation and average size generally well below the 10-hectare threshold.

In turn, Pillar II represents a missed opportunity for Italian regions. Other European nations with comparable levels of complexity such as France, Germany, and Spain, have used the National Framework, an overarching discipline containing elements common to all RDPs, to facilitate coordination between the regions. In Italy, 21 regional RDPs have been approved, as well as a national RDP addressing risk management, the protection of animal biodiversity, and the efficient use of water resources.

Merging the environmental and agricultural policies listed in Table 2 may serve as a practical framework to implement green infrastructure stemming from Pillar II of the CAP, targeting Priority 4—Restoring, conserving, and improving ecosystems that depend on agriculture and forestry—and Priority 5—Promoting the efficient use of resources and supporting progress towards a low-carbon economy that can adapt to climate change in the agricultural, food, and forestry sectors.

Table 2. Environmentally oriented measures within the 2014–2020 RDPs.

Measures	Sub-Measures
Measure 4—Investment in physical assets	4.3 Aid for investment in infrastructures tied to developing, modernizing, and adapting agriculture and forestry 4.4 Aid for non-productive investment in meeting agricultural, environmental, and climate targets
Measure 7—Basic services and village renewal in rural areas	7.3 Aid for creating, improving, and expanding basic local services for the rural population, including recreational and cultural activities, and related infrastructure 7.4 Aid for investment in recreational infrastructure for public use, tourist information, and small-scale tourist infrastructure 7.5 Aid for studies/investments tied to maintaining, reclaiming and restoring the natural and cultural heritage of villages, rural landscapes, and areas with a high natural value, including their social and economic aspects, as well as ecological awareness-raising initiatives
Measure 10—Agri-environment and climate	10.1 Aid for agri-environmental and climate commitments 10.2 Aid for the conservation, sustainable use, and development of genetic resources in agriculture
Measure 12—Payments under Natura 2000 and the Water Framework Directive	12.1 Compensation payments for agricultural areas in the Natura 2000 Network 12.2 Compensation payments for forestry areas in the Natura 2000 Network 12.3 Compensation payments for agricultural areas included in basin hydrological plans
Measure 15—Forestry, environmental and climate services and forest conservation	15.1 Payments for forestry, environmental, and climate commitments
Measure 16—Cooperation	16.2 Aid for pilot projects and developing new products, practices. processes, and technologies 16.5 Aid for joint actions performed to mitigate climate change and adapt to it, and for joint approaches to environmental projects and existing environmental practices

Furthermore, cooperation arrangements supported by measure 16 deemed capable of encouraging technical, organizational, and social innovation have been poorly financed or poorly practiced, missing the objective of prioritizing green networks beneficial for the climate and the environment [25,26].

The CAP reform after 2020 is supposed to change the entire structure of the policy itself through three key elements. Firstly, it foresees the drafting of a single "CAP strategic plan" per Member State expected to mediate between the European policy and regional programming. Such national programming frames of reference will be based on the assessment of needs, targeting and performance monitoring to cover and ensure greater coherence between the two Pillars (direct payments to farmers and support for rural development) [27]. Indeed, the Commission has to approve the strategic plans when satisfied with their quality. As a consequence, each region will develop its own RDP according to the national strategic plan, which will have an operational character [28].

Secondly, a new system of "enhanced conditionality" is expected to streamline current mechanisms of cross-compliance and greening payments in tying income support (and other area- and animal-based payments) to environmentally and climate-friendly farming practices and standards defined as "Good Agricultural and Environmental Conditions" (GAECs) and "Statutory Management Requirements" (SMRs).

Thirdly, a complementary set of (voluntary) tools will be offered to farmers to help achieve the CAP environmental and climate objectives. 'Eco-schemes' represent a new stream of funding aiming at strengthening the ecological baseline. They are payment schemes for caring for the environment and climate that will be funded by member states under the CAP's direct payments. Member states (within state/regional conferences) will have to provide one or more eco-schemes, which will remain voluntary measures testing a reward beyond conditionality. The Commission clarifies that even though eco-schemes have features in common with support for agri-environment climate commitments available under Pillar I, there are significant differences between the two. In particular, eco-schemes offer the possibility to grant direct payments as an incentive to farmers adopting practices beneficial for the environment and the climate (going beyond the costs incurred or the income foregone due to the adoption of these practices) [27].

2.2. From Ecological Networks to Green Infrastructure: An Attempt to Integrate Agriculture and the Environment

Urbanization and urban sprawl pose a strong threat to natural and agricultural areas [29]. In the 1990s, European countries initiated planning and management processes addressing ecological networks according to the Habitats Directive complying with major international strategies for biodiversity [30–32]. Ecological networks, as tools for improving biodiversity and ecological connectivity among habitats, were designed to consider different levels of nature protection (Table 3).

Table 3. Components of Ecological Networks (prepared by the authors).

Component	Description
Core natural areas	Areas of high-quality habitats that are managed primarily for biodiversity conservation, whether or not they are protected.
Buffer zones	Transitional areas, located around highly natural areas where restrictions on resource use and special development measures are undertaken in order to enhance the conservation value of the protected area.
Ecological corridors	Linear and continuous structures in the landscape varying in shape and size, connecting highly natural areas to each other and representing the key element of ecological networks since they allow for species mobility and genetic exchange, an indispensable phenomenon for maintaining biodiversity.These corridors form the "edges" in the network, linking core nature areas and stepping stones.
Stepping stones	Smaller areas of quality habitats that are intended to help the movement of individuals by serving as islands of favorable habitats between larger core nature areas. These also represent "nodes" in the ecological network that, due to their strategic position or composition, may support species passing through a territory or host particular microenvironments in critical habitat situations.

The aim of connecting open spaces with different degrees of naturalness has gradually been combined with the need to also enhance interactions in rural and peri-urban areas [33].

Starting in the early 2000s, eco-agricultural connections have been at the core of seminal debates among planners and policy makers highlighting the multiple benefits of ecological and natural systems for communities and economies. The key role of GI, somehow supplanting ecological networks due to more comprehensive targets in addressing environmental sustainability [34,35], leverages not only ecological functionality but also social, cultural and economic activities involving communities at large [36,37]. In France, the legislative framework sets the ground for regional and urban planning dealing with green infrastructure, merging ecological issues with the basic needs of human communities on different scales [38]. In Germany, cases of urban regeneration are associated with energy and climate challenges. In Italy, which lacks national-level directives to harness CAP measures and ecological connectivity/green infrastructure strategies, implementation follows different paths at different paces based on the specific case. Scant propensity to cooperate and a lack of funds for implementation and management referring to defective multi-level governance are delaying alignment with other countries.

In sum, GI delivers environmental (and social) advantages [39] regarding biodiversity and ecological restoration, adaptation, and mitigation measures to counter the manifold effects of pollution and climate change [40,41]. Furthermore, it lends itself to light, flexible policy frameworks addressing multilevel governance. Hansen and Pauleit have listed three main features: (i) strategic approach (GI planning aims for long-term benefits but remains flexible for changes over time); (ii) social inclusion (GI planning stands for communicational and socially inclusive planning and management); and (iii) trans-disciplinarity (GI planning is based on knowledge from different disciplines, such as: landscape ecology, urban and regional planning, and landscape architecture, and is developed in partnership with different local authorities and stakeholders) [42].

Two further perspectives—"landscape", maximizing perceptive, utility, and recreational value, and "ecosystem services", i.e., direct and indirect contributions of ecosystems to human well-being—feed both ecological networks and green infrastructure, requiring spatial planning to specify features, patterns, and uses (Figure 1), and coupling human and natural science as well as different budget headings.

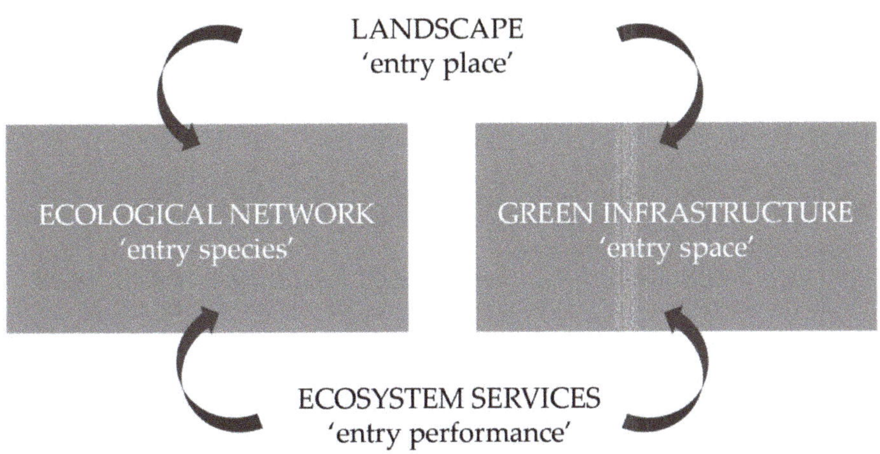

Figure 1. Ecological Networks and Green Infrastructure. The nexus of two paradigms. Coupling biodiversity (entry "species") with human activities (entry "space") may clarify and enable relationships among sectors, policies, and tools conventionally separated (prepared by the authors).

3. Materials and Methods

Without a national strategy integrating agricultural and environmental policies, the Italian case can be read as an array of independent approaches "in the field". In this framework, several particular aspects emerge in terms of:

- legislative references displaying an overlapping of competences and administrative sectors;

- spatial constraints due to poor directions for green infrastructure implementation and weak support for accommodating agri-environmental measures within spatial planning;
- structural constraints such as land-use and land-tenure fragmentation, lack of cooperation between public and private stakeholders.

Considering the policy-oriented character of the processes being analyzed, a qualitative approach in the design of the study cases was selected and performed according to the seminal work of Bruno Dente and Francesco Kjellberg. A wide range of governance issues have been taken into account concerning both nature and function of rules dealing with structural arrangements, decision-making and financial aspects [43] (Table 4).

Table 4. Types of reforms in local governance, adapted by the authors from Dente and Kjellberg, p.11.

Subject of Reforms	Intergovernmental Relationships	Internal Local Government
1. Organizations	1.1 Structural (i.e., change in number of Local Authorities, administrative boundaries)	1.2 Organizational (i.e., changes in relations between political and administrative bodies)
2. Decision-making	2.1 Functional and procedural (i.e changes in the Administrations' functions, procedures)	2.2 Local programming (i.e, implementation/update of programming tools, plans)
3. Financial resources	3.1 Funding (i.e., change in EU, national, regional funding)	3.2 Local funding (i.e., variation in fares regimes, local accounting regulations)

It has to be noted that in the selected study cases, formal agreements among institutional bodies have been made possible with no changes in the administrative structure, while decision-making aspects as well as innovations related to the financial support take on greatest importance. It can be argued that place-specific governance reforms in the public domain have been able to prompt private partnerships by creating a milieu conducive to shared paths. First, ongoing experiences throughout the country were selected using desk analysis and interviews with administrators and professionals committed to environmental and agricultural issues. Subsequently, three study cases were selected in as many Italian regions (Piedmont, Marche, Lazio, Figure 2). The common features of these experiences, which differ in geographic condition, size, promoter, community engagement, and budget headings, are the following (Table 5).

- They are good practices where existing programming tools have been linked to CAP funding for the implementation and management of GI;
- They target contexts differing in terms of environmental, productive, social and regulatory features;
- They are part of regional programming and planning (such as the Environmental Regeneration Program, Provincial Territorial Plan, Regional Landscape Plan) with a specific focus on biodiversity conservation;
- They involve the agricultural sector stakeholders in the policy process development for the definition of objectives, strategies and interventions.

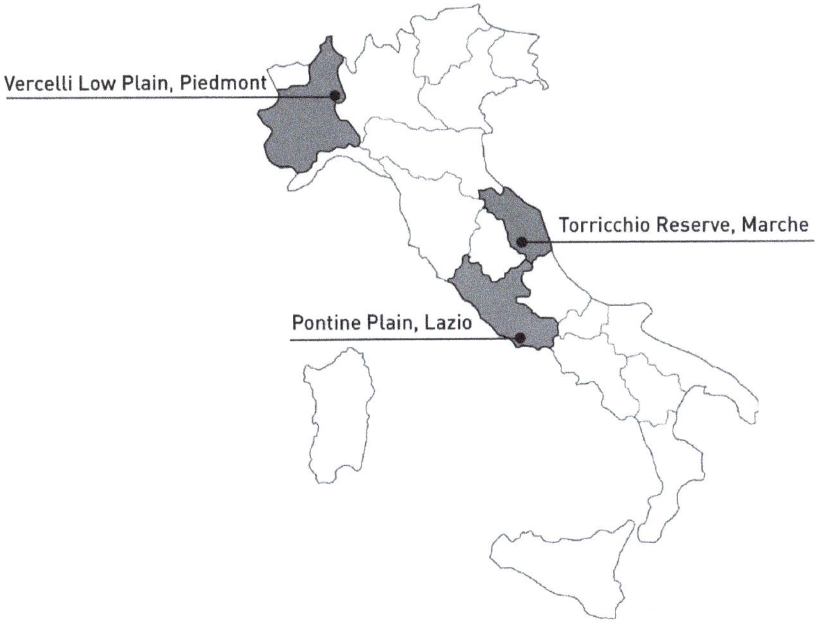

Figure 2. Overview of the regions and selected study cases.

Table 5. Main features of the study cases.

Target Area/Region	Promoter/Coordinator	Productive Features	Policy Process	Target Area Extension	CAP Payments
Vercelli lowlands plain, Piedmont	Province of Vercelli	Submerged rise cultivation	Wetland contract	70,736 hectares	Pillar II
Pontine Plain, Lazio	Province of Latina	Intensive farming	Pact for Biodiversity	77,000 hectares	Pillar I, II
Torricchio Reserve, Marche	Università di Camerino	Grazing	Agri-environmental Agreement	1231 hectares	Pillar II

The following systematized methods were used to collect the data for each study case:

- Study Case 1:
    - Consultation of 7 territorial plans and programs (Hydrological Structure Plan; Po River Basin Management Plan; Piedmont Regional Territorial Plan; Piedmont Rural Development Programme; Regional Water Protection Plan; Provincial Coordination Plan; Regional Guidelines for River/Lake Contracts) and the management plans of 2 protected areas; 1 Memorandum of Understanding for the signing of the Wetland Contract; 5 progress reports (stakeholder analysis; scientific description of the area; Wetland Contract trend and oriented scenarios; participatory process development; Wetland Contract and Action Plan);
    - Open ended and focused interviews to 1 manager of the Vercelli Province, and 1 manager of the Piedmont Region in charge for River/Lake/Wetland Contract coordination;
    - Direct and participant observation to 5 Technical Secretariat meetings and 3 Focus groups [44].
- Study Case 2:
    - Consultation of 5 plans and programs (Regional Water Protection Plan, Lazio Rural Development Programme, Provincial Coordination Plan, Provincial Ecological Network,

    Environmental Restoration Program of Agro Pontino); 1 Memorandum of Understanding for the subscription of the Pact for Biodiversity;
    - Open ended and focused interviews to 1 manager of the Latina Province, 1 manager of the Lazio Region in charge for agri-environmental measures coordination, and 1 director of Confagricoltura association;
    - Direct and participant observation to 4 Technical Secretariat meetings [44].
- Study Case 3:
    - Consultation of 2 plans and programs (Management Plan of the Torricchio Mountain Nature Reserve, Marche Rural Development Programme); 2 scientific papers on the topic; 1 Agri-Environmental Agreement, and 1 Land Stewardship Agreement;
    - 1 focused interview to the delegate of the director for agri-environmental policies in the Torricchio Mountain Nature Reserve.

A key part of the work reported in each study case is related to stakeholders analysis (Tables 6 and 7). Table 6 reports the analysis of stakeholders' typology, resources, objectives and roles. Particular attention has been payed to the identification of roles, where, according to Dente's definition, the *Promoter* is the actor giving the first stimulus to the process, that can correspond to the process *director*, the one that has to pilot the process until the end; the *allied* has content and process objectives congruent to those of the promoter and brings its resources to the process partnership; the *gatekeeper*, because of the resources he controls, has the possibility to use its veto power and block the process [45], contributing to shaping a policy process of regional and/or national interest but with local consequences. Table 7 proposes a matrix revealing the complexity (within each cell and among different cells) of the area-based multi-level and multi-actor decision process at stake, due to heterogeneity of the interests and potential conflicts [45].

Table 6. Stakeholder analysis framework: Typology, Resources, Objectives, Roles.

Stakeholders	Typology	Resources	Objectives	Role
Stakeholder 1	Politician	Political	contents	Promoter & Director (policy entrepreneur)
Stakeholder 2	Administrators	Financial	contents	Gatekeeper
Stakeholder 3	Administrators	Legal	process	Allied
Stakeholder 4	Special Interest	Knowledge	contents	Allied

Table 7. Framing the policy process complexity.

Dimension of the Interest	Stakeholders Typology				
	Politicians	Administrators	Experts	Special Interests	General Interests
National					
Regional					
Local					

The three study cases share multilevel governance schemes calling into question public-private partnerships and forms of agreement that allow to adopt a set of regulations in which criteria of public utility, economic return, social value and environmental sustainability equally take part in the search for effective solutions in an area-based approach. Those agreements, whose specificities will be detailed in the 'Results' paragraph, do not constitute a new planning act or a new decision-making level, but rather bring the specific strategies and competences of the stakeholders involved towards a governance process, respecting the specificities and autonomies, with a flexible updatable, inter-sectoral and inter-scalar approach. At the same time, they should not be understood as mere inter-institutional agreements aimed at sharing government objectives, but rather as a decision-making and operational process that makes up the environmental and socio-economic interests of a water system, implementing the superordinate provisions (territorial and sectorial) [46].

The first study case, led by the Province of Vercelli, in the Piedmont Region, relates to the Vercelli Lowlands plain Wetland Contract in the framework of the Interreg MED WETNET project.

The second study case regards the Pact for Biodiversity of the Pontine Plain, a multilevel governance tool aimed at improving the ecological management of agro-ecosystems. It falls under the EU Life Nature and Biodiversity funded project GREENCHANGE in the Pontine plain.

The third study case concerns the Agri-environmental Agreements of the Marche Region, promoted with the objective to engage and aggregate stakeholders around a specific environmental issue and joint territorial interventions for the conservation of biodiversity and the safeguard of the land and of the waters.

4. Results

4.1. Regional Reward Criteria for Implementing the River/Wetland Contract Strategic Vision in Piedmont

4.1.1. General Features, Problems, and Open Issues

The Vercelli Lowlands are a vast inland area of 70,736 hectares characterized by fresh-water paddy fields, constituting natural and landscape value as a whole. The area includes part of the western floodplain of the Sesia River and the northern floodplain of Dora Baltea and Po Rivers (Figure 3). It is also characterized by a complex system of irrigation canals and artificial waterways guaranteeing crop irrigation and rice cultivation. The area hosts 9 sites protected as Special Nature Reserves, 7 Sites of Community Importance (SCI), 6 Special Protection Areas (SPA) and 3 Natura 2000 sites; it falls within the administrative area of the Province of Vercelli (intermediate administrative body with coordination functions) and 28 municipalities.

Figure 3. Paddy fields, Province of Vercelli, Piedmont Region (Author Stefano Magaudda).

The Vercelli Lowlands Wetland Contract was developed within Piedmont regional strategies for the integrated management of water resources in accordance with the Water Framework Directive (WFD). Through regional guidelines for implementing the River/Lake/Wetland Contracts, the Region institutionalizes the Contracts as a tool for implementing the Regional Water Protection Plan [47].

Formally the River/Wetland Contracts can be defined as voluntary commitments undertaken by various public and private entities in various capacities interested in the wetland system for the environmental restoration and socio-economic regeneration of water systems. The agreements are developed and formalized within a participatory and negotiated decision-making process leading to drafting of the programming act (Action Plan) with a medium-term duration. The latter compiles and integrates the various interests regarding a body of water, defining responsibilities and implementation tools for the governance and sustainable management of the water system. [48].

Initial funding for the development of this process came from the European Commission Interreg MED Program within the WETNET cooperation project which aims to define common priorities for wetland conservation in the Mediterranean area and encourage the adoption of River/Wetland Contracts for integrated governance.

4.1.2. Governance Process Development

The Province of Vercelli is in charge of area administration in all its aspects and promotes the focused process through its Urban and Regional Planning Department. The latter undertakes responsibilities in the fields of environmental protection, sustainable development, and regional planning, as it is in charge of the Provincial Coordination Plan (PCTP). With the Wetland Contract tool, it aims to implement a homogenous vision for the Province's sustainable development in accordance with the Ecological Network Plan (attached to the PTCP).

More specifically, the Province began the process of implementing the Wetland Contract, setting the scope of sustainably managing the Vercelli Lowlands and enhancing the ecological role of GI in paddy fields.

The process has developed consistently with superordinate planning tools, within which those concerning water management are of greater relevance. On the supraregional level, the Po River Basin Management Plan is the coordination tool of reference, aimed at implementing a coherent and sustainable water-protection policy, integrating the management and ecological aspects arising on the river-basin scale. On the regional level, the Piedmont Regional Water Protection Plan sets out the strategies for achieving the quality objectives for water bodies; it also mentions the possibility of activating negotiated tools such as River/Wetland Contracts.

To create a participatory process, the Province elicited 41 stakeholders with influence or interests in the area, of which 29% are public bodies, 20% are private for-profit entities, and 12% are private non-profit entities. Participatory meetings have then been attended by anywhere from 10 to 46 stakeholders (the Region and municipalities; instrumental bodies such as irrigation associations, protected-area managers, river authorities; farms and farmers' organizations; research institutes; environmental associations; educational institutions; trade associations; citizens). The Memorandum of Understanding (MoU) preliminary to the final signing of the Contract (the procedure is required by the regional guidelines) has been signed by the Province, 28 municipalities, 3 park authorities, the Region, and the Po Basin Authority. The stakeholders subscribing to the Contract have been analyzed with reference to their type, available resources, objectives (either content—problem to face, solution to adopt—or process), and roles in the focused process (Table 8). Secondly, a matrix combining the dimension of territorial interest with the type of stakeholder was developed to measure the complexity of the policy process (Table 9). This data shows that towards the end the process increasingly became a policy act aimed at managing legal, administrative, and financial issues for the successful implementation of the Contract and its actions.

The Action Plan for the target Contract was structured around the strategic areas of governance, the environment, and social and economic development. The general objectives were: (i) to improve biodiversity, landscape, and water quality through new shared governance and planning schemes (e.g., Environmental Compensation Plan); and (ii) to mitigate the environmental impacts of rice paddies, strengthen their role as a habitat for wildlife, and improve their landscape quality. Much attention was focused on identifying detailed actions to implement GI on different levels. These include: the promotion of agreements to sustainably create and manage GI between municipalities and public and private entities, such as Land Stewardship Agreements (LSA); the adaptation of local (municipal) planning tools to the PTCP requirements in terms of ecological network, buffers and filter ecosystems in agricultural areas and wooded and green areas in urban context.

Table 8. Stakeholder analysis (main contributors): Type, resources, objectives, roles.

Stakeholder	Type	Resources	Objectives	Role
Vercelli Province	Politician	Political/legal	process/contents	Promoter & Director (policy entrepreneur)
Piedmont Region	Administrators	Political/legal/financial	process/contents	Ally
Po Basin Authority	General Interests	Legal	contents	Ally
Municipalities	Politician	Political/legal	contents	Ally
Park Authorities	Special Interests	Knowledge	contents	Ally
University	Experts	Knowledge	contents	Ally
Regional Agency for Environmental Protection	Experts	Knowledge	process/contents	Ally

Table 9. Analysis of the policy process complexity, combining stakeholder type and the dimension of the interest at stake.

Dimension of the Interest	Stakeholder Type				
	Politicians	Administrators	Experts	Special Interests	General Interests
National			University		Po Basin Authority
Regional		Piedmont Region	Regional Agency for Environmental Protection		
Local	Province, Municipalities			Park Authorities	

4.1.3. Main Outcomes

Some innovative elements for the governance of GI emerge from this study case. The first concerns the adopted tool, the Wetland Contract, which appears to be a flexible tool for embedding medium-term strategies and addressing multilevel, integrated responses for specific issues with a high impact such as GI. While the Wetland Contract does not constitute a new planning tool, it brings the specific strategies and skills of the stakeholders into the governance process, respecting specifics and autonomies with a flexible approach. At the same time, the tool is not merely an inter-institutional agreement aimed at sharing government objectives; rather, it can be considered a decision-making and operational process capable of creating and implementing a strategic vision in accordance with superordinate regulations [46].

The second element of interest concerns the structuring of the Contract's actors and their respective roles. The Region addresses, coordinates, and accompanies, also by defining the basic policies to construct and manage decision-making processes. At turn, the Province, as the body in charge of the environment, drafts and manages the Contracts and the participatory process. Finally, the municipalities act as collectors of local projects, interests and initiatives, conveying to the Province some local options. This structure seems to be applied in most of the River/Lake/Wetland Contracts in Piedmont, since it is also encouraged by the Region [48].

The last aspect is the successful connection of European policies to local context in an area-based approach. Besides integrating River/Wetland Contracts in its policies, the Piedmont Region has developed a reward mechanism for funding initiatives/projects/interventions included in River/Wetland Contracts. This mechanism consists of a specific score reserved for such actions in public calls for Regional Operational Programme (ROP)/RDP funding. In this way, the Region promotes the implementation of actions within a vision based on a wide-area scope potentially integrated with regional funding.

In reference to the types of reforms identified in Table 5, this study case addresses point 2.1, functional and procedural decision-making changes at the intergovernmental level, which also affect the local scale by introducing horizontal subsidiarity principles and by promoting the implementation and update of programming tools and plans, as specified at point 2.2; and point 3.1, concerning changes in funding procedures for a more efficient allocation of RDP funds in reference to the focused field.

4.2. *The Pact for Biodiversity of the Pontine Plain: Towards the Ecological Network in Agricultural Areas*

4.2.1. General Features, Problems, and Open Issues

The Pontine Plain is the result of the "Great Land Reclamation" of the 1920s, which profoundly changed the hydrography, topography, biodiversity, urbanization, economy, identity, and landscape of the area. This peculiar landscape is currently undergoing extensive transformation due to intensive agriculture, industrialization, and urban sprawl (Figure 4).

Figure 4. Pontine Plain, Gricilli area, Province of Latina, Lazio Region. (Author Giovanni Mastrobuoni).

Over the last 20 years, the Province of Latina has carried out a series of activities aimed at restoring the Pontine Plain by taking advantage of the potential connectivity of its green and blue infrastructure (windbreak strips, reclamation of canal banks and their buffer zones). Consistent with the Water Framework Directive (2000/60/EC) and Habitat Directive (1992/43/EEC), the Province developed the Ecological Network within the Provincial Coordination Plan (PTCP), and the Integrated Environmental Restoration Program of the Pontine Plain. This set of previous experiences committed the Province to a new initiative within the context of the EU-funded project LIFE GREENCHANGE, which aims to test intervention models for implementing these plans.

More specifically, GREENCHANGE intends to help halt the loss of biodiversity and enhance the ecological value of the plain by introducing the Pact for Biodiversity committed to the ecological management of agro-ecosystems through an integrated set of actions and tools. It works to systematize actions in agricultural areas and harmonize priorities, methods for intervention, and financing sources towards ecosystem preservation and restoration. To this aim, networking among individual farm-scale restoration actions has been spurred in order to achieve ecological continuity, even relying to the optimization of funding sources.

The Pact is developed based on recognition of the agroecosystems' added values by maintaining certain natural elements within the agricultural pattern, even in terms of productivity and quality products. Awareness about the "services" provided by such agroecosystems supports the identification of best practices and the most adequate tools to maximize them. The pact's actions concern the development of scenarios to maintain the agricultural land through the lens of ecosystem services, the production of shared rules and guidelines for the environmental management of interventions, and the active involvement of farmers in the management through LSA.

4.2.2. Governance Process Development

As mentioned above, the Province of Latina was the promoter of the Pact for Biodiversity of the Pontine Plain and acted also as director of the process, taking responsibility for engaging the relevant stakeholders on both the regional and local levels. The Pact is designed as an open and permanent working table with the goal of encouraging stakeholders to share methods, tools, and practices to manage agricultural areas in order to achieve its objectives.

As shown in Table 10, the Pact involves an articulated set of stakeholders belonging to both the environmental and farming sectors who are interested for political, managerial, and production reasons. These are local and regional authorities (Rural Development Programmes' managing bodies), the Protected Area managing bodies, and farmers and their associations [49]. It is worth mentioning the key role of the Lazio Region as a gatekeeper of the process, since some of the Pact's achievements depend on the cooperation of regional bodies in enabling specific processes (see below for loans for the use of GI) and the successful integration of its actions and objectives with RDP funding.

Table 10. Subscribing stakeholder analysis: type, resources, objectives, roles.

Stakeholder	Type	Resources	Objectives	Role
Province of Latina	Politician/Administrators	Political/legal	contents	Promoter & Director (policy entrepreneur)
Lazio Region	Politician/Administrators	Political/legal/financial	process/contents	Gatekeeper
Land Reclamation Association	Special Interest	Legal	contents	Ally
Municipalities	Politician/Administrators	Political/legal	contents	Ally
Park authorities	Special Interests	Knowledge	contents	Ally
Farmers associations	Special Interests	Political	contents	Ally
Research institutions	Experts	Knowledge	contents/process	Ally

Participation of the listed stakeholders is necessary to establish the integrated objectives and measures of the Pact and to ease the collaborative implementation of such measures. The complexity of interests and issues emerges from Table 11.

Table 11. Analysis of the policy process complexity.

Dimension of the Tnterest	Stakeholder Type				
	Politicians	Administrators	Experts	Special Interests	General Interests
National			Research institutions		
Regional		Lazio Region			
Local	Province of Latina, Municipalities			Park authorities, Farmers associations, Land Reclamation Association	

The direct involvement of farmers in the implementation and management of the ecological network is a key issue. This has been promoted considering the essential role of EU funding for farmers and rural development, and is therefore based on the remuneration of the areas in the context of the CAP.

The Pact relates to financing opportunities offered by the CAP in two ways. With reference to funding under the First Pillar, the Pact develops tools to facilitate farmers' access to it, such as:

(i) a catalogue containing the description of selected agricultural practices and their economic, productive, and environmental impacts in the short and long term; (ii) an information desk to support farmers during the funding application process; and (iii) an information technology tool to help farmers to plan sustainability and assess business efficiency in agronomic, economic, and ecological terms. The Pact acts as a learning environment for long-term management of rural areas (including productivity), also offering training for farmers about ecosystem services and related benefits on the farm business. It presents scenarios about the variation of ecosystem services and their potential impact on yield (quantity/quality/costs), while providing best practices for the proper management of agroecosystems (to maintain/improve ecosystem services).

Moreover, a specific mechanism is established by the Pact to enhance access to CAP funding. This mechanism is based on the possibility for farmers to fulfil the greening requirements on state-owned GI. Through loan-for-use contracts with the Region (owner and body in charge of managing windbreaks and canal banks), such areas become part of the farm dossier and are eligible for funding. The benefit farmers receive is twofold: besides CAP direct payments, the possibility of devoting areas outside of their productive land to greening. In this framework, LSA formalize farmers' commitments to the conservation and restoration of nature on their property and in state-owned areas whose management has been transferred to them.

With reference to the Second Pillar, the Pact works to improve the grant criteria for RDP funding in order to meet the objectives established in the Pact and the results achieved during its testing phase (GREENCHANGE project). Interaction with the Lazio Region will enable the identification of rewarding criteria to favor the allocation of funds to farms adhering to the Pact by signing LSA or carrying out interventions consistent with the measures of the Pact.

4.2.3. Main Outcomes

The Pact proves an innovative governance tool able to streamline regional complexity in an area-based, multilevel vision coordinating actors with specific skills and interests. On the one hand, the Pact prepares a selection of the regional-owned areas suitable for applying restoration measures as a knowledge base for implementing GI in the region. At the same time, the Pact concretely implements GI, by offering the support of regional authorities to farmers associations, thanks to availability of RDP funding and a mechanism for enhancing the capture of CAP direct funding. In this respect, this second study case embeds the type of reforms listed at point 2.1, 2.2 and 3.1 of Table 5, by setting a multilevel governance tool focused on the agri-environmental field capable of implementing the existing tools (i.e., the Integrated Environmental Restoration Program of the Pontine Plain) and effectively integrating Pillar I and II funding for specific local needs.

The Pact approach allows for a tailor-made regional coordination consistent with the next CAP programming period, where the relevance of agri-environmental measures and the cooperative approach are expected to be reinforced.

4.3. The Area-Based Agri-Environmental Agreements of the Marche Region

4.3.1. Context

The third study case focuses on the Agri-Environmental Agreements (AEA) promoted by the Marche Region with the aim of fostering an integrated and area-based approach while implementing sustainable rural practices. According to the Marche RDP, an AEA is defined as a "set of commitments for farmers in a specific area, supported through a mix of RDP measures to achive specific environmental goals". Based on a territorial approach involving public and private actors, AEAs are aimed at implementing collective and coordinated actions to manage and improve the environment.

This study case focused on the type of AEA foreseen by the 2007–2013 RDP—financed for the first time in 2011—which targets the protection of biodiversity in Natura 2000 areas. It considers the area of the Torricchio Mountain Nature Reserve, which was established in 1974 and is owned by the University of Camerino (UNICAM), who is also in charge of its management with WWF patronage (Figure 5). The reserve extends for 500 hectares and targets the protection of priority habitat "6210 Semi-natural dry grasslands and shrubland facies on calcareous substrates (Festuco-Brometalia) (important orchid sites)". UNICAM, together with local mountain communities, has created a 3,000 hectare district corresponding to the surrounding "Valnerina, Montagna di Torricchio, Monti Fema e Cavallo" Special Protection Area (SPA), for a more effective pasture and ecological management.

Figure 5. Torricchio Reserve, Province of Macerata, Marche Region (Author: Andrea Catorci).

4.3.2. Process Development

UNICAM, as an AEA promoter and coordinator, was in charge of facilitating the agreement's development, drafting the general project, coordinating the individual beneficiaries applying for funding, and reporting to management authorities about the implementation process. UNICAM initially had to establish a dialogue with the 40 livestock farms participating in the agreement-development process in order to overcome their skepticism about the Reserve management body. For this purpose, UNICAM not only coordinated an institutional board of local stakeholders (municipalities, farmers associations, business unions), but also formed a multidisciplinary team of scientists and appointed an agronomist facilitator in charge of dialoguing with every farm in the area.

The first step was to launch round tables with the farmers and their technicians in order to present RDP measures and funding opportunities, notably in reference to the protection of pasture-related habitats. Such workshops offered the opportunity to jointly define the objectives and conservation measures of the Natura 2000 and area management plans. The stakeholders involved in this process are listed in Table 12.

Table 12. Participating stakeholder analysis: type, resources, objectives, roles.

Stakeholders	Type	Resources	Objectives	Role
UNICAM/Reserve Manager	Administrators	Legal/Knowledge	contents	Promoter & Director (policy entrepreneur)
Marche Region	Administrators	Legal/financial	process/contents	Gatekeeper
Municipalities/Mountain Communities	Politician/Administrators	Political/Legal	contents	Ally
NGOs	Special Interests	Knowledge	contents	Ally
Farmers	Special interests	Knowledge	contents	Ally
Farmers Associations	Special Interests	Political/Knowledge	contents	Ally

The AEA of the Torricchio Reserve aims to bring the mountain meadow up to a good state of conservation. The solutions to improve the habitat were drafted by the entire community via a participatory process that lasted about a year. Thirty-eight farmers were engaged in the process, and twenty-five signed the agreement. The governance process resulted in a Pasture Plan containing conservation measures and management regulations that are also binding for the farmers that have not endorsed the agreement (Table 13).

Table 13. Analysis of the policy process complexity.

Dimension of the Interest	Stakeholder type				
	Politicians	Administrators	Experts	Special Interests	General Interests
National					NGOs
Regional		Marche Region	UNICAM	Farmers associations	
Local	Municipalities/Mountain Communities	UNICAM/Reserve Manager		Farmers	

After the approval of the AEA, the Region launched a three-year call for funding addressed at the participating farmers who, in order to obtain funding, not only have to respect the agreement, but also have to co-operate to manage the collective goods. The role of UNICAM was pivotal in the discussion with the Region, especially when addressing RDP calls for measures and interventions. The loss of earnings due to the meadows' maintenance as well as experimental interventions for limiting invasive species were supported by regional funds through a system of mobile fences. In three years, € 600,000 have been spent from the RDP for funding applications presented by agronomists, who had previously received training about the specific instrument.

4.3.3. Main Outcomes

In Italy, the Marche is undoubtedly the region that experimented and promoted a collective partnership approach within the RDP, even with training and in-depth information sessions for farmers. The AEAs were already developed in the 2007–2013 programming period and were further strengthened in the 2014–2020 one [12]. The main feature of this study case is the choice made by the Marche Region to allocate a significant share of the RDP resources for agricultural communities and businesses that implement and sign the AEAs, addressing in such way point 3.1 of Table 5 (funding reforms under intergovernmental relations).

In conclusion, AEAs innovated the relationships among the Reserve manager, farmers, and other stakeholders, setting the ground for the development of regional joint initiatives, allowing for further development of LSAs to manage and maintain common goods.

5. Discussion and Conclusions

Within the targeted study cases, the area-based experiences, whether or not depending on administrative boundaries, have demonstrated the effectiveness of governance practices based on both a large-scale vision and strong public leadership. The administrations involved successfully developed a comprehensive strategy for the restoration and enhancement of the territory by considering the different environmental and landscape components as part of a single object of intervention. A main feature to all the experiences is the adoption of voluntary agreements. Regardless of their name, these "Pacts" are currently used as governance tools dealing with environmental protection, biodiversity, climate change, ecosystem services, land use management and sustainable development in rural areas.

The three study cases highlight the capacity to overcome conflicts between sectoral and institutional skill areas thanks to a collaborative and strategic approach, which is appropriate for the spatial context and planning goals [50], and not just for matters of habitat conservation.

In addition, the identification of a promoter in charge of coordinating the collaborative governance tools on the territorial scale has shown significant effectiveness. This organization has a key role in defining, managing, and monitoring the Pact. Moreover, it has to bring local needs and issues to the negotiating table with the State/Region, managing the allocation of EU funds. The study cases have demonstrated that "intermediate bodies" between the upper and the lower administrative level best suit in managing collaborative governance since they combine regional strategy with direct dialogue with local actors. The study cases therefore stress that effectiveness of collaborative tools in reference to GI implementation is closely connected to the promoters' capacity to negotiate and orient regional programmes' measures and funds.

Tools such as these Pacts do not add another planning level to a context already characterized by multiple overlapping regulatory tools. On the contrary, flexibility and administrative simplification allow for greater adaptability to local contexts and EU funding opportunities.

Another successful aspect to highlight is the participatory approach in defining the practices to be adopted, likely to increase the engagement of farmers and their awareness of environmental issues [51].

The study cases of the Pact for Biodiversity and the Torricchio Reserve AEA witnessed the solid involvement of farmers in decision-making processes, and in both cases a cooperative approach was promoted for managing common goods and natural resources. The great relevance given to the social aspect in terms of participation and appreciation of the involved communities are viewed as preconditions for maintaining and correctly managing GI [52].

The European Commission promotes these integrative approaches to strengthen synergies resulting from commitments jointly undertaken by a group of farmers, thereby multiplying the environmental and climate benefits [53]. The advantages of applying a collective approach to agriculture are widely documented and concern not only the interventions' effectiveness, but especially the capacity to encourage technical, organizational and social innovations that allow new knowledge networks to be built [49].

Considering the key role of the agricultural sector in providing GI, the CAP's radical reform is certainly expected to set ambitious objectives for the environment and the fight against climate change. On the one hand, greening would become an additional element of enhanced conditionality, and the introduction of eco-schemes would substantially move some agri-environmental issues from the Pillar II to the Pillar I, with the aim of encouraging sustainable practices in traditional and intensive agricultural production systems. On the other hand, by doubling some environmental measures between eco-schemes (Pillar I) and RDP (Pillar II), a strategic vision in planning and management of these measures at the regional scale might be missing. In the preliminary phase of negotiating of the national "CAP strategic plan" (in state/regional conferences), the challenge has been to make environmental and climatic measures consistent, since they are still on separate tracks. The first drafts of the "CAP strategic plans" lack a territorial vision that, in turn, could lead to the loss of the synergic and cumulative effect of some ecological interventions and measures. It is clear that in Italy, characterized by diverse agricultural realities (type of farms and territorial features), all regions, through the RDP measures (such as the "cooperation" measure), will be charged with ensuring the effectiveness of interventions in a territorial framework. On the one hand, planning tools should be tied by an overall coherence with respect to structural and strategic objectives rather than to rigid compliance rules. On the other hand, governance schemes in decision-making processes should introduce public partnerships and public-private agreements.

Finally, convergence between different disciplines and institutional frameworks can be attained by bridging ecological connectivity with greening and agri-environmental measures provided by the CAP. However, such crucial issue can be settled only through policy measures countering land fragmentation at different scales in tune with planning tools likely to accommodate ecological corridors and green networks. These actions should be framed on a regional scale and through cooperative approaches.

Future research could analyze the most suitable tools to foster and organize the relationships among key stakeholders according to the collaborative governance tools described above.

Author Contributions: All authors have read and agreed to the published version of the manuscript. Issues and contents of this article were largely discussed and shared by the Authors, who have jointly written all Sections, except for Section 2, written by A.L.P. and R.D.A., and Section 4, written by S.M. (Stefano Magaudda) and S.M. (Serena Muccitelli).

Funding: This research received no external funding.

Acknowledgments: The Authors are grateful to Andrea Catorci of UNICAM; Veronica Platinetti of the Vercelli Province and Carlotta Sigismondi of the Research Institute Poliedra for the interviews and the materials made available. The authors would like to thank the reviewers for their careful, constructive and insightful comments

Conflicts of Interest: The authors declare no conflict of interest

References

1. European Commission. *The Fifth EC Environmental Action Programme*; European Commission: Brussels, Belgium, 1992.
2. European Environmental Agency. *Environment in the European Union at the Turn of the Century*; EEA: Copenhagen, Denmark, 1999; ISBN 92-828-6775-7.
3. European Commission. *White Paper on Environmental Liability COM (2000) 66 Final 9 February 2000*; European Commission: Brussels, Belgium, 2000.
4. Kingston, S. (Ed.) *European Perspectives on Environmental Law and Governance*; Routledge: New York, NY, USA, 2013.
5. European Commission. *Communication to the European Parliament, the Council, the European Economic and Social Committee and the Committee of the Regions: Green Infrastructure (GI)—Enhancing Europe's Natural Capital*; European Commission: Brussels, Belgium, 2013.
6. European Commission. *The EU Guidance Document on Integrating Ecosystems and Their Services in Decision-making*; European Commission: Brussels, Belgium, 2013.
7. European Commission. *EU Guidance Document on a Strategic Framework for Further Supporting the Deployment of EU-Level Green and Blue Infrastructure SWD, 193 Final, 2019*; European Commission: Brussels, Belgium, 2019.
8. Millennium Ecosystem Assessment. *Ecosystems and Human Well-Being: Synthesis*; Island Press: Washington, DC, USA, 2005.
9. TEEB. The Economics of Ecosystems and Biodiversity. In *The Economics of Ecosystems and Biodiversity—An Interim Report*; European Commission: Brussels, Belgium, 2008.
10. Allen, B.; Keenleyside, C.; Menadue, H. *Fit for the Environment: Principles and Environmental Priorities for the 2014–2020 Rural Development Programmes*; Report Produced for the RSPB; Institute for European Environmental Policy: London, UK, 2012; Available online: http://minisites.ieep.eu/assets/985/IEEP_2012_Fit_for_the_environment.pdf (accessed on 28 October 2019).
11. Roggero, P.P.; Todari, M.; Seddaiu, G. Agroecologia e politiche agroambientali. *Agriregionieuropa* **2006**, *2*, 35–39.
12. Vanni, F.; Cisilino, F.; Fiorani, F. L'esperienza degli Accordi Agroambientali d'Area nelle Marche. *RRN Mag. 190* **2011**, *60*, 58–59. (In Italian)
13. EC. *Our Life Insurance, Our Natural Capital: An EU Biodiversity Strategy to 2020*; COM, 2011, 244 Final; European Commission: Brussels, Belgium, 2011.
14. De Filippis, F. (Ed.) *La Nuova PAC 2014–2020*; Quaderni del Gruppo: Rome, Italy, 2013. (In Italian)

15. Fabiani, G. *Letture Territoriali Dello Sviluppo Agricolo*; Franco Angeli: Milan, Italy, 1991. (In Italian)
16. Sotte, F. Scenari evolutivi del concetto di ruralità. *Proposte e Ricerche* **2013**, *71*, 122–144. (In Italian)
17. Esposti, R.; Sotte, F. *Le Dinamiche Del Rurale. Letture Del Caso Italiano*; Franco Angeli: Milan, Italy, 2001. (In Italian)
18. European Community. *Single European Act*; Title VII, art. 130 R; European Commission: Brussels, Belgium, 1987.
19. European Commission. *Communication from the Commission to the European Parliament and the Council on "The Future of the Rural World"—COM(88) 501 Final 1988*; European Commission: Brussels, Belgium, 1989.
20. Anania, G. (Ed.) *Policy Vision for Sustainable Rural Economies in an Enlarged Europe*; Studies on Spatial Development; Akademie für Raumentwicklung in der Leibniz-Gemeinschaft: Hannover, Germany, 2003; Volume 4.
21. European Commission. *EU Commission—COM Document (1997) Agenda 2000—For a Stronger and Wider Union*; Document Drawn up on the Basis of COM (97) 2000 Final, 13 July 1997. Bulletin of the European Union, Supplement 5/97; European Commission: Brussels, Belgium, 1997.
22. De Filippis, F. (Ed.) *Oltre il 2013. Il Futuro Delle Politiche Dell'unione Europea Per L'agricoltura e le Aree Rurali*; Quaderni del Gruppo: Rome, Italy, 2013. (In Italian)
23. Boschetto, R.G. Il greening nella PAC quale strumento innovativo per una gestione conservativa del suolo. *Reticula* **2014**, *7*, 36–40. (In Italian)
24. Campli, M.; Pascale, A. Sdoppiare La Pac. 2018. Available online: http://www.alfonsopascale.it/index.php/sdoppiare-la-pac/# (accessed on 22 December 2019).
25. Pupo d'Andrea, M.R. Finestra sulla Pac n. 31. *Agriregionieuropa* **2016**, *44*, 31. Available online: https://agriregionieuropa.univpm.it/it/content/article/31/44/finestra-sulla-pac-n-31 (accessed on 29 November 2019).
26. Cagliero, R.; Monteleone, A.; Varia, F. Le criticità di attuazione dei Psr 2014-20 attraverso le Postazioni regionali della Rete Rurale Nazionale. *Agriregionieuropa* **2018**, *52*. Available online: https://agriregionieuropa.univpm.it/it/content/article/31/52/le-criticita-di-attuazione-dei-psr-2014-20-attraverso-le-postazioni-regionali (accessed on 29 November 2019).
27. European Commission. DG Agriculture and Rural Development: The Post-2020 Common Agricultural Policy: Environmental Benefits and Simplifications. 2019. Available online: ec.europa.eu/info/sites/info/files/food-farming-fisheries/key_policies/documents/cap-post-2020-environ-benefits-simplification_en.pdf (accessed on 6 December 2019).
28. Saraceno, E. Le molte ambiguità delle politiche rurali dell'UE: Come affrontarle dopo il 2020? *Agriregionieuropa* **2017**, *50*, 31–35. (In Italian)
29. European Environment Agency. *Urban Sprawl in Europe—The Ignored Challenge. EEA Report n. 10/2006*; European Environment Agency: Copenhagen, Denmark, 2006.
30. Council Directive 92/43/EEC of 21 May 1992 on the Conservation of Natural Habitats and of Wild Fauna and Flora. Available online: https://eur-lex.europa.eu/LexUriServ/LexUriServ.do?uri=OJ:L:1992:206:0007:0050:EN:PDF (accessed on 12 December 2019).
31. United Nations. Convention on Biological Diversity. 1992. Available online: https://www.cbd.int/doc/legal/cbd-en.pdf (accessed on 26 November 2019).
32. SCBD. *Decision Adopted by the Conference of the Parties to the Convention on Biological Diversity at Its Tenth Meeting. X/2. The Strategic Plan for Biodiversity 2011–2020 and the Aichi Biodiversity Targets*; SCBD: South Jakarta, Indonesia, 2010; Available online: https://www.cbd.int/doc/decisions/cop-10/cop-10-dec-02-en.pdf (accessed on 26 November 2019).
33. D'Ascanio, R.; Palazzo, A.L. Molteplicità Territoriali: Il periurbano tiburtino. *Spec. Issue Urban. Inf.* **2018**, *272*, 651–654. (In Italian)
34. Sandstrom, U.G. Green Infrastructure Planning in Urban Sweden. *Plan. Pract. Res.* **2002**, *17*, 373–385. [CrossRef]
35. Davies, C.; McGloin, C.; MacFarlane, R.; Roe, M. Green Infrastructure Planning Guide Project: Final Report. *Annfield Plain NECF* **2006**. [CrossRef]
36. Benedict, M.A.; McMahon, E.T. Green infrastructure: Smart conservation for the 21st century. *Renew. Resour. J.* **2002**, *20*, 12–17.
37. Benedict, M.A.; McMahon, E.T. *Green Infrastructure: Linking Landscapes and Communities*; Island Press: Washington, DC, USA, 2006.

38. Clergeau, P.; Blanc, N. (Eds.) *Trames Vertes Urbaines. De la Recherche Scientifique au Projet Urbain*; Le Moniteur Editions: Paris, France, 2013.
39. Andreucci, M.B. *Progettare Green Infrastructure. Tecnologie, Valori e Strumenti Per la Resilienza Urbana*; Wolters Kluwer: Milano, Italia, 2017. (In Italian)
40. Lafortezza, R.; Davies, C.; Sanesi, G.; Konijnendijk, C.C. Green Infrastructure as a tool to support spatial planning in European urban regions. *For. Biogeosci. For.* **2013**, *6*, 102–108. [CrossRef]
41. Grădinaru, S.R.; Hersperger, A.M. Green infrastructure in strategic spatial plans: Evidence from European urban regions. *Urban For. Urban Green.* **2019**, *40*, 17–28. [CrossRef]
42. Hansen, R.; Pauleit, S. From multifunctionality to multiple ecosystem services? A conceptual framework for multifunctionality in green infrastructure planning for urban areas. *Ambio* **2014**, *43*, 516–529. [CrossRef] [PubMed]
43. Dente, B.; Kjellberg, F. (Eds.) *The Dynamics of Institutional Change. Local Government Reorganization in Western Democracies*; Sage: London, UK, 1988; p. 191.
44. Yin, R.K. *Case Study Research. Design and Methods*; Sage: Thousand Oaks, CA, USA, 1994.
45. Dente, B. *Le Decisioni Di Policy*; Il Mulino: Bologna, Italy, 2011. (In Italian)
46. Ministero dell'Ambiente, Codice dell'Ambiente, D.Lgs 152/2006 (mod. 2016) art.68-bis. Available online: https://www.gazzettaufficiale.it/dettaglio/codici/materiaAmbientale. (accessed on 12 December 2019).
47. Ires Piemonte. *PI Contratti di Fiume e di Lago in Piemonte. Politiche per la Tutela e il Mantenimento Della Risorsa Acqua*; Rapporto di Ricerca; Centro Stampa Regione Piemonte: Turin, Italy, January 2012; Available online: www.ires.piemonte.it/pdf/Rapporto-Contratti-fiume-di-lago-Piemonte.pdf (accessed on 6 November 2019).
48. Gusmaroli, G.; Dodaro, G.; Schipani, I.; Perin, C.; Alberti, F.; Magaudda, S. Wetland Contracts: Voluntary-based agreements for the sustainable governance of Mediterranean protected wetlands. In *Proceeding of the 2nd Euro Mediterrenean Conference for Environmental Integration, Sousse, Tunisia, 10–13 October 2019*; Springer Nature: Berlin/Heidelberg, Germany, 2019. (In Italian)
49. Magaudda, S.; Muccitelli, S.; Pozzi, C. La governance ambientale dell'Agro Pontino: Dalla rete ecologica ai contratti di fiume. *Reticula* **2019**, *22*, 49–59. (In Italian)
50. Ahern, J. Green infrastructure for cities: The spatial dimension. In *Cities of the Future Towards Integrated Sustainable Water and Landscape Management*; Novotny, V., Brown, P., Eds.; IWA Publishing: London, UK, 2007; pp. 267–283.
51. Chiodo, E.; Vanni, F. La gestione collettiva delle misure agro-ambientali: Oltre le esperienze pilota? *Agriregionieuropa*. 2014, 36. Available online: https://agriregionieuropa.univpm.it/it/content/article/31/36/la-gestione-collettiva-delle-misure-agro-ambientali-oltre-le-esperienze-pilota (accessed on 6 November 2019).
52. Malcevschi, S. Infrastrutture verdi e buone pratiche. *Valutazione Ambient.* **2013**, *24*, 21–23. (In Italian)
53. European Commission. *Regulation of the European Parliament and of the Council No 1305/2013 of 17 December 2013 on Support for Rural Development by the European Agricultural Fund for Rural Development (EAFRD) and Repealing Council Regulation (EC) No 1698/2005*; European Commission: Brussels, Belgium, 2013.

sustainability

MDPI

Article

Integrating Green-Infrastructures Design in Strategic Spatial Planning with Geodesign

Michele Campagna *, Elisabetta Anna Di Cesare and Chiara Cocco

Department of Civil and Environmental Engineering and Architecture (DICAAR), University of Cagliari, 09123 Cagliari, Italy; elisabetta.dicesare@unica.it (E.A.D.C.); chiara.cocco@unica.it (C.C.)

* Correspondence: campagna@unica.it; Tel.: +39-070-675-5203

Received: 31 December 2019; Accepted: 24 February 2020; Published: 28 February 2020

Abstract: In the last decades green infrastructure planning, design, and management have been widely recognized as a way to contribute to reach higher levels of sustainability of development. However, often green infrastructures are considered in a sectoral way, while their design should be more integrated within comprehensive planning and design. The paper proposes the use of geodesign methods and technology to support the early phases of integrated strategic territorial planning, in order to enrich the relationships between the design of green infrastructure and of the other relevant systems via more comprehensive planning and design, and by applying systems thinking. A case study developed with architecture and engineering students under the umbrella of the International Geodesign Collaboration is used, to demonstrate how with intensive geodesign workshops it is possible to create spatially explicit design scenarios which take into account the relationships between green infrastructure and other territorial systems and dynamics. A set of analyses on the case study results of the two scales is used to demonstrate the assumption. It is also argued that geodesign intensive workshops can, in a very short time, contribute to raising the awareness among the participants of collaborative design to the importance of green infrastructure in strategic territorial planning.

Keywords: spatial green infrastructures; geodesign; spatial planning

1. Introduction

Over the last decades Green Infrastructure (GI) has stimulated growing interest in research policy-making and planning [1]. Since the concept was first introduced in the second half of the 1990s, in contrast with the one of "grey infrastructure", it has been recognized as an important element for contributing to the sustainability of development [2]. Green infrastructure can be defined as a strategically planned network of high-quality natural and semi-natural areas, either in terrestrial, freshwater and marine zones, which contribute to enhancing environmental protection and biodiversity preservation, while delivering multiple valuable ecosystem goods and services [3]. The importance of GI lies in its capacity to provide a wide range of benefits, such as environmental benefits (e.g., protection against soil erosion, provision of clean water, rainwater retention), biodiversity benefits (e.g., ecological corridors, landscape permeability), social benefits (e.g., food production, diversification of local economy, human wellbeing, additional tourism opportunities) and climate change mitigation and adaptation benefits (e.g., carbon storage and sequestration, strengthening ecosystems resilience, flood alleviation).

GI can be found both in natural and green artificial areas and it may take different forms. The following main typologies of green infrastructure elements can be distinguished at various scales: core areas of high biodiversity value; restoration zones such as reforestation areas; ecosystem service zones oriented to provide a range of ecosystem service benefits; green urban and peri-urban areas such as urban parks and green roofs; and natural and artificial connectivity features to assist species movement [4]. Connectivity and multi-functionality are considered key characteristics for the GI to remain resilient to change [1,5]. The first characteristic is related to the need of biotic functional groups to have not only high-quality living and restoration space (i.e., core areas), but also to be able to move across patches in order to support genetic diversity [6]. The second characteristic concerns the ability of a GI to perform several functions in the same spatial area related to the provision of a variety of Ecosystem Services (ESs), serving a range of functions for both nature and society.

Due to its multi-sectorial nature, GI requires a holistic and cross-sectorial approach to spatial planning in order to both stimulating possible synergies and coordinating initiatives, and to avoid the consequent risk of conflicts between objectives related to different goals [1]. Munoz-Criado et al. [5] recognize the central role of GI in guiding the early stage of the planning process across the different levels (i.e., regional, municipal and project scale) and proposed a reform of the legal framework of landscape and urban planning at the Autonomous Region of Valencia by identifying a unique GI as an ecological-based tool to overcome fragmentation and to interconnect all the different policies concerning landscape protection. Over the last decades a wide range of GI projects have been carried out on local, regional, national and trans-boundary levels, highlighting the necessity for projects defined in different scales to be closely interconnected and coordinated in order to maximize the GI benefits.

Green infrastructure can be reinforced through strategic initiatives oriented to maintaining, restoring and connecting existing, as well as creating new, areas and features. Nowadays there is not a single globally recognized working framework for supporting GI design, however, in literature it is possible to find some different approaches trying to integrate GI in decision-making processes concerning spatial planning and design. Lanzas et al. [7] define an operational GI planning framework to reconcile the need of biodiversity maintenance and the need of maximizing benefits deriving from ESs both for humans and for natural conservation, by using a systematic planning approach. Cannas et al. [8] propose a methodology to support planners in spatially identifying multifunctional GI through (i) the combination and mapping of four values, specifically calculated, representing four main functions of a GI (i.e., biodiversity conservation, supply for ecosystem services, recreation and cultural heritage), and (ii) the identification of the most suitable land parcels to be included in ecological corridors on the basis of their connectivity. Liquete et al. [9] propose a methodology, useful at different spatial scales, to design green infrastructure networks at landscape level based on the delivery of multiple ecosystem services, essential habitats and connectivity.

At the institutional level, the European Union (EU) recognize the contribution of GI in achieving some of the EU's key policy objectives, such as protecting and restoring natural capital and combat the impacts of climate change [10]. The EU Biodiversity Strategy, adopted by the European Commission in 2011, sets out 6 targets and 20 actions with the aim of putting an end to the loss of biodiversity by 2020. In particular, target 2 explicitly refers to GI and requires that "ecosystems and their services are maintained and enhanced by establishing green infrastructure and restoring at least 15% of degraded ecosystems" [11]. The EU Biodiversity Strategy also included a commitment for the Commission to develop an EU-wide Strategy on GI, later adopted on May 2013, which identifies some urgent issues to be addressed in order to encourage GI projects and investments to protect and improve natural capital. The need for GI to become a standard part of spatial planning and territorial development is recognized, together with the need for consistent and available data concerning ecosystems and the development of innovative approaches. In this regard, a significant role in the definition of a strategically planned EU-level GI is Natura 2000, which is the main instrument for biodiversity maintenance within the European Union. Specifically, it is an ecological network extended for over 700,000 km^2 on terrestrial areas and for over 500,000 km^2 on marine areas [12] established by Directive 92/43/CEE (Habitats Directive), which identifies a series of regions where long-term conservation of natural habitats and flora and fauna species need to be ensured. The Natura 2000 network includes over 27,000 sites, classified as: Sites of Community Importance (SCI)—subsequently designated as Special Areas of Conservation (SAC) pursuant to the Habitats Directive— and of Special Protection Areas (SPA) in accordance with 2009/147/CE (Birds Directive). Natura 2000 sites should be not only preserved, by avoiding the development of deleterious activities within their boundaries, but they should also be properly interconnected to allow the genetic renovation of the species populations [4]. Within this context green infrastructure contributes to the full implementation of both the Birds and Habitats Directives.

To date, despite that several GI design best practices can be found in Europe [13], a more integrated approach to their design within the spatial planning practice is desirable. This would require to inform design taking into account the tight relationship between GI and the other natural (e.g., blue infrastructure, agriculture) and artificial (e.g., grey infrastructure, commerce, housing) systems according to comprehensive approach at least in the early phases of strategic territorial planning. According to this line of thought, several geodesign case studies have been recently carried out [14–16] which, within their broader scope, somehow included the synergic design of GI as well as of the other systems stemming from local development priorities applying systems thinking, so ensuring a contextual territorial coordination of future development scenarios.

With a similar approach, this paper presents a recent geodesign study carried out at the University of Cagliari, Italy involving undergraduate and graduate students in developing strategic planning alternatives for the Metropolitan City of Cagliari (MCC). The objective is to present an operative example focusing on applying systems thinking to planning and design, and to demonstrate how the geodesign approach, and more specifically geodesign workshops can facilitate the integration of GI design within early phases of regional strategic planning. To this end, in Section 2, a brief discussion on the geodesign approach is given as methodology background, and the Cagliari case study is presented, including the description of the study area, the design process, and the general results of the study. In the second part of the paper, in Section 3 a detailed focus to the GI system modelling is given in order to clarify how knowledge about GI can be structured to inform integrated design. Lastly, Section 4 focuses on the analysis of the results of the geodesign study in order to clarify the relationships between the GI and the other systems along the design process. The conclusions summarize the overall results and propose issues for further research.

2. The Geodesign Approach to Spatial Planning and Design

Geodesign applies systems thinking [17] to spatial planning and design [18], that is seeking to understand the big picture in territorial transformation dynamics in relation to short-and long-term consequences. It would be reductive to think green infrastructure performance is only due to the correct design of its inner parts. From a systems thinking perspective it would be rather more correct to assume GI performance is affected by the complex functioning of the surrounding environment. If this is true, green infrastructure planning and design should be handled applying a systems approach and designing GI and other interacting subsystems in a single design endeavor.

The landscape architecture and planning tradition were since long aware of this principle, however, until recently, traditional analogue methods had some limitations inherent to the supporting media. With the introduction of computation in geography, and with most recent advances in digital (geographic) technologies, it is for the first time currently possible to build user-friendly multiscale interactive computation environments to support collaborative planning and design [19]. This is what characterizes geodesign, as opposite to traditional landscape architecture and planning methods, by offering concepts and methods and enabling technologies supporting the creation of design proposals with real-time dynamic impact simulations informed by geographic contexts.

In the last decades, the shift from hand-drawing, to Computer Aided Design (CAD), to Geographic Information Systems (GIS), enabled the creation of Planning Support Systems (PSS) [16] which can offer real-time computational feed-back in terms of visualization, simulation, and impact assessment. Some available PSS are structured to help designers to better understand the effects of a design choice in one system not only within itself, but within all the other interacting systems, supporting systems thinking. From the larger to the smaller design scale, Building Information Modelling (BIM), GIS and PSS nowadays enable all that to an unprecedented level. However, until recently, even more advanced PSS somehow failed to address important parts of the planning and design workflow, resulting in an implementation gap which limited their widespread diffusion [20]. Among the issues to be addressed for improving PSS diffusion, Geertman and Stillwell [21] argued their design should be more tightly related to the contextual planning and design process, in order to better meet the user's needs. Along the same line, Campagna [22] argued the need for more process orientation in PSS design. Geodesign methods, and in particular the Steinitz's framework for geodesign [23], may help in this respect as it supplies a guide for the organization of the full planning and design process in a way that is adaptable to contextual local conditions.

The Steinitz's framework for geodesign [23] entails the creation of six models. The *representation* model provides the description of the study area in its historical evolution until the present, and it is based on available data. The *process* model simulates the evolution of the system under the assumption of no action (i.e., do-nothing design alternative) and it is based on modelling and forecasting. The *evaluation* model may express the values of the experts, of the decision-makers and/or of the stakeholders involved in planning and design, and it is based on their assessment of the output of the process model. While the development of the representation and especially of the process models is often carried on by system experts, the evaluation model is the base for action (i.e., linking the analysis to the design of different planning and design alternatives) and it may require, to a variable degree depending on the local context, the expression of values by a wider number of stakeholders, granted the guidance of experts. The process of implementing the models is not strictly linear though, and several iteration loops may be required during their construction. The *change* model uses the evaluation model as input to orient actions, locational choices and avoidance of risks of hazard in the creation of design alternatives. In this sense, the construction of the evaluation model is a critical step in the application of the geodesign framework for it bridges the gap between knowledge (i.e., analysis) and action (i.e., design and decision-making). Once design alternatives are created, their performance can be assessed with the *impact* model. Lastly, the *decision* model makes the decision-making context explicit, defining who acts in which step of the workflow, and on the base of what power-relationships and values.

In the last five years, since its availability to the public, the PSS Geodesignhub faced a growing interest by the geodesign community for it was designed to support the second part of the Steinitz' framework, that is the iterative implementation of the change, impact and decision models. Despite its short life, Geodesignhub was widely used to support geodesign workshops worldwide and its usage have been widely documented in literature [14,24–29].

In the light of the above premises, in the next paragraph, the MCC case study is introduced, and in Section 3 a detailed discussion on the creation of the evaluation model for the GI system is given.

The Intenational Geodesign Collaboration and the Cagliari Case Study

The International Geodesign Collaboration (IGC) is a recent global research initiative involving almost one-hundred partners from five continents [30]. In early 2018, a group of scholars active in geodesign coordinated a major research project, which, in one year only, produced fifty-six studies following common guidelines. The main research objective was to gain insights in a systematic and comparable way on how global territorial dynamics affect local regions and cities around the world, and to explore how technology innovations with regards to different interrelated territorial systems may help to address current sustainability challenges. Before the local studies started, IGC working groups developed preliminary studies to define sets of (i) global assumptions; (ii) system innovations; and (iii) common working and reporting rules [31], so that at the end of the geodesign studies the results could be easily compared. The results were presented in February 2019 at the first IGC meetings in Redlands California, hosted by the leading GIS industry ESRI.

Among other common working rules, a set of nine systems structuring the study was agreed beforehand by the coordination team. They included green, blue (BI), energy (EI) and grey infrastructures (TRANS), low-(LDH) and mix high-density housing (MIX), industry and commerce (IND), agriculture (AG), institutional services (INST). A tenth system was left to be defined locally by the study teams on the base of the contextual conditions.

Most of the IGC studies were undertaken following the format of intensive geodesign workshop (WS), and the majority relied in extensive use of supporting digital technologies such as GIS and PSS. According to the IGC common rules for comparability, the study areas were squared, with variable size spanning from 0.5 to 160 km. The case study of the Metropolitan City of Cagliari in Sardinia (Italy) reported in this paper was carried out at two nested scales: (1) a square 80 by 80 km including all the MCC; (2) a square 20 by 20 km focusing on the South-Eastern edge of the MCC, with a shift of the design scale (Figure 1). As tenth system for the MCC case study, historical and cultural heritage was chosen as the area is rich in resources.

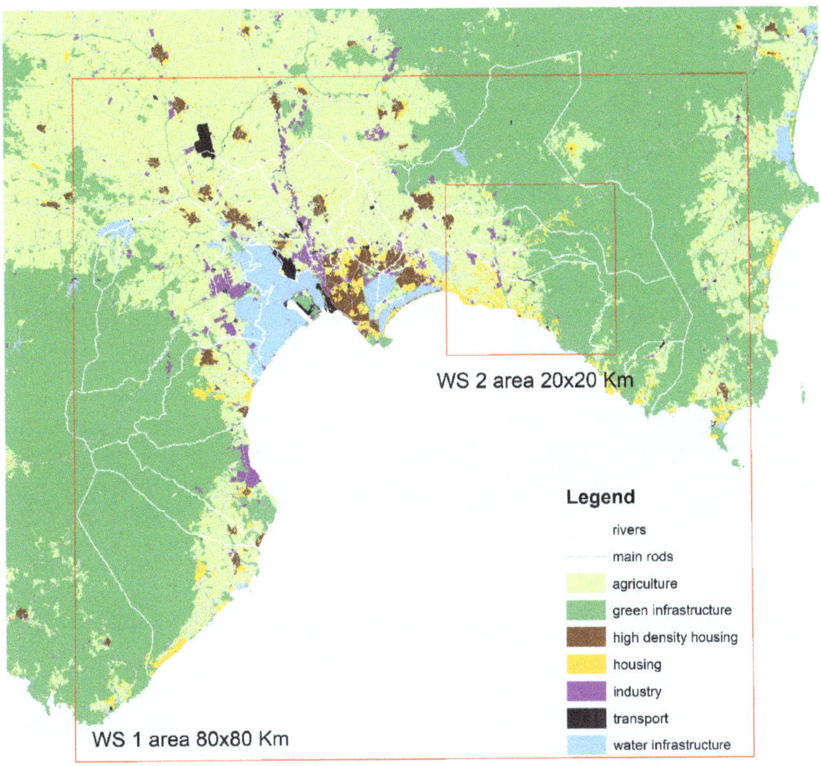

Figure 1. The Metropolitan City of Cagliari geodesign study areas.

The common IGC workflow for the study entailed the development of three alternative development scenarios considering two time-stages, namely year 2035, and 2050 [30]. The design teams had to design according to different assumptions with regard to the timing of introduction of technology innovations in their design. Accordingly, three main scenarios were to be created: (1) Early Adopter (EA) introducing technology innovations as early as in 2035; Late Adopter (LA) introducing innovations after 2035; and Non-Adopter (NA) not considering technology innovations at either of the time stages.

While the IGC study teams were free to use the digital (or analogue) design workflow they deemed appropriate on the base of locally available data and skills, many of the study followed the Steinitz's framework to inform their workflow for the design process. In our study, thanks to the availability of large scale geographic information on the Sardinian Regional geoportal, the process was fully digitally supported, relying on ESRI ArcGIS software for the development of the representation, process and evaluation models (and for the representation of the results at a later stage), and on the web-based Geodesignhub collaborative PSS for implementing the change, impact and decision models during two intensive geodesign workshops (i.e., one for each design scale) held in temporal sequence.

Overall the workshop workflow in both cases included the following main steps:

1. Creation of the representation, process and evaluation models by the study coordination team with GIS software. In this step the evaluation maps were produced for the ten systems to be used as input to inform design in the following phases;
2. Set-up of the Geodesignhub PSS by the coordination team, including the creation of a matrix recording impact scores among systems, as input for the impact model;
3. Creation of project and policy *diagrams* by the study participants. The participants in this phase acted as system experts and were asked to produce diagrams representing projects and policies considering technology innovations at each of the given time stages. The result of this phase was in both workshops the creation of a matrix of diagrams arranged by system, to be shared by the participants in the following phase;
4. Creation of integrated comprehensive design alternatives, or *syntheses*, by design teams according to the IGC scenarios (i.e., EA, LA, and NA). The syntheses were composed using the shared project and policies matrix;
5. Final negotiation among design teams to reach consensus on the final design for each given IGC scenario.

In general, in both cases the workshop was a first experience in applying systems thinking to address the MCC development challenges for the participants, who were architecture and engineering students with little or no spatial planning and design (and GIS) experiences. The scenario-driven design alternatives produced during the two workshops are shown in Figures 2 and 3.

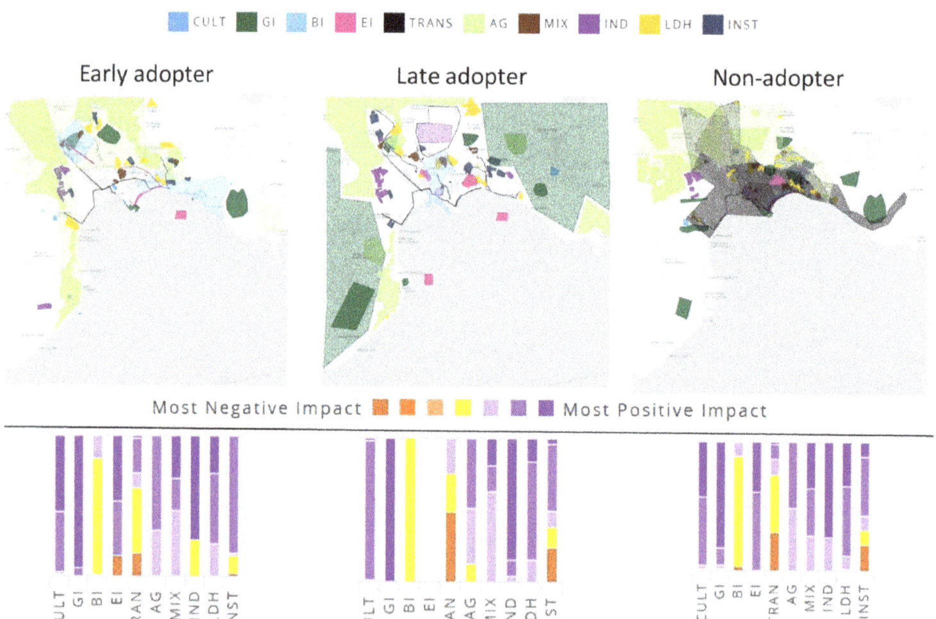

Figure 2. Final syntheses of the International Geodesign Collaboration (IGC) Metropolitan City of Cagliari (MCC) geodesign workshop 2018 (area 80 × 80 km) and related impact summary graph showing the direct impact of selected diagrams in one system on itself.

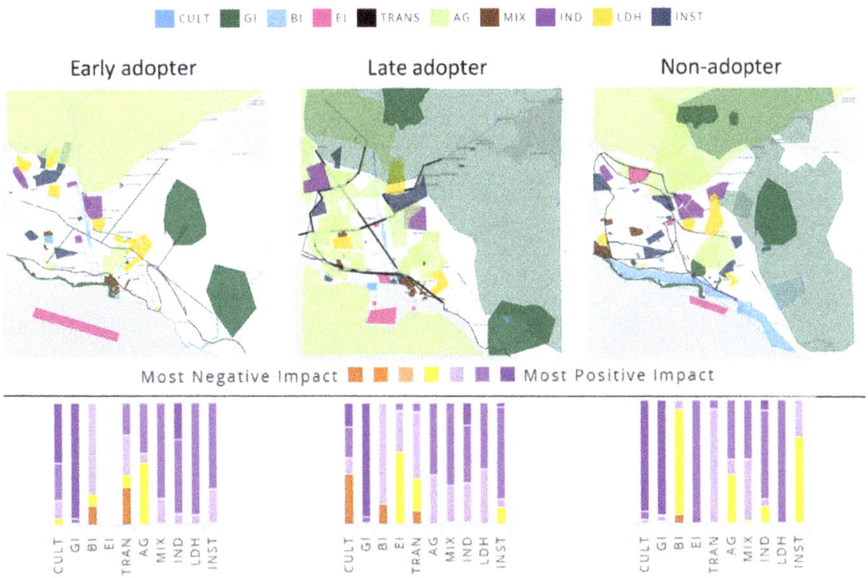

Figure 3. Final syntheses of the IGC South-East MCC geodesign workshop 2018 (area 20 × 20 km) and related impact summary graph showing the direct impact of selected diagrams in one system on itself.

The geodesign approach, in the very limited time of the intensive workshops (i.e., 15 h each), enable them to grasp the complexity of studying comprehensive strategic planning alternatives dealing with the interaction of ten systems at once. As shown in Figures 2 and 3, the impact models of the final syntheses were overall positive, as substantially was the impact on the GI system in all the alternatives. In the light of these general results, in the following sections a closer focus is given on the GI system modelling and design within the MCC study in order to highlight the adopted level of detail in the analysis and in the design.

3. The Green Infrastructure System in the Metropolitan City of Cagliari

The Metropolitan City of Cagliari, recently established with Regional Law n. 2/2016, is located in Southern Sardinia (Italy) and includes 17 municipalities. The area is located on the fertile plain of Campidano and overlooks the Gulf of Cagliari to the South. The area is surrounded to the East and to the West by mountains.

The MCC has currently a total population of 430,798 inhabitants distributed over an area of 1248.71 km^2, with a density of 345 inhabitants per km^2. The population distribution is uneven as it is more concentrated around the regional capital, Cagliari and its neighboring municipalities belt, Quartu Sant'Elena, Selargius, Quartucciu and Monserrato. The territory is characterized by the presence of heterogeneous areas of naturalistic interest, since the geomorphology, lithology, pedology, land cover and landscape are extremely variable.

In particular, the MCC includes sixteen Natura 2000 sites: twelve SACs, and four SPAs both in marine and terrestrial zones (Table 1, Figure 4) covering an area of 52,000 hectares of SACs (31,000 ha of which fall exclusively within the territory of the local authority) and 49,000 hectares of SPAs (18,000 ha of which fall exclusively within the territory of the local authority). Despite the presence of a quite number of core areas characterized by a rich biodiversity, the sites, often located in or near densely populated urban areas, are not always adequately connected to each other. Within this context, one of the main priorities for the creation of a metropolitan level GI is to enhance connectivity among protected areas and counter habitat fragmentation.

Table 1. List of Natura 2000 sites (Special Areas of Conservation – SAC; Special Protection Areas – SPA) in the Metropolitan City of Cagliari.

Site Name	Site Code	Type	Area [ha]
Stagno di Cagliari, Saline di Macchiareddu, Laguna di S. Gilla	ITB042207	SAC	8.567
Costa di Cagliari	ITB040021	SAC	2623.851
Bruncu de Su Monte Moru—Geremeas (Mari Pintau)	ITB040051	SAC	139.000
Monte dei Sette Fratelli e Sarrabus	ITB041106	SAC	9295.794
Canale su Longuvresu	ITB042207	SAC	8.567
Riu S. Barzolu	ITB042241	SAC	281.341
Torre del Poetto	ITB042242	SAC	9.371
Tra Forte Village e Perla Marina	ITB042231	SAC	0.320
Monte Sant'Elia, Cala Mosca e Cala Fighera	ITB042243	SAC	27.448
Foresta di Monte Arcosu	ITB041105	SAC	30,369.312
Stagno di Molentargius e territori limitrofi	ITB040022	SAC	1275.232
Capo di Pula	ITB042216	SAC	1576.379
Stagno di Cagliari	ITB044003	SPA	3756.385
Foresta di Monte Arcosu	ITB044009	SPA	3132.074
Monte dei Sette Fratelli	ITB043055	SPA	40,473.932
Saline di Molentargius	ITB044002	SPA	1307.155

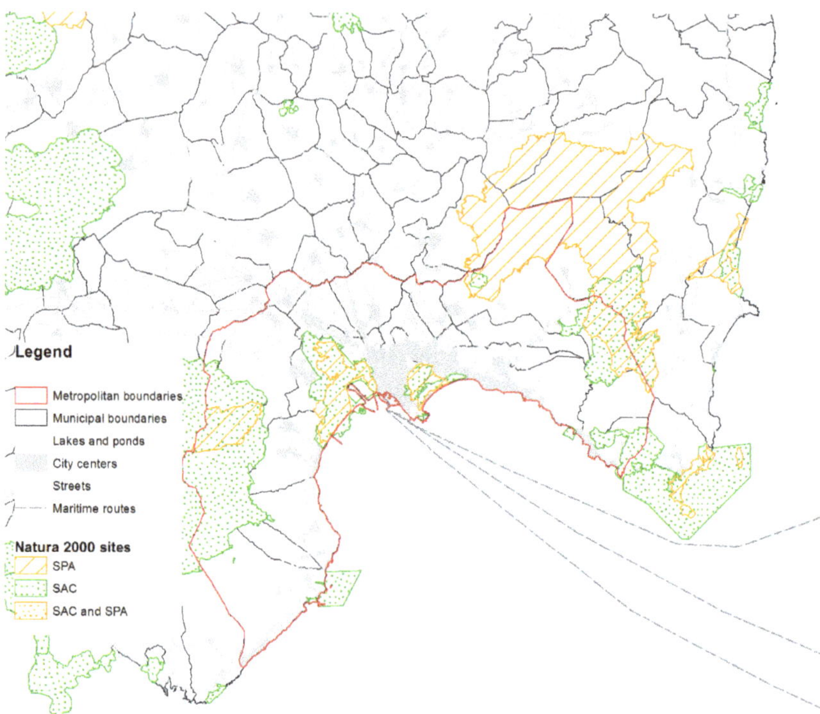

Figure 4. The Metropolitan City of Cagliari and the Natura 2000 sites.

As noted by Slätmo et al. [1] "the land uses that conflict most in terms of habitat fragmentation with GI preservation and development are transport and energy infrastructures and agriculture". In addition, low-density land use for built-up areas, commonly referred to as 'urban sprawl' is an important factor of habitat fragmentation, as well as, "denser urban structure and land cover changes in urban areas are often at the expense of green areas and the ecosystem services they preserve [1]". These statements also apply for the study area, and they have been taken into account in defining the input values of the impact matrix in Geodesignhub.

Green-Infrastructure Evaluation Model

In the knowledge building phase of the study, an evaluation model was prepared for each system by the study coordination team. The construction of the evaluation model is a critical step in the geodesign workflow. On the one hand, the output of the evaluation model (i.e., an evaluation map for each system) needs to include several criteria in order to suggest suitable location choices for an action in the system, as well as, to identify areas were changes are not needed or unfeasible. On the other hand, in doing so the evaluation model needs to be transparent to the user, which may otherwise not accept the recommendations represented in the evaluation maps, due to the possible occurrence of the well-known black-box effect. In addition, the number of criteria, and their spatial patterns, used to identify more or less suitable areas may vary in space. The degree of uncertainty involved in the preparation of the evaluation maps may vary accordingly, sometimes to a substantial degree. Therefore, a step by step modeling process, to be documented in detail, may help to avoid the black-box effect, and to better deal with uncertainties, while making steps based on the recommendations. Formalizing the evaluation model for each system may help in dealing with complexity and in making the model transparent, hence more reliable, to the users. Below, the evaluation model for the GI system is described in detail.

The modelling starts with short but comprehensive verbal description of the system (i.e., the "rationale" row in Table 2). The specification of categories of possible actions helps to make the description more operable.

Table 2. Modeling table of the green infrastructure evaluation model.

<table>
<tr><td rowspan="3">Rationale</td><td colspan="2">GREEN: Green Infrastructures.The Green Infrastructure system focuses on protection and connection of high quality natural and semi-natural areas located in urban, suburban and rural areas, in order both to protect biodiversity and enhance nature's ability to deliver multiple valuable ecosystem goods and services. Sardinia includes 125 Natura 2000 Sites of Community Interest and Special Protection Zones, which are part of Natura 2000, a network of core areas instituted by EU Directive 92/43/EEC (Habitat Directive) and EU Directive 2009/147/EC (Birds Directive), with the aim to ensure the long-term survival of the local most valuable and threatened species and habitats, listed under the body of the Directives. In the Metropolitan City of Cagliari, Natura 2000 sites are situated near densely populated areas and they are not always adequately connected to each other.</td></tr>
<tr><td>Possible actions on landscape (regional/metropolitan scale):
• New Protected areas
• Ecological corridors
• Green Infrastructure for storm water management
• Creek Restoration & Naturescaping
• Multi-functional farming
• Wildlife overpass
• Beehives
• Other</td><td>Possible actions on settlements (urban scale):
• Green roofs and walls
• Green streets
• Rain gardens
• Urban agriculture
• Urban forest
• Wildflower verge
• Other</td></tr>
<tr><td colspan="2">Possible integrated actions
• Green belt
• Other</td></tr>
</table>

Table 2. *Cont.*

	Existing	Not appropriate	Capable	Suitable	Feasible
Classification (Courtesy of Carl Steinitz)	Is where the system is "existing" already and in a healthy state, meaning that it is feasible to remain a constraint in terms of information but not a total Constraint.	Is lowest priority for change "not appropriate" or not capable of supporting the system, meaning don't put it there, e.g., too wet or steep or unless you provide change to the basic area conditions e.g., fill in the ocean for new land, regrade the mountain, etc. (all very risky projects). This is also a constraint in terms of information.	Is low but higher priority "capable", meaning that you can place it here IF you also provide the technology and market to make it feasible, e.g., water and sewers, access roads for mechanical harvesting, etc., and the market comes	Is higher priority "suitable", meaning that the area is capable of supporting the project and it already has the appropriate technologies to support the activity taking place e.g., septic tank soil or sewers, access roads for mechanical harvesting, etc. BUT there may not yet be a market for the change.	Is the highest priority for change "feasible", meaning that it is suitable AND there is a demand or market to provide the new land use change, e.g., that someone wants to buy the product or new house (and at a profit) OR that the government wants to protect and improve an historical landscape.
Description	This class includes: • Core areas of high biodiversity value which act as hubs for GI (i.e., Natura 2000 sites) • Natural features containing healthy functioning ecosystems, acting as wildlife corridors or stepping stones	This class includes: • Urbanized areas (due to the design scale)	This class includes: • Artificial areas that, if sustainably managed, enhance ecosystem services or assist wildlife movement	This class includes: • Natural and artificial areas that, if sustainably managed, could help to improve the general ecological quality and permeability of the landscape to biodiversity	This class includes: • Not protected natural and semi-natural areas • Buffer zones around natural protected areas and natural features that help to reconnect existing natural areas
Data	Layers • Natura 2000 sites (Sites of Community Interest and Special Protection Zones) • Corine Land Cover 2008 (CLC)	Layers • Corine Land Cover 2008	Layers • Corine Land Cover 2008	Layers • Corine Land Cover 2008	Layers • Natura 2000 sites • Corine Land Cover 2008
Operators	• Natura 2000 sites • CLC 1.4: Artificial, non-agricultural vegetated areas • CLC 4: Wetlands • CLC 5: Water bodies	• CLC 1.1: Urban fabric • CLC 1.2: Industrial, commercial and transport units • CLC 1.3: Mine, dump and construction sites	• CLC 2.1: Arable land • CLC 2.2: Permanent crops	• CLC 2.3: Pastures • CLC 2.4: Heterogeneous agricultural areas	• CLC 3: Forest and semi natural areas • Natura 2000 buffer (2000 m) • CLC 4 buffer (500 m) • CLC 5 buffer (500 m)

It should be noted that as evaluation map legend, the classification applied in the Georgia Coastal Region geodesign study [16] based on Carl Steinitz's adaptation of the G. Angus Hills work [32] was used. The description of each class is given in the row "classification" in Table 2. This classification was widely adopted by the authors, as well as, by other geodesign experts and proven to be reliable. Nevertheless, a carefully modelling stands crucial in order to reduce uncertainties in the final evaluation maps.

For each class of the evaluation map legend, a list of spatial criteria to identify them is given first in natural language (row "description", Table 2), then as list of spatial data layers (row "data", Table 2). As a complement to increase transparency, the row "operators" (Table 2) specify the conditions applied to spatial data layers through GIS functions (e.g., query, geoprocessing, and so forth). Table 2 presents the modelling table used for the GI system in the MCC workshop, and it is proposed here with the aim to illustrate the rationale of the creation of the evaluation map. In general terms, the model itself is indeed context-dependent and may vary substantially in different study areas, at different scale, and with the objectives and scope of the design.

The modelling table should be, and it usually is, distributed to the workshop participants to familiarize with the model. In addition, the modelling experts for each system should be present during the workshop in order to give the participants appropriate recommendations in case of needed. Unlike more traditional suitability maps [33,34], which are usually built for a given well-defined land use (e.g., a given type of crops for agriculture, or a given type of building typology for lower density housing, and so forth), depending on the system and on the scale, the model underlying an evaluation map should represent the level of suitability—or, on the contrary, the lack of necessity or feasibility at various locations—of more or less complex sets of development actions (i.e., conservation or development projects or land-use changes). As such, the results should be considered as general recommendation and not in a regulatory way. The importance of understanding the model and its level of uncertainty, which may vary locally depending on the class, is fundamental in order to make informative recommendations for locational choices of design actions during the geodesign workshop.

In our study, the modeling exercise was made even more complicate for it concerned both the regional (workshop 1) and the urban (workshop 2) scale at once. However, it should be noted that the scope of the workshop was exploratory in nature, and undertaken in educational context. As such, the level of the uncertainty was considered acceptable for the working context. The main objectives of the two workshops were understanding the influence in the use of advanced design technologies for the change actions, and educating early career students in architecture and civil and environmental engineering in geodesign methods and techniques for spatial planning and design. Analyzing the results of the design exercise, and the learning curve of the participants observed during the workshops, both objectives could be considered reached. Nevertheless, it should be noted that in a real world geodesign study, such a level of uncertainty might not be considered acceptable, and one model for each scale would be recommendable.

Figure 5 shows the GI evaluation map obtained by geoprocessing implementation of the modelling table (Table 2). It should be noted that the modelling workflow is not strictly linear. Several iterations were carried on by the study team before the results presented here were considered appropriate in terms of model and level of details for this study.

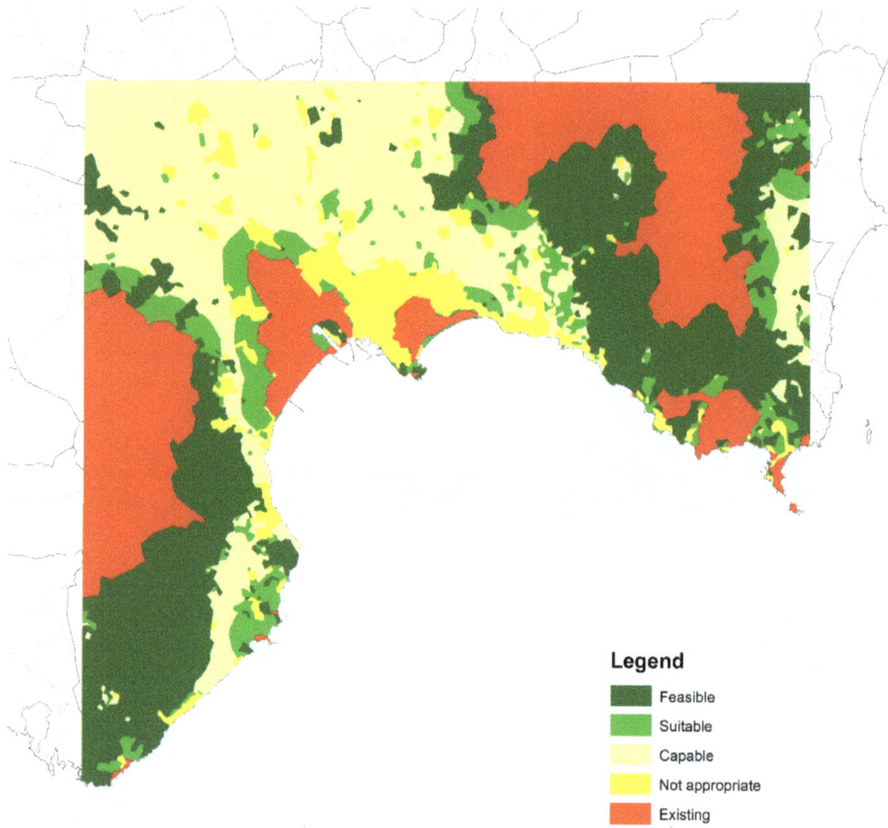

Figure 5. Green Infrastructure (GI) evaluation map classified according to the standard 5-class color code.

4. Analysis of the GI Design Results

The geodesign study was carried out in the context of two studio courses involving a large number of architecture undergraduate students and engineering graduate students (master's degree) respectively. Five three-hour sessions were fully dedicated to the collaborative development of design alternatives for each of the two planning scales (i.e., regional and urban scale). While the design in the workshops were carried out by students with little or no previous knowledge, the analysis of the designs they produced collaboratively provided interesting results, contributing to demonstrate that the overall geodesign workflow effective in supporting the integrated design of GI with other systems. A detailed analysis of the green infrastructure design results was carried out in a de-briefing phase (Figure 6).

During the step 3 (i.e., diagram creation phase) of workshop 1 on the smaller design scale (i.e., larger area) the students grouped in ten system expert teams produced overall 353 diagrams with a mean of 35 proposals in each system. In the system green infrastructure 2 policies (polygons with hatch pattern in Figure 6) and 30 projects (polygons with solid color in Figure 6) were created representing the 9% of the diagram matrix. Among them 18 were selected in the final designs of the three scenario-driven groups: NA (Figure 6b), LA (Figure 6c), EA (Figure 6d). All the design teams included further GI management down the mountain areas on the East and on the West.

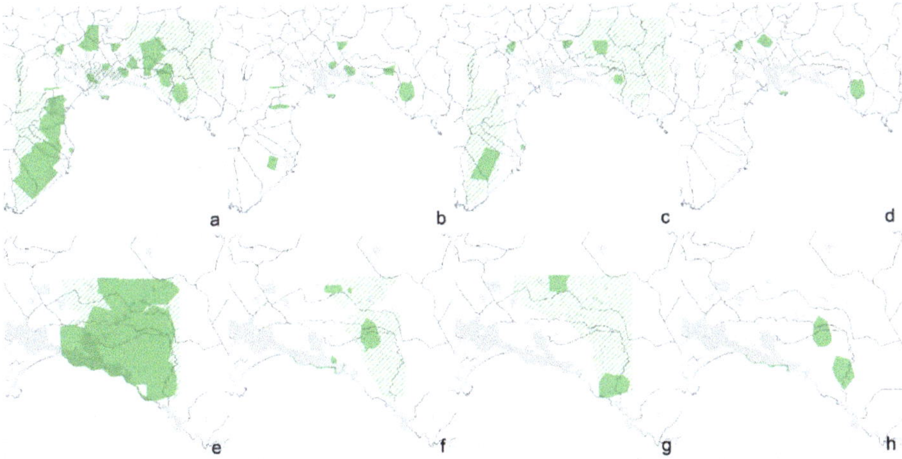

Figure 6. All the GI diagrams created during the workshop 1 on the (**a**) smaller design scale and only those selected in the final syntheses: (**b**) Non-Adopters, (**c**) Late Adapters, (**d**) Early Adapters; and the GI diagrams created during workshop 2 on the (**e**) smaller design scale and selected in the final syntheses: (**f**) Non-Adapters, (**g**) Late Adapters, (**h**) Early Adapters. Polygons with hatch pattern are policies while polygons with solid color represent projects.

Focusing more on individual projects, the EA team proposed a project representing an extended green park, connecting two existing core areas, one in the coastal and one in the inland zone (Figure 7a). The park has been designed both to create new habitats and to allow movement of animals between existing areas of high biodiversity value. Other diagrams, both located in the Northern part of the MCC near urban centers, represent respectively new residential areas designed with a modular system of green facades in order to contribute to the reduction of the urban heat island effect and new public gardens (Figure 7b) with the aim to stabilize the concentration of CO_2.

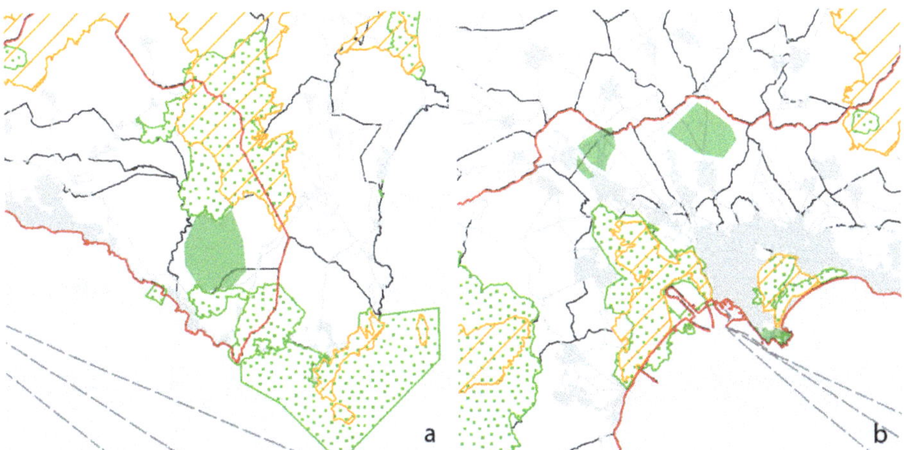

Figure 7. Example of integrated GI projects proposed by EA group in the MCC geodesign workshop 1 in the (**a**) South-Easter edge and in the (**b**) central area. Polygons with hatch pattern are policies while polygons with solid color represent projects.

The LA team proposed two wide area policies, one in the East and one in the West side of the MCC, both proposing multifunctional green areas (Figure 8a,b). Inside these areas they planned reforestation zones and green parks. Moreover, near densely populated areas they planned residential areas with green roofs and walls (Figure 8c).

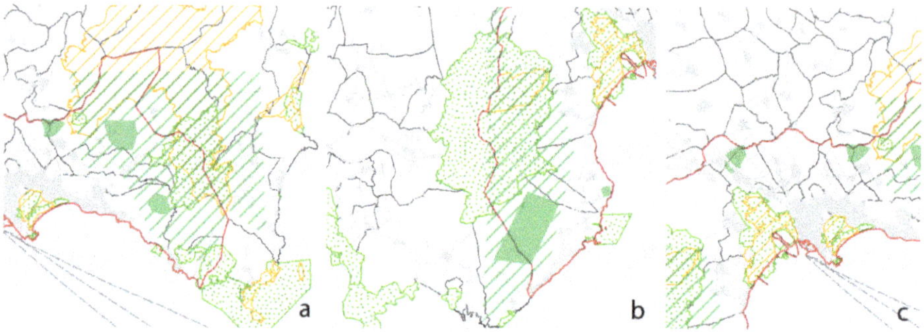

Figure 8. Example of GI projects proposed by LA group in the MCC geodesign workshop 1 in the (**a**) South-Easter edge, (**b**) in the South-Western edge and in the (**c**) central area.

The NA team proposed smaller and spatially widespread projects, most of which represent ecological corridors connecting existing green areas, which support the restoration and enable wild animal populations to move more freely (Figure 9a,b).

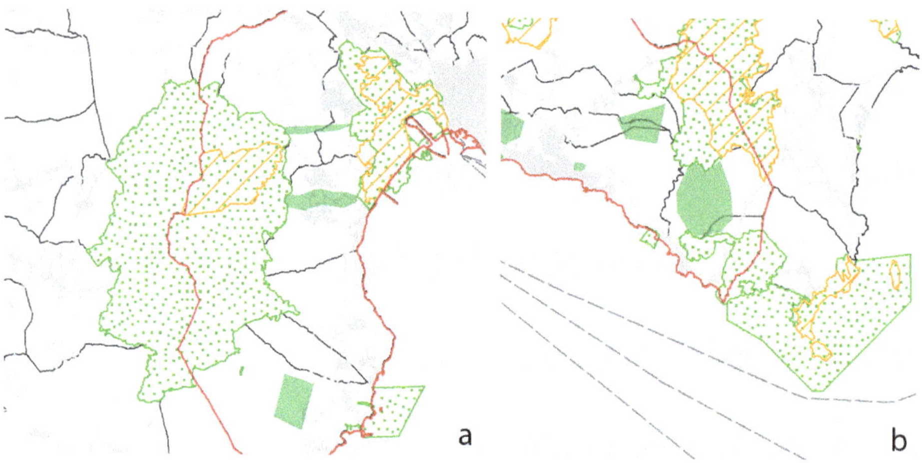

Figure 9. Example of GI projects proposed by NA group in the MCC geodesign workshop 1 in the (**a**) South-Western edge and in the (**b**) South-Easter edge.

The PSS Geodesignhub facilitates the green infrastructure design at different scales and in relation to the other systems, rather than focusing on single-level sectoral planning. Running two geodesign workshops in sequence allowed to experiment the change of scale in the design of future alternatives for the MCC. The architecture students who attended the second workshop in the sequence had to comply with the design earlier developed by the engineering students in the first workshop. Geodesignhub enabled the shift from the metropolitan level to the local level providing the projects and policies from workshop 1 as input in the larger scale study (i.e., smaller area). The GI design of the South-Eastern edge of the MCC was, thus, informed by the whole metropolitan strategy developed in workshop 1.

The architecture students designed 356 diagrams including 12 policies and 21 projects in the GI system (Figure 6e). It is worth noting that the GI diagrams included in the final syntheses of workshop 2 (NA, Figure 6f; LA, Figure 6g; EA, Figure 6h) spatially coincide in part with the small-scale GI proposals with an area of overlapping of 20%, 100% and 5% respectively. For example, the LA group in workshop 2 located an outdoor recreation area represented by a project of 9 km^2 within the multifunctional green area identified as a metropolitan-level policy by LA group in workshop 1.

The EA team proposed two main types of intervention. The first consists of two extended green parks near a Natura 2000 site, which improve the general ecological quality of the landscape and enable nature conservation and human recreation (Figure 10a). The second project is located in the Southern area along the coast and consists of a flora coastline protection green infrastructure (Figure 10b), which protects coastline and residential areas from sea level rise resulting from climate change, while creating new habitats for marine fauna species.

Figure 10. Example of GI projects proposed in the MC by EA group C geodesign workshop 2 in the (**a**) inland zone of the South-Easter edge and in the (**b**) coastal area of the South-Easter edge.

The LA team proposed a big policy in the Eastern rural area, next to a Natura 2000 site, which offers an attractive outdoor area, near the residential one, for recreation (e.g., trekking, horse riding, picnic areas) and wildlife (Figure 11a). The second project is a biodiversity-rich park, which falls under the wide area policy mentioned above and it is located in the Northern edge, planned to help to absorb CO_2 emissions and improve air quality (Figure 11b).

The NA team proposes a big policy in the Easter zone, planned to support multiple land uses in the same spatial area: human recreation and food production (e.g., wildlife-friendly agriculture and farming, which enable the permeability of the landscape to biodiversity) (Figure 12a). The second project is the flora coastline protection green infrastructure (Figure 12b) with the aim to protect from sea level rise.

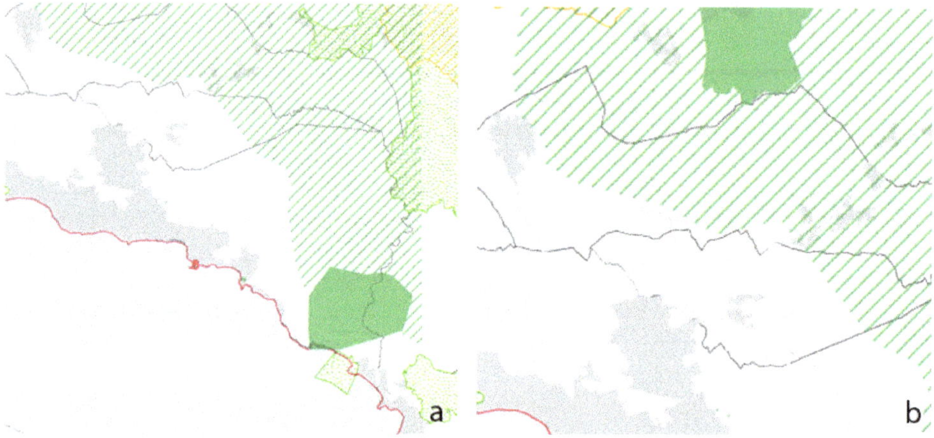

Figure 11. Example of GI projects proposed by LA group in the MCC geodesign workshop 2 in the (**a**) coastal area of the South-Easter edge and in the (**b**) inland zone of the South-Easter edge.

Figure 12. Example of GI projects proposed by NA group in the MCC geodesign workshop 2 in the (**a**) inland zone of the South-Easter edge and in (**b**) the coastal area of the South-Easter edge.

5. Discussion

The PSS Geodesignhub, which was designed to implement the second part of the framework for geodesign [23], encourages the users in applying systems thinking. The cross-system approach is supported by the possibility to compute the impacts of projects hence to evaluate the cumulative performance of design proposals in relation to both the systems the projects belong to, and to the other systems. Geodesignhub overlays diagrams with the evaluation maps and checks their intersections to compute impacts. The relationships between systems are computed thanks to values imputed in the impact matrix beforehand. With reference to the example shown in Figure 13, the two diagrams GI 2 and GI 3 represent the implementation of green corridors on the Eastern and Western edges respectively (Figure 13a). Their impacts on the system Green Infrastructure are positive, meaning that the existing natural resources are in need of protection (Figure 13b). Conversely, the cross-system analysis (Figure 13c) displays negative impacts in the system Commerce and Industry. The area in orange is, in fact, also highly attractive for locating commercial or industrial facilities, thus possible conflicts among different land-uses may occur in the area. The establishment of a green corridor would considerably restrict the possibility in the area for commercial and/or industrial development. Hence the spatial overlay analysis allows to automatically identify possible conflicts of interest, and provides an efficient tool for understanding how systems mutually influence each other in rapid real-time design iterations. Such kind of assessment can be repeated real-time many times during the design providing for an efficient trial and error aid when studying multi-system locational coherence.

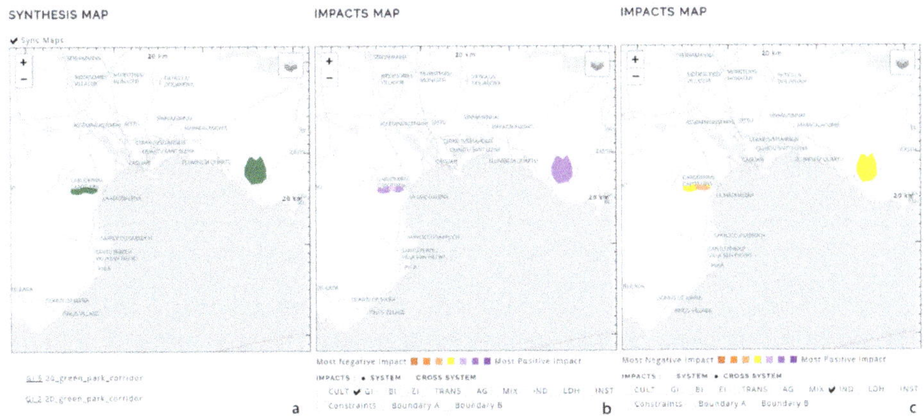

Figure 13. The tool "Compute Detailed Impact" in Geodesignhub. The impact of the two diagrams (**a**) GI 2 and GI 3 (**b**) on the system Green Infrastructure and (**c**) on the system Commerce and Industry.

The "Display Overlaps" function identifies overlapping features in a selection of diagrams by a single design team. Once again, the goal is to represent conflict in design, within a single synthesis. Figure 14a shows the overlapping projects of the final synthesis of the group Early Adopter 2050. Some of the projects may coexists (e.g., GI 27 "New botanic garden" and LDH 44 "Nursing homes with robotic assistants"), while others include potentially conflicting land-uses (e.g., INST 26 "Expansion of the harbor capacity" and CULT 14 "Protection of underwater ancient relict"). Likewise, the function "Combined Analysis" displays overlapping projects in two or more syntheses (Figure 14b). Mismatching dark colors identifies conflicts between teams while consistent colors confirm agreement on a certain land use.

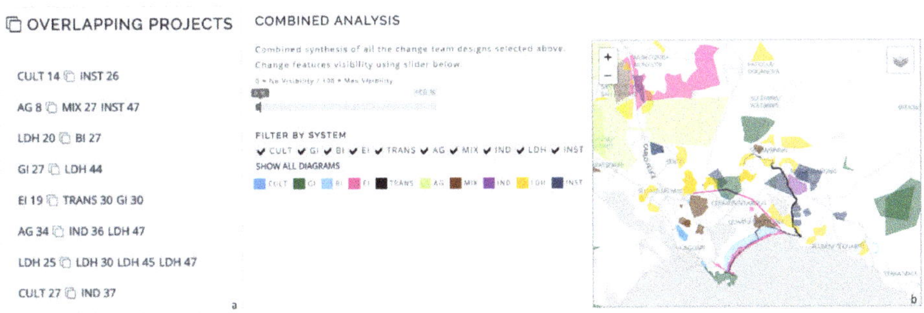

Figure 14. The "Display Overlaps" (**a**) and "Combined Analysis" (**b**) tools in Geodesignhub.

In addition to the available built-in tools in Geodesignhub, the Geodesign Process Analytics (GDPA) approach [35] can be used to process Geodesignhub log-data to monitor and real-time display in a e-dashboard the process dynamics, including the spatial overlaps between projects, the temporal sequence of activities and tasks, the user behavior and productivity, and the evolution of the design in space and time. The set of indicators proposed in these analytics support the application of systems thinking during the rapid design iterations of a typical geodesign workshop [14,15], highlighting in real-time the relationships between GI projects and those of the other systems. Making projects synergies or conflicts more explicit it encourages systems thinking and help design teams in reaching consensus on common proposals. Following on previous studies [36] GDPA exploit the spatial functions of the PostgreSQL database to automatically identify topological relations between diagrams. If the overlapping area represents more than or is equal to 80% of the total area of the two diagrams, they are considered similar. Other types of relations can be identified: diagram A "contains" diagram B when the area of intersection represents more than or is equal to 80% of the total area of the second diagram; diagram A is "within" diagram B when the area of intersection represents more or is equal to 80% of the total area of the first diagram; diagram A "intersects" diagram B when the area of intersection is less than 80% of the area of both diagrams. Table 3 shows the results of the application of the Freitas' and Moura's method [36], that is not only that there is an overlap but also the type of relation that characterize the overlapping diagrams in the final synthesis of the group Early Adopter 2050 (EA50). The same indicator can be used to bring out spatial relations between diagrams of two different syntheses (Table 4). Analyzing the relation between two overlapping diagrams, their titles and related systems, it is possible to disclose potential conflicts resulting from incompatible land-uses, which were earlier identified only graphically in the map in Figure 14b.

Table 3. Excerpt from the output generated by the Extract, Transform and Load (ETL) transformation to measure the topology similarity between the diagrams selected in the last synthesis of the group EA50.

Title	Diagram A	% (A∩B/A)	Relation	% (A∩B/B)	Diagram B	Title
Protection of underwater ancient relict	CULT 14	93.80	within	1.56	INST 26	Expansion of the harbor capacity
New botanic garden	GI 27	47.88	intersects	57.29	LDH 44	Nursing homes with robotic assistants
Drone laser scanning for precision agriculture	AG 34	1.50	contains	99.98	IND 36	Hydrogen fuel cell power plant

Table 4. Excerpt of the output generated by the ETL transformation to measure the topology similarity between the diagrams selected in the last synthesis of the group EA50 and EA35 respectively.

Title	Diagram A (EA50)	% (A∩B/A)	Relation	% (A∩B/B)	Diagram B (EA35)	Title
High-density 3D-printed housing	MIX 21	100	within	0.64	AG 43	Precision agriculture with drones
Smart farming with drones	AG 34	47.88	intersects	57.29	AG 34	Precision agriculture with drones
Poetto beach—solar sidewalk	EI 19	1.59	contains	100	TRANS 6	Viale Poetto—solar road

The implementation of the participants' performance and design evolution indicators in a digital dashboard provides an efficient tool to identify systems of major interest and those not sufficiently considered in the design from the perspective of a comprehensive plan design. Among the first set of indicators proposed, the "Diagram creation by system" (Figure 15a) shows the total number of diagrams created during the workshop. While, the indicator "Diagram selection by system in a synthesis version" offer the possibility to measure how many diagrams were selected in a group synthesis among those initially proposed (Figure 15b). The drop-down menus allow to select the team and synthesis version and display the histogram accordingly. It should be noticed that the number of diagrams created/selected can act as a rapid alert system for the workshop coordinator, but is not in itself a measure able to assess whether all the systems were properly taken into consideration. Local challenges in some systems could be addressed with few diagrams, conversely, the peculiar characteristics of some other could require a greater number of diagrams to face up the changes taking place or envisioned. Nevertheless, it provides useful information for the workshop coordinator when compared to other measures. To better evaluate these design dynamics the number of diagrams selected can be assessed with regards to expected targets. There may be a target associated with some of the systems (e.g., for this study area, 4000 hectares of new green infrastructures were required). Histograms in Figure 15c shows in the left bar the target to be achieved and in the right bar the amount of area the change team devoted to that land use. This comparative analysis can support the workshop coordinator to best focus their assistance to design teams where an increasing/reducing in complexity (i.e., number of diagrams selected) is required to achieve the objectives and their targets. Furthermore, typically at the beginning of a geodesign study participants are asked to formulate a minimum number of design proposals in each of the systems (e.g., for this study area, a minimum of ten diagrams had to be created in each system). In order to ensure that this task has been accomplished a reference line was applied to the "Diagram creation by system" showing the target value (Figure 15a).

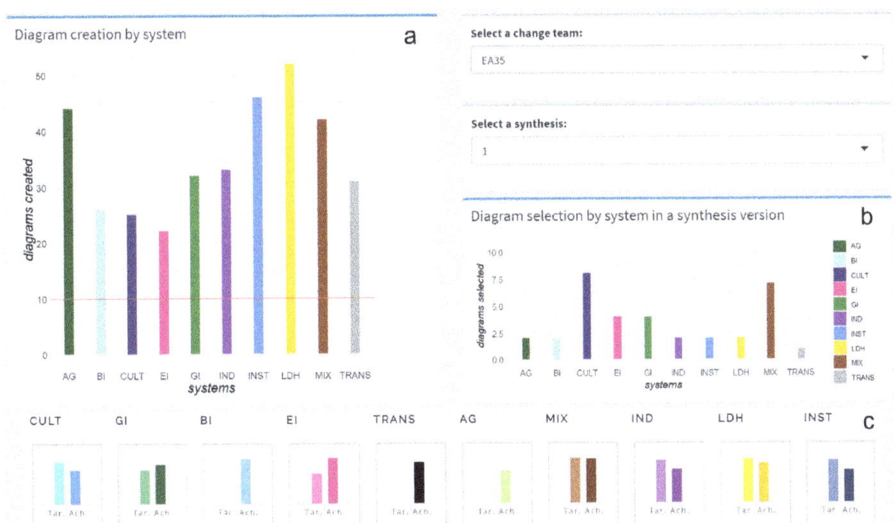

Figure 15. Excerpt from the Geodesign Process analytics dashboard used to analyze the log data of the MCC case study. The figure shows the "Diagram creation by system" (**a**), the "Diagram selection by system in a synthesis version" (**b**) and the histograms (**c**) with the target objectives (in hectares) to be achieved (**left bar**) and already achieved (**right bar**) by the change team in the selected synthesis.

6. Conclusions

The main underlying assumption of this paper is that the design of GI should not be a sector design act, but GI should be considered as one of the main pillars in comprehensive regional strategic planning. Nevertheless, in current practices this is not often the case. This may be due to the fact that the importance of GI -in the current meaning of the term—in contributing to sustainable development has been widely recognized only recently. Hence, GI planning and management often may result as a sector endeavor, and an integration with more comprehensive spatial planning and policy-making is far from be common practice. Possible hindering factors may be institutional, as the recent concept may require time to be structurally introduced in planning systems with long traditions, but also methodological, for it may require innovation through reliable methods and tools to handle the systemic relationships between GI and the other territorial development system and dynamics. In order to address the latter issue, a geodesign study developed within the IGC is presented in order to propose a practical example of how applying systems thinking through geodesign methods and tools may help to tightly couple the design of GI with other important territorial systems.

The value of the application of the Steinitz's framework within intensive workshop with Geodesignhub is to addressing complex problems (in our case the integration of GI with comprehensive territorial planning) in a very short time, enhancing the participant understanding and knowledge, and fast reaching consensus. The limitation is that it may be applied to strategic planning mostly due to limitation in accuracy, and further work will be needed afterword to develop implementation plans.

The geodesign workflow applied in this study was implemented in two design workshops on the MCC. The results contribute to demonstrate how it is operationally possible to include GI in spatially-explicit strategic territorial planning. While this level of design may require a shift in scale to further develop the design in implementation plans, it is very likely to contribute to grow the awareness of decision-makers and stakeholders about the importance of GI, and of its relationships with other systems since the early phases of spatial planning, in order to avoid possible conflicts at a later stage. The built-in analytical tools in Geodesignhub and the set of indicators proposed in the geodesign process analytics facilitate the multi-sectorial approach to GI planning identifying and quantifying in real-time possible synergies and conflicts. The overall approach adopted allowed us to affirm that Geodesignhub effectively supports planning GI at different scales, making the coordination between territorial levels more robust, however, the quality of the results obtained with this academic effort does not allowed us to comment more on the outcomes.

The study presented in this paper was developed for research purposes and in educational settings, and further testing in real-world planning studies may be required in the future to further confirm the robustness of this approach in the real practice. However, while it may be considered only a first example towards more integrated GI planning, we hope the method, techniques and tools presented in this paper may provide a useful aid to planners towards reaching this aim.

Author Contributions: Conceptualization, M.C.; data curation, M.C., C.C. and E.A.D.C.; formal analysis, C.C. and E.A.D.C.; investigation, M.C., C.C. and E.A.D.C.; methodology, M.C., C.C. and E.A.D.C.; software, C.C.; supervision, M.C.; validation, C.C. and E.A.D.C.; visualization, M.C., C.C. and E.A.D.C.; writing— original draft, M.C., C.C. and E.A.D.C.; writing—review and editing, M.C., C.C. and E.A.D.C. This work is an outcome of the joint efforts of the authors. All authors have read and agreed to the published version of the manuscript.

Funding: This research received no external funding.

Acknowledgments: The authors wish to thank Carl Steinitz for sharing the classification used for the evaluation maps in the Metropolitan City of Cagliari geodesign study. Chiara Cocco gratefully acknowledges Sardinia Regional Government for the financial support of her PhD scholarship (P.O.R. Sardegna F.S.E. Operational Program of the Autonomous Region of Sardinia, European Social Fund 2014–2020—Axis III Education and training, Thematic goal 10, Priority of investment 10ii.).

Conflicts of Interest: The authors declare no conflict of interest.

References

1. Slätmo, E.; Nilsson, K.; Turunen, E. Implementing green infrastructure in spatial planning in Europe. *Land* **2019**, *8*, 62. [CrossRef]
2. Wang, J.; Banzhaf, E. Towards a better understanding of Green Infrastructure: A critical review. *Ecol. Indic.* **2018**, *85*, 758–772. [CrossRef]

3. Naumann, S.; McKenna, D.; Kaphengst, T.; Pieterse, M.; Rayment, M. *Design, Implementation and Cost Elements of Green Infrastructure Projects*; Final Report to the European Commission; Ecologic Institute and GHK Consulting: Berlin, Germany, 2011.
4. Bennett, G.; Mazza, L.; de Nocker, L.; Gantioler, S.; Losarcos, L.; Margerison, C.; Kaphengst, T.; McConville, A.J.; Rayment, M.; Brink, P.T.; et al. *Green Infrastructure Implementation and Efficiency*; Institute for European Environmental Policy: London, UK, 2011.
5. Muñoz-Criado, A.; Domenech, V. Green Infrastructure Planning at Multiple Levels of Scale: Experiences from the Autonomous Region of Valencia, Spain. In *Scale-Sensitive Governance of the Environment*; Wiley-Blackwell: Hoboken, NJ, USA, 2014; pp. 283–301.
6. Olds, A.D.; Pitt, K.A.; Maxwell, P.S.; Connolly, R.M. Synergistic effects of reserves and connectivity on ecological resilience. *J. Appl. Ecol.* **2012**, *49*, 1195–1203. [CrossRef]
7. Lanzas, M.; Hermoso, V.; de-Miguel, S.; Bota, G.; Brotons, L. Designing a network of green infrastructure to enhance the conservation value of protected areas and maintain ecosystem services. *Sci. Total Environ.* **2019**, *651*, 541–550. [CrossRef] [PubMed]
8. Cannas, I.; Lai, S.; Leone, F.; Zoppi, C. Green infrastructure and ecological corridors: A regional study Concerning Sardinia. *Sustainability* **2018**, *10*, 1265. [CrossRef]
9. Liquete, C.; Kleeschulte, S.; Dige, G.; Maes, J.; Grizzetti, B.; Olah, B.; Zulian, G. Mapping green infrastructure based on ecosystem services and ecological networks: A Pan-European case study. *Environ. Sci. Policy* **2015**, *54*, 268–280. [CrossRef]
10. *European Commission Communication from the Commission to the European Parliament, the Council, the European Economic and Social Committee and the Committee of the Regions. Green Infrastructure (Gi)—Enhancing Europe's Natural Capital*; COM(2013) 249 Final; European Commission: Brussels, Belgium, 2013; Volume 2013.
11. *European Commission Communication from the Commission to the European Parliament, the Council, the Economic and Social Committee and the Committee of the Regions Our Life Insurance, Our Natural Capital: An Eu Biodiversity Strategy to 2020*; European Commission: Brussels, Belgium, 2011.
12. Natura 2000 Barometer. Available online: https://www.eea.europa.eu/data-and-maps/dashboards/natura-2000-barometer (accessed on 30 December 2019).
13. Green Infrastructure—Environment—European Commission. Available online: https://ec.europa.eu/environment/nature/ecosystems/studies/index_en.htm (accessed on 29 December 2019).
14. Nyerges, T.; Ballal, H.; Steinitz, C.; Canfield, T.; Roderick, M.; Ritzman, J.; Thanatemaneerat, W. Geodesign dynamics for sustainable urban watershed development. *Sustain. Cities Soc.* **2016**, *25*, 13–24. [CrossRef]
15. Pettit, C.; Hawken, S.; Zarpelon, S.; Ticzon, C.; Afrooz, A.; Steinitz, C.; Ballal, H.; Canfield, T.; Lieske, S. Breaking Down The Silos Through Geodesign—Envisioning Sydney's Urban future. *Environ. Plan. B Urban Anal. City Sci. (Spec. Issue Urban Syst. Des. Sci. Des. Des. Sci.)* **2019**. [CrossRef]
16. Rivero, R.; Smith, A.; Orland, B.; Calabria, J.; Ballal, H.; Steinitz, C.; Perkl, R.; McClenning, L.; Key, H. Multiscale and multijurisdictional Geodesign: The Coastal Region of Georgia, USA. *Landscapes* **2017**, *19*, 42–49.
17. Meadows, D.H. *Thinking in Systems: A Primer*; Chelsea Green Publishing: White River Junction, VT, USA, 2008; ISBN 978-1-60358-055-7.
18. Gu, Y.; Deal, B.; Larsen, L. Geodesign Processes and Ecological Systems Thinking in a Coupled Human-Environment Context: An Integrated Framework for Landscape Architecture. *Sustainability* **2018**, *10*, 3306. [CrossRef]
19. Lee, D.J.; Dias, E.; Scholten, H.J. *Geodesign by Integrating Design and Geospatial Sciences*; Springer: Berlin/Heidelberg, Germany, 2014; ISBN 978-3-319-08299-8.
20. Bottlenecks Blocking Widespread Usage of Planning Support Systems—Guido Vonk, Stan Geertman, Paul Schot. 2005. Available online: https://journals.sagepub.com/doi/10.1068/a3712 (accessed on 27 December 2019).
21. Geertman, S.; Stillwell, J. Planning support systems: An inventory of current practice. *Comput. Environ. Urban Syst.* **2004**, *28*, 291–310. [CrossRef]
22. Campagna, M. Metaplanning: About designing the Geodesign process. *Landsc. Urban Plan.* **2016**, *156*, 118–128. [CrossRef]
23. Steinitz, C. *A Framework for Geodesign: Changing Geography by Design*; Esri Press: Redlands, CA, USA, 2012.

24. Rivero, R.; Smith, A.; Ballal, H.; Steinitz, C. Promoting collaborative Geodesign in a multidisciplinary and multiscale environment: Coastal Georgia 2050, USA. *Digit. Landsc. Archit.* **2015**, *1*, 42–58.
25. Campagna, M.; Carl, S.; Di Cesare, E.A.; Cocco, C.; Hrishikesh, B.; Tess, C. Collaboration in planning: The Geodesign approach. *Rozw. Reg. I Polityka Reg.* **2016**, *35*, 55–72.
26. Araújo, R.P.; Moura, A.C.M.; Nogueira, T. Creating Collaborative Environments for the Development of Slum Upgrading and Illegal Settlement Regularization Plans in Brazil: The Maria Tereza Neighborhood Case in Belo Horizonte. *Int. J. E-Plan. Res. (IJEPR)* **2018**, *7*, 25–43.
27. Patata, S.; Paula, P.L.; Moura, A.C.M. The application of geodesign in a Brazilian illegal settlement. Participatory planning in Dandara occupation case study. In *Environmental and Territorial Modelling for Planning and Design*; FedOA—Federico II University Press: Naples, Italy, 2018; ISBN 978-88-6887-048-5.
28. Ernst, F.B.; Erdoğan, S.; Yılmaz, M.; Ulukavak, M.; Şenol, H.İ.; Memduhoğlu, A. A New Master Plan for Harran University Based on Geodesign. *ISPRS—Int. Arch. Photogramm. Remote Sens. Spat. Inf. Sci.* **2017**, *42W6*, 43–45. [CrossRef]
29. Lanfranchi, E.; Fonzino, F. Co-creation of Alternative Futures Using Technologies of Geoinformation Structured in a Geodesign Method: Contributions to the State-of-the-Art. *DISEGNARECON* **2018**, *11*, 4.1–4.14.
30. Orland, B.; Steinitz, C. Improving our Global Infrastructure: The International Geodesign Collaboration. In Proceedings of the Journal of Digital Landscape Architecture, Wichmann: Anhalt University, Dessau, Germany, 22–25 May 2019; Volume 4, pp. 213–219.
31. Home International Geodesign Collaboration. Available online: https://www.envizz1.com/ (accessed on 30 December 2019).
32. Harvard University. Department of Landscape Architecture. Research Office. In *A Comparative Study of Resource Analysis Methods*; Harvard University: Cambridge, MA, USA, 1969.
33. Hopkins, L.D. Methods for Generating Land suitability Maps: A Comparative Evaluation. *J. Am. Inst. Plan.* **1977**, *43*, 386–400. [CrossRef]
34. Malczewski, J. GIS-Based Land-Use Suitability Analysis A Critical Overview. Progress in Planning, 62, 3-65.—References—Scientific Research Publishing. 2004. Available online: https://www.scirp.org/(S(351jmbntvnsjt1aadkposzje))/reference/ReferencesPapers.aspx?ReferenceID=2122500 (accessed on 27 December 2019).
35. Cocco, C.; Jankowski, P.; Campagna, M. An Analytic Approach to Understanding Process Dynamics in Geodesign Studies. *Sustainability* **2019**, *11*, 4999. [CrossRef]
36. Freitas, C.R.; Moura, A.C.M. ETL Tools to Analyze Diagrams' Performance: Favoring Negotiations in Geodesign Workshops. *DISEGNARECON* **2018**, *11*, 15.1–15.23.

sustainability

MDPI

Article

Territorial Energy Decentralisation and Ecosystem Services in Italy: Limits and Potential

Paolo De Pascali *, **Saverio Santangelo**, **Francesca Perrone** and **Annamaria Bagaini**

Department of Planning, Design and Technology of Architecture, Sapienza University, 00185 Rome, Italy; saverio.santangelo@uniroma1.it (S.S.); francesca.perrone@uniroma1.it (F.P.); annamaria.bagaini@uniroma1.it (A.B.)

* Correspondence: paolo.depascali@uniroma1.it; Tel.: +39-348-3962856

Received: 31 December 2019; Accepted: 12 February 2020; Published: 14 February 2020

check for updates

Abstract: This article focuses on the complex relationships between energy processes and ecosystem services. It highlights the conflicts between them due to the anthropocentric value that characterizes their interrelationship. The article reports the initial results of ongoing research on energy decentralization processes in Italy, examining the Italian districts heating performance, concerning ecosystem provisioning and regulating services. The analysis is based on a sample of more than 150 Italian district-heating systems. Contrary to studies that positively evaluate processes of energy decentralization, the results of the research show some critical factors and impacts. An efficiency gap between districts heating and traditional energy systems emerged. The data processed show a critical situation in the development of local networks, highlighting that the decentralized energy model is not deeply rooted in the local area and is poorly characterized by shared governance, which instead would benefit from the integration of ecosystem services. The significant presence of large energy groups and the considerable use of fossil sources in Italy reduces the effectiveness of the decentralization of energy systems. The article presents some conclusive considerations, which outline some general guidelines for proceeding towards a more correct relationship with ecosystem services and greater integration with the territories.

Keywords: ecosystem services; energy decentralization; district heating; urban planning; energy transition

1. Introduction

The interest in ecosystem services arose following an increasing awareness of the interdependence between human beings and natural elements [1]. On one hand, communities receive essential elements both for their personal (e.g., food, reproductive spaces, raw materials, air purification, drinking water, etc.) and social (e.g., places, communities and economic and environmental conditions, which directly impact human well-being) survival, through (direct and indirect) ecosystem services; on the other hand, anthropic interventions are able to alter and irreparably compromise the balance and quality of ecosystem features and components [2].

The production and consumption processes of energy sources, including their complicated internal systems, are strictly related to human activities. Useful energy, which comes from primary sources and goes through the processes of transformation, transport, storage and distribution, up to the so-called "end-use energy", is strictly created by man for man. It starts by using existing natural resources, both exhaustible and renewable. The idea of "useful energy" itself, therefore, has an anthropic, or rather anthropocentric, connotation. It highlights the strategy of energy processes that are focused directly or indirectly on the human being and his social, cultural and economic organisations.

The useful energy processes now influence, as they did in the past, all human activities, behaviours, socio-economic organisations and settlements, with increasing overall intensity over time. It is foreseen that the growth of this trend will increase in the near future.

Useful energy is one of the main interface elements with ecosystems and it uses ecosystem services [3,4], by affecting them in ways that have become critical for their vitality itself [5].

Monitoring energy production in relation to the features of ecosystems is important because it allows us (i) to keep the entire "ecological infrastructure" intact; (ii) to evaluate environmental issues deriving from anthropogenic pressure on ecosystems, due to the exploitations of energy sources; (iii) to measure the dependence on exhaustible energy sources, taking into account the strong ecosystem stress; (iv) to examine the balance between the different types of energy resources and Ecosystem Services; (v) to control the energy demand and the definition of energy plans, on the basis of a thorough knowledge of the resulting ecosystem problems; (vi) to reduce environmental impacts, which occur in presence of local energy systems, of directional flows (e.g., hydroelectric energy, etc.), of extended technological systems (e.g., wind farms, biomass systems, etc.) [6].

The two trends (growth of useful energy and preservation of ecosystem services) move in the same anthropocentric direction but increasingly conflict with each other. The social model—especially of advanced industrial societies, as well as the emerging economies—demands increasing quantities of useful/commercial energy, that affects the energy sources themselves of ecosystem services and, in the near future, they threaten the existence of these invaluable resources.

2. State of the Art

The production of useful/commercial energy involves and, both directly and indirectly, influences all the ecosystem services categories [7–9]. Thus, complex interactions between energy processes and ecosystem services emerged. The dynamics can hardly be defined within a logical framework; the complex conceptual mapping of these interactions is still incomplete due to the vastness of the subject, as reported in Holland [10]. The most relevant direct impacts of useful energy, which affect us more than the four classic classification categories [11–14], involve the ecosystem services classified as "provisioning services", which refer to the production function of ecosystems and—as for our case study—they refer to the primary energy sources (fossil sources, biomass, geothermal, solar, hydraulic, wind energy, etc.) [15] as well as the secondary energy sources (electricity, fuels and biofuels, fuel oil, etc., that derive from the transformation of the primary sources). The most relevant impacts of useful energy that mainly involve us, are also the classification of the category defined as "regulation services", which provide for the balance conditions of ecosystems and, as for our case study, they concern the control of climate-changing emissions and polluting factors deriving from anthropic energy processes.

They pursue the maintenance of the levels of the two components of the ecological source and sink model [16], which seems to be very difficult to achieve. Indeed, the gap from the balance system increases.

Therefore, the two categories of ecosystem services are severely compromised in their function, due to the intensity of production and consumption, which considerably exceed the ability of ecosystem services to cover energy demand without affecting existing natural resources (see fossil sources), as well as natural heritage. The two categories are also compromised in their function due to the inability of the ecosystem services to properly provide for the absorption and/or the metabolization of the negative effects resulting from the energy processes.

Looking ahead, the convergence of guidelines between the transformation of the energy model and the enhancement of ecosystem services is required. Thus, an explanation of the latter is essential for the pursuit and stabilisation of the former, as well as the virtuous change of the former, which must align and integrate with the features of the latter, also following strategies and time programs.

In this context, an important strategic role is played by the local dimension—that is the territories as providers of ecosystem services in relation to energy processes. The local dimension could lay the basis for the wide (shared and inclusive) governance of these strategies and integration programs, as well as for a socio-cultural growth, and the strengthening and development of virtuous models regarding ecosystems.

The literature [17,18] positively evaluates energy decentralization (both in terms of governance and technological systems) in terms of increasing environmental protection, combatting climate change and boosting sustainable development. Local energy grids/networks (smart grids; heat/cold grids; electricity grids; micro-grids, etc.) represent an important development path for energy decentralization, supported by the energy liberalization process promoted in Europe since 1990 [18]. District heating is based on the local production and distribution of thermal fluid, which can have a significant impact on the physical and functional organization of settlements—particularly on the economy, on the enhancement of endogenous energies (biomass and biofuel, waste-to-energy, the energy use of urban waste, geothermal, solar thermal uses) and on environmental protection. Local energy grids can also have positive impacts on the ecosystem services [10]. Comparted to traditional centralised energy systems, local district-heating systems have greater energy efficiency—and therefore less environmental impact—as well as better effects on the socio-political-cultural dimension. The research conducted shows that all good implications related to local energy networks are not always achieved, and that it is necessary to recalibrate the energy policies for a better relationship with the ecosystem services.

In Italy, the local district-heating systems do not seem to maintain a good relationship and integration with the ecosystem services.

The article aims to verify this consideration, through a study on district-heating systems in Italy. This study aims to outline two aspects of the relationship between local energy networks and ecosystem services: 1) how the ecosystem services are taken in consideration for the development of local energy networks, with regards to provisioning services and regulating services, in terms of greater or lesser efficiency compared to traditional energy systems; 2) how the energy decentralization characterized by a shared management and governance in relations with the ecosystem services.

The study outlines an overall critical situation in Italy, whose critical factors may characterize the relationship with the ecosystem services. In conclusion, the paper provides a list of opportunities and guidelines for a better local energy policy.

3. Materials and Methods

The methodology applied is shaped by the four following points:

1. The management of "energy" within the framework of ecosystem services refers to the production and consumption of both fossil and renewable sources, and the impact they have upon the environment. The "anthropogenic energy" topic, in relation to the ecosystem services, is still underestimated in literature. It considers only some man-made energy supply chains in relationship with ecosystem services (e.g., the biomass supply chain). Other, even more important sources, such as fossil sources, which display straight involvement within ecosystem services, are not fully described, probably due to the complexity of the relationship. This research analyzes the topic literature, in order to highlight and identify the ecosystem services involved. Energy management following forms of decentralization and energy localism are positively considered in relation to ecosystem services. It allows us to get close the demand and the supply, and therefore it demonstrates the efficiency achievement. This occurs when energy management is deep-rooted in territories and settlements and promotes participation and inclusive governance possibilities concerning the ecosystem dimension. The research analyzes the topic literature. This point aims to answer the following question: 1) in literature, is the relationship between decentralized energy systems and ecosystem services considered particularly important?
2. The third step of the research is an operative assessment of the Italian context. Does energy decentralization (especially in Italy) create virtuous conditions in relation to the ecosystem services? A detailed analysis of local energy networks in Italy has been conducted and the first results are reported. The criteria used are the consumption efficiency (energy consumption and monitoring systems) and the impact on the social cohesion and inclusion. The results of this

elaboration provide negative results. The district-heating systems analyzed in Italy do not seem to develop an effective relationship with the ecosystem services. The analysis also reveals elements of crisis in the development of the field itself, which must be analyzed in depth. The elaboration process, data used and the interpretations from which the results were obtained are reported below. This step aims to answer the question: In Italy does the energy decentralization process establish a positive relationship between the ecosystems?

3. The study outlines some policy guidelines to consider in order of improving the non-positive trend for the development of Italian local energy networks. The question to be answered in the research is: Which are the prospects for increasing the state of Italian districts hating, in terms of institutional frameworks review for promoting the sector in relation to the ecosystem services?

The research is based on the study on the Italian district-heating systems, which also includes many cogeneration plants (production of electricity and heat), and less cold grids, based on a trigeneration plant (production of electricity, heat and cold), which is still few presents.

It is estimated that in Italy, district-heating systems amount to over 350 units, concentrated mainly in Northern Italy. The district-heating systems, built from the 1970s, had a great development between the 2010 and 2015, after that, they show a slowdown. This was due to the expansion of existing networks (especially the larger ones, located in Turin, Milan, and Brescia, rather than the small ones). The main document reporting the general status of the Italian district-heating systems [AIRU] shows a little increase of new local networks, which concerns small systems. This evolution path is explained by operators as an effect of the absence of a stable institutional framework for supporting the sector. Other critical factors exist, some of which are reported below. For the study, two main data sources were used:

- AIRU report [19] which reports the data of about 340 local networks, of which 150 (including the biggest ones) are in a detailed and complete form, and for the remainder few data are unavailable. The analysis is based on a sample base variable within this range, which is the most complete and significative sample;
- ISTAT (National Institute of Statistics) data [20] which provides information on the district-heating systems located in the provincial cities—42 district-heating systems of a total of 109 Italian capitals of the province. The ISTAT source has also been used for population and housing data that have been cross-referenced with the local energy networks data.

For the statistical analysis, three groups of energy sources were considered which feed the local networks. The three groups concern fossil sources (now almost exclusively refer to natural gas), geothermal energy, and biomass. In many cases, the district-heating systems are supplied by a mix of energy sources, such as residual parts of urban waste, recovery heat, solar source, industrial waste, etc.). The criterion adopted is that of the prevailing source used according to the official statistics.

4. Results

Considering only the district-heating systems fed by fossil sources (in general natural gas), it is found that 28% of the networks considered have higher consumption than that of the traditional energy systems and 33% of them have lower consumption of just 10% (Figure A1). Overall, more than 60% of the district-heating systems have a consumption of fossil sources approximately equal or even higher than traditional systems. This percentage generally regards local grids with small energy production plants. The local networks with medium-large energy production plants have higher fossil consumption avoided (between 10% to 30%). The largest district heating, in Turin, reaches 23.4% of fossil consumption avoided. The local networks that exceed 30% of fossil consumption avoided represent about 10% of the total sample. The total set of district-heating systems considered and based on fossil fuels do not seem to reach high percentages of fossil fuel consumption avoided. Figure A1 is provided in Appendix A.

Figure A2 shows the percentages of fossil consumption avoided by the district-heating systems (compared to the traditional energy systems), broken down the energy source in the power plant (fossil, geothermal, biomass source), by the plant's first year of operation. For fossil-fuel plants, there is an approximation of overtime of the consumption of fossil fuels towards parity with traditional energy systems; especially in the last few years, in which the energy-saving range has decreased considerably, concentrating approximately within −10% and + 10%. The trend line represents this slightly upward trend which tends to cross the abscissa axis. It shows very low values, considering that individual energy production systems, such as natural gas condensing boilers—now widely used—easily save 15%–20% of energy compared to traditional boilers, and that heat pumps and micro-cogeneration can reach even higher values. The few plants with geothermal sources indicate a virtuous trend towards a lower consumption of supplementary fossil sources, approaching the maximum percentage value (100%). District-heating systems supplied by biomass show a trend with a discrete upward inclination which indicates an increase over time of the use of fossil sources. The district-heating performance tends to worsen over time with respect to the consumption of fossil fuels and therefore in the relationship with the ecosystem services, probably due to the greater presence of mixed plants (fossil + renewable). Figure A2 is provided in Appendix A.

Many international studies [17,21,22] evaluate the efficiency of a district heating network considering two main indicators: the linear density (MWh/m) and the demand density (kWh/m3). The study of the IEA [21] on a large number of plants proposes a limit value of 1.8 MWh /m for the linear density; the Woods work [22] reports the minimum demand density at 20 kWh/m2, which can correspond about 7 kWh/m3. The linear density (Figure A3) indicates the energy distributed per linear meter of the network. It should be noted that the trend line of the fossil-source based networks remains substantially constant over time, with a slight downward inclination, with average values between 2 and 3 MWh /m, close to the values coming from the international literature [17,18,21,22]. The local energy networks supplied by renewable sources (geothermal + biomass) have a very marked downward trend, which may indicate a decrease in the efficiency of renewable energy networks. This phenomenon can present a critical factor and should be examined in detail to highlight its causes. Figure A3 is provided in Appendix A.

Figure A4 shows the local networks by extension and by type of ownership (public, private, co-owners, private-public partnership, etc.) according to the initialization. The owners are grouped into three classes: energy groups, local companies, and local authorities. The analysis aims to understand the grade of shared governance and management linked to energy decentralization. Figure A4 shows the size of the networks with regard to new plants which tend to decrease over time. In recent years, the start-up of large plants in large cities has been decreased. The reduction in size does not only concern the networks but also the size of cities involved, analysed in terms of the size of settlements (millions of inhabitants) [20] (Figure A5). Figure A5 also shows that the realization of small local networks particularly affects energy groups, even those with national importance. Figures A4 and A5 are provided in Appendix A.

Figure A6 highlights how fossil-fuel supplied networks are mostly managed by large energy groups. To give some examples, the presence of the Hera group, with EUR 6.5 billion in total turnover, and the A2A group, with EUR 6.6 billion in turnover, and the Iren group, with a turnover of EUR 4 billion [23]. Local companies and local authorities mostly manage renewable energy plants, especially biomass plants. The small presence of local administrations is also highlighted in Figure A7, which reports the situation referred to in the provincial cities with district-heating systems. Figures A6 and A7 are provided in Appendix A.

Figure A8 examines the corresponding energy values in terms of MWh supplied per year. The Figure shows the large gap between the quantity of energy supplied by the local energy networks managed by large energy groups and that supplied by local companies. Big energy private groups mostly own energy plants that use fossil sources (in general, natural gas), while renewable sources are less common. Figure A8 is provided in Appendix A.

5. Discussion

The energy decentralization process in Italy seems to negate the statements made in the literature, both in terms of sharing management—with the prevalence of centralized corporate economic entities—and in terms of sources used, with a large presence of fossil fuels resources being used, while other, endogenous sources (renewable energy sources) are less frequently used.

Furthermore, from the policymaking side, is evident that the structural difficulty encountered in transforming energy processes to incorporate the ecosystem services with which they are related, such as those relating to the [11,12] relating to the provisioning and regulating services categories.

In our study, conducted on the Italian situation, a very fragmented picture of these relationships emerges—without a generalized institutional framework—able to drive and better inform the numerous stakeholders involved in these processes.

The energy decentralization path emerged from the ongoing energy liberalization process; however, it encounters resistance and obstacles to implementing virtuous models and even forms of transition to virtuous models which would incorporate measures for the protection of the ecosystem services.

Contrary to the widespread opinion that considers energy decentralization to always be positive, the data collected on the Italian situation do not seem to confirm this assumption. The study carried out on district-heating systems in Italy highlights several important critical factors in relation to ecosystem services and, at the same time, creates the opportunity to suggest guidelines for the sector. The data processed on the Italian district-heating systems present a problematic picture, from which the contradiction with the common, internationally favourable judgments is clear.

District-heating in Italy has only grown by a very limited extent in recent years, unlike in other European countries [24], and only due to the expansion of a few existing local networks and the construction of limited new networks in small towns.

The explanation coming from businesses is that there is lack of a regulatory and institutional framework, which would favour investment with a degree of security and stability for the future. The expansive prospects generated by some studies [25,26]—which reveal conditions for great development potential at the national level—have not found fulfilment.

In addition, other critical factors emerged from the study related to ecosystem services.

The first critical issue concerns the efficiency of district-heating systems, assessed in terms of the avoided consumption of fossil sources compared to the use of traditional energy systems, which corresponds directly to the emissions avoided. The study shows a substantially negative situation with regards the low values of fossil fuel consumption avoided. The critical situation mainly regards the networks fed by fossil sources, which are the majority and show very low overall values and—in many cases—even negative values, which means that they consume more than traditional plants.

Even traditional efficiency indicators (linear density and demand density) are located at the lower limits of the values considered to be effective at the international level, even with regard to the networks powered by renewable sources that have a trend line in a rapid descent.

This critical situation also has a negative impact in the future forecast for ecosystem services regarding the regulation and provision categories in terms of resource consumption and emission production.

The second critical issue concerns the social dimension of energy. The processes of energy decentralization should bring opportunities for the implementation of participatory social models. Energy Communities [27,28], starting from the shared management of energy, aim to foster a democratic process: where energy is a vehicle for the establishment of new social models with shared and inclusive governance, which also incorporate socio-economic development and cultural/behavioural change in the ecological-environmental field. From the conducted study, this form of political–cultural decentralization, connected to local networks, is not particularly noticeable. Indeed, it would appear that the overall picture is not quite favourable, due to the significant presence of large energy groups and the scarcity of local companies/cooperatives and municipal administrations. Finally, large groups are mainly active in plants supplied by fossil sources, primarily natural gases, which utilize a centralized system of management/distribution. Few of them are involved in using renewable endogenous sources that benefit local ecosystem services and for which the integration between energy and territorial development is more relevant. The lack of a common institutional framework probably tends to undermine any certainties, especially in corporate governance models with strong "business-oriented" push.

6. Conclusions

Energy production and management, as well as energy end-use, have complex and high-impact relationships with ecosystem services. The anthropocentric dimension of the energy chain, triggered by man for man, has an intensive, often irreversible, impact on the ecosystem.

An increased awareness of the interaction between energy systems and ecosystem services is required. This would encourage better policymaking and management, with a view to address the new environmental and human challenges (land/sea use, overexploitation, climate change, etc.) [10].

The study provides some general guidelines for improving the sector towards a more effective relationship with ecosystem services and a greater integration with the territories:

- There is a need for a broad program to promote the redevelopment of the inefficient plants and local networks that consume more than traditional systems, and of those that only achieve a low avoided consumption. In many cases, it is necessary to rethink the entire network or parts of it, in relation to the territorial changes that have occurred, to improve the physical–functional organization of a settlement, and encourage a greater diversification of outputs (cogeneration, heat and cold production). A stable institutional framework must also be produced to facilitate the financing of interventions with innovative forms.
- The energy transition from fossil fuels to renewable sources is becoming central to European and even global policies. It is necessary to facilitate the progressive, but rapid, conversion to renewable sources, as well as the construction of new local networks supplied by renewable sources. For endogenous energies, it is necessary to operate through the creation of production chains, in which the district-heating system and the energy production are part of a sustainable territorial development (for instance, the use of agricultural and forest biomass, waste-to-energy, the use of heat recovery from industrial processes, etc.) which can have a virtuous impact on ecosystem services and integration with the territories.
- Energy production and management at the local level has important potential for the creation of social cohesion. The process of energy decentralization can foster innovative bottom-up initiatives and projects, including collaborative projects (inhabitants + energy companies)—often promoted by Local Administrations. The process of the liberalization of the energy market is not only open to large operators, but also to potential prosumers and local communities. The most interesting line is the creation of corporate-collective entities (e.g., small public companies or local cooperatives) in which citizenship directly participates, called Local Energy Communities (LECs) or Sustainable Energy Communities, or Green Communities [27–33]. The institutional framework should combine the promotion of business-oriented lines with community-based values rooted in the territories.
- Local energy networks can find an important role in socio-economic development plans in the inland areas and medium-sized cities, and they can be integrated into urban regeneration plans. Along this systemic line, the energy–environmental value can be a driver for settlement redevelopment and social inclusion. In this context, policies and programs can act as starters for projects with an eco-energy aim, with the stable and progressive recognition of ecosystem services.

Author Contributions: This paper is the result of the joint work of the authors. All authors have read and agreed to the published version of the manuscript.

Funding: This research received no external funding.

Conflicts of Interest: The authors declare no conflict of interest.

Appendix A

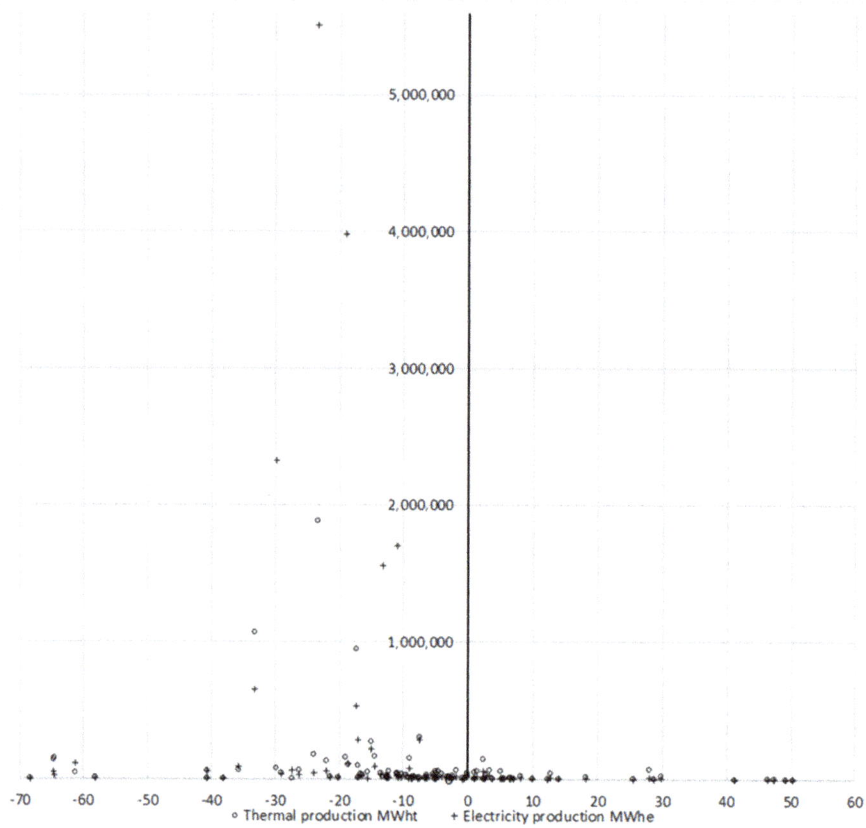

Figure A1. Avoided consumption of fossil fuels in district-heating systems using fossil sources, by the quantity of energy produced (MWht, MWhe) [19].

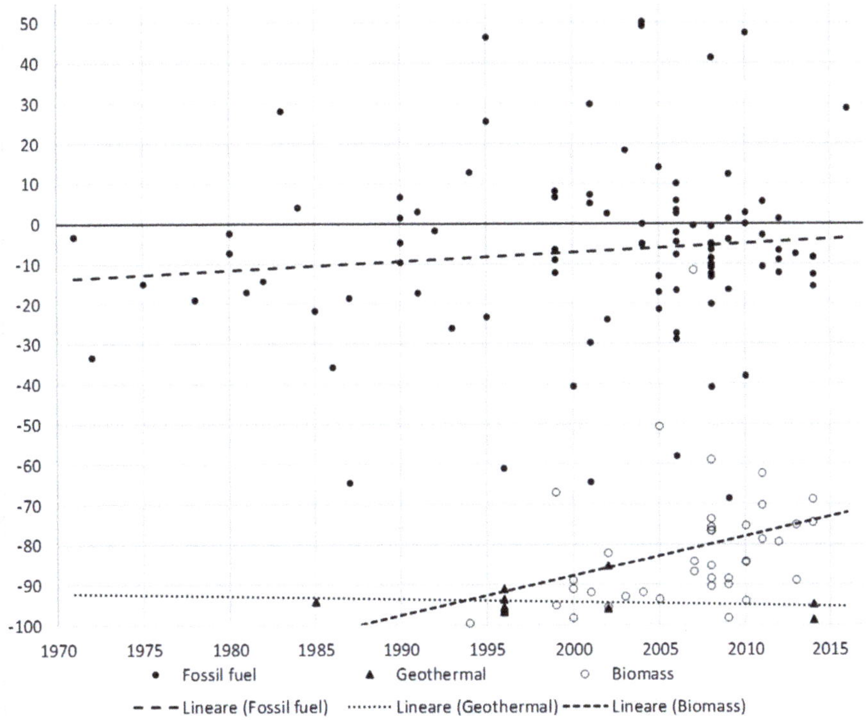

Figure A2. Reduction in the consumption of fossil fuel (%) of the local networks powered by fossil, geothermal and biomass) sources, by initial year of production [19].

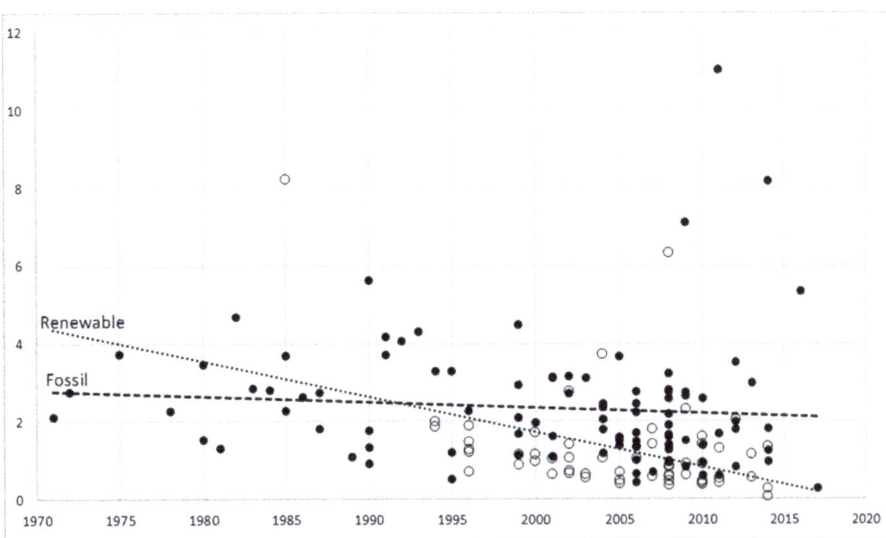

Figure A3. Linear Density (MWh/m) by start date (dark fossil, bright renewable) [19].

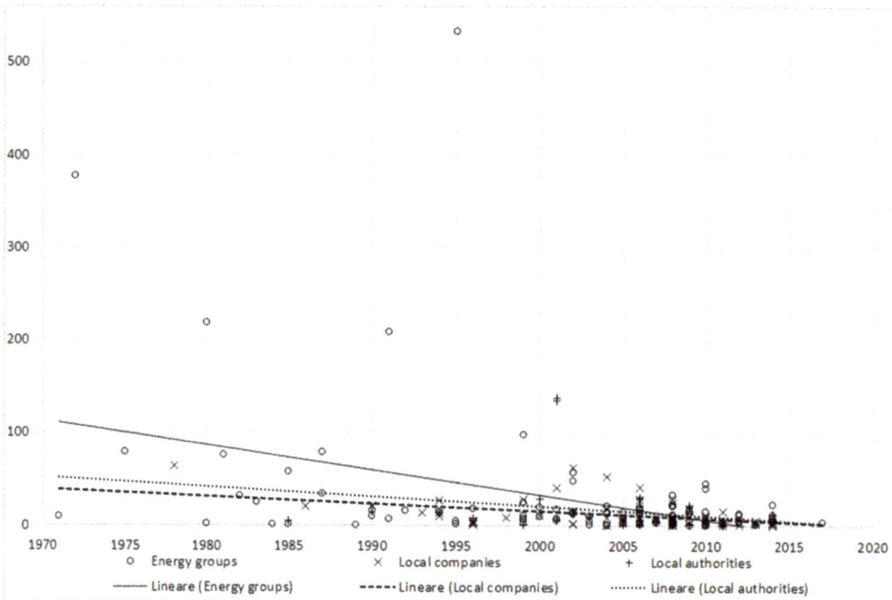

Figure A4. Network ownership by initialization and extension (km) [19].

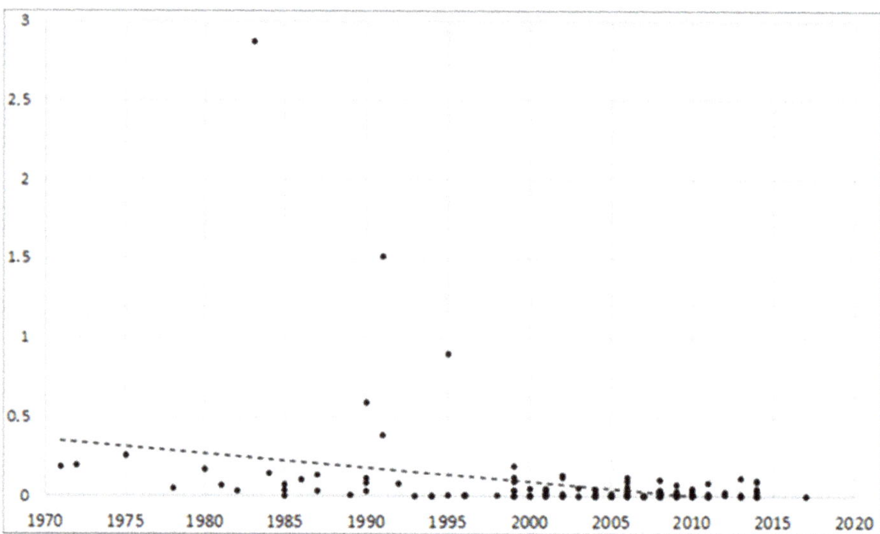

Figure A5. Local networks per initialization year and size of settlements (millions of inhabitants) [19,20].

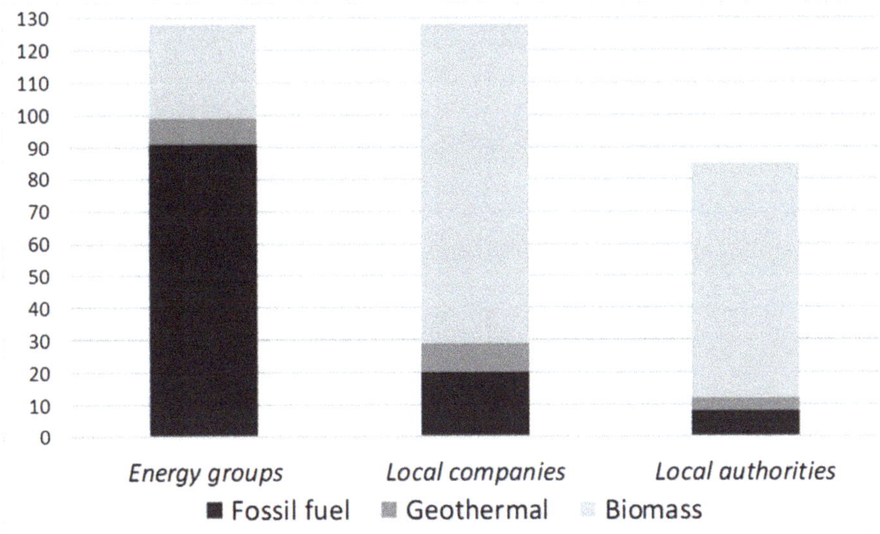

Figure A6. Number of local networks per ownership and energy source [19].

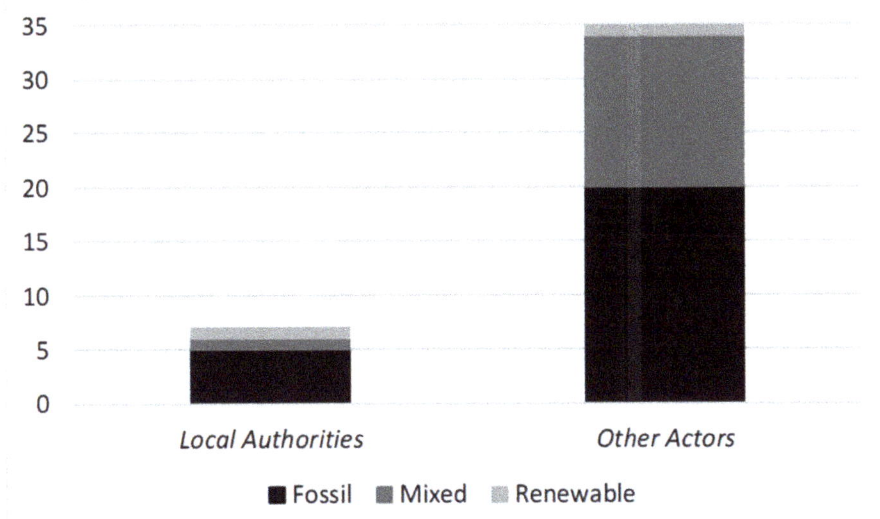

Figure A7. Number of local energy networks in capitals of province per ownership and energy source [20].

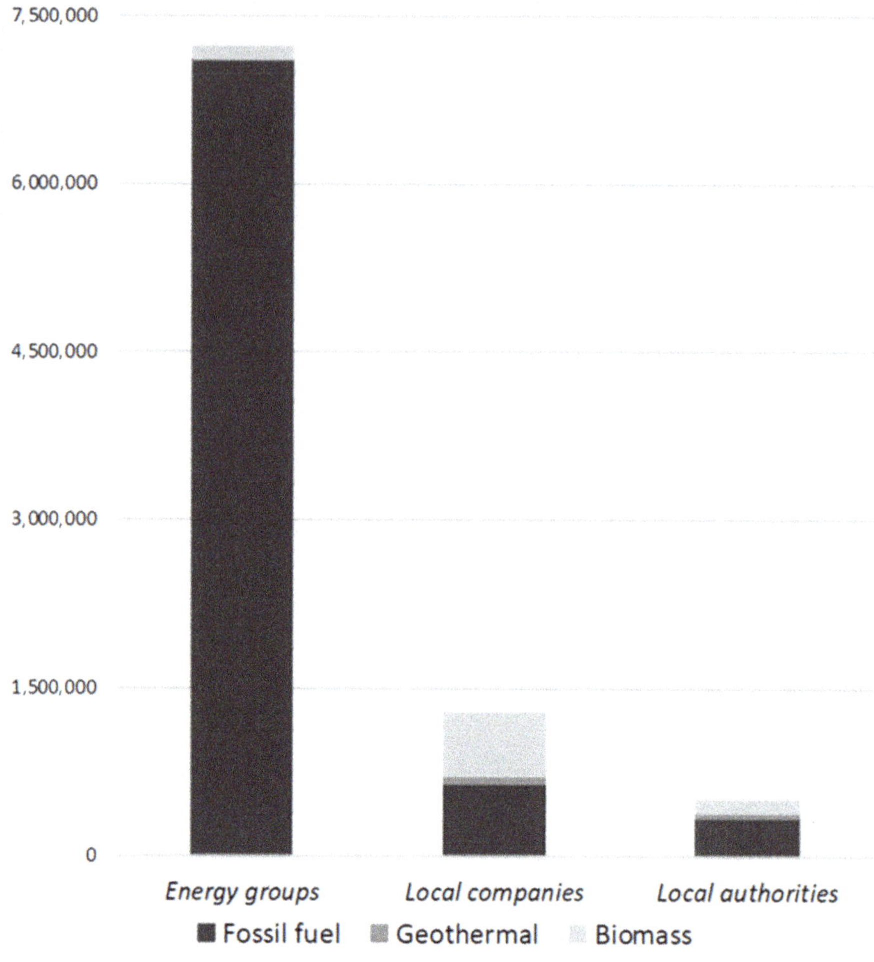

Figure A8. Thermal energy supplied (MWh/year) per ownership and energy source [19].

References

1. Gómez-Baggethun, E.; De Groot, R.; Lomas, P.L.; Montes, C. The history of ecosystem services in economic theory and practice: From early notions to markets and payment schemes. *Ecol. Econ.* **2010**, *69*, 1209–1218. [CrossRef]
2. Daily, G.C. *Nature's Services: Societal Dependence on Natural Ecosystems*; Island Press: Washington, DC, USA, 1997; ISBN 1559634758.
3. Braat, L.C.; De Groot, R. The ecosystem services agenda: bridging the worlds of natural science and economics, conservation and development, and public and private policy. *Ecosyst. Serv.* **2012**, *1*, 4–15. [CrossRef]
4. Bouwma, I.; Schleyer, C.; Primmer, E.; Winkler, K.J.; Berry, P.; Young, J.; Carmen, E.; Špulerová, J.; Bezák, P.; Preda, E.; et al. Adoption of the ecosystem services concept in EU policies. *Ecosyst. Serv.* **2018**, *29*, 213–222. [CrossRef]
5. Howells, M.; Roehrl, R.A. Perspectives on Sustainable Energy for the 21st Century. In *Sustainable Development in the 21st Century (SD21) Project*; Department of Economic and Social Affairs, Division for Sustainable Development: New York, NY, USA, 2012.

6. Hastik, R.; Basso, S.; Geitner, C.; Haida, C.; Poljanec, A.; Portaccio, A.; Vrščaj, B.; Walzer, C. Renewable energies and ecosystem service impacts. *Renew. Sustain. Energy Rev.* **2015**, *48*, 608–623. [CrossRef]
7. Boyd, J.; Banzhaf, S. What are ecosystem services? The need for standardized environmental accounting units. *Ecol. Econ.* **2007**, *63*, 616–626. [CrossRef]
8. Fisher, B.; Costanza, R.; Turner, R.K.; Morling, P. Defining and classifying ecosystem services for decision making. In *CSERGE Working Paper EDM*; University of East Anglia, The Centre for Social and Economic Research on the Global Environment (CSERGE): Norwich, UK, 2007; ISSN 0967-8875.
9. De Groot, R.S.; A Wilson, M.; Boumans, R.M. A typology for the classification, description and valuation of ecosystem functions, goods and services. *Ecol. Econ.* **2002**, *41*, 393–408. [CrossRef]
10. Holland, R.A.; Scott, K.; Hinton, E.D.; Austen, M.C.; Barrett, J.; Beaumont, N.; Blaber-Wegg, T.; Brown, G.; Carter-Silk, E.; Cazenave, P.; et al. Bridging the gap between energy and the environment. *Energy Policy* **2016**, *92*, 181–189. [CrossRef]
11. MEA—Millennium Ecosystem Assessment. *Ecosystems and Human Well-being: Synthesis*; Island Press: Washington, DC, USA, 2005; ISBN 1597260401.
12. TEEB—The Economics of Ecosystems and Biodiversity. *Guidance Manual for TEEB Country Studies*; UNEP TEEB Office: Geneva, Switzerland, 2013; ISBN 9780415501088.
13. Haines-Young, R.; Potschin, M.B. *Common International Classification of Ecosystem Services (CICES) V5.1 and Guidance on the Application of the Revised Structure*; Fabis Consulting: Nottingham, UK, 2018.
14. Biodiversity Information System for Europe—BISE. *Ecosystem Services Categories*; European Environment Agency: Copenhagen, Denmark, 2018; Available online: https://biodiversity.europa.eu/maes/ecosystem-services-categories-in-millennium-ecosystem-assessment-ma-the-economics-of-ecosystem-and-biodiversity-teeb-and-common-international-classification-of-ecosystem-services-cices (accessed on 20 October 2019).
15. IUCN. Energy, Ecosystems and Livelihoods: Understanding Linkages in the Face of Climate Change Impacts. In Proceedings of the 4th World Conservation Congress, Barcelona, Spain, 9 October 2008.
16. Pulliam, H.R. Sources, Sinks, and Population Regulation. *Am. Nat.* **1988**, *132*, 652–661. [CrossRef]
17. Linares, P.; Altmann, M.; Brenninkmeijer, A.; Ellison, D.; Crisan, A.; Hugyecz, A.; Koreneff, G.; Hänninen, S. *Decentralized Energy Systems*; European Parliament: Brussel, Belgium, 2010.
18. Adil, A.M.; Ko, Y. Socio-technical evolution of Decentralized Energy Systems: A critical review and implications for urban planning and policy. *Renew. Sustain. Energy Rev.* **2016**, *57*, 1025–1037. [CrossRef]
19. AIRU. Il Riscaldamento Urbano. In *Annuario 2018*; AIRU: Milano, Italy, 2018.
20. ISTAT. Energia. 2018. Available online: https://www.istat.it/it/files//2018/12/tavole-1.zip (accessed on 14 February 2020).
21. Nussbaumer, T.; Thalmann, S. *Status, Report on District Heating Systems in IEA Countries*; Bioenergy Task 32 and the Swiss Federal Office of Energy: Zürich, Switzerland, 2014; Available online: http://task32.ieabioenergy.com/wp-content/uploads/2017/03/IEA_Task32_DHS_Status_Report.pdf (accessed on 14 February 2020).
22. Woods, P. *ETI Macro Distibuted Energy Project*; Energy Technologies Institute: Loughborough, UK, 2012.
23. Pagni, L. Luce, energia, acqua e rifiuti: il potere soft delle multiutility. *La Repubblica*. 2019. Affari & Finanza. Available online: https://rep.repubblica.it/pwa/affari-e-finanza/2019/09/02/news/luce_energia_acqua_e_rifiuti_il_potere_soft_delle_multiutility-234779073/ (accessed on 2 September 2019).
24. Aa. Vv. *Nuova Energia*; n.2/2019; Alkes: Milano, Italy, 2019.
25. ENEA-FIRE. Indagine Sulla Conoscenza e Diffusione del Teleriscaldamento nei Comuni in Zone E ed F. 2014. Available online: http://www.fire-italia.org/prova/wp-content/uploads/2015/04/RdS_PAR2013_132.pdf (accessed on 14 February 2020).
26. GSE. Valutazione del Potenziale Nazionale e Regionale di Applicazione della Cogenerazione ad Alto Rendimento e del Teleriscaldamento Efficiente. 2016. Available online: https://ec.europa.eu/energy/sites/ener/files/documents/it_potenziale_car_tlr_nazionale_e_regionale_dic_2016.pdf (accessed on 14 February 2020).
27. Van Der Schoor, T.; Scholtens, B. Power to the people: Local community initiatives and the transition to sustainable energy. *Renew. Sustain. Energy Rev.* **2015**, *43*, 666–675. [CrossRef]
28. Koirala, B.P.; Koliou, E.; Friege, J.; Hakvoort, R.A.; Herder, P.M. Energetic communities for community energy: A review of key issues and trends shaping integrated community energy systems. *Renew. Sustain. Energy Rev.* **2016**, *56*, 722–744. [CrossRef]

29. Rescoop position paper. What Local Energy Communities Need from the Clean Energy Package. 2016. Available online: https://www.duurzameenergie.org/f/files/download/rescoop-position-paper-on-clean-energy-package-final.pdf (accessed on 14 February 2020).
30. Public Affairs Bruxelles. What Role for 'Local Energy Communities' in the EU Energy Transition? 2017. Available online: https://www.pubaffairsbruxelles.eu/event-highlights-what-role-for-local-energy-communities-in-the-eu-energy-transition/ (accessed on 14 February 2020).
31. Tavazzi, L. Le Energy Community: Inquadramento e Potenziale di Diffusione in Italia. 2016. Available online: https://www.ambrosetti.eu/whats-hot/citta-e-territori/lo-sviluppo-delle-energy-community-in-italia/ (accessed on 14 February 2020).
32. CoR1. *Unlocking the Potential of Local Energy Communities*; European Committee of the Regions: Bruxelles, Belgium, 2018; Available online: https://cor.europa.eu/it/news/Pages/unlocking-the-potential-of-local-energy-communities-.aspx (accessed on 14 February 2020).
33. CoR2. *Models of Local Energy Ownership and the Role of Local Energy Communities in Energy Transition in Europe*; European Committee of the Regions: Bruxelles, Belgium, 2018; Available online: https://cor.europa.eu/en/engage/studies/Documents/local-energy-ownership.pdf (accessed on 14 February 2020).

sustainability

MDPI

Article

Military Training Areas as Semicommons: The Territorial Valorization of Quirra (Sardinia) from Easements to Ecosystem Services †

Ginevra Balletto [1], Alessandra Milesi [1], Nicolò Fenu [1], Giuseppe Borruso [2,*] and Luigi Mundula [1]

[1] Department of Civil and Environmental Engineering and Architecture, University of Cagliari (DICAAR), 09123 Cagliari, Italy; balletto@unica.it (G.B.); alessandra.milesi@gmail.com (A.M.); nicolofenu@gmail.com (N.F.); luigimundula@unica.it (L.M.)

[2] Department of Economics, Business, Mathematics and Statistics "Bruno de Finetti" (DEAMS), Trieste University, 34127 Trieste, Italy

* Correspondence: giuseppe.borruso@deams.units.it; Tel.: +39-040-558-7008

† This article is part of the work of drafting the Municipal Urban Plan of Villaputzu (MUPV) and in the related coastal use plan (LCP) and strategic environmental assessment (SEA): Scientific coordinator Ginevra Balletto. Participated in: MUPV, LCP, and SEA Alessandra Milesi and LCP Nicolò Fenu.

Received: 24 October 2019; Accepted: 8 January 2020; Published: 15 January 2020

check for updates

Abstract: The paper addresses the issue of the concurrent use of coastal areas for military training and civil activities, namely tourism. In the paper, starting from the consideration of publicly owned assets as 'semi-commons', we propose a method based on the comparison of planning instruments related to the different uses, and try to model them in a grid, where different weights and degrees of evaluation can be considered, in order to promote, rather than blocking, possible activities, compatible with concurrent use. The military areas in Sardinia (region and island, Italy) are around 234 km^2, which constitutes 60% of the national surface affected by military easements. This situation is due to its geographic position, considered centrality in the Mediterranean for strategic reasons. This contribution evaluates the performance of the Local Coastline Plan (LCP) and the Site management plan of Community Interest (SCI) in conditions of military constraint. The case study is the municipality of Villaputzu South Sardinia, Italy), where an important coastal military easement and the use of the coast for recreational tourism purposes coexist together through specific planning, a consequence of institutional agreements between the Municipal Administration of Villaputzu and the Ministry of Defense. The idea is considering the concurrent possible land uses guaranteed by the different planning instruments, instead of focusing, as it is generally the rule, on the sum of constraints provided by the laws. The local coastline plan has been identified as the ideal planning tool, which addresses the co-existence of apparently opposite land uses and interests, as those expressed by the local municipal planning and those expressed by the military. An evaluation of the congruence of the specific objectives of the LCP and SCI shows how their combined action favors the environmental enhancement of Sardinia, contributing to the formation of ecosystem services, even in particular conditions arising from military easements. These are sites that evolve from 'anticommons' to 'semicommons'. In fact, the military release process in Sardinia, together with the promiscuous military and civil use, activates unique governance policies of their kind that find a significant field of application in Sardinia to guarantee sustainable renewal of economic development of the 'semi-commons' awaiting to become 'commons'.

Keywords: landscape connectivity; Natura 2000 network; strategic environmental assessment; protected areas and spatial planning; semi-commons; ecosystem services

1. Introduction

State property indicates a set of publicly owned assets very different from each other. The state property 'Demanio' (as in the Italian Civil Law, art. 822 and following) consists, among others, of the maritime and water state properties (all water bodies and part of lands touched by water), military property (land and ad hoc infrastructure for national defense) and 'accidental' state property (transport infrastructure, cultural buildings and sites). It finally consists of the other assets that are by law subject to the regime proper to the public domain. Such assets can also belong to the regions, metropolitan cities, provinces, or municipalities—the intermediate administrative partitions in Italian system - equally subject to the state property regime. Public use is exercised on state property [1], i.e., the community can enjoy its benefits directly, as in the case of beaches or museums [2], or indirectly, in the case of ports or airports. The main characteristic of these goods is their inalienability, as they can neither be sold—except by a specific new law—nor can they be subject to rights in favor of third parties, except in the ways and limits established by the laws concerning them (Nav. Code 30 et seq.). They always remain state owned even if abandoned for a long time. Among the assets owned by the public, those relating to military uses also take on particular importance in relation to the fact that military defense is configured, in strictly economic terms, as a pure public good—non rival and non-excludable—and therefore, as an asset, the benefits of which fall indistinctly on the totality of the population, a fact that partly explains the huge amount of financial resources that the various states allocate [3]. After a descending phase in the 1990s, starting from 2000, military spending has in fact been continuously growing.

More specifically as in Figure 1, in 2018, the United States of America was the leading country in military spending of the sector, with 649 billion dollars - worth 36% of the world military spending and 3.2% of its Gross Domestic Product (GDP), followed by China with 250 billion dollars—14% of the world military spending and 1.9% of its GDP.

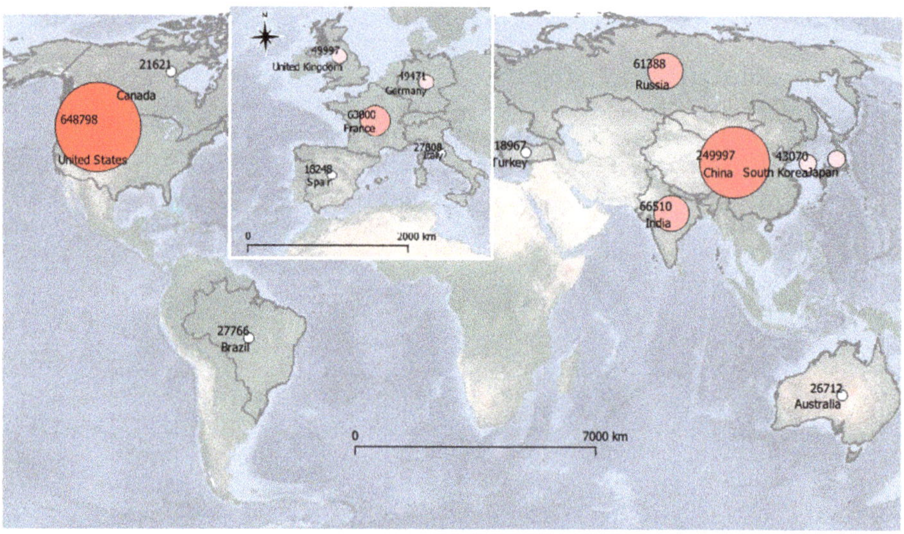

Figure 1. World military expenditure, by region, 1988–2018 (Source: Sipri, 2019. Available online: https://www.sipri.org/research/armament-and-disarmament/arms-transfers-and-military-spending/military-expenditure).

In this ranking, Italy occupies 11th place, with 27.8 billion dollars (equal to 1.5% of the World spending and 1.3% of its GDP). The state properties, on the one hand, can be traced back to the so-called 'enclaves', which are closed areas defined by administrative or cultural characteristics different from the surrounding territory and on the other hand, they have characteristics typical of the so-called 'anticommons', first introduced by Michelman in 1982, in contrast to that well-known common [4,5]. This concept was then widely developed by Heller [6] and Eisenberg through the theory concerning the under-use of a resource caused by the right of ownership, and therefore of exclusion, legitimately attributed to a multitude of subjects. Since it is sufficient for only one of the subjects to exercise their right, it is clear that it is very probable, due to the high transaction times and costs necessary to reach a satisfactory agreement for all, that the fruition of that will, lead to a blocking situation in which 'no one has the actual privilege of use' [7]. This is what happens in a by now recurrent form, in the occasion of the dismissal of the state property, both in the urban areas and in extra-urban ones [8]. This brief summary also includes military proving grounds (MPGs), which have the following European locations as shown in Figure 2, with a total of 284 MPGs, mainly concentrated in the United Kingdom (55), the Netherlands (27), France and Germany (26 in each country), Switzerland (22), Spain (21), Italy and Poland (13 in each country).

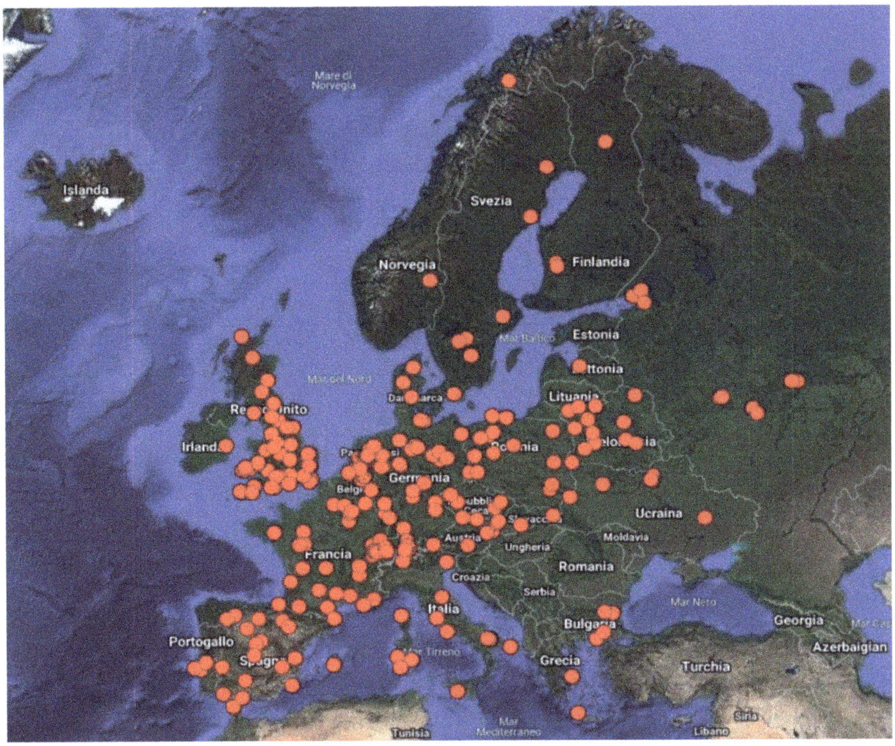

Figure 2. Maps of European military training areas.

The multitude of activities that are carried out there. Activities range from the training of national and foreign units to testing missile prototypes and targets, from quality tests in cooperation with industries and organizations in the aerospace electronics sector and activities related to scientific research, testing, and experimenting of naval ammunition and medium- and long-range terrestrial, including the testing of missile systems, shooting practices, even interforce and for out-of-area operations. They have a significant negative impact [9,10] not only for the territory directly concerned, but also for the local communities, also given by a series of constraints (restriction on flying navigation, bathing, etc.) that they express themselves through easements—According to the Italian Civil Code (art. 1027), an easement (or "praedial servitude" or land easement in the case of land), in the legal lexicon, indicates a minor real right of enjoyment over something else, consisting in the weight or limitation imposed on a fund (called servant) for the usefulness of another fund (called dominant) belonging to another person—which vary according to the security that is required.

In particular, military easement involves the set of limitations or prohibitions that can be imposed both on private assets and on public assets located near military installations. The state of servitude can, for example, impose the prohibition of building buildings higher than a given height, and the evacuation of land and housing in conjunction with the operations of exercises—In Italy, military servitude is an institution governed by law n. 898, promulgated December 24, 1976 and subsequent amendments. https://www.normattiva.it/uri-res/N2Ls?urn:nir:stato:legge:1976;898, access 15 September 2019.

If, on the one hand, it is true—as evidenced by several studies [11–16]—that military goods and servitudes perform a substantial function of indirect landscape protection that, in many cases, has prevented or severely limited speculative appetites on territories of great naturalistic value, on the other hand, the intensity and concentration of fire drills, as well as the testing of armaments with the use of fuels and propellants, have a significant impact on the environment and biodiversity [17], whose possible redevelopment may require expensive and difficult land reclamation works. The alleged correlation between military activities and the anomalous appearance of damage to health is still not completely defined. With regard to the economic aspects related to the presence of military bases and Military Proving Grounds (MPG), it should be noted that the territories concerned see potential forms of economic development linked to the exploitation of land for agricultural use and of many sea areas suitable for fishing damaged. To this, it must be added the failure to take off the tourist industry, both due to the unavailability of the sites, and to the interference that military activities have with the normal process of tourist settlement. It is also true, however, that this 'cumbersome' presence is in any case connected to an induced, also industrial, which relies mostly on local labor, as well as the local civilian staff serving in the MPG, which is often local [18].

The problems arising from the presence of military easements are therefore manifold and this feeds the debate on the search for possible solutions. The current orientation of several European countries is aimed at the rebalancing of military easements and the closing down of buildings [19]. In Italy, these procedures have often turned out to be cumbersome and slow due to the frequent regulatory interventions that have repeatedly changed the discipline. Also, in terms of environmental protection, similar considerations apply, in the sense that overcoming the exclusively military destination of certain areas may require guarantees of naturalistic protection [20], especially if such sites are included in the trade-in and exchange agreements program with local authorities. In this framework, the concept of semicommon [21] is well suited to be used for a new interpretation of those areas that present a mixture of civil and military uses in time and space. This concept, in fact, interposed between that of commons and anticommons, establishes an interaction between public and private property [1], and 'allows the right holders to benefit from the joint use of the resource' [22]. The semicommons, in fact, incorporate all private collective rights, but at the same time attribute a series of public rights to other subjects, such as those arising from military servitude. In the presence of this regime, the following two effects arise, deriving from the combined use [22]:

1. Economies of scale are achieved because public and private interests are combined; and
2. Environmental protection—the interaction of private and common use reduces the phenomenon of over-use.

In other words, there is a balance between the right of use and exclusion (Vanneste, Van Hiel, Parisi and Depoorter, 2006; Brede and Boschetti, 2008), typical of semicommons with the following characteristics:

1. A multitude of subjects (public and private) are involved;
2. There is the co-presence of public and private rights; and
3. There is the simultaneous presence of divergent public and private interests.

To this end, the second section of the contribution, after a brief classification of military easements in Sardinia and the main memoranda of understanding signed over the years between the Ministry of Defense and the Autonomous Region of Sardinia (Section 2.1), is as follows:

1. Classification of the case study 'MPG' of Villaputzu (Section 2.3);
2. Evaluation of coastal planning and management tools (Sections 2.4 and 2.4.1, Sections 2.4.2 and 2.4.3);
3. Assessment of the consistency of the objectives and actions of the coastal planning and management tools (Section 2.4.4).

In the third section, the results are shown and discussed, and finally, in the fourth section, the conclusions and future activities are presented.

2. Materials and Methods. The Case Study of Villaputzu 'Military Enclave'

The present contribution analyzes the case study of the coast of the Municipality of Villaputzu in Sardinia, with particular reference to the beach of Murtas, which constitutes the main case of Sardinia in the management and use of the areas subjected to military easements. This easement covers approximately 144.4 km^2 and accounts for 62% of all military easements in Sardinia (Section 2.2). The need to renew the role of military easements in relation to changed geopolitical conditions requires a transition from anticommons to semicommons. In this sense, the proposed approach is that of the transition through ecosystem services within integrated environmental planning. The proposed approach, although to be improved and refined, can be replicated in similar contexts, in coherence with the paradigm underlying ecosystem services.

The case study is interesting as the authors have been involved in the preparation of the local coastline plan (LCP), a compulsory planning tool for municipalities for addressing the issues related to coastal management and planning. Given the peculiarity of the area, being characterized by the important presence of a military proving ground, it was considered as an ideal case study where find possible co-existence features of civilian and military uses could be found.

In particular, the area under study is set up as an enclave within the Military Proving Ground Salto di Quirra (MPGSQ), which has effectively prevented its use by transforming the area into an anticommon. Furthermore, the beach of Murtas is included within a site of community interest (SCI) identified with code ITB040017 and name Stagni di Murtas e S'Acqua Durci. In other words, it is a site where divergent strategic objectives (environmental and military) converge, but which for a long time has been totally prevented from entering. The possibility of making the coastline accessible was reached with the recent state–region agreements (2014–2017), which allowed the use of the coast for tourism purposes in a limited period of the year (1 June–30 September), thus allowing the municipal administration of Villaputzu to equip itself with the appropriate littoral use plan (LCP). The authors, after framing the main state–region and municipality agreements and protocols (Section 2.1), the military easements in Sardinia (Section 2.2), and the analysis of the Villaputzu case study (Section 2.2), proposed the following methodology organized according to the following phases:

- Description of the case study (Section 2.3);
- Evaluation of the LCP and SCI (Sections 2.4.1–2.4.3); and
- Evaluation of the congruence of the specific objectives (qualitative and quantitative) and of the (qualitative) actions of the LCP and SCI (Section 2.4.4)

2.1. *Agreements and Memorandum for the Reorganization of Military Areas in Sardinia*

In Italy, military easements date back to the 1950s and derive from international agreements signed by Italy as a defeated country, at the end of World War II, in particular by the bilateral 'Mutua Sicurezza' (Reciprocal Security) agreement (1952) under which the United States imposed military bases in Italian territory. These agreements provide for the limitation of the right of ownership in the areas adjacent to installations of military interest. In 1976, the first law governing all matters of military servitude was issued (L. December 24, 1976, No. 898 'New Regulation of military easements), which provided for the establishment, for each region, of a joint commission with the task of assessing the compatibility of military programs with territorial development plans. With the National Law 104/1990, a list of the regions most affected by military easements was introduced to provide for the provision of compensation with the protocol of an understanding between the Ministry of Defense and ARS of 09.08.1999. Below is the list of the main agreements and memorandum of the Ministry of Defense and ARS as in Table 1.

Table 1. Main agreements and memorandum for the reorganization of military areas in Sardinia.

Date	Document Type	Object
9 August 1999	Memorandum of Understanding between the Ministry of Defense and the Autonomous Region of Sardinia	Regulation of compensation to economic operators for the removal of water bodies affected by military exercises
8 September 2005	Memorandum of Understanding between the Ministry of Defense and the Autonomous Region of Sardinia	Integration of the 1999 Memorandum of Understanding with the calculation of additional compensation due to the economic operators of the Teulada and Sant'Anna Arresi marinas
10 November 2006	Agreement between the Ministry of Defense and the Autonomous Region of Sardinia	Reorganization of the military presence on the island
28 March 2007	Memorandum of Understanding between the Ministry of Defense and the Autonomous Region of Sardinia	Disposal by the ministry of the buildings listed in the attached tables and the contextual commitment of the Region to the relocation of the functions performed in the buildings for which the disposal will take place
7 March 2008	Program Agreement between the Ministry of Defense, the Autonomous Region of Sardinia, the State Property Agency	Definition of procedures, methods, and timing of disposal of the properties listed in the annexes
21 December 2011	Resolution n. 45/5 Autonomous Region of Sardinia	Integration to the 2008 Program Agreement
18 December 2017	Memorandum of Understanding between the Ministry of Defense and the Autonomous Region of Sardinia	Coordination of military activities present in the territory of Sardinia with which in particular the suspension of fire activities at the Sardinian Military Proving Ground (MPG) is formalized from June 1 to September 30 of each year

Finally, with the memorandum between the Ministry of Defense and the ARS dated 18 December 2017, the suspension of the exercises within the MPG from 1 June to 30 September was formalized in order to guarantee the exploitation and use of the coast, also consistent with the environmental defense policies deriving from the Natura 2000 Network and the regional landscape plan (PPR).

2.2. Military Proving Ground (MPG) in Sardinia

In Sardinia, the military bases were installed in 1956 with the construction of the three coastal military proving grounds: Capo Frasca, Teulada, and Salto di Quirra (Military Proving Ground Salto di Quirra, MPGSQ, case study, Section 2.3), with a total area of approximately 234 km^2.

These MPGs together with the remaining military easements [23] account for about 1.5% of the total surface area of Sardinia, which in addition to blocking the air spaces during the exercises, determine that in some periods of the year almost the whole island is concerned as in Figure 3.

In particular, the MPG are intended for military services, such as training, experimentation with new weapon systems, simulated wars, deposits, etc.

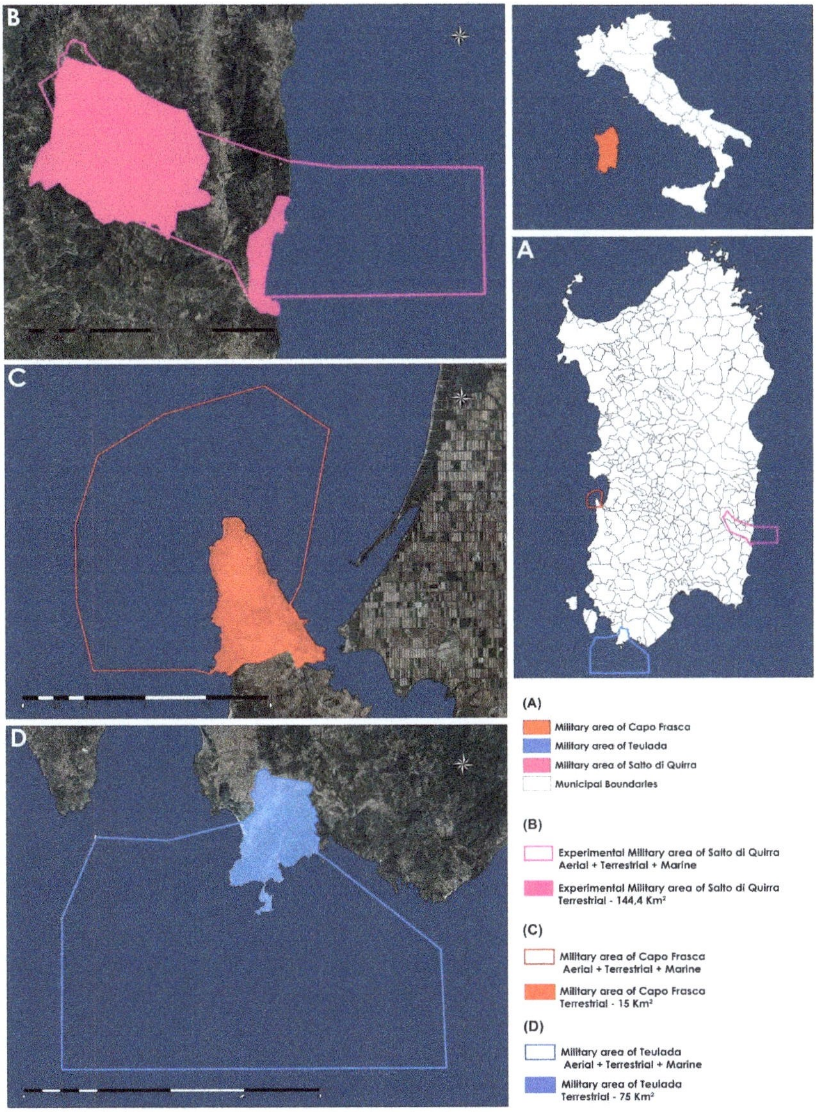

Figure 3. Maps of military training areas in the Sardinia region (Italy).

These services have always confirmed the strategic role of Sardinia in the context of the North Atlantic alliance of the political–military system related to control in the Mediterranean. These military services require a complex system of easement, with variable limitations, both in the inland and coastal areas and in the airspace. In particular, the limitations consist in the prohibition of the internal areas, in the permanent and/or temporary limitations in the coastal areas, and in the prohibition of flight in the air spaces. In this framework, the system of military services, which is expressed through the relative state property, land, sea, and air servitude, determines the complex and articulated system of military constraints that exist in Sardinia. Only in the last decade has the military state property been the object of specific state federalism, particularly in the coastal area, partly reducing military constraints. In fact, with the state-owned federalism process starting with Legislative Decree 85/2010, the transfer of part of the State's assets to the municipal administrations was envisaged. However, this process has not yet had the desired effects. In fact, in this situation, on the one hand, the ARS pushes for the non-onerous transfer of state property based on statutory regulations (while the decree provides for costs); on the other hand, the municipal administrations and local authorities are pushing for ownership to activate the development [24].

2.3. The Case of Military Proving Ground of Villaputzu (MPGSQ)

In this regional framework fits the case study of Murtas beach, in the Quirra coast, municipality of Villaputzu, an 'enclave' within the MPGSQ [25]. It represents the first case in Sardinia of planning for tourist use in the summer months and uses military in the remaining months of the year, the result of the complex process of state federalism (Sections 2.1 and 2.2). The MPGSQ is located in the southeastern part of Sardinia and develops mainly on two distinct areas: An area characterized by a plateau called Salto di Quirra [25]. The MPGSQ involves the municipalities of Villaputzu, Perdasdefogu, Tertenia, Ballao, Osini, Ulassai, Jerzu, and Arzana, as in Figure 4.

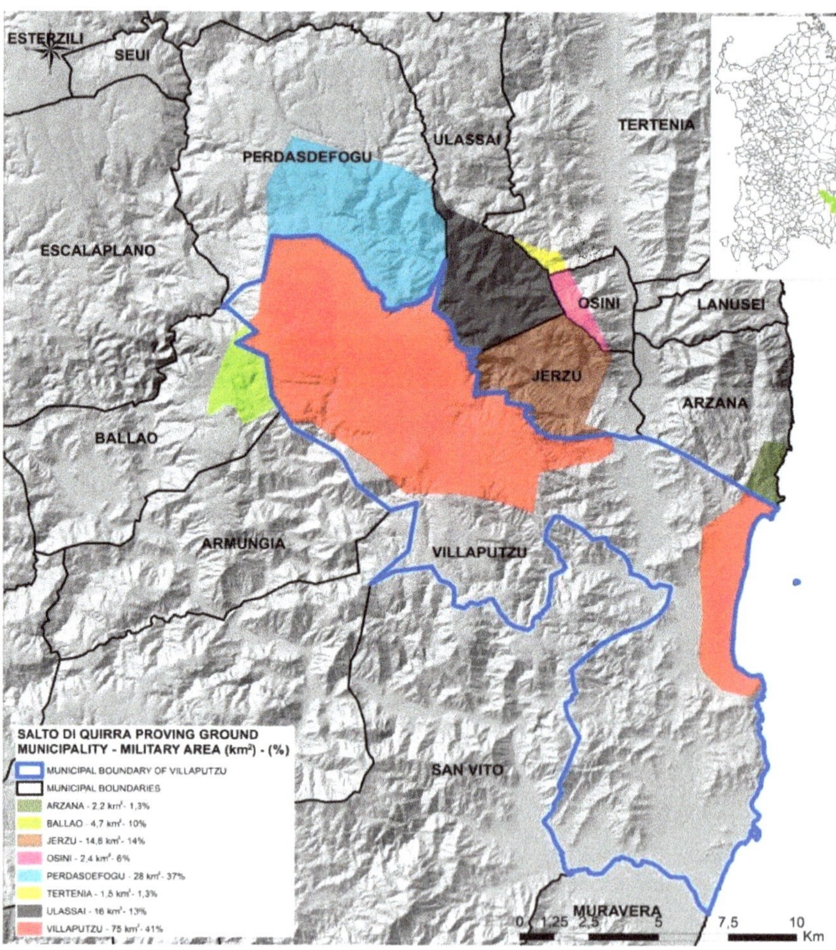

Figure 4. Municipalities affected by Military Proving Ground Salto di Quirra (MPGSQ). Percentage compared to the total municipal area.

The municipality of Villaputzu covers about 181.25 km^2 and is the most affected by the MPGSQ, with about 41% of the land area occupied by military easements. Furthermore, the MPGSQ at sea falls within a site of community interest (SCI) classified with code ITB040017 'Ponds of Murtas and S'Acqua Durci' [26] as in Figure 5.

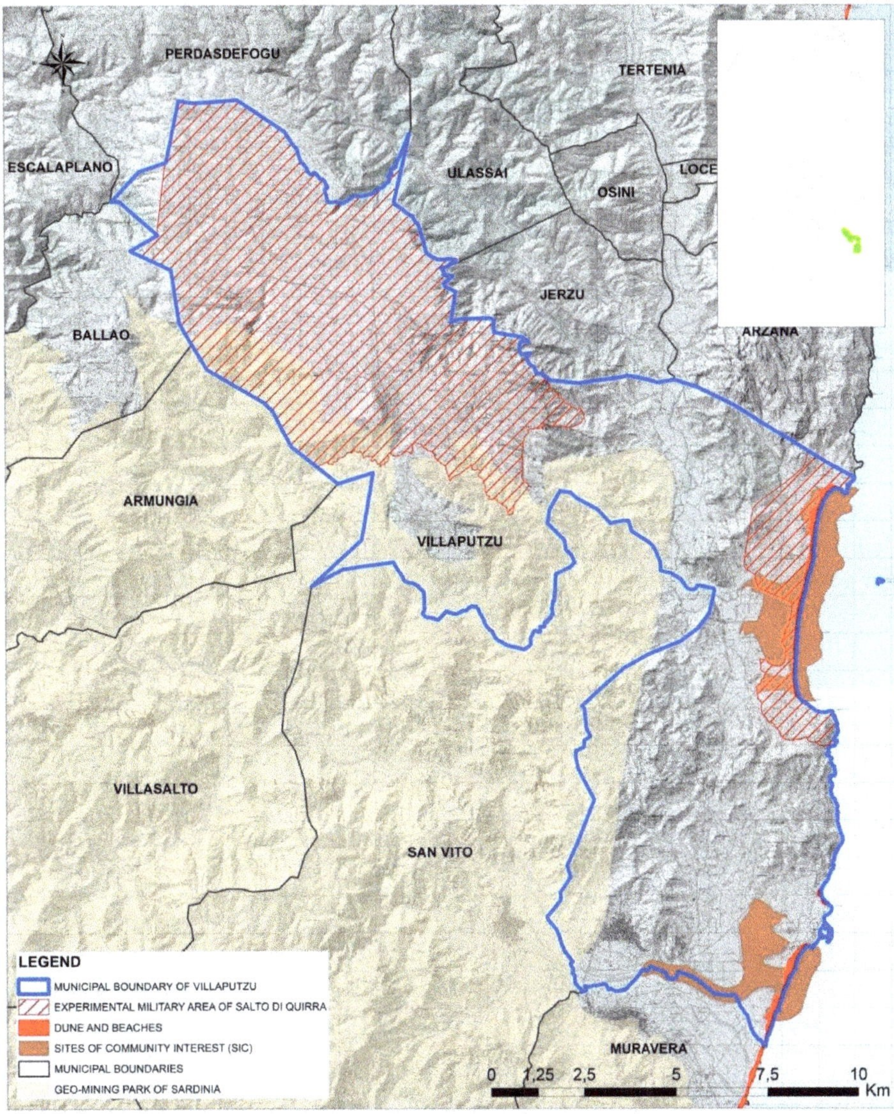

Figure 5. Military Proving Ground Salto di Quirra (MPGSQ) and the Site of Community Interest (SCI) of Villaputzu.

This situation deriving from military constraints has stimulated an impressive action from the bottom coming from the local community that has offered a sensitization of political opinion so that through the agreement of the Ministry of Defense and the municipality of Villaputzu of 27 June 2013, from June to September, tourist use on the beach of Murtas is allowed.

In 2017, the beach of Murtas was then included in the regional list of bathing beaches, thus allowing the municipality of Villaputzu to include this beach within its local coastline plan (LCP).

2.4. Planning and Management Tools for Coastal Uses

The planning and management of the coastal environment requires careful assessment of both coastal evolutionary trends, such as marine and hydrodynamic weather phenomena, and the phenomena of increasing anthropic pressure connected to tourist use. This in order to integrate the planning and management of the coastal environment between risk mitigation and conscious use [27,28]. In this sense, the authors have identified the main instruments for the planning and coastal management of Villaputzu: The coastal use plan (LCP) and the SCI management plan (SCIMP) [29]. In particular, the need to equip coastal municipalities with the LCP stems from the need to regulate the use of maritime state-owned areas, such as beaches, for recreational tourism purposes in order to protect and enhance the coastal environmental heritage.

The Law of 4 December 1993 n. 494 (article 6 paragraph 3) gives the faculty to allow the regions to have plans for the use of state-owned maritime areas. Subsequently, with the Legislative Decree 112/98, the delegation from the state to the regions was transferred for the functions related to 'the release of property concessions of the state of the inland navigation, of the maritime land and of areas of the territorial sea for purposes other than supplying of energy sources. This transfer of powers does not operate in ports and areas of national interest - SIN - identified by the decree of the President of the Council of Ministers of 21 December 1995. The Sardinia region—Regional Law of 12 June 2006, n. 9, the 'Assignment of functions to Local Authorities'—holds the task of adopting the general guidelines for the preparation of LCPs. Municipalities are assigned different functions on the subject of processing and approval of LCPs; concessions, on the assets of the maritime domain or inland navigation, for tourist recreation purposes, on uncovered areas, or those which involve easy removal facilities; and other administrative functions concerning maritime land ownership and the territorial sea not reserved for the region or the state.

The regulatory system where the LCP is inserted concerns the reorganization of maritime state property concessions.

According to the regional landscape plan (Regional Government Decree - DGR 36/7 of 2006), the municipalities are obliged, within the process of adaptation of the Municipal Urban Plans (MUPs) to the Regional Landscape Plan (RLP) [30], to draw up the LCP as an integral and substantial part of the municipal urban plan, and then the LR 8/2015 recognizes the LCP as an implementation plan. The general guidelines for the preparation of the LCP are currently represented by the 'guidelines' (Regional Council Resolution No. 10/5 dated 21 February 2017). In particular, the LCP regulates the use of the coasts and immediately contiguous territories for recreational tourism functions, dividing the coast in relation to the specific environmental characteristics, establishing the use and the related support services.

The LCPs in Sardinia often act within the territories of the Natura 2000 Network. Such a network represents the main instrument of the European Union policy regarding the conservation of biodiversity—established pursuant to Directive 92/43/CEE 'Habitat' and subsequent national transpositions—to guarantee the long-term maintenance of natural habitats and threatened flora and fauna species, divided into Sites of Community Interest (SCI), special protection areas (SPAs), and special areas of conservation (SAC). According to the recent literature, these are recognized as ecosystem services [31,32]. The LCP planning proposals must therefore take into account the SCI on which they act, so that the respective specific objectives are congruent.

2.4.1. The Coastal Use Plan (LCP): Murtas Beach, Villaputzu

In this framework, the Villaputzu LCP was developed, which, in addition to the state-owned maritime areas, also governs the contiguous areas, regulating road and pedestrian access to the areas in order to create an integration between the coast and the areas not immediately close to the coasts, thus also directing tourist flows towards less privileged areas as in Figure 6.

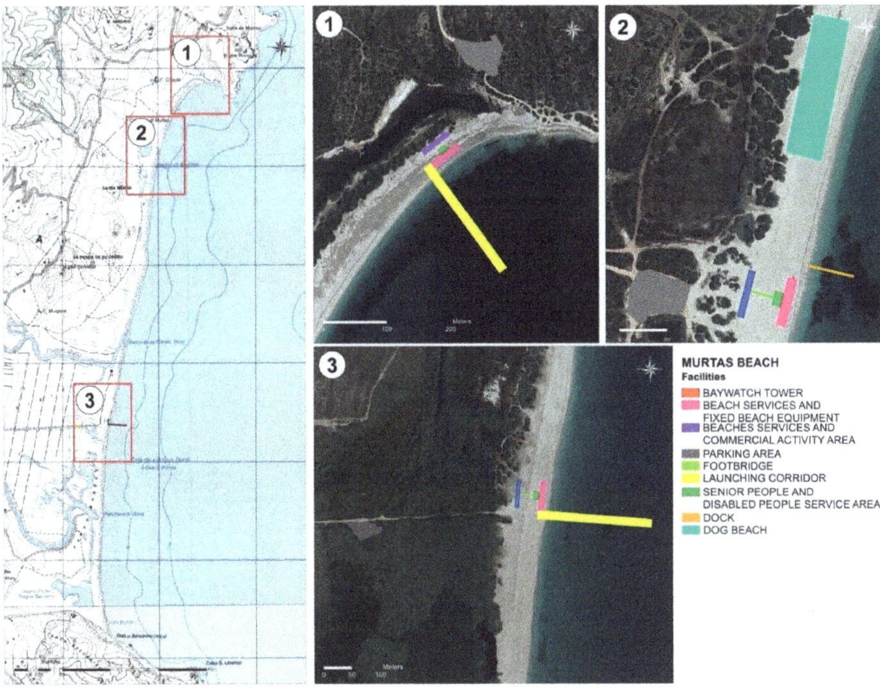

Figure 6. LCP concept Murtas beach, municipality of Villaputzu (2019).

In particular, the LCP proposes the following strategic objectives:

- Guarantee the conservation and protection of local coastal ecosystems, with particular reference to habitats—Directive 21 May 1992, 92/43/CEE and subsequent amendments;
- Harmonize actions on the territory for sustainable development, in particular by promoting measures for the reduction of degradation and land consumption processes;
- Promote and encourage environmental redevelopment through re-naturalization projects;
- Guarantee the continuity between the sand dunes and the ecological plant corridor, as well as improve the accessibility of the state-owned maritime areas;
- Promote innovation and diversification of the tourist offer; and
- Regulate the various activities for the purposes of integration and complementarity between them.

The Villaputzu LCP also affects the enclave of the MPG for the Salto di Quirra section (MPGSQ). This constitutes the first case in Sardinia of coastal planning that intends to reconcile the military activities of the winter period and the tourist ones recreating the summer period, in compliance with the current legislation in terms of health and safety and in the context of environmental monitoring [33].

2.4.2. The Management Plan of the SCI Municipality of Villaputzu

The SCI called 'Ponds of Murtas and S'Acqua Durci' (code ITB040017) located in the coastal sector of the mouth of the Rio Quirra, in central-eastern Sardinia, includes the portion of territory that from the promontory of Torre Murtas reaches Capo San Lorenzo, extending for an area of 7.4 km^2, Figure 7.

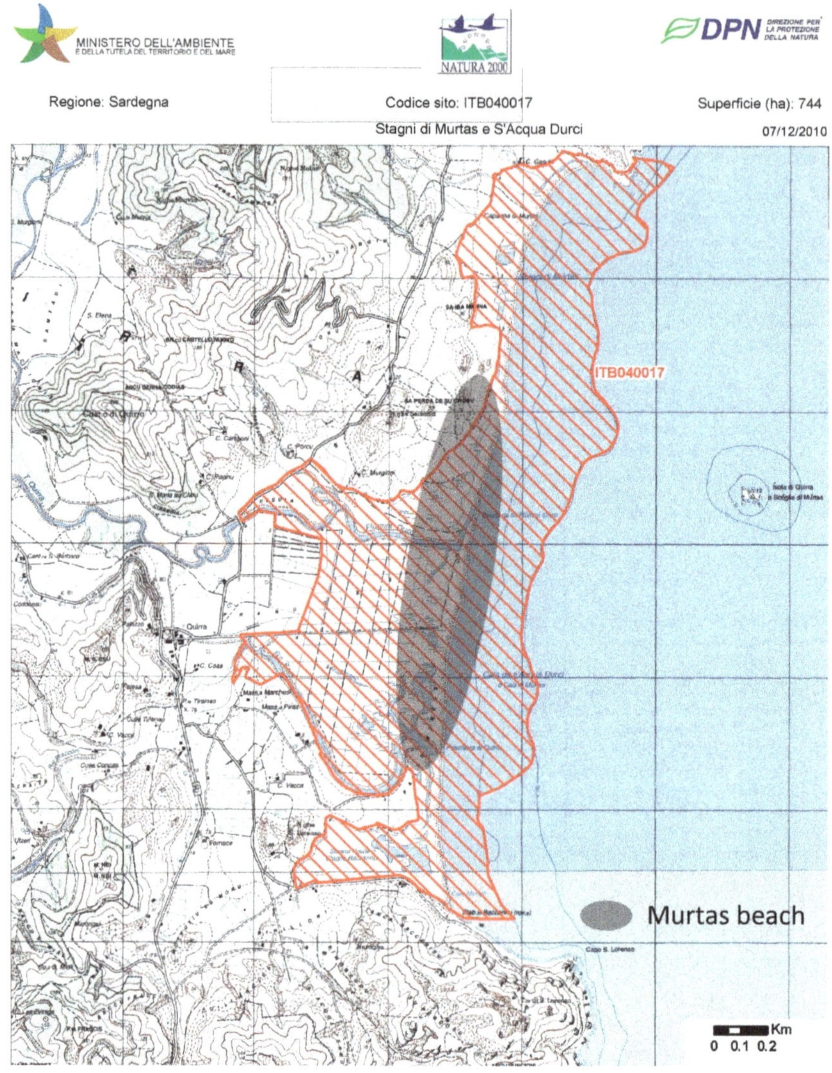

Figure 7. Sites of Community Interest (SCI) ITB040017 'Stagni di Murtas e S'Acqua Durci' (Source: https://www.minambiente.it/pagina/sic-zsc-e-zps-italia).

The management plan (SCIMP) regulates the SCI, which is consistent with the provisions of the art. 6 of the 'Habitat' Directive and of the art. 4 of Presidential Decree 120/2003, and has the objective of guaranteeing the conservation and protection of the habitats and species of fauna and flora, implementing protection and management strategies capable of allowing the maintenance of areas in optimal conditions, even in the presence of human activities, and thus favoring the maintenance of ecosystem services.

2.4.3. Ecosystem Services, LCP (Murtas Beach), and Management Plan of the SCI of Villaputzu

According to the definition given by the Millennium Ecosystem Assessment (2005), "the multiple benefits recorded by ecosystems for mankind" are expressed through four main categories: Life support, such as the nutrient cycle, soil formation, and primary production; procurement, such as the production of food, drinking water, and materials or fuel; regulation, such as climate and tide regulation, water purification, etc.; and cultural values, such as educational, recreational, etc.

In this sense, the importance of including ecosystem services in planning processes is supported by a vast literature. In fact, the overcoming of the sectoral approaches (Geneletti, 2011; Baker et al. 2013, Zoppi et al. 2018) [34–37], oriented to the coherence between the economic and environmental dimension (Danley et al. 2016) [37], build the prerequisites for the integration of ecosystem services within environmental planning. In particular, the transition to anticommons and semicommons is activated through the integrated environmental planning of ecosystem services. Therefore, to evaluate this integration referred to the case study, the methodology proposed by Mascarenhas et al. (2015) [38] was applied to the management plan of the SCI and to the Local Coastline Plan (LCP) of Murtas beach by searching for the following keywords, as in Table 2. Table one shows how the selection of the keywords related to the ecosystem services present in the LCP Murtas beach are highly representative in quantitative terms with respect to the management plan of SCI code ITB 040017. This confirms the role of the LCP strategy (Murtas beach) in the transition between anticommons and semicommons of the military proving ground of Villaputzu (MPGSQ).

Table 2. Number of keywords in Local Coastline Plan (LCP) Murtas beach (Villaputzu, Sardinia) and management plan of SCI (code ITB040017, Sardinia).

Keywords	LCP Murtas Beach Villaputzu	Management Plan of SCI Code ITB040017
ecosystemic | ecosystem	3	2
ecosystem service (s)	2	2
environmental service (s)	2	-
ecosystem balance	2	-
ecosystem functionality	1	3
ecosystem value	-	-
ecosystemic benefit (s)	2	-
regulating service (s)	1	-
Provisioning service (s)	1	1
cultural service (s)	-	-
Ecotourism	2	-
sustainable tourism	1	2
natural tourism | naturalistic tourism	2	-
recreational service (s)	2	-

2.4.4. Evaluation of the Congruence of Specific Objectives: LCP and SCIMP

The authors then proceeded with the evaluation of the congruence of the specific LCP and SCIMP objectives, building the following logical framework, as in Table 3.

This congruence was evaluated both in qualitative terms in Figure 8 and in quantitative terms in Figure 9, as well as by examining the actions of the two instruments in Figure 10.

From the qualitative analysis in Figure 8, we can extrapolate an overall coherence between LCP and SCIMP, given by the prevalence of the congruence relations, in line with the assumptions of the integrated management of the coasts in Sardinia [39].

The authors subsequently built a matrix with the attribution of weights (quantitative evaluation) to the congruence relations to obtain the degree of convergence between the specific objectives of the LCP and SCIMP, Figure 9.

Table 3. Logical framework: Specific Objectives of LCP and SCIMP.

Specific Objectives [SO] of Local Coastal Plan [LCP]	Specific Objectives [SOs] of Sites of Community Importance—Management Plan [SCIMP]
LCP_SO_01—Promoting innovation and diversification of the tourism offer, also through an integrated advertising-offer circuit	SCIMP_SO_01—Improve the quality and effectiveness of the organization responsible for the implementation, verification and updating of the Management Plan.
LCP_SO_02—Establish a framework for the harmonization of the actions of public and private subjects on the coastal strip.	SCIMP_SO_02—Improve the quality and effectiveness of communication and territorial control activities.
LCP_SO_03—Promote and encourage environmental redevelopment of the areas identified in the PUL	SCIMP_OS3—Improve the quality and effectiveness of monitoring activities
LCP_SO_04—Adopt recognition and monitoring systems of the coastlines in order to activate actions aimed at reducing the degradation and consumption processes of the territory.	SCIMP_SO_04—Restore and promote the expansion of all the surfaces that can be potentially occupied by habitats and habitats of species thanks to the involvement of stakeholders.
LCP_SO_05—Promote the decongestion of some stretches of coastline where the greatest load of bathing users is usually concentrated.	SCIMP_SO_05—Removal of landfills, exotic and invasive species and stray dogs.
LCP_SO_06—To guarantee the conservation and protection of local coastal ecosystems in harmony with the development of tourist activities and the free use of stretches of coast.	SCIMP_SO_06—Implementation of internal nature trails, adaptation of parking areas, and ecological connection of the SCI with the other neighboring SCIs and SPAs.
LCP_SO_07—To guarantee the continuity between the sandy shore and the dune system, improving the accessibility of the state-owned maritime areas	

	SCIMP_SO_01	SCIMP_SO_02	SCIMP_SO_03	SCIMP_SO_04	SCIMP_SO_05	SCIMP_SO_06
LCP_SO_01	✓	✓	✓	✓	×	✓
LCP_SO_02	✓	✓	✓	×	✓	✓
LCP_SO_03	✓	✓	✓	✓	✓	✓
LCP_SO_04	×	✓	✓	✓	✓	✓
LCP_SO_05	×	✓	✓	✓	✓	✓
LCP_SO_06	✓	✓	✓	✓	✓	✓
LCP_SO_07	×	✓	✓	✓	✓	✓

✓ relationship of congruence between objectives
× no relationship

Figure 8. Qualitative assessment of the specific objectives of LCP and SCIMP.

From Figure 9, it can be extrapolated that there is a maximum convergence for about 88% of cases, a weak convergence in 23.8% of cases, and no convergence in 11.9% of cases.

The authors also proceeded to assess the adequacy of the individual actions envisaged by the LCP and the ICS management plan to achieve their specific objectives.

The results of the consistency assessment of the actions (LCP and SCI) are shown in Figure 8. The white squares indicate that there are no interferences between the actions of the two instruments, the + symbol indicates a positive congruence between the actions, and the ++ symbol indicates highly positive congruence between actions.

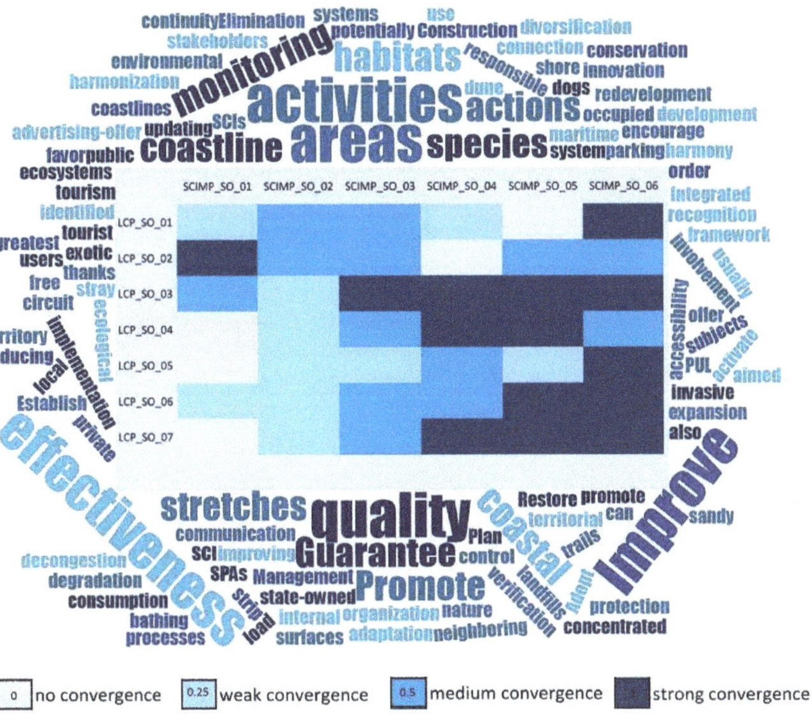

Figure 9. Convergence matrix (qualitative evaluation) of the objectives of the PUL and the SCIMP.

LCP ACTIONS	SCIMP ACTIONS																											
	AI1	AI2	AI3	AI4	AI5	AI6	AI7	AI8	AI9	AI10	AI11	AI12	AI13	AI14	AI15	RE1	IN1	IN2	MR1	MR2	MR3	MR4	MR5	MR6	MR7	EP1	EP2	EP3
LCP_Act_01	+			+																								
LCP_Act.02		++																										
LCP_Act.03													++															
LCP_Act.04			+							+						+												
LCP_Act.05																												
LCP_Act.06			+				+	+	+		++			++		+												
LCP_Act.07																												
LCP_Act.08												+							+	+	+				+			
LCP_Act.09												+			+	+	+											
LCP_Act.10																										+	++	
LCP_Act.11			++						+	++						+												

AI1-15 = Active interventions RE1 = Regulation IN1-2 = Incentives MR1-7 = Monitoring programs EP1-3 = Educational Programs

+ positive congruence ++ highly positive congruence

Figure 10. Evaluation of the congruence between the actions of LCP and SCIMP.

Figure 10 shows that there are no negative interferences between the actions of the respective plans, therefore the pursuit of objectives that have a low degree of convergence or no convergence does not derive from an interference between actions, but from actions that act on levels and different themes without conflict.

3. Results and Discussion

The proposed approach highlights how an integrated environmental planning of ecosystem services can allow the transition from anticommons to semicommons, where multiple interests coexist. This transition is in fact a fundamental phase, often neglected in environmental planning, but which requires the utmost attention during changes in the use of the territory and in particular in the occasion of the disposal of public assets. This transition is favored by the LCP, local coast planning, in coherence with the environmental protection policies (SCIMP) and the military regulations.

In particular, in the specific case study of the Villaputzu municipality, Quirra beach, the coexistence of multiple military, public, and private interests found a synthesis in the LCP. In fact, the authors realized a specific grid of converging specific objectives that can converge, particularly in terms of the overall convergence of the specific objectives of the Villaputzu LCP with the agreements and protocols of understanding between the Ministry of Defense and the Sardinia region (Table 1), and with the specific objectives of the SCIMP. An attempt was made here to respond to the debate of the local communities that they oppose military servitude, especially in the coastal area. Furthermore, this methodological application referred to the military enclave of Villaputzu and can be replicated in the remaining contexts of Sardinia subjected to military servitude, as reported in Figure 1, and more generally in the occasion of a transition of public goods from anticommons to semicommons. In this paper, we tackled a little covered issue, that of allowing the co-existence of military zones with civil uses, and particularly those related to leisure and tourism, by the application of a spatial planning policy instrument—the LCP— as the framework in which the different interests on a specific territory were considered together. The example cited and analyzed is not a trivial one and implied profound reflections on the concept, planning aspects, and opportunities deriving from a concurrent use of areas presenting military bonds and tourism.

With reference to the concept, it is worth noting that considerations have been made on the nature of the goods as semicommons, as those particular goods holding all the private collective goods, but at the same time attributing a set of public rights to other subjects, as those deriving from being territories subject to easements for military purposes. Such a concept seems nearly straightforward and obvious; nonetheless, it is not explicitly cited when considering this kind of use.

When moving to planning, the consideration of the delimitation of the areas is central, and from a visual and graphical (and cartographical) point of view, the precise localization of the areas, which are the bonds in terms of both military and environmental uses, as well as the possible destinations to a civil use, are important. That point alone required a deep and articulated survey on the planning instruments as well as on the official documents and materials to correctly identify the areas and different aims, bonds, and opportunities on a given territory.

A last, innovative, and interesting point deals with the evaluation process, which aimed at developing new techniques for resolving issues in terms of bonds, and proposing common and concurrent planning capable of complying with most of the different aims and orientations of the planning issues of origin. In the case study tackled here, the authors, as scholars and professionals involved in studying spatial planning issues, as well as working with local authorities on spatial planning issues, proposed an instrument as an evaluation matrix for putting together the different pieces of evaluation elements belonging to the different levels of planning, as military purposes, local spatial and urban planning, and environmental constraints.

Among the benefits expected by the novel approach presented here, there is that of allowing a wider awareness and acceptance of bond and easements over environmentally sensitive and protected areas, as well as a fair exploitation of resources.

4. Conclusions

The methodology proposed by the authors, consisting in the evaluation of congruence and convergence between the specific objectives of LCP and SCIMP and in the evaluation of congruence of the respective actions, allowed us to demonstrate how a military enclave, such as that highlighted in the case study of Villaputzu (Sardinia, Italy), can be configured as a semicommon [40], which is able to configure services' ecosystems [36,41] and use.

The transition from anticommon to common of the Villaputzu military enclave was in fact possible due to ecosystem services. In particular, the condition of a semicommon good incorporates the private sector related to tourism and beach services. In the presence of this regime (semicommons) the following two effects arise, deriving from the combined use: 1. Economies of scale are obtained thanks to the combination of public interests (military activities) and private (services seasonal tourist); 2. environmental protection: The interaction between private use (coastal use plan (LCP) and the common (management plan of the SCI of Villaputzu) reduces the phenomenon of excessive use. In this sense, ecosystem services offered in the LCP, Murtas beach allow responsible use based on the dynamism of the natural context. In fact, the military enclave has led to an anticommons, but it is also true that it has preserved the coast, prohibiting the fruition. This prohibition allowed the natural beach to evolve without human interference in the Murtas beach. In this sense, the LCP is inserted and proposes tourist use compatible with conservation and valorization of the natural heritage. It should be noted, however, that the research carried out appears to be at an initial stage, and with little evidence of other cases in sensible areas around the world. To the authors' knowledge, no other cases have been widely tackled in the literature and in spatial planning actions. The authors in particular highlighted, in theory and practice, how a concurrent approach on planning can be adopted, allowing de facto multiple targets to be addressed: Military operations, environmental protection and management, and tourist recreational use. The basic idea, extended for the consideration of the goods and services considered as 'semicommons', is that their use, observed by means of a fit-for-purpose matrix, can be in some sense concurrent. If rights of use can be present on a theory, it is extremely unlikely that the same territory is subjected to different uses at the same time, or at least this can be regulated and planned. A right of use should not translate into a perpetual use of a given portion of territory, as that can be in some particular moments of the year and/or of the day. This is valid for proving grounds—generally used in certain circumstances and moments—as well as tourism, which is generally subject to strong levels of seasonality in the presence of tourists and in the use of areas. Appropriate and calibrated planning can act also in terms of urgency or special cases in which a dominant use can overcome the other ones but adopting adequate announcements and communication.

The study developed here it is also interesting because it can be replicated in other similar contexts and where concurrent planning for sustainable use is also necessary in the territories used for military defense. In changing times in which military needs change the ways in which operations and training are performed, a convergence, at least in periods of time and for a certain amount of space, is actually needed. Some Italian regions, and Sardinia in particular, present a vast heritage of military grounds and easements, the offspring of the Cold War period, which are now still persisting. However, while the military presence has reduced its size and operating methods, in terms of personnel and resources, it did not reduce the occupied spaces and easement. In such sense, the authors have proposed an approach that takes into account the change, in the interests of both military and civilian uses. In this sense, little research seems to have tackled such an issue. Further research will be carried out in the same region and in other ones, where the heritage of military forces is still strong and pushes for a reuse are ongoing, with little delayed responses. The authors therefore intend to continue the research by evaluating the performance of ecosystem services on the occasion of semicommons connected to military proving grounds (MPGs).

Author Contributions: The paper derives from the join reflections of authors. However, Section 1 can be attributed to G.B. (Ginevra Balletto) and L.M.; Sections 2 and 2.1, Sections 2.2 and 2.3 to G.B. (Ginevra Balletto) and A.M.; Sections 2.4, 2.4.1 and 2.4.2 to G.B. (Ginevra Balletto), A.M. and N.F.; Sections 2.4.3 and 2.4.4 to G.B. (Ginevra Balletto); Abstract and Section 3 to G.B. (Giuseppe Borruso); Section 4 can be attributed to G.B. (Ginevra Balletto). All authors have read and agreed to the published version of the manuscript.

Funding: This research received no external funding.

Conflicts of Interest: The authors declare no conflict of interest.

Glossary

LCP Local coastline plan: Municipal plan that defines the structure of the coastlines through an integrated and systemic management that allows a guarantee of the preservation and enhancement of the integrity of the state property for tourist uses.

SCI Site management plan of community interest: It is the management tool for specific measures, required by the 'Habitat' directive 92/43/EEC, for the conservation of natural habitats and species of wild flora and fauna of community interest, taking into account the particularities of the specific site.

MPG Military proving ground: The training area is a military installation or reservation where weapons or other military technology are experimented with or are tested, or where military tactics are tested.

MA Military area: A facility that houses military equipment and personnel.

References

1. Angiolini, C.S.A. La Questione Dei Beni Comuni: Un Itinerario FRA Regimi Dei Beni E Diritti Fondamentali. Ph.D. Thesis, Università Ca' Foscari, Venezia, Italy, 2018.
2. Imperiale, F. *Il Valore Dei Beni Culturali Nel Patrimonio Dello Stato Italiano*; FrancoAngeli: Milano, Italy, 2018.
3. Caruso, R. *Chiamata Alle ARMI: I Veri Costi Della Spesa Militare in Italia*; EGEA spa: Burago di Molgora, Italy, 2018.
4. Hardin, G. The Tragedy of the Commons. *Science* **1968**, *162*, 1243–1248. [PubMed]
5. Marciano, A.; Frischmann, B.M.; Ramello, G.B. Tragedy of the Commons after 50 Years. 2019, 3451688. Available online: https://ssrn.com/abstract=3451688 (accessed on 15 May 2019).
6. Heller, M.A. The Tragedy of the Anticommons: Property in the Transition from Marx to Markets. *Harv. Law Rev.* **1997**, *111*, 621–688. [CrossRef]
7. Heller, M.A.; Eisenberg, R.S. Can Patents Deter Innovation? The Anticommons in Biomedical Research. *Science* **1998**, *280*, 698–701. [CrossRef] [PubMed]
8. Gastaldi, F.; Camerin, F. Processi di dismissione degli immobili militari. Temi e problemi aperti per la rigenerazione urbana in Italia. *Sci. Reg.* **2017**, *16*, 103–120.
9. Odoni, M. Degrado ambientale prodotto da esercitazioni militari tra interessi individuali e sicurezza nazionale alla luce del principio di non discriminazione. *Diritti Um. E Dirit. Internazionale* **2018**, *2*, 269–306.
10. Esu, A.; Maddanu, S. Military pollution in no war zone: The military representation in the local media. *Journalism* **2018**, *19*, 420–438. [CrossRef]
11. Hreško, J.; Petrovič, F.; Mišovičová, R. Mountain landscape archetypes of the Western Carpathians (Slovakia). *Biodivers. Conserv.* **2015**, *24*, 3269–3283. [CrossRef]
12. Špulerová, J.; Dobrovodská, M.; Izakovičová, Z.; Kenderessy, P.; Petrvič, F.; Štefunková, D. Developing a strategy for the protection of traditional cultural landscapes based on a complex landscape-ecological evaluation (the case of a mountain land scape in Slovakia). *Morav. Geogr.* **2013**, *21*, 15–26. [CrossRef]
13. Zentelis, R.; Lindenmayer, D. Bombing for Biodiversity—Enhancing Conservation Values of Military Training Areas. *Conserv. Lett.* **2015**, *8*, 299–305. [CrossRef]
14. Silva Arimoro, O.A.; Reis Lacerda, A.C.; Tomas, W.M.; Astete, S.; Roig, H.L.; Marinho-Filho, J. Artillery for Conservation: The Case of the Mammals Protected by the Formosa Military Training Area, Brazil. *Trop. Conserv. Sci.* **2017**, *10*. [CrossRef]
15. Ellwanger, G.; Müller, C.; Ssymank, A.; Vischer-Leopold, M.; Paulsch, C. *Management of Natura 2000 Sites on Military Training Areas*; Landwirtschaftsverlag: Münster, Germany, 2016; p. 152.
16. Reinhardt, I.; Kluth, G.; Nowak, C.; Szentiks, C.A.; Krone, O.; Ansorge, H.; Mueller, T. Military training areas facilitate the recolonization of wolves in Germany. *Conserv. Lett.* **2019**, *12*, e12635. [CrossRef]
17. Jenni, G.D.L.; Peterson, M.N.; Cubbage, F.W.; Jameson, J.K. Assessing biodiversity conservation conflict on military installations. *Biol. Conserv.* **2012**, *153*, 127–133. [CrossRef]

18. Paloyo, A.R.; Vance, C.; Vorell, M. The regional economic effects of military base realignments and closures in germany. *Def. Peace Econ.* **2010**, *21*, 567–579. [CrossRef]
19. Camerin, F.; Gastaldi, F. Il Ruolo dei Fondi di Investimento Immobiliare Nella Riconversione del Patrimonio Immobiliare Pubblico in Italia. Working papers. Rivista online di Urban@it2018, 2/2018. Available online: https://www.urbanit.it/wp-content/uploads/2018/10/BP_Camerin_Gastaldi-002.pdf (accessed on 15 May 2019).
20. Ellwanger, G.; Reiter, K. Nature conservation on decommissioned military training areas—German approaches and experiences. *J. Nat. Conserv.* **2019**, *49*, 1–8. [CrossRef]
21. Smith, H.E. Semicommon property rights and scattering in the open fields. *J. Leg. Stud.* **2000**, *29*, 131–169. [CrossRef]
22. Bertacchini, E.; De Mot, J.P.B.; Depoorter, B. Never two without three: Commons, anticommons and semicommons. *Rev. Law Econ.* **2009**, *5*, 163–176. [CrossRef]
23. Perelli, C. Geografie militari e nuovi movimenti di opposizione alle basi in Sardegna. In Proceedings of the International Conference Military Landscapes. A future for Military Heritage, La Maddalena, Italy, 21–24 June 2017; Skira: Milan, Italy, 2017; pp. 1210–1217.
24. Colavitti, A.M.; Serra, S.; Usai, A. Demanio Militare e "Beni Comuni": regolazione dei rapporti interistituzionali e coinvolgimento della società civile nei processi di valorizzazione della Regione Sardegna. In *Commons/Comune: Geografie, Luoghi, Spazi, Città., Giornata di Studio Della Società di Studi Geografici Rome, Italy, 11 dicembre 2015 2016*; Società di Studi Geografici: Firenze, Italy, 2016; pp. 613–620.
25. Codonesu, F. *Servitù Militari Modello di Sviluppo e Sovranità in Sardegna*; CUEC Editrice: Cagliari, Italy, 2013; Volume 23.
26. SardegnaAmbiente Web Site. Available online: http://www.sardegnaambiente.it/documenti/18_183_20140507151419.pdf (accessed on 15 May 2019).
27. Marx, J.T.; Liebenberg, I. Into the future: Donkergat Military Training Area and the Langebaan Ramsar site. *J. Transdiscipl. Res. S. Afr.* **2019**, *15*, 1–11. [CrossRef]
28. Seidel, V.; Dourte, D.; Diamond, C. Applying Spatial Mapping of Remotely Sensed Data to Valuation of Coastal Ecosystem Services in the Gulf of Mexico. *Water* **2019**, *11*, 1179. [CrossRef]
29. Leone, F.; Zoppi, C. Local Development and Protection of Nature in Coastal Zones: A Planning Study for the Sulcis Area (Sardinia, Italy). *Sustainability* **2019**, *11*, 5095. [CrossRef]
30. Sardegna Territorio Web Site. Available online: http://www.sardegnaterritorio.it/paesaggio/pianopaesaggistico2006.html (accessed on 15 May 2019).
31. Ministero dell'Ambiente e Della Tutela del Territorio e del Mare Web Site. Available online: https://www.minambiente.it/pagina/rete-natura-2000 (accessed on 15 May 2019).
32. Santolini, R.; Morri, E.; Scolozzi, R. I servizi ecosistemici nelle aree protette e oltre i loro confini. *Urban. Inf.* **2015**, *263*, 41–43.
33. Regione Sardegna Web Site. Available online: http://82.85.20.38/profili/stazione/profilo/id/665 (accessed on 15 May 2019).
34. Geneletti, D. Assessing the impact of alternative land-use zoning policies on future ecosystem services. *Environ. Impact Assess. Rev.* **2013**, *40*, 25–35. [CrossRef]
35. Blouin, M.; Hodson, M.E.; Delgado, E.A.; Baker, G.; Brussaard, L.; Butt, K.R.; Cluzeau, D. A review of earthworm impact on soil function and ecosystem services. *Eur. J. Soil Sci.* **2013**, *64*, 161–182. [CrossRef]
36. Lai, S.; Leone, F.; Zoppi, C. Strategic environmental assessment and enhancement of ecosystem services: a study concerning spatial planning in Sardinia (Italy). *Curr. Politics Econ. Eur.* **2018**, *29*, 393–412.
37. Danley, B.; Widmark, C. Evaluating conceptual definitions of ecosystem services and their implications. *Ecol. Econ.* **2016**, *126*, 132–138. [CrossRef]
38. Mascarenhas, A.; Ramos, T.B.; Haase, D.; Santos, R. Ecosystem services in spatial planning and strategic environmental assessment—A European and Portuguese profile. *Land Use Policy* **2015**, *48*, 158–169. [CrossRef]
39. Lai, S.; Leone, F. La pianificazione nei paesaggi costieri: l'applicazione del concetto di integrazione nelle esperienze di due aree protette italiane. In Proceedings of the Seventh International Symposium Monitoring of Mediterranean Coastal Areas: Problems and Measurement Techniques, Livorno, Italy, 19–21 June 2018; Benincasa, F., Ed.; Firenze University Press: Firenze, Italy, 2018.

40. Ling, G.H.T.; Ho, C.S.; Tsau, K.Y.; Cheng, C.T. Interrelationships between Public Open Space, Common Pool Resources, Publicness Levels and Commons Dilemmas: A Different Perspective in Urban Planning. *Int. J. Built Environ. Sustain.* **2019**, 6, 13–21. [CrossRef]
41. Cannas, I.; Lai, S.; Leone, F.; Zoppi, C. Green infrastructure and ecological corridors: A regional study concerning Sardinia. *Sustainability* **2018**, 10, 1265. [CrossRef]

MDPI
St. Alban-Anlage 66
4052 Basel
Switzerland
Tel. +41 61 683 77 34
Fax +41 61 302 89 18
www.mdpi.com

Sustainability Editorial Office
E-mail: sustainability@mdpi.com
www.mdpi.com/journal/sustainability

www.ingramcontent.com/pod-product-compliance
Lightning Source LLC
Chambersburg PA
CBHW061607220926
51850CB00005B/445
* 9 7 8 3 0 3 6 5 1 7 1 6 2 *